1998
anuf

Ergonomics in Manufacturing

Ergonomics in Manufacturing

Raising Productivity
Through Workplace Improvement

Edited by
Waldemar Karwowski
and
Gavriel Salvendy

Society of
Manufacturing Engineers
Dearborn, Michigan

Engineering & Management Press
a division of the
Institute of Industrial Engineers
Norcross, Georgia

Copyright © 1998 by Society of Manufacturing Engineers

987654321

All rights reserved, including those of translation. This book, or parts thereof, may not be reproduced in any form or by any means, including photocopying, recording, or microfilming, or by an information storage and retrieval system, without written permission from the copyright owners.

No liability is assumed by the publishers with respect to the use of information contained herein. While every precaution has been taken in the preparation of this book, the publisher assumes no responsibility for errors or omissions. Publication of any data in this book does not constitute a recommendation or endorsement of any patent, proprietary right, or product that may be involved.

No responsibility is assumed by the publisher for any injury and/or damage to persons or property as a matter of product liability, due to negligence or otherwise, or from any use or operation of any methods, products, instructions, or ideas contained in the material herein.

Library of Congress Catalog Card Number: 98-060113
ISBN: 0-87263-485-X

Additional copies may be obtained by contacting:

Society of Manufacturing Engineers
Customer Service
One SME Drive, P.O. Box 930
Dearborn, MI 48121-0930
(800) 733-4763; fax (313) 271-2861

Engineering & Management Press
Institute of Industrial Engineers
Order Department
25 Technology Drive
Norcross, GA 30092
(800) 494-0460 or (770) 449-0460; fax (770) 441-3295
e-mail cs@iienet.org

SME staff who participated in producing this book:
Larry Binstock, Senior Editor
Rosemary Csizmadia, Production Team Leader
Jennifer Courter, Production Assistant
Frances Kania, Production Assistant
Dorothy Wylo, Production Assistant
Dave McWilliams, Production Assistant
Judy Munro, Cover Designer

Printed in the United States of America

CONTENTS

PREFACE xv

1 ERGONOMICS IN PLANT OPERATIONS 1
    Gavriel Salvendy, NEC Professor of Industrial Engineering,
    Purdue University, West Lafayette, IN

2 PARTICIPATORY ERGONOMICS—A PRACTICAL GUIDE
    FOR THE PLANT MANAGER 5
    Donald Day, Consultant in Ergonomics and Health Promotion,
    Greenwood Village, CO
    What is Participatory Ergonomics? 5
    The Benefits 6
        Is Participatory Ergonomics Best?, Assess Readiness, Moving
        Toward Participatory Ergonomics, Getting Started, Analysis Process
    Keeping it Going 20
        Monitoring Progress and Evaluating the Process, Methods
    Opportunities 26

3 LOWERING COSTS THROUGH ERGONOMICS 29
    Hal W. Hendrick, Emeritus Professor of Human Factors,
    University of Southern California
    Los Angeles, CA
    Case Studies 30
        Forestry Industry, Material Handling Systems, Workstation
        Redesign, Reducing Work-related Musculoskeletal Disorders
    Human Factors Test and Evaluation 38
        Macroergonomics

4 MANUFACTURING WORKSTATION DESIGN 43
    Biman Das, Professor,
    Department of Industrial Engineering,
    Dalhousie University, Halifax, Nova Scotia, Canada
    Applying Engineering Anthropometry to Workstation Design ... 44
    Adjusting Anthropometric Data 45
    Determining Workstation Design Parameters 45
    Determining Workstation Dimensions 48
        Work Height, Normal and Maximum Reaches, Lateral Clearance,
        Angle of Vision and Eye Height
    Case Studies 50
        Supermarket Checkstand Workstation
    Computerized Human Modeling Programs for Workstation
    Design 54
        CYBERMAN, COMBIMAN, CREW CHIEF, JACK, SAMMIE, MANNEQUIN

v

5   **A DESIGN AND SELECTION GUIDE FOR HAND HELD TOOLS** 65
    F. Aghazadeh, Associate Professor,
    S.M. Waly, Assistant Professor,
    Industrial and Manufacturing Systems Engineering Department,
    Louisiana State University, Baton Rouge, LA

    Evolution of Tools 65
       Two Categories, Proper Matching, Anatomy of the Upper Extremities

    Injuries and Illnesses from Hand Held Tools 71
       A Survey, Injury Parameters

    Principles of Hand Tool Design 75
       Effects of Grip Type, Size, and Shape; Effects of Gloves; Effects of Wrist Position; Effects of Tool Weight and Muscle Group

    Safety Considerations 81

6   **COMPUTER-AIDED DESIGN FOR ERGONOMICS AND SAFETY** 87
    Markku Aaltonen, Project Manager,
    Department of Occupational Safety,
    Finnish Institute of Occupational Health, Vantaa, Finland

    Markkus Mattila, Professor, Vice-Rector,
    Occupational Safety Engineering Department,
    Tampere University of Technology, Tampere, Finland

    Gathering Safety Data 87
       Standardization

    Safety Management Information Systems 89
       Supportive Safety Information; Job Risk Assessment and Participatory Safety and Health; Training and Instructional Software Packages; Simulation Softwares for Ergonomics; Job Design Tools for Ergonomics; Integrated Systems for Occupational Safety, Health, and Ergonomics; Examples of Computer-aided Design for Ergonomics; Interactive Ergonomic-oriented Production System Design; TRANSOM JACK™ Human Modeling; IGRIP Software

    Integration of Safety and Ergonomics 102

7   **ERGONOMICS TRAINING AND EDUCATION FOR WORKERS AND MANAGERS** 107
    Marilyn Joyce, Director,
    The Joyce Institute/A Unit of Arthur D. Little, Seattle, WA

    Characteristics of Good Training 107
       Nature of the Adult Learner, Role of the Trainer/Facilitator, The Environment

    Five Phases 110
        Planning, Development, Delivery, Measuring the Impact, Integration and Improvement
    Job-specific Training 117
        Managers; Health and Safety Professionals; Manufacturing Engineers; Supervisors, Technicians, Ergonomics Team Leaders, Labor Representatives; Employees; Medical and Human Resource Professionals

8 ASSESSING PHYSICAL WORK LOAD 121
Veikko Louhevaara, Professor, Regional Institute of Occupational Health and University of Kuopio, Kuopio, Finland

Juhani Smolander, Senior Researcher, Department of Physiology,
Tatiana Aminoff, Researcher,
Juhani Ilmarinen, Professor,
Finnish Institute of Occupational Health, Vantaa, Finland

    Physical Load at Work 121
        Physiology of Muscular Work, Static Muscle Work, Effect of the Size of Working Muscle Mass, Muscular Overload, Field Methods for Assessing Physical Overload, Heavy Dynamic Muscle Work, Manual Material Handling, Static Muscle Work, Repetitive Work, Physical Work Load Assessment in Practice
    Preventing Physiological Overload 131

9 STATIC WORK LOAD AND ENDURANCE TIMES 135
Nico J. Delleman, Senior Researcher/Consultant,
Jan Dul, Head
Department of Innovation, NIA TNO, Amsterdam, The Netherlands

    Maintaining Working Postures 135
        Perceived Discomfort, Maximum Holding Time, Maximum Holding Time versus Discomfort, Maximum Acceptable Level of Discomfort
    Workstation Design and Adjustment 139
        Sewing Machine Operation, Press Operation, Hand Positions
    Standards 145
        Scope, Contents
    Work-rest Model 146
        Software, Example

10 WORKER STRENGTH EVALUATION: ERGONOMIC AND LEGAL PERSPECTIVES 153
Patrick G. Dempsey, Researcher, Liberty Mutual Research Center for Safety and Health, Hopkinton, MA

    Musculoskeletal Disorders 153
        Ergonomic Job Design

Methods of Strength Evaluation 154
   Isometric Testing, Isokinetic Testing, Isoinertial Testing, Comparing Strength Evaluation Methods, Selecting a Strength Evaluation Method
Legal Implications 158
   The Americans with Disabilities Act, Examples

11 METHODS FOR EVALUATING POSTURAL WORK LOAD 167
W. Monroe Keyserling, Professor, Industrial and
Operations Engineering, The University of Michigan, Ann Arbor, MI

Work Posture 167
   Health Effects of Awkward Posture, Productivity Effects of Awkward Posture, Posture Analysis Methods, Exposure Assessment Methods, Root Cause Analysis Methods
A Case Study 181
   Pre-change Analysis, Ergonomic Changes, Post-change Analysis

12 MANUAL MATERIAL HANDLING: DESIGN DATA BASES 189
Christin L. Shoaf, Doctoral Student,
Ashraf M. Genaidy, Associate Professor,
Mechanical and Industrial and Nuclear Engineering Department,
University of Cincinnati, Cincinnati, OH

Models 189
Examples 200
Reducing Exposure to Manual Material Handling Hazards ... 203

13 ASSESSMENT OF MANUAL LIFTING— THE NIOSH APPROACH 205
Thomas R. Waters, Research Physiologist,
Vern Putz-Anderson, Ergonomist,
Applied Psychology and Ergonomics Branch,
National Institute for Occupational Safety and Health, Cincinnati, Ohio

The Body of Work 205
   Recommended Weight Limit (RWL), Measurement Requirements, Lifting Index (LI), Miscellaneous Terms, Equation Limitations, Horizontal Component, Vertical Component, Distance Component, Asymmetry Component, Frequency Component, Special Frequency Adjustment Procedure, Coupling Component
Procedures 223
   Step 1: Collect Data, Step 2: Single- and Multi-task Procedures
Applying the Equations 230
   Using the RWL and LI to Guide Ergonomic Design, Rationale and Limitations for LI, Job-related Intervention Strategy
Example Problems 231

14 PERSPECTIVE ON LIFTING BELTS FOR MATERIAL HANDLING 243

Malgorzata J. Rys, Assistant Professor,
Department of Industrial and Manufacturing Systems Engineering,
Kansas State University, Manhattan, KS

Luis Rene Contreras, Professor,
Department of Industrial and Manufacturing Engineering,
Institute of Engineering and Technology,
Autonomous University of Ciudad Juarez, Ciudad Juarez, Chihuahua, Mexico

Prevention Attempts 244
   Personnel Training, Personnel Selection, Job Design

External Support Devices 245
   Back Support Belts, Weight-lifting Belts, Industrial Back Support Belts, Unknown Factors

Rationale for Using Back Support Belts 249
   Effect of Back Support Belts, Epidemiological Research

Physical Research 253

15 TRAINING AND EDUCATION IN BACK INJURY PREVENTION 265

Glenda L. Key, President, KEY Method, Minneapolis, MN

Impact Points 265
   Prior to the Hire, After the Hire

Prevention Programs 267
   Audit Assessment, Back Belts, Education, Ergonomics, Exercise, Fitness

Functional Capacity Assessments 273
   FCA Principles, FCA Reporting and Outcomes, FCA Standardization

Functional Therapy 280
   Job Analysis

Job Placement Assessments 282
   JPA Outcomes, "Tool Box Talks"

16 AN OVERVIEW OF UPPER EXTREMITY DISORDERS 287

J. Steven Moore, Professor,
Department of Nuclear Engineering,
Texas A&M University, College Station, TX
Co-director, NSF I/UCRC in Ergonomics, College Station, TX

Five Disorders 287
   Trigger Finger and Trigger Thumb, de Quervain's Tenosynovitis, Peritendinitis, Lateral Epicondylitis, Carpal Tunnel Syndrome

17 CUMULATIVE TRAUMA DISORDERS IN INDUSTRY 305
Fadi A. Fathallah, Researcher,
Patrick G. Dempsey, Researcher,
Barbara S. Webster, Researcher,
Loss Prevention Department,
Liberty Mutual Research Center for Safety and Health, Hopkinton, MA

What are CTDs? 305
   CTD Incidence and Costs, Common CTDs in Industry, CTD Risk Factors
Surveillance Methods 311
   Passive Surveillance, Active Surveillance
Prevention and Control of CTDs in Industry 314

18 ANSI-Z365 STANDARD: CONTROL AND PREVENTION OF CUMULATIVE TRAUMA DISORDERS 317
Marvin J. Dainoff, Professor, Psychology, and Director,
Center for Ergonomic Research, Miami University, Oxford, OH

Developing ANSI-Z365 317
   The ANSI Role
The Process 318
   The Accredited Standards Committee, Compliance
Structure and Content 320
   Components
Issues and Concerns 323
   Additional Factors
Other Standards 324

19 MANAGING WORK-RELATED MUSCULOSKELETAL INJURIES 327
Gary A. Mirka, Assistant Professor,
Carolyn M. Sommerich, Assistant Professor,
Department of Industrial Engineering,
North Carolina State University, Raleigh, NC

Three-Tier System 327
   Widespread Commitment
Management's Role 328
Human Resources' Role 331
Engineering's Role 340
The Supervisor's Role 342
The Operator's Role 344
Medical Management's role 346
The Safety Specialist's Role 347
Roles for Other Production Support Groups 348
   Integration

20 ERGONOMICS: PART OF CONTINUOUS IMPROVEMENT 351
Steven L. Johnson, Professor of Industrial Engineering,
University of Arkansas, Fayetteville, AR
Terminology 352
  Right and Wrong Terms
Effective Implementation of Ergonomics 353
  Policy and Procedures Document, Employee Involvement,
  Ergonomics Committee, Surveillance Methods, Job-site Analysis
  Methods, Ergonomics Training
Medical Management Program 365
  Prevention

21 VIBRATION-INDUCED CUMULATIVE TRAUMA DISORDERS 369
Donald E. Wasserman, Human Vibration Consultant,
Cincinnati, OH
The Nature of Vibration 369
Whole-body Vibration 370
Hand-arm Vibration 371
Vibration Measurements 373
  Standards
Controlling Vibration in the Workplace 376

22 EVALUATING ERGONOMICS PROGRAMS 381
Gary B. Orr, Industrial Engineer, OSHA,
Office of Ergonomic Support,
Department of Labor-Occupational Safety and Health Administration,
Washington, DC

David C. Alexander, President,
Auburn Engineers, Auburn, AL
The Ergonomics Process and Its Evaluation 382
  Management Commitment, Employee Involvement
Control and Prevention of Occupational Health Hazards 386
  Work Site Analysis, Hazard Prevention, Medical Management,
  Training, The Bottom Line
Control of Conditions Affecting Performance 391
Evaluating Products of the Organization 392
Outreach within the Trade 393
Contributions to the Technical/Legislative Community 394

23 AUDITING ERGONOMICS 397
Colin G. Drury, Professor, Department of Industrial Engineering,
State University of New York at Buffalo, Buffalo, NY
Measurement, Change, and Auditing 397
  Why Audit Ergonomics?

Choices for an Audit System 399
    Types of Checklists, Audit Design
How to Audit Ergonomics 402
An Ergonomics Audit Example 407
Lessons from Auditing 409

24 ECONOMIC EVALUATIONS OF ERGONOMIC INTERVENTIONS 413

James R. Buck, Professor,
Department of Industrial Engineering,
University of Iowa, Iowa City, IA

Benefits and Costs 414
    Interest Calculations and Discounted Cash Flows
Inspection Economics 425
    Inspection Costs, Rate Considerations in Inspection, Location of the Inspection Station
Economics of Learning 430
    Training and Transfer
Lifting Belts and Economics 436

25 WORLDWIDE CORPORATE ERGONOMICS EFFORTS—USA 439

Brian Peacock, Manager, Manufacturing Ergonomics Laboratory,
General Motors Corporation, Warren, MI

The Growth of Ergonomics 440
    The Profession
Polarization of Physical and Cognitive Ergonomics 442
Physical Ergonomics in Industry 443
    Organizational Factors, Ergonomics Programs, Cumulative Trauma Disorders
Commercial Opportunities 447
    Training, Job Analysis, Physical Devices
The Problem of Science and Standards 453
    Consensus, Ergonomics Standards
Exposure—The Time Factor 458
    Teams
Objectives of Ergonomics 460
    Fitting the Tasks, Preventing Unwanted Outcomes
Anthropology of Work 462

26 CORPORATE ERGONOMIC EFFORTS IN GERMANY 465

Hans-Jörg Bullinger, Professor and Head,
Fraunhofer Institute for Industrial Engineering (IAO), Stuttgart, and Head
Institute for Human Factors and Technology Management (IAT),
University of Stuttgart, Germany

Martin Braun, Research Scientist,
Institute for Human Factors and Technology Management (IAT),
University of Stuttgart, Germany

Rainer Schopp, Head Product Design,
Fraunhofer Institute for Industrial Engineering (IAO),
Stuttgart, Germany

    Evolution of Ergonomics 465
    Motives of Ergonomics 466
        Economic Efficiency, Humanization, Normative and Legal Aspects, Sociodemographic Development
    Institutions Involved in Ergonomics 468
        Public Institutions and Programs, Science and Research Institutions
    Ergonomic Design 472
        Goals, Applications, Design Requirements and Methods, Examples of Ergonomic Design

27 CORPORATE ERGONOMIC EFFORTS IN SWEDEN 485

Åsa Gabrielsson, Project Leader, Quality and Process Development,
Jörgen Eklund, Associate Professor,
Division of Industrial Ergonomics,
and Center for Studies of Humans, Technology and Organization,
Linköping University, Sweden

Gunnela Westlander, Professor, Division of Industrial Ergonomics,
Linköping University, Sweden

    Strategies on the National Level 486
        Internal Control, Impact of the Swedish Foundation of Work Life
    Ergonomics as a Professional Field 490
    Ergonomic Equipment for Industrial Production 490
    Cases from the Manufacturing and Service Industries 491
        The Case of ABB, Automotive Cases—Volvo and Saab, Sociotechnical Job Design in Sawmills, Service Industry—Tools for Ergonomic Improvements

28 ERGONOMICS AND TQM 505

Holger Luczak, Professor and Director,
Kai Krings, Research Assistant,
Stefan Gryglewski, Research Assistant,
Georg Stawowy, Research Assistant,
Institute of Industrial Engineering and Ergonomics, Aachen, Germany

    TQM Philosophy 505
    Objectives of Ergonomics 508
        Ergonomics as a Change Agent
    Health Promotion in TQM 514
        Occupational Safety and Health as a Quality Target, Economics and Health Promotion

Organizational Development through Health Promotion 523
Areas of Conflict, The Problem of Applying TQM

INDEX 533

PREFACE

This book is concerned with the role of people in complex manufacturing systems, the design of equipment and facilities for effective human use, and the development of environments for productivity and safety in a variety of contemporary industries. The field of human factors and ergonomics in the manufacturing sector has developed and broadened considerably since its inception more than 50 years ago. Recently, a substantial body of knowledge has been generated in many areas of ergonomics applications, including the four main issues that we focus on in this book:

- Methods for ergonomics job design and evaluation;
- Cumulative trauma disorder and low back injury prevention;
- Organization and management of ergonomics efforts; and
- Worldwide corporate ergonomics activities.

The foregoing clearly shows how broad the field of manufacturing ergonomics has become today. Thus, this book should be of great value to all ergonomics and human factors specialists, manufacturing engineers, industrial hygienists, plant managers and supervisors, team leaders, labor unions, human resource specialists, and health and safety professionals.

The 28 chapters of this book have been written by the foremost authorities from industry, government, and academia, representing eight of the world's leading manufacturing countries.

Each chapter is heavily tilted toward manufacturing applications, and a significant number of case studies, examples, figures, and tables are utilized to facilitate usability of the material.

The effective use of ergonomics knowledge presented in this book should help the reader achieve such contemporary manufacturing objectives as increased productivity, better workplace design, decreased cost of work-related accidents, and higher quality of working life.

We had the privilege of working on this book project with Larry Binstock, the SME Editor, who significantly facilitated our editorial work. We would also like to acknowledge the invaluable assistance provided to us during preparation of this book by our most able assistants, Laura Abel and Kim Gilbert.

Waldemar Karwowski *Gavriel Salvendy*
Louisville, Kentucky *West Lafayette, Indiana*

CHAPTER 1

ERGONOMICS IN PLANT OPERATIONS

Gavriel Salvendy
NEC Professor of Industrial Engineering
Purdue University
West Lafayette, IN

Ergonomics focuses on the role of humans in complex systems, the design of equipment and facilities for human use, and the development of environmental comfort and safety. Thus, ergonomics must be part of, and not a supplement to, effective design and operation of facilities. Table 1-1 shows the broad application of ergonomics in the workplace and the powerful impact it can have when effectively managed.

To achieve this objective, the ergonomics function in an organization must be placed under a reasonably high management position so that decisions about ergonomics intervention can be made promptly and effectively, with complete and full support of all management levels and the total workplace. The manager of the ergonomics function must have complete authority and the necessary financial resources to take full responsibility for effective implementation of ergonomics.

The ten steps toward successful ergonomics are listed in Table 1-2. They must be effectively integrated with the ten steps on how to win in a competitive marketplace, which appear in Table 1-3.

Where the ergonomics function is placed in the organization will dictate its function, effectiveness, and overall usefulness to the organization. It can be part of product design, manufacturing, or personnel. When it is part of product design, the main function of the ergonomist is to make the product more user-friendly, which may increase customer satisfaction and, depending on other factors, increase market share. In contrast to these product ergonomists, the process ergonomists are concerned with designing the process to increase safety and health of workers, as well as productivity, quality, and profitability of the corporation. When ergonomics is part of the personnel function, the main emphasis of the

Table 1-1. Ergonomics in the workplace (Salvendy 1997)

Job design
- Allocation of functions
- Task analysis
- Mental work load
- Job and team design
- Participatory ergonomics
- Models in training and instruction
- Computer-based instruction
- Organizational design and macroergonomics
- Socially centered design

Equipment, workplace, and environmental design
- Visual displays
- Controls
- Nonconventional controls
- Biomechanical aspects of workplace design
- Noise
- Vibration
- Illumination
- Toxicology and thermal comfort
- Climate and clothing
- Design for macrogravity and microgravity environments
- Architecture and interior design

Design for health and safety
- Occupational risk management
- Work schedules and sustained performance
- Psychosocial approach in occupational health
- Manual materials handling
- Work-related musculoskeletal disorders (WMSDs) of upper extremeties
- Warnings and risk perception

Performance modeling
- Decision making
- Feedback control models: manual control and tracking
- Supervisory control
- Cognitive modeling
- Computer modeling and simulation
- Decision support systems

Evaluation
- Data—collection and evaluation of outcome measures
- Exploratory sequential data analysis: qualitative and quantitative handling of continuous observational data
- System effectiveness testing
- Usability testing
- Maintainability
- Human factors audits
- Assessing cost/benefits of human factors

Human-computer interaction
- Design of computer terminal workstations
- Software-user interface design
- Virtual environments
- Social computing: computer supported cooperative work and groupware
- Human factors in information access of distributed system
- Multi-media

Selected applications of human factors
- Human factors in manufacturing
- Human factors in automation
- Human factors in process control
- Human factors in transportation
- Design for people with functional limitations due to disability, aging, or circumstance

Table 1-2. Ten elements for ergonomics contributions to win in a competitive marketplace

• Flat organizational design	• Optimizing psychological stress
• Continuous training	• Lowering back injuries
• Compensation for knowledge	• Reducing carpal tunnel syndrome
• Job design	• Personnel selections
• Workplace design	• Ergonomics audits

Table 1-3. Ten commandments on how to win in a competitive marketplace (Salvendy 1992)

- Get people on your side
- Emulate the best
- Use short interval scheduling
- Diversification: products and customers
- Design and manufacture for customers' needs
- Investment in research and development
- Use human resource accounting
- Design products for human use
- Use adaptive products and processes
- Use system integration

ergonomist is the health and safety of the workers. Whereas when ergonomics is a part of the manufacturing function, the major thrust of the ergonomist is to increase quality, productivity, and profitability. Frequently health and safety come into consideration if their inclusion helps increase quality, productivity, or profitability.

The number of ergonomists needed per facility depends on the number of people at each facility, homogeneity and heterogeneity of the facility's population, complexity and diversity of products manufactured, complexity and diversity of the manufacturing systems, and social, cultural, and environmental factors affecting product design and work processes. The total number of engineers employed at a facility is a good indicator for the variables. The number of ergonomists needed per facility to achieve safe and healthy work environments, together with high quality and productive work with increased profitability, is based on the diversified set of the author's experiences in more than 20 countries with different facility sizes and cultures (Table 1-4).

Table 1-4. Number of ergonomists needed per facility

Category of evaluation	Ergonomists per engineer
• All facilities • Best 10% of facilities	• One ergonomist per 210 engineers • One ergonomist per 80 engineers

It is important to note that the better managed, most profitable facilities where employees have the highest motivation level, employ the most ergonomists per engineer; the least profitable facilities, with the lowest morale, employ the fewest ergonomists per engineer.

How is your facility managed? How profitable is it? How high is worker morale? If you are in doubt, the first effective step is to conduct an ergonomics audit, which, depending on the facility size, can take from one to five days. The outcome of the audit will pinpoint potential areas for improvement.

REFERENCES

Salvendy, G., ed. *Handbook of Human Factors and Ergonomics*, 2nd Edition. New York: John Wiley & Sons, 1997.

Salvendy, G., ed. *Handbook of Industrial Engineering*, 2nd Edition. New York: John Wiley & Sons, 1992.

CHAPTER 2

PARTICIPATORY ERGONOMICS—A PRACTICAL GUIDE FOR THE PLANT MANAGER

Donald Day
Consultant in Ergonomics and Health Promotion
Greenwood Village, CO

In the late 1970s and early 1980s, Eastman Kodak began an endless journey on the road of quality and employee-based processes. Anyone who has been in business the last 20 years knows the various philosophical changes corporate management and leadership have initiated and, in some cases, implemented. Some of these programs came and went relatively fast, often dubbed program of the quarter or month. The only programs still remaining as part of the overall process are those that 1) evolved with the culture or blended with other processes and kept pace with management and leadership philosophies, 2) had good basic processes to support them, and 3) had some form of participation from affected employees. By nature, these programs generally are a process.

That was very much the philosophy of the Eastman Kodak Ergonomics Group from 1970-80 (Rodgers 1983). It made sense. And better than that, it worked! What was done at Eastman Kodak as a best practice was *participatory ergonomics*, even though it did not have a label at that time.

WHAT IS PARTICIPATORY ERGONOMICS?

Participatory ergonomics is a hybrid of other organizational efforts designed to accomplish more than just ergonomics (Fig. 2-1). Employees from all levels, functions, and organizations work and communicate collectively in functional or natural groups or teams, using ergonomics as a forum. Through the participatory ergonomics process, a commitment is made to agree upon and attain desirable outcomes for:

- Organizations and work systems (or macroergonomics); and

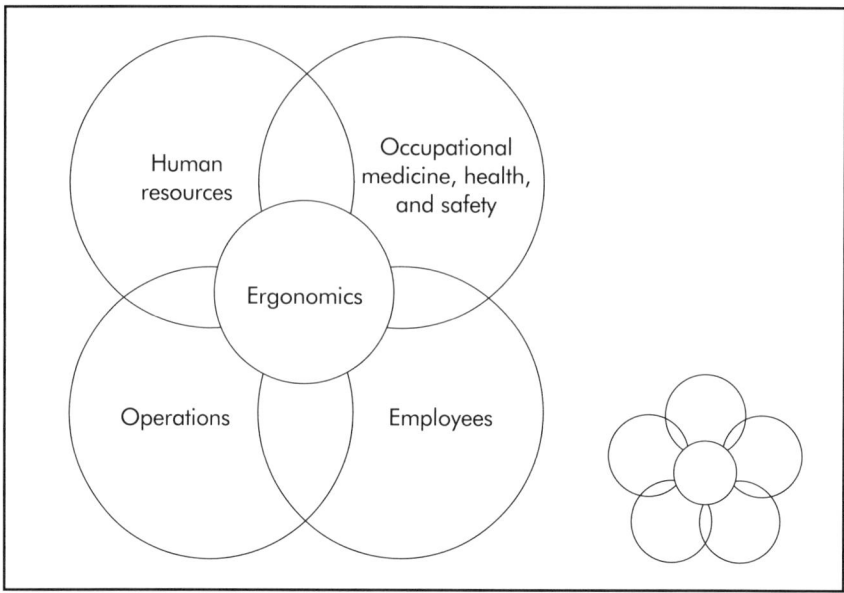

Figure 2-1 Organizational and functional considerations and needs.

- Workstations and specific tasks and jobs associated with them (or microergonomics) (Hendrick 1987; Imada and Nagamachi 1995; Nagamachi 1995; Wilson 1995).

THE BENEFITS

One common question—what can ergonomics, more specifically, the participatory ergonomics approach, do to enhance the organization and work for our business and employees? The answer depends upon where your company, functional groups, employees, and you as the leader are when considering multiple factors that influence participatory efforts.

The demands of work must be matched to worker capabilities and capacities, and the influence the organization and work systems have on the work and worker. The overall benefit, as seen in Figure 2-2, is a balance of work demands to employee capabilities and capacities within the company culture, organizational style, and work systems. Benefits of general participatory programs have been well established (Sherwood 1988; Pasmore 1990; Allen 1991; Proctor 1986; Auguston 1989), and participative ergonomics programs have some of those same benefits.

Many benefits of participation efforts, and therefore participatory ergonomics, are difficult to measure and substantiate due to some of the following concerns or situations:

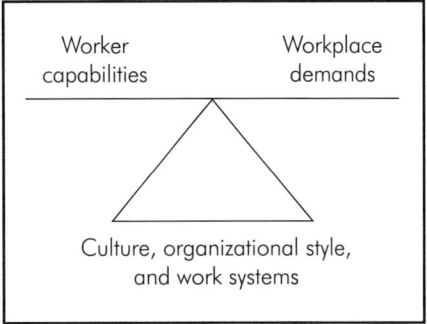

Figure 2-2 The participatory ergonomics balance.

- Being unaware of the effect or relationship of how work is organized and its influence on the possible benefits;
- Variations in work and individuals lead us to assume benefits can't be measured;
- Change occurs, adding variance to the study;
- Poor control or other influences may cause the outcome to be unclear;
- Not wanting to share information on these studies because it did or didn't work; and
- Not having good measurement processes for such parameters as quality or productivity.

Benefits may be difficult to measure, but there are certain points that leadership will gravitate to when considering specific benefits of participatory ergonomics (Rodgers 1984):

1. More employees can view and understand each job, which in turn creates a better work process through a more knowledgeable work force.
2. Ergonomics is an excellent communications platform. In a participative process, ergonomics provides more and improved communications to all levels, functions, and organizations.
3. There are more satisfied, committed, and motivated employees who accept responsibility and ownership of their work and workplace. Participation is perceived by the employee as improving his or her quality of work life (Della-Giustina and Della-Giustina 1989).
4. A better understanding of injuries and illness parameters by employees results in decreased injury and illness incidents associated with work and nonwork situations. This can result

in reduced workers' compensation costs, as well as health care costs and hidden costs associated with these incidents.
5. Employee knowledge is enhanced about the realities of work and work systems, including business, technology, and cost issues through application of participatory processes in the ergonomic process.
6. There are decision-making and process-setting capabilities realized by involving the employees in participatory ergonomics.
7. Increased organizational health, resulting in decreased absenteeism, less turnover, enhanced employee performance, shorter and more effective job training, and fewer complaints or grievances.
8. Participatory ergonomics is an excellent vehicle for beginning an ergonomics initiative as well as to implement general participatory processes.

Gains from technical advances to increase productivity and quality goals may have gone as far as they can without utilizing a company's main resource—people (Sherwood 1988). More companies

---

Work- and Nonwork-related Health Care Interaction

An example of this mixing of work and nonwork related cases comes from the findings of a review of health care cost utilization and workers' compensation cases of a plant with 3,200 employees. The results showed that musculoskeletal incidents accounted for the largest percentage of cases relating to the work environment with psychosocial cases being the second highest. A parallel review of the general health care cost utilization indicated that the largest percentage of cases involved the personal health of the employee as a result of psychosocial issues and the second highest percentage was musculoskeletal incidents.

There very well may be some sort of relationship between the musculoskeletal disorders in these two categories that affects the employee as well as the employer, whether it is work-related or nonwork-related, considering the overall health care costs. As a result, programs were developed to address both work- and nonwork-related health and safety issues.

are moving toward employee participative programs to get more involvement and cooperation from the available work force.

Utilizing a participatory approach with the ergonomic process itself can help in attaining some of the following benefits:

1. Meeting existing and proposed regulatory requirements (ADA, OSHA, ANSI 365) for participation of employees in health and safety programs and ergonomics, and accommodation of disabled employees.
2. Removing the mystery of ergonomics.
3. Generating more effective and longer lasting ergonomic changes with practical ergonomic analysis and sound problem solving methodology (Rodgers 1984).

Participatory ergonomics should not be used as a panacea. It does have limitations. But participatory ergonomics can be a more effective approach to this multifunctional science and should be used as a tool common to safety, productivity, and quality-enhancing efforts in all business and industrial functions.

IS PARTICIPATORY ERGONOMICS BEST?

You can have an ergonomics program without it being participatory. There are those who believe the longer lasting benefits of participatory ergonomics far outweigh any downside. But it may not be appropriate in all cases. So, don't feel compelled to adopt participatory ergonomics, especially if you or your leadership staff, supervision, employees, or union (if applicable) meet some of the following criteria:

Employee Participation in Proposed Regulatory Programs

The 1989 OSHA Voluntary Safety and Health Program Management Guidelines set expectations for company safety and health programs based on the degree of employee participation in the following elements:
- Development of the program including conducting training and education;
- Workplace audits including collecting data;
- Program reviews; and
- The authority to stop activities when deemed hazardous.

1. May stand to lose or gain to a substantial degree, especially when compared to other groups, with the adoption of participatory ergonomics. It should be a win situation for all groups. See Table 2-1.

Table 2-1. Potential gains and losses in developing a participatory approach

Affected groups	Gains	Losses
Production	• Driving force • Awareness	• Measured for value • Increased resource allocation
Union	• Seen as "team" player	• Less need as employee representative • Measured for value
Management	• Makes job easier • More commitment	• Less (different) control • More direct reports • More commitment
Supervisor	• Makes job easier • Can delegate more • More as a "coach"	• Less (different) control • Less need as level • More as a "coach"
Engineer	• More assessable • More interaction	• Additional work generated • Less (different) prestige • More assessable
Employees	• More control in their job • More input	• More responsibility • Measured for value

2. Not willing or able to commit to a participatory ergonomics effort.
3. Want to adopt participatory ergonomics for the wrong reasons (weaken the union, lessen management influence, exploit workers).
4. Want to adopt participatory ergonomics knowing that it will fail by their own actions, actions of others, or circumstances (Garrigou et al. 1995).

ASSESS READINESS

A preliminary assessment of management, social, and cultural aspects of the employees and company regarding ergonomics and participatory processes is required when a facility is considering participatory ergonomics. This can be facilitated by an outside or internal consultant or by self-assessment tools.

First, consider your thoughts about ergonomics in general. Even if your facility has a program, go to seminars or visit other facilities doing ergonomics to see if they use the participatory type. Talk to experts and consultants to gain their perspective about ergonomics and participatory processes.

Second, assess your values and beliefs toward participation programs. You need to be personally committed. If you don't want participation or aren't willing to nurture the effort, it will not begin to develop.

Third, consider the culture of your company and your facility and how it will react to participatory programs and ergonomics. Gain as much information as possible to allow you to make this important decision.

Table 2-2 may assist you in evaluating your management style and the culture of your plant. Your management style will influence employee participation in ergonomics and other health and safety efforts in your facility.

Following are some additional thoughts for determining whether you and your plant are ready for participatory ergonomics:

- If you and your company already subscribe to the philosophy of participatory efforts, adding an ergonomics component will be relatively simple.
- If you and your company already have an ergonomics effort in place, transforming the effort to participatory should be a natural and effective enhancement to your program.
- If you believe in ergonomics and the participative efforts, but the company culture doesn't, the change will be a challenge and might be more successful with limited pilot efforts.
- If you believe in ergonomics, but your company doesn't, you will have to measure the efforts and processes to gain their trust and show benefits to gain their commitment.

MOVING TOWARD PARTICIPATORY ERGONOMICS

Generally, until your company's culture is very aware and trusting as well as accepting of change, utilizing an outside source, such as a consultant, to facilitate and guide change will probably be most effective. But this depends upon the factors that influence change toward adopting participatory efforts. If conditions are not right for change, or there isn't some driving force facilitating the change, you may want to first use pilot programs or another ergonomic model before evolving to participatory ergonomics.

Table 2-2. Management styles and their characteristics
(Kavianian, Roe, and Sanchez 1989)

Characteristic	Style			
	Participative	Consultative	Authoritative	
			Benevolent	Exploitive
Confidence in subordinates	Yes	Substantial	Condescends	None
Free to communicate upward?	Yes	Mostly	Not much freedom	None
Ideas sought?	Yes, always	Usually	Sometimes	Seldom
Communication	Down, up, and sideways	Down and up	Mostly down	Down
Accuracy of upward communication	Accurate	Limited accuracy	Censored for boss	Often wrong
Decision making	Throughout organization	To some extent throughout	Top and middle management	Top management
Setting goals	By group action	Orders, after some discussion	Orders, some comment invited	Orders issued
Control of work and processes	Widely shared	Moderate delegation	Relatively high at top	At top only
Informal organized resistance?	None	Sometimes	Usually	Always
Commitment	High at all levels	Some	Little except for individuals	Very little
Training	Many aspects of business, at various levels	Some across functions with guidelines	Some with much justification	Very little outside of job responsibilities
Integration of processes functions	At operator level	Given to specialist	Guarded within department	Very guarded, no crossover

> **Use Other Experiences to Help Your Participatory Ergonomics Efforts**
>
> In a plant of 800-plus employees, a major effort was made to implement a total quality management (TQM) process in a participatory manner. Because the middle management and leadership, as well as department heads, did not generally believe in the principles and tools of TQM, or because they were not measured as to implementation and longevity of the process, TQM died a slow but certain death. The production management levels were not involved in developing the overall TQM process. The TQM process was being presented and directed by the human resource organization rather than the production organization. These factors all contributed to the demise of the TQM effort in this company. Whether conscious or unconscious, the employees at all levels felt they and the TQM process were set up to fail. Passive subversiveness can overcome even the wishes of company presidents.

The plant manager's role in the ergonomics process is to champion the ergonomics goals to ensure buy-in and commitment at levels. He or she has the power to subtly help mold the direction and level of importance of the ergonomics effort.

As many functional aspects of your company as possible should participate in developing the overall ergonomics effort. All employees, directly or indirectly involved with the effort, should be aware of the limitations and pitfalls. Participatory ergonomics may not always work the exact same way for all facilities and companies. Everyone will be more accepting, committed, and realistic when beginning the ergonomics effort if they know there is some structure and flexibility in the overall process.

GETTING STARTED

As with most programs and process efforts, participatory ergonomics should be implemented with a well-thought-out plan. Plan the introduction process and follow it, but be flexible and have milestones to help measure the implementation process. All stakeholders and parties should be involved and prepared for the participation process as much as possible to ensure a smoother implementation.

Efforts to establish fertile groundwork and strong cornerstones would include the following concepts and steps:

- Determine what value participatory ergonomics can deliver to each of the stakeholders and involved parties and communicate those points. (What is in it for me?)
- Introduce to leadership staff for discussion, planning, and ownership. Upper level management should be educated and persuaded to begin the commitment towards participatory ergonomics (Brown 1990). Objectives for educating the management and leadership levels should include the following:
    1. Know what participatory ergonomics is.
    2. Know the benefits of participatory ergonomics.
    3. Know the structure of participatory ergonomics.
    4. Understand the resource (time and personnel) requirements for participatory ergonomics.
    5. Know what is required to maintain participatory ergonomics.
    6. Know the basic signs and symptoms of musculoskeletal disorder (MSD).
    7. Know the response and tracking mechanisms for occupational health for your facility, including medical case management, transitional work, and return to work.
    8. Know the process for identifying and solving problems using ergonomics.
    9. Know and understand management's role and responsibility in the participatory ergonomics process.
    10. Know and understand how management will be measured and held responsible for the participatory ergonomics process.

Most of these same points should be applied in educating all levels of the organization about the participatory ergonomics process. Select a member of the staff to sponsor the process. Good project management skill is a desirable capability for this champion.

- Ideally, the champion and core leadership team should set goals, roles, responsibilities, boundaries, and accountabilities for each involved party. Steps must be taken to involve management, leadership, and supervision in developing and implementing such efforts, and to hold them accountable for these efforts. This can be difficult for certain individuals as they may feel they have lost their power.

In assisting top management in this effort, plant managers should consider the following steps (Sherwood 1988):

1. Take the road blocks down.
2. Have patience. In change processes, slower change is generally more complete and longer lasting.
3. Do not let the process drag on forever. Appropriate measurement intervals should be established to help facilitate change. Set goals and timelines for the effort.
4. Education efforts should be delivered in different amounts to all levels.
5. Intraorganizational and interorganizational issues should be identified and addressed.
6. Agree on the process, structure, boundaries, budget, resources, etc.
7. Assess current communications issues for all levels and organizations, and establish enhancement strategies and efforts to aid the ergonomics process.
8. Select a usable ergonomics analysis system and concentrate on practical and economical ergonomic solutions. The most effective and long lasting results often stem from problem solving efforts during the process.
9. Work with key levels in leadership as well as functional support groups to establish commitment and support for the overall process.
10. Find key people—facilitators, steering committee members, and process coordinators. At times, gentle persuasion or outright sneaky efforts should be applied to get the best individuals, strategically selected, to become involved. However, be cautious; many times these individuals are already overloaded with other team duties.
11. Establish seed money and a budget to account for the resources and needs for travel, training, measurement devices, and tools for the ergonomics team as well as the initial fixes they determine. Track and review this effort. Don't let it continue so long that you create a dependent process.

12. Establish communication processes early, including written, verbal, and review processes to inform and gain buy-in from employees on the implemented ergonomic solutions.
13. As often as possible, review your plans for avoiding pitfalls and develop back-up plans should pitfalls occur.

Ergonomics Process and Structure

Several key elements should be folded into an ergonomics process. First and foremost there should be structure (Fig. 2-3), which gives the ergonomics effort legitimacy in the eyes and minds of the leadership and employees.

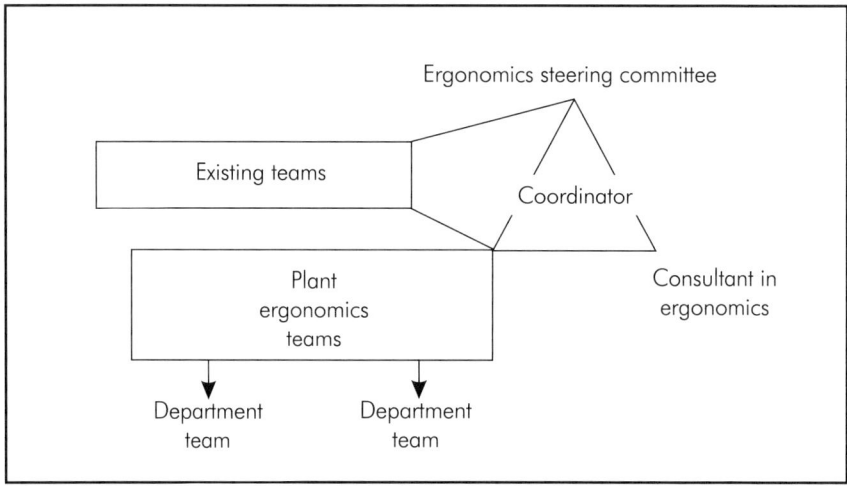

Figure 2-3 Participatory ergonomics process structure.

The structure helps facilitate the participative effort as a change agent. The structure should be flexible, allowing continuous improvement to evolve as a long lasting initiative for the company. If there is no structure, these initiatives will not be long lasting.

Although it is beneficial to have many workers directly represented in a participative team effort, it may not be feasible. However, workers could be involved in varying levels by participating as ad hoc members when actually analyzing their workstation in the problem-solving analysis session.

Develop proactive communication with designers, planners, and schedulers and encourage them to apply ergonomics principles in

> **Basic Structural Considerations for a Participative Ergonomics Model**
>
> - Ergonomics Steering Committee (ESC)
>   - Focus on macro versus microergonomics.
>   - Has and can transfer authority.
>   - Can break barriers, run interference.
>   - Sets boundaries.
>   - Plan for long-term program needs and direction of ergonomics process.
>   - Considers business issues and the relationship with ergonomics.
> - Coordinator and facilitators should be good listeners, delegators, and project managers and should understand other business issues and needs.
> - Plant Ergonomics Team (PET) or Departmental Ergonomics Team (DET)
>   - Focus on micro versus macroergonomics.
>   - Has been given authority.
>   - Regular progress reports made to ESC.
>   - Responsible for effort on a day-to-day basis.
>   - Makes recommendations to ESC.
>   - Integrates with other business initiatives.
>   - Overall responsibility for direction of organizational/departmental ergonomics.
>   - Involves employees off the line as ad hoc members to help PET or DET with specific ergonomics projects or when analyzing their workstation.

designing for manufacturability to enhance work. This should be a separate, but integrated, effort from the reactive ergonomics efforts to redesign work and workplaces. Allow the ergonomics team (Fig. 2-4) to participate with this effort as well.

Team Parameters

The following points should be kept in mind when considering team boundaries and parameters.

1. Use other teams, such as quality and injury reduction, that have already been established to help supplement the ergonomics effort. Even if these other teams are not given direct

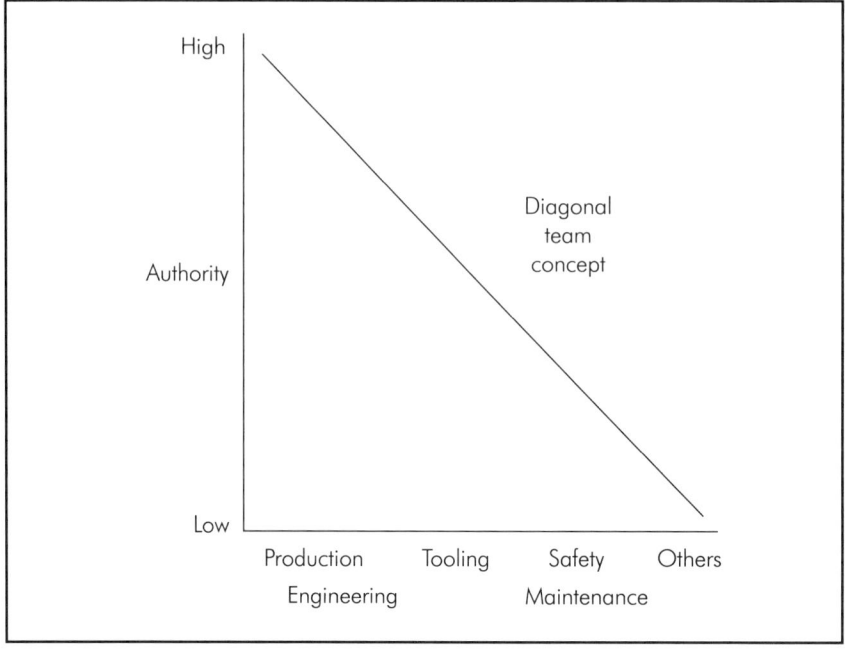

Figure 2-4 Participative team effort. Teams should contain a variety of functions and authority levels.

ergonomics projects, at least involve them as often as possible, perhaps through regular meetings.
2. Keep in mind all the initiatives to better integrate ergonomic efforts and teams into business and the bottom line.
3. In establishing meeting lengths and frequencies, keep the following in mind:
   - Initial efforts should be more frequent, with meetings held at least every other week;
   - Give the team enough time for practice and practice to gain confidence;
   - One hour every week may not be a long enough time frame to meet, but it does keep the ergonomics effort fresh in the minds of leadership;
   - Two hours allow for enough concentrated effort, but don't detract too much from production duties;

- When processes mature, the team should still meet at least once a month;
- Special needs teams can meet more often, or as needed.
4. Set and keep meeting agendas, take and keep notes, and maintain an ergonomics project file.
5. Team size should be 4-6 regular team members. Too small a team is viewed as a special interest project; too large and it becomes unmanageable.
6. The team can include two members from the affected work area and can recruit additional members from various other support departments.
7. The team should be primarily employees/operators that do not hold leadership positions.
8. Develop specific technical assistant networks for the team, including EHS, engineering, purchasing, and finance.
9. Have the teams report out to area or department leadership and plant leadership either after projects are completed or on a regular basis (at least once a month).
10. Train as a team to build capability as well as compatibility. Work as a team to facilitate the team learning curve.

The team should have a process or road map for analyzing work and workstations, clarifying and focusing on the problems, and arriving at better, longer-lasting solutions.

ANALYSIS PROCESS

The analysis process should be a road map—giving consistency and a standard methodology in approaching as many workplace tasks and situations as possible across all organizations in your facility. Whatever the analysis process, it can be very important for initiating and maintaining an effective ergonomics effort. An effective analysis process will incorporate some of these steps (Rodgers 1989; Day and Rodgers 1992):

1. Focus the team's efforts.
2. Enhance leadership awareness and appeal to the logic of doing business.
3. Be simple enough for nontechnical individuals to understand.
4. Basic problem-solving steps, including a) identification of the concern or problem, b) identification and integration of contributing factors (possible root cause), and c) identification of alternative solutions.

The analysis process should not leave out the scientific principles of ergonomics.

If you overanalyze or overcontrol, you reduce participation. Generally, this also takes longer and creates a negative response from leadership and technical support. Likewise, if the analysis process is very technical and inflexible, it will be seen by the employees as not trustworthy. If the analysis process is oversold or subjective, it will be viewed as untrustworthy by all personnel. If the analysis is just brainstorming or just a quick fix, it may not address the root cause, thus bypassing the science of ergonomics.

There are different ergonomic analysis models as well as avenues to consider for the analysis. The "Workplace Analysis Process Flow" (Fig. 2-5) should be relatively easily understood by the ergonomics team. Participation occurs throughout the process, but particularly during the observation, fatigue analysis, and problem solving steps. This particular flow is primarily for repetitive work.

Basic considerations for analysis (Fig. 2-6) would be the following:

1. Due to all the different situations and contributing factors addressed by the science of ergonomics, integration of the various issues and analysis methodologies must be considered. Ergonomics can influence 80-90% of the quality and safety factors of the problems analyzed.
2. Have options for different workplace analysis needs such as quality, safety, nonrepetitive and repetitive work.
3. After the data collection and review, survey/workplace measurements, and initial analysis have been completed, a problem solving approach should be applied by the analysis teams.
4. For those situations that require more technical study, experts or consultants can be brought in to help facilitate or perform the analysis. However, the expert must work closely with the analysis teams.

KEEPING IT GOING

Ways and means of keeping an ergonomics process going and adding fresh ideas for the longevity of the process are many, but they won't work in all situations and cultures. The following points are specific to keeping an ergonomics process going:

1. Make a plan, review it, work it, and measure against it.
2. Occupational health, ergonomics, safety, and industrial hygiene need to show value added to the business.

Participatory Ergonomics—A Practical Guide for the Plant Manager

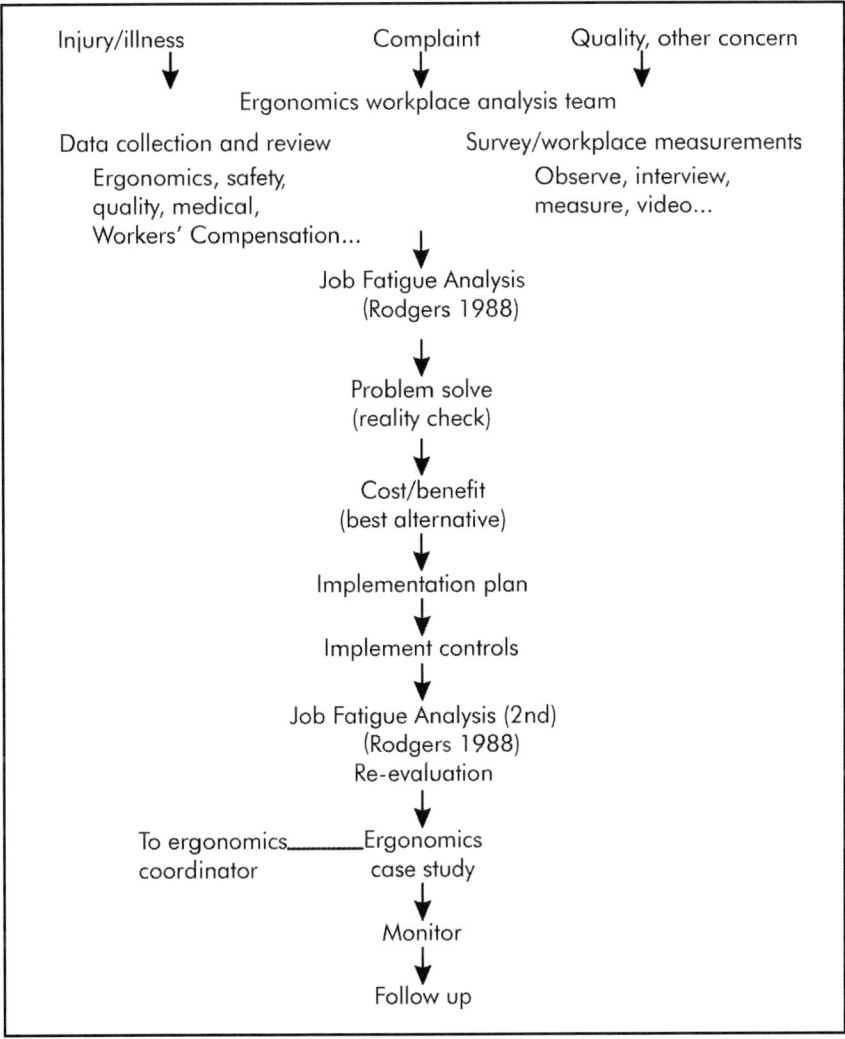

Figure 2-5 Ergonomics workplace analysis process flow (Adapted from S. H. Rodgers 1988 and 1992).

3. Once the team has the capability, allow members to train the rest of the plant employees in general ergonomics awareness.
4. These same team members can train other teams or new team members to problem solve.

Ergonomics in Manufacturing

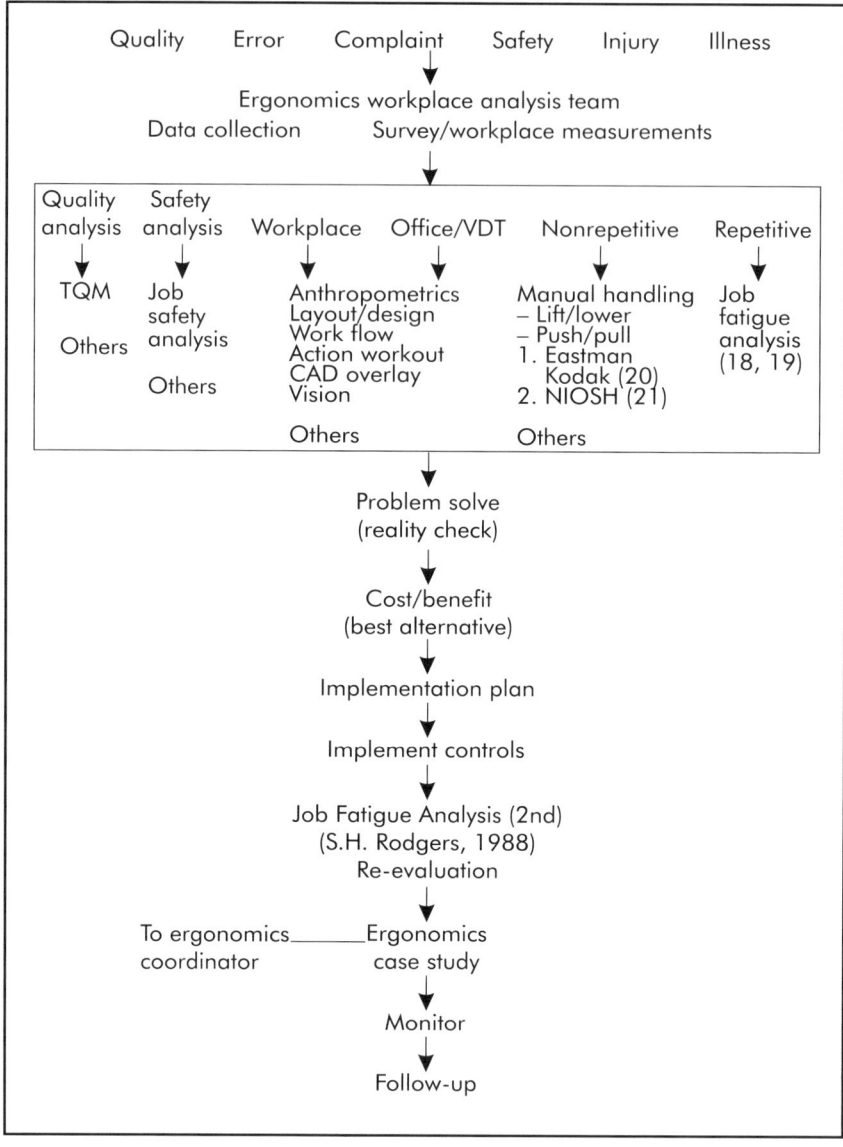

Figure 2-6 Considerations for the workplace analysis process flow.

5. Implementation should be the responsibility of the engineering/facility/maintenance.
6. Rotate team members so they do not stagnate or burn out.

7. Stagger or overlap when team members leave the team so that a team is not comprised of just new members.
8. Experienced members can coach new members.
9. Allow the team to be involved with special projects such as occasional on-site vendor shows, trips to other companies and plants, and trips to seminars and shows such as those with a focus on tooling or manual handling devices.
10. Share successes through general communication tools, (maybe with a special " ergo day" attended by all plant personnel).
11. Audit or review accomplishments and needs against the internal ergonomics program on a regular basis.

Be patient. These types of cultural changes can take three to four years to develop. One major company has seen it take 10 years to develop the culture needed to implement design for manufacturability. Some companies have stated that it is a never-ending journey of continuous improvement.

MONITORING PROGRESS AND EVALUATING THE PROCESS

Procedures and mechanisms should be developed to evaluate implementation of the program and monitor progress. Programs to address the concerns identified within the data are an integral part of ergonomics and safety performance management at both the organizational and departmental levels. This effort in ergonomics and safety performance measurement is an important step in the goal of providing:

- An ergonomics and safety program based on quality management principles and business needs; and
- An aid toward empowering working teams with the knowledge and tools to better perform their ergonomics and safety responsibilities.

You may want to utilize an expert in the participatory sciences or in participatory ergonomics to assist you with the employee survey design and administration.

Data

Evaluation may include the following data:

Injury and illness record analysis. Identifies high risk jobs and monitors trends over time, utilizing OSHA 200 injury and illness records, workers' compensation claims and costs, and safety management reports.

Employee surveys. All employees or targeted groups could be surveyed for symptoms of musculoskeletal discomfort and other organizational health concerns. These surveys can be done on a more frequent basis early in the initial phases of the ergonomics process, and then less frequently as time goes on.

Facility plant organization evaluations. Mechanisms should be developed to follow through on identified problem areas. Time lines should be established and used to ensure ergonomics projects are completed on a timely basis.

Ergonomics process and projects log/management. Each organization could maintain a log that tracks identified ergonomic issues by project management processes, such as responsibilities, needs, milestones, time lines, and effectiveness.

Some measurement bias will affect the evaluation results due to such factors as: 1) previous exposure to ergonomics and participatory efforts; 2) corporate culture; 3) importance placed in the program; and 4) behavior in the target population and areas to be studied. Since there is generally some increase in reporting incidents and concerns when a program is initiated, these points should be considered when planning, developing, and maintaining an ergonomics management program.

METHODS

Descriptive, as well as goal fulfillment, methods should be the preferred means to measure effectiveness of the ergonomics management program.

Descriptive Measurement Criteria

Current rates and numbers should be compared to past rates and numbers prior to implementing the ergonomics process. Fixes and projects for specific ergonomic issues should be identified within the organization or workplace and evaluated for effectiveness. Periodic analysis, monthly and/or quarterly and annually, should be done using past rates and numbers before any new ergonomics program is implemented. Acting as a benchmark for measuring improvements and program effectiveness, the analysis should include:

1. Injury and illness categories, including nature of incident, body part involved, activity being done, type of activity, and source of the incident. Give other employee details such as years with company, shift, time of day, and supervisor to identify specific ergonomic concerns.

2. OSHA incident rates compared to similar SIC codes, number of cases, lost time, and restricted cases.
3. Worker compensation incident numbers, time to maturation and time to close.
4. Worker compensation costs, including total dollars as well as dollars per case.

Goal Fulfillment Measurement Criteria

Goal fulfillment measurement criteria may be the best method to assess the participatory aspect of participatory ergonomics. Much of the desired change will be related to the culture of the company, perception, and behaviors of the employees. Such goals would include the improvement of:

- Job satisfaction—teamwork, self management;
- Quality of product—less poor product missed, or good product reworked;
- Productivity—increase throughput, etc.;
- Quality of work life;
- Accommodation of injured/ill worker—quicker and more successful returns to work;
- Design and redesign efforts; and
- Communication within and across organizational lines.

Evaluation Methods

Analysis should be done using two or more factors, typically before and after the program's initiation. At first, look for simple changes that keep the process open and flexible. More in-depth effects can be made by comparisons, using statistical analysis.

1. Job satisfaction—teamwork, self management (survey).
2. Quality of work life (survey, complaints [grievances if appropriate]).
3. Quality of product (numbers, costs, percent scrap, rework, changes).
4. Productivity—decrease throughput (numbers, costs, time, training time).
5. Accommodation of injured/ill worker (time, numbers, costs, process improvement).
6. Design and redesign efforts (numbers and costs, process improvement).

7. Psycho-physiological discomfort (survey).
8. Number of employees participating, number of active teams, team activities, meetings, completed (implemented) projects.

OPPORTUNITIES

There are many opportunities afforded by participatory ergonomics. Probably the most advantageous opportunity is to better utilize the most valuable asset in business and industry today, the employee. The opportunity to have input and some degree of control over the employee's working life is very motivating and rewarding. This opportunity is not just limited to the hourly worker. Supervisors will also have more time to devote to business needs. Perhaps most important, employees will feel more freedom to do their jobs because of the attention to ergonomics and safety issues since they will be able to glean for input from others who are actually more knowledgeable. There is also great opportunity for enhanced communications through a sound structure and analysis process.

Another major opportunity that may not always be appreciated is in improved customer relationships. In the customer's view, the image of having workers help in making the work and workplace more streamlined and efficient through a sound structure and process can be a sales motivator. The decrease in injuries and illnesses and the associated costs of making a product could result in lower overhead cost, a savings that could be passed on to the customer. When customers visit the production site, they will have an image of a safe, clean, and more efficient workplace.

REFERENCES

Allen, L.A. " The Care Crisis: Hospitals Need New Leadership." *Management Review* January 1991: 46-49.

Amoroso, C.R. "Optimal Work Systems." *Personal Communications*, 1988.

Auguston, K.A. "Polaroid's Journey to Materials Handling Excellence. Part I: Getting Started." *Modern Materials Handling* July 1989: 60-63.

Brown, O. " Marketing Participatory Ergonomics: Current Trends and Methods to Enhance Organizational Effectiveness." *Ergonomics* 1990, 33(5): 601-604.

Day, D.E., and Rodgers, S.H. " Problem-solving Methodologies in Ergonomics." Springfield, TN: Brouha Work Physiology Symposium, 1992.

Della-Giustina, J.L., and Della-Giustina, D.E. "Quality of Work Life Programs Through Employee Motivation." *Professional Safety* May 1989: 25-28.

Garrigou, A., Daniellou, F., Carballeda, G., and Ruaud, S. "Activity Analysis in Participatory Design and Analysis of Participatory Design Activity." *International*

Journal of Industrial Ergonomics, Special Issue: Participatory Ergonomics 1995, 155: 311-329.

Hendrick, H.W. "Macroergonomics: A Concept whose Time has Come." *Human Factors Society Bulletin* 1987: 30(2): 1.

lmada, A.S., and Nagamachi, M. "Introduction to a Special Issue on Participatory Ergonomics." *International Journal of Industrial Ergonomics, Special Issue: Participatory Ergonomics* 1995: 15.

Kavianian, J.K. Rao, and Sanchez, V.F. "Management Thinking and Decision-making Styles: Their Effect on Occupational Safety and Environmental Health." *Professional Safety* September 1989: 24-27.

Nagamachi, M. "Requisites and Practices of Participatory Ergonomics." *International Journal of Industrial Ergonomics, Special Issue: Participatory Ergonomics* 1995, 15(5): 371-379.

Pasmore, B. *Designing Effective Organizations*. New York: John Wiley & Sons, 1990.

Proctor, B.H. "A Sociotechnical Work-Design System at Digital Enfield: Utilizing Untapped Resources." *National Productivity Review*, Summer 1986: 262-270.

Rodgers, S.H., ed. *Ergonomic Design for People at Work,* Vol. 1. New York: Van Nostrand Reinhold, 1983.

Rodgers, S.H. Rochester, New York: Brouha Work Physiology Symposium, 1984.

Rodgers, S.H. "Job Evaluation in Worker Fitness Determination." *Occupational Medicine: State of the Art Reviews*. Philadelphia: Hanley & Belfus, 1988.

Rodgers, S.H. St. Malo, Colorado: Brouha Work Physiology Symposium, 1989.

Rodgers, S.H. "A Functional Job Analysis Technique." *Occupational Medicine: State of the Art Reviews*. Philadelphia: Hanley & Belfus, 1992.

Rodgers, S.H., ed. *Ergonomic Design for People at Work*, Vol. 2. New York: Van Nostrand Reinhold, 1986.

Sherwood, J.J. "Creating Work Cultures with Competitive Advantage." *Organizational Dynamics* Winter 1988: 5-27.

Waters, T.R., Putz-Anderson, V., Garg, A., and Fine, L.J. " Revised NIOSH Equation for the Design and Evaluation of Manual Lifting Tasks." *Ergonomics* 1993, 36(7): 749-776.

Wilson, J.R., and Corlett, E.N., eds. " Ergonomics and Participation In." *Evaluation of Human Work: A Practical Ergonomics Methodology*, 2nd Edition. London: Taylor & Francis 1995, 1071-1096.

CHAPTER 3

LOWERING COSTS THROUGH ERGONOMICS*

Hal W. Hendrick
Emeritus Professor of Human Factors
University of Southern California
Los Angeles, CA

One of the clearest ways to delineate a discipline is by its unique technology. The Human Factors and Ergonomics Society (HFES) of the United States has noted, as have others internationally, that the technology of human factors/ergonomics is human-system interface technology (Hendrick 1996). Thus, the discipline of human factors can be defined as the development and application of human-system interface technology.

Human-system interface technology deals with the interfaces between humans and the other system components, that include hardware, software, environment, jobs, and organizational structures and processes. Like the technology of other design-related disciplines, it also includes principles, specifications, guidelines, methods, and tools.

As noted by the HFES, ergonomists use this technology for improving the quality of life, including health, safety, comfort, usability, and productivity. Regarding it as a science, ergonomists study human capabilities and limitations for the purpose of developing human-system interface technology. As a practice, the technology is applied by ergonomists to the analysis, design, evaluation, standardization, and control of systems. It is this technology that clearly defines human factors/ergonomics as a unique, stand-alone discipline that identifies what ergonomists do and what ergonomics offers for the betterment of society.

Human factors/ergonomics professionals have long recognized the tremendous potential of their discipline for improving the health, safety, and comfort of persons, and human and system pro-

---

*Reprinted with modifications and additions from Proceedings of the Human Factors and Ergonomics Society 40th Annual Meeting. Santa Monica, CA: Human Factors and Ergonomics Society, with permission.

ductivity. Indeed, through the application of human-system interface technology, ergonomics has the potential to truly make a difference in the quality of life for virtually all persons on this globe.

In light of this potential, why don't more organizations—with their urgent need to obtain employee commitment, reduce expenses, and increase productivity—give greater attention to ergonomics? Although various reasons could be cited, a major one is often lack of managerial exposure to the discipline of ergonomics. Of particular note, managers often are not aware of the cost benefits that can result from effective use of ergonomics in industrial settings. Put simply, when there is managerial commitment to effectively apply ergonomics to the design or modification of work systems and environments, not only are improvements in health, safety, comfort, and production possible, but so are considerable cost benefits (Hendrick 1996).

CASE STUDIES

FORESTRY INDUSTRY

A coordinated series of projects were undertaken by the Forest Engineering Technology Department of the University of Stellenbosch (South Africa) along with Ergotech, an ergonomics consulting firm in South Africa, to improve safety and productivity in the South African forestry industry. Many aspects of the following forest industry examples are comparable to the field of manufacturing.

Leg Protectors

An anthropometric survey of foresters was conducted to extract the basic data needed for redesigning leg protectors for this very heterogeneous work force.

The South African forestry industry is populated with a wide variety of ethnic groups, having varying anthropometric measurements. The original protector, obtained from Brazil, was modified to ergonomically improve the fastening and anthropometric dimensions, and to incorporate improved materials. Part of the ergonomic design modification process was an extensive series of usability tests over a six-month period. In a well-designed field test, the modified leg protector was introduced for use by persons responsible for ax/hatchet debranching at a eucalyptus plantation. Among the 300 laborers, an average of ten injuries per day was occurring with an average sick leave of five days per injury. During the one-year period of the test, not one single ax/hatchet leg injury occurred, resulting not only in considerable savings in human pain and suffering, but also in a direct net cost savings of $250,000. Use of the leg protectors throughout the South African hardwood forestry industry is conservatively calculated to save $4 million annually (Warkotsch 1994).

Tractor-trailer Design

A second study involved ergonomically improving the seating and visibility of 23 tractor-trailer forwarding units with an investment of $300 per unit for a logging company. This resulted in a better operating position for loading, and improved vision and operator comfort. Downtimes caused by accident damage to hydraulic hoses, fittings, etc., went down by $2,000 a year per unit. Daily hardwood extraction was increased by one load a day per vehicle. So, for a total investment of $6,900, a hard cost savings of $65,000 per year was achieved—a 1:9.4 cost/benefit ratio (Warkotsch 1994).

Other Projects

Other innovations by this same collaborative effort between Stellenbosch University, Ergotech, and various forestry companies include (a) development of a unique, lightweight, environmentally-friendly pipe type of timber chute for transporting logs down slopes more efficiently and safely, (b) redesign of three-wheeled hydrostatic loaders to reduce excessive whole-body vibration and noise, (c) classifying different terrain conditions—ground slope, roughness, etc.—and determining the most effective tree harvesting system (method and equipment) for each, and (d) developing ergonomic checklists and work environment surveys tailored to the forestry industry. All of these are expected to result in significant cost savings, as well as greater employee satisfaction and improved quality of work life.

These series of ergonomics applications provide a good example of what ergonomics can potentially contribute to any given industry when there is a true collaborative effort and commitment.

MATERIAL HANDLING SYSTEMS

One group that does a good job of documenting the costs and benefits of its ergonomic interventions is the faculty of the Department of Human Work Sciences at Lulea University of Technology in Sweden. The following examples are from their Division of Environment Technology's work with steel mills. The basic approach to ergonomic analysis and redesign in these projects was to involve employee representatives with the Lulea faculty. For each project, the economic payoff period was calculated jointly with the company's management.

Steel Pipe and Rod Handling and Stock-keeping System

A semiautomatic material handling and stock-keeping system for steel pipes and rods was ergonomically redesigned, because it had an unacceptably high noise level and rejection rate. The redesign reduced the noise level in the area from 96 dB to 78 dB, increased production by 10%, dropped rejection from 2.5% to 1%, and paid back the redesign and development costs in approximately 18 months. After that, it was all profit.

In a tube manufacturing facility, a tube handling and storage system had an unacceptably high noise level, high rejection rate from damage, inefficient product organization, a poor safety record, and required heavy lifting. Ergonomic redesign eliminated stock damage, improved stock organization, reduced lifting forces to an acceptable level, reduced the noise level by 20 dB, resulted in less accidents, and increased productivity with a payback period of only 15 months.

Forge Shop Manipulator

In a forge shop, the old manipulator was replaced with a new one having an ergonomically designed cabin and overall better workplace design. Whole body vibration was reduced, noise was reduced by 18 dB, operator sick leave dropped from 8% to 2%, productivity improved, and maintenance costs dropped by 80%.

WORKSTATION REDESIGN

Food Service Stands

Using a participatory ergonomics approach, Dr. Andy Imada of the University of Southern California and George Stawowy, a visiting ergonomics doctoral student from the University of Aachen in Germany, redesigned two food service stands at Dodger Stadium in Los Angeles (Imada and Stawowy 1996). The total cost was $40,000. Extensive before and after measures demonstrated a reduction in average customer transaction time of approximately eight seconds. In terms of dollars, the increase in productivity for the two stands was approximately $1,200 per baseball game, resulting in a payback period of 33 games, or 40% of a single baseball season. Since modification of these two stands served as the prototype for developing all vendor stands in the ballpark, the effort was costly and time consuming. Modifying the other 50 stands in Dodger Stadium, however, can now be done at a price of $12,000 per stand, resulting in a payback period of only 20 games.

The new design increases productivity by reducing customer waiting time, thereby increasing customer satisfaction. This modification effort is only one part of a macroergonomics intervention project to improve productivity. Ongoing work to improve the total system process, including packaging, storage, and delivery of food products and supplies, and managerial processes, eventually will result in an even greater productivity increase.

Fine Assembly Workstations

Typical workstations at one major electronics assembly plant resulted in poor postures leading to musculoskeletal disorders. Valrie Venda of the University of Manitoba designed a new type of fine assembly workstation that utilizes a TV camera and monitor. Not only does the TV camera provide a greatly enlarged image of the assembly work, but it enables the worker to maintain a better posture and more dynamic motion. Based on extensive comparative testing, a 15% higher productivity rate was obtained with the new workstation.

Venda reports that the average value of products assembled per worker per shift at these types of workstations varies between $15,000 and $20,000. Thus, the additional value produced by one worker per day using the new workstation is $2,250 to $3,000 per day. Although it is too early to say precisely, Venda expects that the new workstations eventually will decrease occupational injuries for these jobs by 20%.

Workstation CRT Display

The CRT display used by the directory assistants at Ameritech were ergonomically redesigned by Scott Lively, Richard Omanson, and Arnold Lund to reduce average call processing time. The redesign replaced all upper-case display with a mixed-case display and added a highlighting feature for listings selected by the directory assistant. Extensive before-and-after measurements showed a 600 millisecond reduction in average call operating time after introduction of the ergonomically redesigned CRT display. Although seemingly small, this reduction represents an annual savings of approximately $2.94 million across the five-state region served by Ameritech, according to Lively and Lund.

Training System Redesign

In a related Ameritech effort, done jointly with Northwestern University's Institute for Learning Sciences, the traditional lecture and practice training program for new directory assistants was replaced by an ergonomically designed computer-based training program that incorporated a simulated work environment and error feedback. As a result, operator training time was reduced from five days to one and a half days.

Workstation Job Aids

Soon after IBM started shipping its Displaywriter™ product to customers, a report came back that customer setup of the product was failing. Follow-up by ergonomist Daniel Kolar, president of Info Xfer, a usability consulting firm in Austin, Texas, determined that the problem was caused by frequent errors in the packing line. The packers had no idea what they were doing because they had inappropriate documentation. Kolar conducted a task analysis, then used it to develop highly pictorial story boards that detailed the specific packing steps at each station. Following installation of the story boards, the shipping error rate dropped from 35 per hundred to less than one in a thousand. IBM's cost effectiveness personnel calculated the savings at $2 million over a two year period.

Workstation Tools

A conventional deboning knife used at a poultry packaging plant did a poor job of deboning and resulted in high rates of carpal tunnel syndrome, tendonitis, and tenosynovitis. A new, ergonomically designed pistol-shaped knife was introduced by ergonomist Ian Chong, principal of Ergonomics, Inc. of Seattle, Washington. Less pain and happier cutting crews were reported almost immediately. Upper extremity work-related musculoskeletal disorders were greatly reduced, line speeds increased by 2% to 6%, profits increased because of more efficient deboning, and, over a five-year period, $500,000 was saved in workers' compensation premiums. This is a good example of how a simple, often inexpensive ergonomic solution can sometimes have a very high cost/benefit payoff.

REDUCING WORK-RELATED MUSCULOSKELETAL DISORDERS

AT&T Global Information Solutions

AT&T Global Information Solutions in San Diego, with 800 employees, manufactures large mainframe computers. Following analyses of their OSHA 200 logs, the company identified three types of frequent injuries caused by lifting, fastening, and keyboarding. The company next conducted extensive worksite analyses to identify ergonomic deficiencies. As a result, the company made extensive ergonomic workstation improvements and provided proper lifting training for all employees. In the first year following the changes, workers' compensation losses dropped by more than 75% from $400,000 to $94,000.

In a second round of changes, conveyor systems were replaced with small, individual scissors-lift platforms, and heavy pneumatic drivers with lighter electric ones. This was followed by moving from an assembly line process to one where each worker builds an entire cabinet, with the ability to readily shift from standing to sitting positions. A further reduction in workers' compensation losses to $12,000 resulted. In terms of lost work days due to injury, in 1990 there were 298; in both 1993 and 1994 there were none. These ergonomic changes have reduced workers' compensation costs at AT&T Global over the 1990-1994 period by $1.48 million. The added costs for these ergonomic improvements represent only a small fraction of the savings (Center for Workplace Health Information 1995a).

Red Wing Shoes

The Red Wing Shoes Company made a commitment to reducing work-related musculoskeletal disorders (WMSDs) via ergonomics. Steps included (a) initiating of a safety awareness program that encompassed basic machine setup and operation, safety principles and body mechanics, information about CTDs, and monthly safety meetings, (b) a stretching, exercise and conditioning program, (c) hiring of an ergonomics advisor, and (d) specialized training on ergonomics and workstation setup for machine maintenance workers and industrial engineers.

The company:

- Purchased adjustable ergonomic chairs for all seated operators and antifatigue mats for all standing jobs;
- Instituted continuous flow manufacturing, which included operators working in groups, cross training, and job rotation;
- Ergonomically redesigned selected machines and workstations for flexibility and elimination of awkward postures, and greater ease of operation; and
- Modified production processes to reduce cumulative trauma strain.

As a result of these various ergonomic interventions, workers' compensation insurance premiums dropped by 70% from 1989 to 1995, for a savings of $3.1 million. During this same period, the number of OSHA-reportable lost time injury days dropped from 75 for 100 employees working a year, to 19. The success of this program is attributed to upper management's support, employee education and training, and having everyone responsible for coordinating ergonomics (Center for Workplace Health Information 1995b).

Reducing WMSDs via Ergonomics Training

In 1992, Bill Brough of Washington Ergonomics conducted a one-day seminar for cross disciplinary teams of engineers, human resource management personnel, and safety/ergonomics committee members from seven manufacturing companies insured by Tokyo Marine and Fire Insurance Company, Ltd. The seminar taught basic principles of ergonomics and provided materials to implement a participatory ergonomics process. The training focused on techniques for involving workers in evaluating present workplace conditions and making cost-effective improvements.

The class materials provided the tools for establishing a baseline, setting improvement goals, and measuring results. In six of the companies, the seminar data and materials used by the teams to implement a participatory ergonomics program received funding from management and support from labor. (The seventh company did not participate in implementing the training). Follow-up support was provided by a senior loss control consultant for Tokyo Marine. For the six companies that did participate, strain-type injuries dropped progressively from 131 in the six months prior to the training to 42 for the six month period ending 18 months later. The cost of these injuries for the six months prior was $688,344. For the latter period, the injury costs dropped to $72,600, for a net savings over 18 months of $1.34 million using the prior six months as the baseline.

Tokyo Marine traces these reductions in strain-type injuries directly back to Bill Brough's participatory ergonomics training program and related materials. A good example of what can happen when you couple collaborative management and labor commitment with professional ergonomics. Worker involvement reportedly created enthusiasm and encouraged each individual to assume responsibility for the program's success. According to Bill Brough, the reduction of injuries resulted from a commitment to continuous improvement and was fostered by many small changes, not a major singular event.

For the one company that did not participate in implementing the training, the number of reported strain injuries was 12 for the six months prior to training, and 10, 16, and 25 respectively the next three six-month periods. In short, things got worse rather than better even with management and labor's active support.

Deere and Company

One of the best known successful industrial safety ergonomics programs is at Deere and Company. In 1979, Deere recognized that traditional interventions like employee lift training and conservative medical management were, by themselves, insufficient to reduce injuries. So, the company began to use ergonomic principles to reduce the physical stresses of the job.

Eventually, ergonomics coordinators were appointed in all of Deere's U.S. and Canadian factories, foundries, and distribution centers. These coordinators, chosen from the industrial engineering and safety departments, were trained in ergonomics.

Today, job evaluations and analyses are done in-house by part-time ergonomics coordinators as well as wage-employee ergonomics teams and committees. The company has developed its own ergonomics checklists and surveys. The program involves extensive employee participation. Since 1979, Deere has recorded an 83% reduction in back injuries, and by 1984 had reduced workers' compensation costs by 32%. According to Gary Lovestead, each year, literally hundreds to thousands of ergonomics improvements are implemented. Today, ergonomics is built into Deere's operating culture (Center for Workplace Health Information 1995c).

Union Pacific

In the early 1980s, the Palestine Car Shop near Dallas, Texas, had the worst safety statistics of the Union Pacific Railroad's shop operations. Of particular note was the high incidence of back injuries. For example, in 1985, 9 out of 13 lost-time injuries were back injuries, and 579 lost and 194 restricted or limited work days were accumulated. Only 1,564 cars were repaired that year, and absenteeism was 4% (American Association of Railroads 1989).

The University of Michigan Center for Ergonomics' computer model for back compression was modified and expanded for easy application to the railroad environment. It was packaged by the Association of American Railroads (AAR). The AAR back model was introduced at the Palestine Car Shop to identify job tasks that exceeded acceptable back compression values and equipment that supported various lifting jobs was redesigned. For example, a coupler knuckle storage table was designed for storing the 90-pound knuckles, rather than piling them on the ground and lifting them from there. In addition, a commercial back injury training program was adopted, and every employee was taught how to bend and lift safely.

Finally, management attitude and priorities about safety were conveyed through weekly meetings with safety captains from each work area and quarterly meetings with all shop employees. From 1985 to 1988, the total incidents of injury went from 33 to 12 per year. Back incidents went from 13 to 0; lost days from 579 to 0; restricted days from 194 to 40 (all from minor, nonback injuries); and absenteeism from 4% to 1%. The number of cars repaired per year went from 1,564 in 1985 to 2,900 in 1988, an increase in dollar value of $3.96 million. Union Pacific calculated the cost/benefit ratio as approximately 1:10.

HUMAN FACTORS TEST AND EVALUATION

NYNEX, one of the "Baby Bells," developed a new workstation for its toll and assistance operators who help customers complete their calls and record the correct billing. The primary reason for the newly designed workstation was to help operators reduce average time per customer.

The existing workstation (in use for several years), employed a 300-baud, character-oriented display and a keyboard on which functionally related keys were color coded and spatially grouped. This functional grouping often separated common sequences of keys by a large distance on the keyboard. In contrast, the new workstation, ergonomically designed with sequential as well as functional considerations, incorporated a graphic, high-resolution 1,200-baud display, used icons, and, in general, was a good example of a graphical user interface because the designers paid careful attention to human-computer interaction issues.

Wayne Gray and Michael Atwood of the NYNEX Science and Technology Center, and Bonnie John of Carnegie Mellon University designed and conducted a comparative field test, replacing 12 of the current workstations with 12 new ones (Gray, John, and Atwood 1993). In addition, they conducted a goals, operators, methods, and selection rules (GOMS) analysis (Card, Moran, and Newell 1980) in which both observation-based and specification-based GOMS models of the two workstations were developed and used.

Contrary to expectations, the field test demonstrated that average operator time was 4% slower with the new workstation than with those to be replaced. The GOMS analyses accurately predicted this outcome, demonstrating the validity of the GOMS models for efficiently and economically evaluating telephone operator workstations. Had this test and evaluation not been conducted, and had the presumably more efficient workstation been adopted for all 100 operators, the performance decrement cost per year would have been $2.4 million. This is good example of the value of performing careful human factors tests and evaluations before decisions are made.

MACROERGONOMICS

Petroleum Distribution Company

Several years ago, Andy Imada of the University of Southern California began a macroergonomic analysis and intervention program to improve safety and health in a company that manufactures and distributes petroleum products. The key components of this intervention included an organizational assessment that generated a strategic plan for improving safety, equipment changes to improve working conditions and enhance safety, and three macroergonomic classes of action items. These items included improving employee involvement and integrating safety into the broader organizational culture. The program utilized a participatory ergonomics approach involving all levels of the division's management and supervision, terminal and filling station personnel, and truck drivers.

Over the course of several years, many aspects of the system's organizational design and management structure and processes were examined from a macroergonomics perspective and, in some cases, modified. Employee-initiated ergonomic modifications were made to some of the equipment. New employee-designed safety training methods and structures were implemented. Employees were given a greater role in selecting new tools and equipment related to their jobs.

Two years after the program's initial installation, industrial injuries were reduced by 54%, motor vehicle accidents by 51%, off-the-job injuries by 84%, and lost work days by 94%. Within four years, further reductions occurred for all but off-the-job injuries, which climbed back 15% (Nagamachi and Imada 1992). The company's area manager of operations reports that he continues to save one-half of one percent of the annual petroleum delivery costs every year as a direct result of the macroergonomics intervention program. This amounted to a net savings of approximately $60,000 per year over three years and the savings is expected to continue. Imada reports that perhaps the greatest reason for these sustained improvements has been the successful installation of safety as part of the organization's culture (Nagamachi and Imada 1992).

L.L. Bean

The use of macroergonomics was reported as an approach and methodology for introducing total quality management (TQM) at the L.L. Bean Corporation (Rooney, Morency, and Herrick 1993).

Using methods similar to those described for Imada's intervention (Nagamachi and Imada 1992), but with TQM as the primary reason, the production and distribution division experienced more than a 70% reduction in lost time accidents and injuries within a two year period. Other benefits, such as greater employee satisfaction and improvements in additional quality measures, also were achieved. Given the present emphasis in many organizations on implementing ISO 9000, these results take on an even greater significance.

CONCLUSION

The foregoing case studies are only limited examples of the variety of ergonomic interventions that the human factors/ergonomics profession can provide to improve the human condition, as well as the bottom line. Truly good ergonomic interventions are almost always beneficial especially when managers use effective criteria in allocating resources.

Ergonomic interventions can offer a wonderful common ground for labor and management collaboration. Invariably, both benefit. Managers benefit in reduced costs and improved productivity and employees in improved safety, health, comfort, and usability of tools and equipment. Both groups derive increased competitiveness leading to long-term organizational survival.

REFERENCES

American Association of Railroads. "Research Pays Off: Preventing Back Injuries." AAR program adopted at Union Pacific. *TR News* 1989, 140: 16-17.

Card, S.K., Moran, T.P., and Newell, A. "Computer Text Editing: An Information Processing Analysis of a Routine Cognitive Skill." *Cognitive Psychology* 1980, 12: 32-74.

Center for Workplace Health Information. "An Ergonomics Honor Roll: Case Studies of Results-oriented Programs." AT&T Global. *CTD News Special Report: Best Ergonomic Practices* 1995a: 4-6.

Center for Workplace Health Information. "An Ergonomics Honor Roll: Case Studies of Results-oriented Programs, Red Wing Shoes." *CTD News Special Report: Best Ergonomic Practices,* 1995b: 2-3.

Center for Workplace Health Information. "An Ergo Process that Runs Like a Deere." *CTD News 8,* August 1995c: 6-10.

Gray, W.D., John, B., and Atwood, M. "Project Ernestine: Validating a GOMS Analysis for Predicting and Explaining Real-world Task Performance." *Human-Computer Interaction* 1993, 8: 237-309.

Hendrick, H.W. "Presidential Address: The Economics of Ergonomics is the Ergonomics of Economics." *Proceedings of the Human Factors and Ergonomics*

Society 40th Annual Meeting. Santa Monica, CA: Human Factors and Ergonomics Society, 1996: 1-10.

Imada, A.S., and Stawowy, G. "The Effects of a Participatory Ergonomics Redesign of Food Service Stands on Speed of Service in a Professional Baseball Stadium." *Human Factors in Organizational Design and Management-V*. Amsterdam, The Netherlands: North-Holland, 1996.

Nagamachi, M., and Imada, A.S. "A Macroergonomic Approach for Improving Safety and Work Design." *Proceedings of the 36th Annual Meeting of the Human Factors and Ergonomics Society*. Santa Monica, CA: Human Factors and Ergonomics Society, 1992: 859-861.

Rooney, E.F., Morency, R.R., and Herrick, D.R. "Macroergonomics and Total Quality Management at L.L. Bean: A Case Study." *Advances in Industrial Ergonomics and Safety-V*. London, England: Taylor & Francis, 1993: 493-498.

Warkotsch, W. "Ergonomic Research in South African Forestry." *Suid-Afrikaanse Bosboutydskrif* 1994, 171: 53-62.

CHAPTER 4

MANUFACTURING WORKSTATION DESIGN

Biman Das
Professor
Industrial Engineering
Department of Industrial Engineering
Dalhousie University
Halifax, Nova Scotia, Canada

An ergonomics approach to the design of a workstation for the manufacturing industry endeavors to obtain an adequate balance between worker capabilities and work requirements. The aim is to optimize worker productivity and the total system, as well as enhance worker physical and mental well-being, job satisfaction, and safety.

A manufacturing workstation is often designed in an arbitrary manner, with little consideration to the anthropometric measurements of the anticipated user. The situation is aggravated by the lack of usable design parameters or dimension (Das and Grady 1983a; Das 1987). The physical dimensions in the design of a manufacturing workstation are of major importance from the viewpoint of production efficiency and operator physical and mental well-being. Small changes in workstation dimension can have a considerable impact on worker productivity and occupational health and safety. Inadequate posture caused by an improperly designed workstation causes static muscle efforts, eventually resulting in acute localized muscle fatigue. Consequently, it decreases productivity, and increases possibility of operator-related health hazards (Corlett et al. 1982).

For a successful industrial workstation basic design, rules or guidelines have been provided by Khalil (1972), Tichauer (1975), Eastman Kodak Company (1983), Corlett (1988), Konz (1995), and Corlett and Clark (1995). Corlett (1988) has proposed the most com-

---

The contributions made by Dr. Arijit K. Sengupta in the preparation of this chapter are gratefully acknowledged.

prehensive guidelines that include task design and workplace layout. These guidelines, emphasizing the requirement of healthy operator posture, are directed toward improving the operator's physiological efficiency. Many painful afflictions of the musculoskeletal system, known as cumulative trauma disorders (CTDs), are associated with working posture. These are sometimes caused and aggravated by the repeated forceful exertions connected with awkward work postures of the upper extremities (Armstrong et al. 1986).

An attempt is made in a workstation design to achieve an optimum compromise between the variable anthropometry of the targeted operator population and the physical size and layout of the workstation components. An ergonomic analysis for a workstation design is concerned with spatial accommodation, posture, reaching abilities, clearance, interference of the body segments, field of vision, available strength of the operator, and biomechanical stress. The appropriate anthropometric data regarding body size, strength, segment masses, and inertial properties from the established data bases are typically used in the analysis.

An obstacle in implementing the ergonomic recommendations in a real-world design situation is human variability in size and capability. It is a challenge to designers to come up with solutions that optimally fit the diverse anthropometry of the users and satisfy their task demands (Das 1991).

APPLYING ENGINEERING ANTHROPOMETRY TO WORKSTATION DESIGN

For the design of a workstation, Das and Grady (1983a, 1983b) determined design dimensions by using existing anthropometric data, so that these dimensions could be readily employed by a designer. Workspace design dimensions were determined for industrial tasks in sitting, standing, and sit-stand positions. Worker populations consisted of a combination of male and female workers and the individual male and female workers for the 5th, 50th, and 95th percentiles based on existing anthropometric data.

The normal and maximum reach dimensions were based on the most commonly used industrial operations, which require a grasping movement or thumb and forefinger manipulations. However, appropriate allowances were provided to adjust reach dimensions for other types of industrial operations. The normal and maximum horizontal and vertical clearances and their reference points were established to facilitate the design. The concepts developed by Farley (1955) and Squires (1956) were used to describe the workspace envelope for the individual worker. The dimensions of smaller (5th percentile) and larger (95th percentile) workers were used to determine the limits of reach and clearance requirements, respectively.

ADJUSTING ANTHROPOMETRIC DATA

The existing anthropometric data were derived on the basis of measurements from nude subjects. Therefore the data were adjusted for clothing and shoe allowances. Since the data for the standing, sitting, and eye heights were based on erect positions at work or rest, the data were adjusted to account for the "slump" posture involved in the normal standing and sitting positions. Also, necessary adjustments were made for the reach dimensions used for performing various industrial operations (Hertzberg 1972).

The corrected or adjusted anthropometric measurements to account for clothing, shoe and slump posture are presented in Table 4-1. The measurements were used subsequently to generate mathematically, the pertinent design dimensions of workplace layout (Das and Grady 1983a).

DETERMINING WORKSTATION DESIGN PARAMETERS

To design a workstation for manufacturing, it is necessary to obtain relevant information or data on task performance, equipment, working posture, and environment. For a new workstation design, it is advantageous to obtain such information from a similar existing task/equipment situation. Several methods such as direct observation, one-to-one interviews with experienced operators, videotaping, and questionnaires can be used for this purpose. Before redesigning an existing workstation in industry, often it is desirable to conduct a worker survey to determine the effect of the equipment or system design on employee comfort, health, and ease of use (of the equipment). The survey response is useful in reinforcing the recommended modifications to an existing workstation design based on ergonomic principles and data. However, frequently it is necessary to design an entirely new industrial workstation. Even then, it may still be desirable to obtain feedback from the operators engaged in performing a similar type of industrial task. The feedback may generate heightened awareness of workstation design problems and issues.

In the beginning, decisions are formalized regarding the task sequence, available space, equipment and tools. Work methods need to be established before embarking on the new design.

Table 4-1. Corrected anthropometric measurements to account for clothing, shoe, and slump posture

Sex	Body feature	Percentiles in. (cm)		
		5th	50th	95th
Male	Total height (slump)	65.4 (166.2)	69.3 (176.1)	73.3 (186.3)
	Body height (sitting, slump)	32.3 (82.1)	34.5 (87.6)	36.5 (92.7)
	Eye height (slump)	63.0 (160.0)	64.9 (164.9)	68.8 (174.8)
	Eye height (sitting, slump)	27.9 (70.9)	30.0 (76.2)	32.0 (81.3)
	Shoulder height	54.3 (138.0)	58.1 (147.7)	61.7 (156.8)
	Shoulder height (sitting)	21.5 (54.7)	23.5 (59.8)	25.4 (64.4)
	Body depth	10.5 (26.7)	11.9 (30.2)	13.4 (34.0)
	Elbow-to-elbow	15.7 (40.0)	17.8 (45.1)	20.4 (51.7)
	Thigh clearance	5.6 (14.2)	6.4 (16.2)	7.3 (18.5)
	Forearm length	14.6 (37.1)	15.9 (40.4)	17.2 (43.7)
	Arm length	26.9 (68.3)	29.6 (75.2)	32.3 (82.0)
	Elbow height	41.6 (105.6)	44.5 (113.0)	47.4 (120.4)
	Elbow height (sitting)	7.4 (18.8)	9.1 (23.1)	10.8 (27.4)
	Popliteal height (sitting)	16.7 (42.4)	18.0 (45.7)	19.2 (48.7)
Female	Total height (slump)	60.6 (153.8)	64.3 (163.2)	68.4 (173.8)
	Body height (sitting, slump)	30.9 (78.5)	32.6 (82.8)	34.3 (87.1)
	Eye height (slump)	56.5 (143.6)	60.4 (153.5)	64.3 (163.4)
	Eye height (sitting, slump)	27.0 (68.6)	28.5 (72.4)	30.1 (76.5)
	Shoulder height	50.4 (127.9)	54.1 (137.5)	57.7 (146.6)
	Shoulder height (sitting)	20.6 (52.2)	22.5 (57.2)	24.3 (61.8)
	Body depth	8.6 (21.8)	9.7 (24.6)	10.9 (27.6)
	Elbow-to-elbow	14.1 (35.7)	15.0 (38.2)	17.2 (43.8)
	Thigh clearance	4.9 (12.4)	5.7 (14.4)	6.5 (16.5)
	Forearm length	12.8 (32.5)	14.4 (36.6)	16.0 (40.7)
	Arm length	23.7 (60.2)	26.0 (66.0)	28.5 (72.4)
	Elbow height	39.0 (99.0)	41.4 (105.1)	43.8 (111.2)
	Elbow height (sitting)	7.4 (18.8)	9.1 (23.1)	10.8 (27.4)
	Popliteal height (sitting)	14.7 (37.3)	16.0 (40.6)	17.2 (43.6)

Determination of the workstation dimensions usually proceeds according to the steps outlined in Table 4-2 (Das and Sengupta 1996).

The workstation design procedure commences with the collection of relevant data through direct observation, video taping, and input from experienced operators and supervisors (step 1, Table 4-2). It is necessary to identify the appropriate user population based on such factors as ethnic origin, gender, and age (step 2). The necessary anthropometric dimensions of the population are obtained or approximated from the results of the available anthropometric surveys that reasonably represent the user group. As these dimensions are taken from nude subjects in erect posture, they need to be corrected appropriately for the effect of clothing, shoe, and normal slump posture during work (Das and Grady 1983a).

In developing a manufacturing workstation, a designer should take into account the height (step 3). The work table height must

Table 4-2. A systematic approach for determining workstation design

1. Obtain relevant information on the task performance, equipment, working posture, and environment through direct observation, video recording, and/or input from experienced personnel.
2. Identify the appropriate user population and obtain the relevant anthropometric measurements or use the available statistical data from anthropometric surveys.
3. Determine the range of work height based on the type of work to be performed. Provide an adjustable chair and a foot rest for a seated operator and an adjustable work surface or platform for a standing operator.
4. Layout the frequently used hand tools, controls, and bins within the normal reach space. Failing that, they may be placed within the maximum reach space. Locate control or handle in the most advantageous position, if strength is required to operate it.
5. Provide adequate elbow room and clearance at waist level for free movement.
6. Locate the displays within the normal line of sight.
7. Consider the material and information flow requirements from other functional units or employees.
8. Make a scaled layout drawing of the proposed workstation to check the placement of individual components.
9. Develop a mock-up of the design and conduct trials with live subjects to ascertain operator-workstation fit. Obtain feedback from these groups.
10. Construct a prototype workstation based on the final design.

be compatible with worker height, whether standing or sitting. Konz (1967) found that the best working height is about 1 in. (2.5 cm) below the elbow. However, he found that the working height can vary without any significant effect on performance. The nature of the work to be performed must be taken into consideration in determining work height. For seated operators, provide an adjustable chair and foot rest, and for standing operators, an adjustable work surface or platform.

Frequently used hand tools, controls, and bins need to be located within the normal reach spaces (step 4). The items used occasionally may be placed beyond normal reach, but they should be placed within the maximum reach space. For positioning a control that requires strength, give consideration to the human strength profile in the workspace. Extreme reach space, involving twisting of trunk, should be avoided at all times. Adequate lateral clearance must be provided for the large (95th percentile) operator for ease of entry and exit at the workstation and to provide ample elbow room for ease of work (step 5).

Placement of the displays should not impose frequent head and eye movement on the operator. The optimum display height for the normal (slump) eye height is 15° downward gaze (step 6). Appropriate personnel from other functional units or departments should be consulted regarding material and information flow requirements (step 7). It is beneficial to consider the physical size of the individual components and make a scaled layout drawing of the proposed workstation to check the placement of the individual components within the available space (step 8).

Operator/workstation fit should be evaluated with a workstation mock-up and an appropriate user population (step 9). This will ensure the task demand and layout do not impose an undesirable working posture. It is desirable to check the interference of body members with the workstation components. From the interest groups, such as users, equipment manufacturers, marketing, and health and safety committees, feedback should be obtained regarding the new design. If necessary, the design should be modified. Finally, it is beneficial to construct a prototype workstation based on the final design (step 10). All steps shown in Table 4-2 may not be applicable in every industrial workstation design situation.

DETERMINING WORKSTATION DIMENSIONS

For the physical design of manufacturing workstations, the four essential design dimensions are: work height, normal and maximum reaches, lateral clearance, and angle of vision and eye height.

WORK HEIGHT

Height of the working surface should maintain a definite relationship with the operator's elbow height, depending on the type of work. The standing U.S. population work heights for the 5th, 50th, and 95th percentile female operators for performing different types of work are presented in Table 4-3. The table provides guidelines especially for the design of delicate, manual, and forceful work. Similar data for males can be obtained from Ayoub (1973), and Das and Grady (1983a).

NORMAL AND MAXIMUM REACHES

The normal reach is defined by the tip of the thumb while the forearm moves in a circular motion on the table surface. During this motion, the upper arm is kept in a relaxed downward position. The maximum reach can be considered as the boundary on the work surface in front of an operator that he/she can reach without flexing his/her torso. For performing repetitive tasks, the hand movement should preferably be confined within the normal working area. The controls and items of occasional use may be placed beyond the normal working area. Nevertheless, they should be placed within the maximum working area.

The concept of normal and maximum working areas describes the working area in front of the worker in a horizontal plane at the elbow level; the areas are expressed in the form of mathematical models (Das and Grady 1983b; Das and Behara 1995). The most frequently used area of the workstation preferably should be within the normal reach of the operator. The reach requirements should not exceed the maximum reach limit to avoid leaning forward and bad posture. The maximum working area at the elbow level is determined from the data provided in Table 4-4. The adjusted anthropometric measurements for arm length (K), shoulder height (E), and elbow height (L) are used to calculate arm radii (R) for the 5th, 50th, and 95th percentiles for females.

LATERAL CLEARANCE

A well-known approach is to design the reach requirements of the workstation corresponding to the measurements of the 5th percentile of the representative group and the clearance corresponding to the 95th percentile measurements, so as to make the workstation compatible for both small and large persons. The minimum lateral clearances at waist level are determined by adding 2 in. (5 cm) on both sides or 4 in. (10 cm) to hip breadth (standing). For determining the clearance at elbow level, Squires'

(1956) concept of the normal horizontal working area is used. The concept postulates that in describing the area, the elbow moves out away (half of body depth) from the body, in a circular path, as the forearm sweeps. Considering the elbow-to-elbow distance and the sweep of both the elbows within the normal horizontal working area and adding 2 in. (5 cm) on both sides, minimum lateral clearance at elbow level is determined. The values for lateral clearances are shown in Table 4-5.

ANGLE OF VISION AND EYE HEIGHT

Das and Grady (1983a, 1983b) have provided the eye height for standing female operators: 56.5 in. (143.6 cm) for 5th percentile, 60.4 in. (153.5 cm) for 50th percentile, and 64.3 in. (163.4 cm) for 95th percentile. For males, similar eye height data can be obtained from the previously stated source. Using trigonometry, the angle of sight can be calculated from the horizontal distance of the display from the operator's eye position.

These parameters form the basis of ensuring human/machine fit of a workstation design for the users. Nevertheless, the aforementioned factors direct the designer's attention towards the ergonomic requirements in workstation design, and form the basis of compromise in the layout design.

CASE STUDIES

SUPERMARKET CHECKSTAND WORKSTATION

The steps outlined in Table 4-2 were applied in designing a supermarket checkstand workstation. The working posture and work methods of the cashiers were recorded through direct observation (step 1). Based on the direct observation, the major shortcomings in the design of the supermarket checkstand workstation were identified. The design problems associated with the present checkstands were: reach, work height, bagging, lifting, and visual requirements.

A survey of female cashiers was conducted in three superstores to determine the effect of 1) environmental factors, 2) general fatigue induced by the task, 3) physical demand of tasks, and 4) the postural discomfort of operators during a regular working day (step 2).

The salient findings from the survey were: 1) one store had rated the temperature as unacceptable, 2) the product bin handling task and prolonged standing posture were perceived to be the most strenuous of all tasks, and 3) the mean postural discomfort rating was found to be increasing as time on the work shift increased.

Table 4-3. Female operator standing work surface height in. (cm)

Type of work	Population percentile		
	5th	50th	95th
Delicate work with close visual requirement	39.0-40.9 (99-104)	43.3-45.3 (110-115)	45.7-47.6 (116-121)
Manual work	33.1-35.0 (84-89)	35.4-37.4 (90-95)	37.8-39.8 (96-101)
Forceful work aided by upper body weight	23.2-33.1 (59-84)	25.6-35.4 (65-90)	28.0-37.8 (71-96)

Table 4-4. Female anthropometric measures for maximum reach in. (cm)

Population percentile	Arm length [K]	Shoulder height [E]	Elbow height [L]	Maximum reach [R]
5th	23.6 (60)	50.4 (128)	39.0 (99)	20.9 (53)
50th	26.0 (66)	54.3 (138)	41.3 (105)	22.8 (58)
95th	28.3 (72)	57.9 (147)	43.7 (111)	24.8 (63)

Table 4-5. Female anthropometric measurements for lateral clearances in. (cm)

Population percentile	Hip breadth [W]	Elbow-to-elbow [H]	Body depth [G]	Clearance at waist level $C1 = W+10$	Clearance at elbow level $C2 = H+G+10$
5th	15.7 (40)	14.2 (36)	8.7 (22)	19.7 (50)	26.8 (68)
50th	17.7 (45)	15.0 (38)	9.8 (25)	21.7 (55)	28.7 (73)
95th	20.5 (52)	17.3 (44)	11.0 (28)	24.4 (62)	31.9 (81)

Significantly, high postural discomfort ratings were found in the lower back, back, neck, ankle, foot, knee, and leg regions.

The dimensional compatibility between the cashiers' structural anthropometry and the workspace dimensions was evaluated using a scaled drawing of the existing checkstand, which was prepared from actual measurements taken at the site.

By employing an engineering anthropometry approach, the supermarket checkstand workstation design parameters or dimensions were determined for: 1) optimum work height, 2) normal and maximum reaches, 3) lateral clearance, and 4) angle of vision and eye height (steps 3, 4, 5, and 6). The appropriate data were used subsequently for the design of the checkstand workstation (Figure 4-1) (Das and Sengupta 1996). The superimposition of the normal and maximum horizontal and vertical working areas on the supermarket checkstand drawing facilitated the design and placement of the checkstand components (step 8). This procedure enabled placement of the components within the normal working area when possible, and failing that, within the maximum working area. Several alternative checkstand layout drawings were considered using a computer-aided design (CAD) package with human modeling capability. An adjustable and padded floor platform was provided for the cashiers. The platform height could be lowered by 2 in. (5 cm) at a time to accommodate taller cashiers. The padding was provided to reduce foot fatigue from prolonged standing.

The components that needed frequent operation were placed in front of the cashier to reduce twisting of the torso and neck. The width and depth of the laser scanner was reduced to accommodate the printer and code catalog right in front of the cashiers. The reduced scanner width improved the cashiers' reach over the conveyor belt and bag handling area. A deflector was provided on the conveyor belt to ensure that the products were within the maximum reach of cashiers. The original product bins at the left of the cashiers were eliminated and replaced by plastic bag hangers at appropriate heights and locations.

A group of cashiers, a management representative, and a representative from the equipment supplier were consulted through the various design stages. In the final stage, a mock-up wooden model was constructed to fine tune the design (step 9). Additionally, representative female subjects were used to compare the designed checkstand with the existing one. A time study of the simulated

Manufacturing Workstation Design

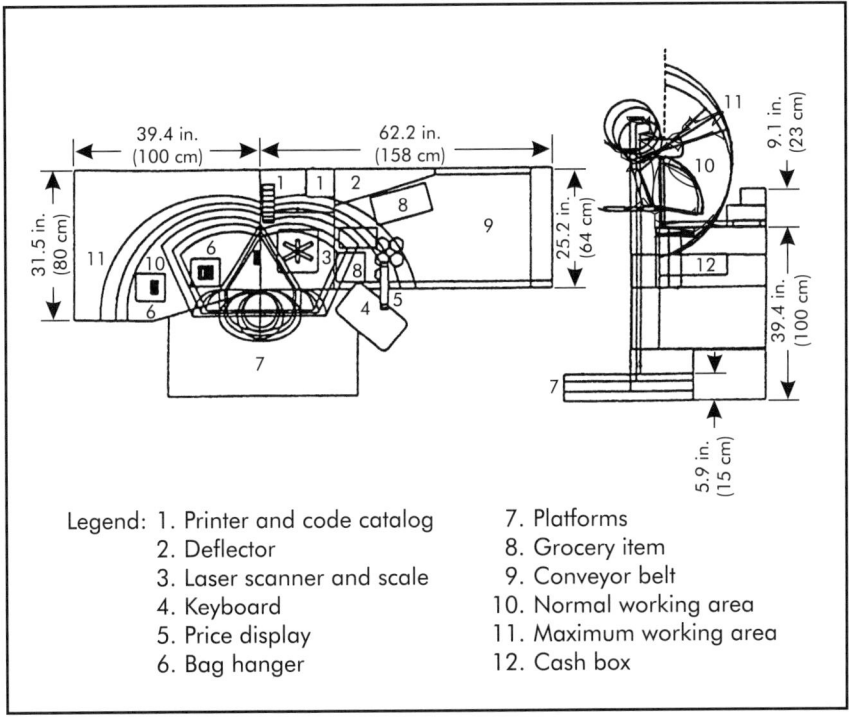

Legend:
1. Printer and code catalog
2. Deflector
3. Laser scanner and scale
4. Keyboard
5. Price display
6. Bag hanger
7. Platforms
8. Grocery item
9. Conveyor belt
10. Normal working area
11. Maximum working area
12. Cash box

Figure 4-1 The normal and maximum working areas for a supermarket checkstand workstation comprised of the 5th, 50th, and 95th percentiles for the female in the horizontal and vertical planes.

cashiers' task in the laboratory showed a 15% improvement in worker productivity.

The main improvements in the proposed design were: 1) forward facing work posture and eliminating the required torso twisting, (forward facing location of laser scanner, weigh scale, bag hangers, keyboard and printer/code cataloging), 2) increased area on the conveyor belt within the normal working area, 3) placing the visual display (item price) within the normal line of sight of the operator, and 4) adjustable height platforms to accommodate 5th, 50th, and 95th percentile female operators. The proposed design would improve working posture, provide flexible work height, reduce reach requirements, improve visual display requirements, and enhance productivity of the cashiers.

COMPUTERIZED HUMAN MODELING PROGRAMS FOR WORKSTATION DESIGN

The computerized human modeling programs for a manufacturing workstation design provide a convenient interface for the user to generate and manipulate true-to-scale, 3-dimensional (3D) images of human and workstation characteristics graphically on the video display terminal (VDT). Through their use, the designer can construct a large number of anthropometric combinations to represent the human. The programs give the user complete control over the development of the human model. They provide a comprehensive package to evaluate human/machine interaction through easy-to-understand programmed commands. The user does not need to be a computer specialist to work with such programs.

To illustrate the current state of development, Das and Sengupta (1995) selected six representative programs: CYBERMAN, COMBIMAN, CREW CHIEF, JACK, SAMMIE, and MANNEQUIN. The important features of these human modeling programs follow.

CYBERMAN

This model was developed by Chrysler Corporation (Waterman and Washburn 1978) in the late 1970s and later it was marketed by Control Data Corporation (Szmrecsanyi 1982). The basic anthropometry of the human model is based on the SAE 2D mannequins. The size of the mannequin can be changed by varying 11 scale factors. If all the scale factors are equated to one, the mannequin will have 95th percentile stature of male population.

Through system prompts, the user is asked to supply all necessary positions and postural angles of the individual segments. Upon correct input of all these data and the scale factors, the program displays the human model. The workstation model is first generated in an external CAD system and then imported into this program for analysis.

To manipulate the posture of the human model, the postural data table is called back, and the necessary angles and position data are entered. On entering the display mode, the new posture of the human model is displayed. The human model can be displayed either in a stick figure form or with a wireframe outline. However, it does not offer any external profiling with surfaces (Figure 4-2).

COMBIMAN

The COMBIMAN (computerized biomechanical man) program went through many revisions and the current version, reported by McDaniel (1990), was developed by Armstrong Aerospace Medical

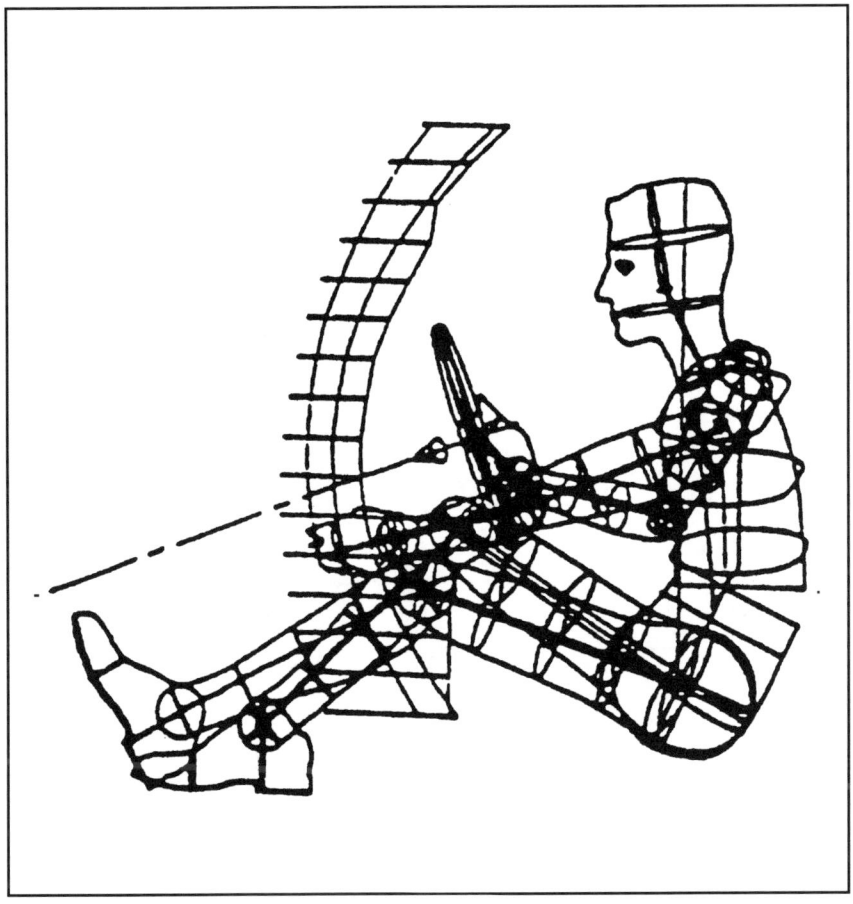

Figure 4-2 CYBERMAN in front of a control panel (Szmrecsanyi 1982).

Research Laboratory. It is used to analyze the seated workplace for airplanes and helicopters. The program contains six anthropometric data bases for males and females. The individual body dimensions of the human models can be defined in terms of any percentile (1st-99th) from any of these data bases. Alternatively, actual measurement data of an individual in terms of 12 external body dimensions can be used for sizing the human model.

The program allows analysis of reach capability with arms and legs, visual limitations, and strength for operating controls. The human model can be manipulated by simply defining the task to be performed.

Reach analysis includes reach to a single control and computing a reach envelope in the plane of an instrument panel. Figure 4-3 shows a typical reach analysis screen in the COMBIMAN program. Reach originates either from the torso or shoulder joint, depending on the selected type of shoulder restraints. To evaluate strength, if a point or control in the forward hemisphere can be reached, the program displays the amount of force that can be exerted by an operator in that position.

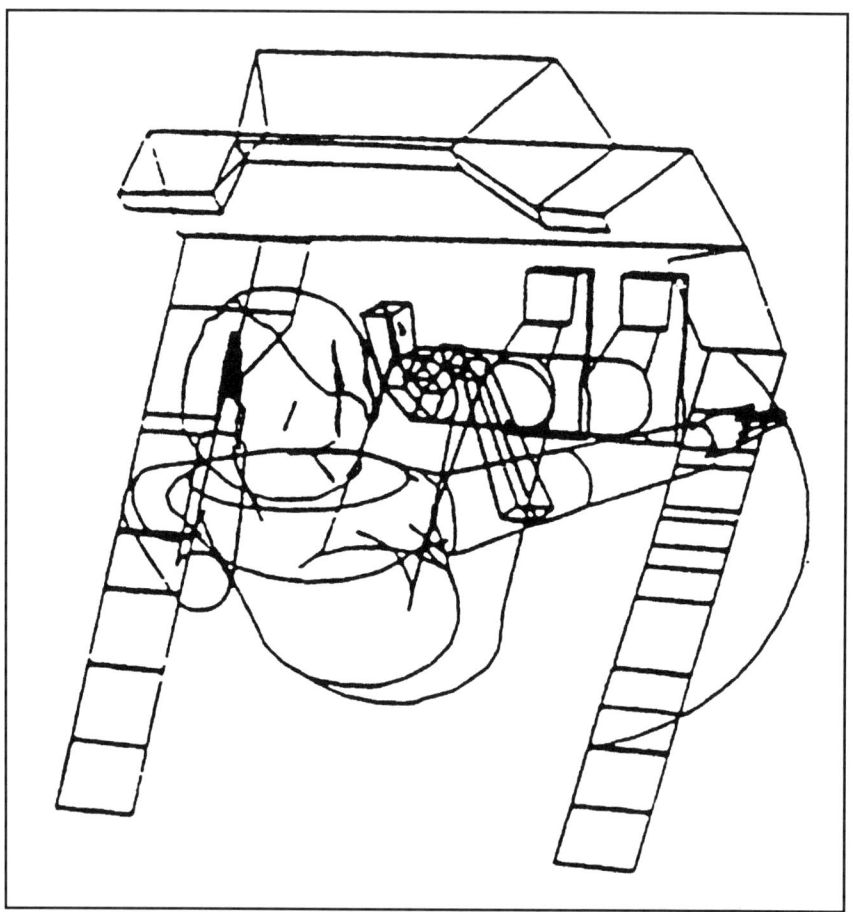

Figure 4-3 COMBIMAN human model showing the reach area on the right side console of a helicopter cockpit (McDaniel 1990).

CREW CHIEF

This program was also developed at Armstrong Aerospace Medical Research Laboratory (McDaniel 1990). It was designed to evaluate the maintainability of aircraft and other complex systems in terms of physical access for reaching into confined areas with hand tools and other objects, visual accessibility, and strength assessment for using both hand tools and manual material handling.

The CREW CHIEF human model represents a maintenance technician whose data base includes discrete 1st, 5th, 50th, 95th, and 99th percentile body segment sizes of male and female technicians. It provides 12 default postures, such as standing, sitting, stooping, squatting, crawling, in which the model can be imported into an environment and can be manipulated in task-oriented command. The user can select from four clothing types that affect the joint mobility while performing automated reaches. Similar to a COMBIMAN model, CREW CHIEF has the capabilities of reaching to a point, visual ability assessment, and image presentation. It contains 105 types of hand tools in its data base. Figure 4-4 shows a female model using a wrench to tighten a bolt. It contains strength data regarding lifting, carrying, pushing, and pulling.

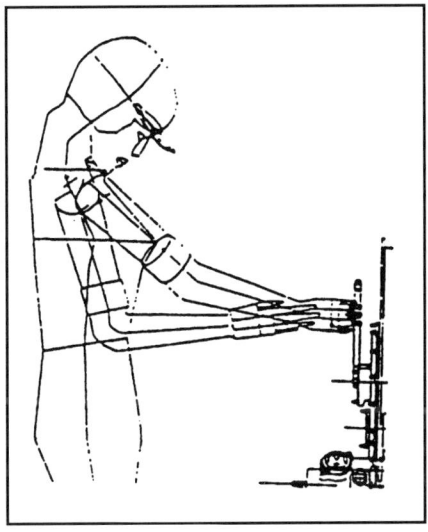

Figure 4-4 Female CREW CHIEF human model using a ratchet wrench to remove a bolt from a jet engine (McDaniel 1990).

JACK

This program was developed at the Computer Graphics Laboratory of the University of Pennsylvania (Badler 1991). The human model consists of 88 articulated joints and has a 17-segment flexible torso, which allows closer simulation of the human torso movement. Either population statistics or an actual data base of individuals can be used to provide the percentile data to carry the size of the human model. By selecting an appropriate somatotype, the external flesh profile can be altered. Default size of the model is based on the anthropometry of civilian astronauts at NASA. Additionally, it provides a data base of biostereometric-scanned body surface data of 76 subjects. Through a spreadsheet-type interface, the user can try out and modify the percentile, gender, joint limits, or individual segment lengths. Reach can be analyzed by interactive positioning of the desired hand or foot. Human vision is analyzed through a view as seen by the model or by placing translucent view cones at the model's eye point. JACK provides on-screen strength data display. Figure 4-5 shows JACK within a typical workstation.

SAMMIE

SAMMIE (system for aiding man/machine interaction evaluation) was developed at Nottingham University, U.K. (Bonney et al. 1979). The program is marketed now as a self-sufficient CAD program (Porter 1992). The SAMMIE human model is based on the dimensions derived from Dreyfus (1967). It contains 17 joints and 21 segments contoured with plane surfaces. The anthropometry menu allows the user to access and change the segment lengths independently or collectively using either a percentile value or an actual dimension. The external flesh profile can be altered by using the options available under the somatotype menu. The joint movements are realistically constrained. By moving its individual segments, the model can assume any posture. Figure 4-6 shows the use of SAMMIE in a computer workstation design. The program can produce a view of the workstation and the human model from any position. In a similar manner, it can produce the view of the workstation as seen from the human model's eye.

MANNEQUIN

The program was developed by Humancad (1991), a division of Biomechanics Corporation of America. This human modeling program runs on the personal computer (PC) within the DOS operating system. The program contains a data base of anthropometry from ten countries. Body type may be selected from three somatotypes; heavy, normal, and thin. The human model can be shown in any of the three forms with increasingly complex outer profiles, stick figure, robot, and human form. The initial posture of the human model is selected from the eight commonly adopted postures described by graphic icons.

The human model consists of 46 body segments with each hand having five individual fingers and each finger divided into three joints. The program is provided with the function to produce a perspective view that can be seen from any viewpoint and direction.

Additionally, for the analysis of the human model's vision, it provides translucent eye cones for foveal and peripheral vision. Its torque function calculates the movements of different body parts resulting from an applied load and moment on the hand and foot. Figure 4-7 shows the MANNEQUIN-generated human model within a meat-cutting workstation.

Comparative Analysis

The individual programs were compared under four broad criteria: 1) usability in terms of hardware and software; 2) anthropometry and structure of the human model; 3) model manipulation,

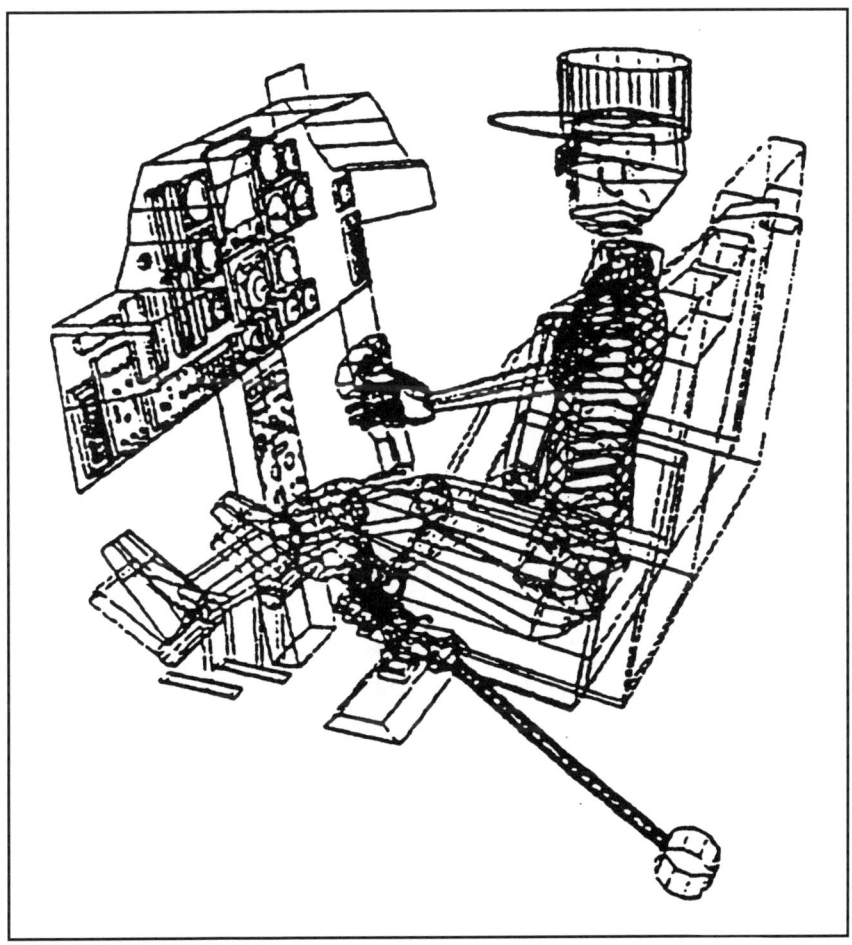

Figure 4-5 JACK human model seated within a workstation (Badler 1991).

Figure 4-6 A computer workstation and human model developed by SAMMIE (Porter 1992).

reach, and visual analysis functions; and 4) other ergonomic evaluative functions (Das and Sengupta 1996). The programs differ considerably in terms of system requirements, operating characteristics, applicability, and the various ergonomic evaluation functions available in the human modeling programs.

Conventional versus Computerized Workstation Design Methods

Computerized workstation design methods permit a convenient and low-cost way of constructing and/or modifiying alternative workstation designs and easily evaluating many such iterations with the help of the variable anthropometry human model. The method allows for 3D evaluations of human/machine interface from the early stage of design. In the conventional workstation design process, this type of operator compatibility check is only possible at the later stage of the design, through a mock-up construction of the workstation. The computerized models, with their ability to produce a multitude of variable anthropometry human

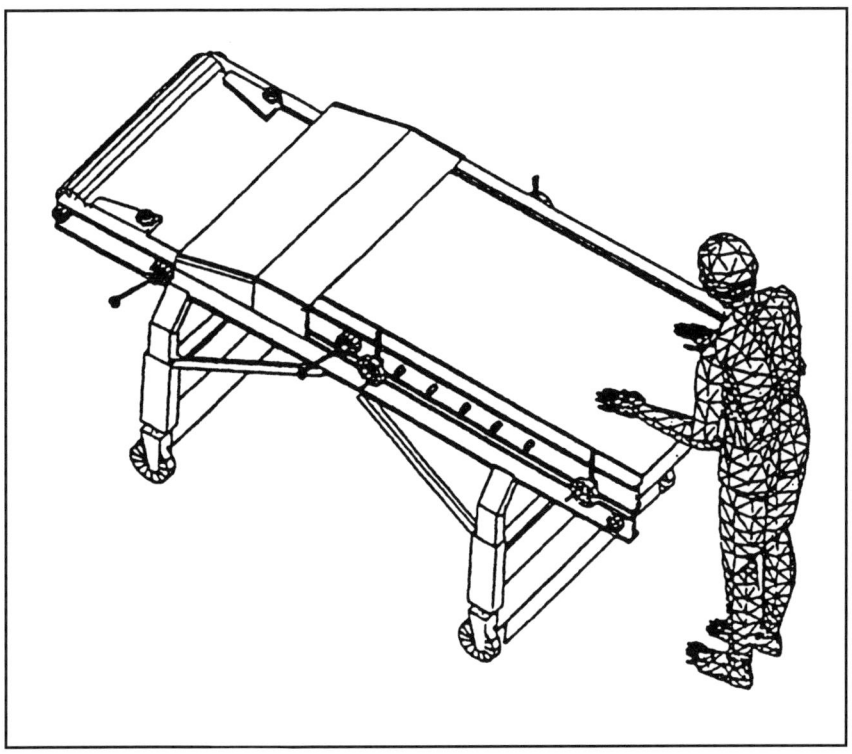

Figure 4-7 MANNEQUIN–generated human model in front of a meat-cutting workstation (Humancad 1991).

models in standard user-defined postures, can provide a relevant and standardized fitting trial, eliminating the difficulty of selecting representative personnel.

The conventional design aids, such as line drawings, mannequins, 3D dummy human models, and mock-up workstations, can virtually be replaced by the 3D computerized human workstation models. As the human attributes—structural size, volume, joint movement ranges, strength, mass, and inertial properties—are embedded within the computer program, the necessary calculations required to evaluate posture, clearance, spatial compatibility, visual requirements, biomechanical loads, and body balance can be handled by programmed routines. However, contrary to common belief, the computerized workstation design method does not act like an expert system. It cannot by itself generate an optimum workstation design from a set of given conditions. The designer must determine posture, working height, preferred line of sight, body clearances, and other pertinent physiological factors. Thus, for effective use of such computer programs in workstation design, the designer must be knowledgeable of the relevant ergonomic principles and guidelines.

With all its virtues, computerized human models are seldom used for workstation design in manufacturing industry. An estimated 7% of the surveyed designers in the field use CAD models, whereas nearly half of them build mock-ups (Kern and Bouer 1988). CAD human modeling programs have been criticized in the past for their scarcity of documentation in literature and being allegedly appraised only by their developers. Many of the programs have been developed for a specific application (Rothwell and Hickey 1986).

REFERENCES

Armstrong, T.J., Radwin, R.G., Hansen, D.J., and Kennedy, K.W. "Repetitive Trauma Disorders: Job Evaluation and Design." *Human Factors* 1986, 28: 325-336.

Ayoub, M.M. "Workplace Design and Posture." *Human Factors* 1973, 15(3): 265-268.

Badler, N.I. "Human Factors Simulation Research at the University of Pennsylvania." Department of Computer and Information Science, University of Pennsylvania. *Computer Graphic Research Quarterly Progress Report No. 38*, Fourth Quarter 1991: 1-17.

Bonney, M.C., Blunsdon, C.A., Case, K., and Porter, J.M. "Man/Machine Interaction in Work Systems." *International Journal of Production Research* 1979, 6: 619-629.

Corlett, E.N. "The Investigation and Evaluation of Work and Workplaces." *Ergonomics* 1988, 31(5): 727-734.

Corlett, E.N., Bowssenna, M., and Pheasant, S.T. "Is Discomfort Related to the Postural Loading of the Joint?" *Ergonomics* 1982, 25: 315-322.

Corlett, E.N., and Clark, T.S. *The Ergonomics of Workspaces and Machines: A Design Manual*, 2nd Edition. London, England: Taylor & Francis, 1995.

Das, B. "An Ergonomics Approach to the Design of a Manufacturing Work System." *International Journal of Industrial Ergonomics* 1987, 1: 231-240.

Das, B. "Industrial Workstation and Workspace Design: An Ergonomic Approach." *Industrial Ergonomics: Case Studies*. Norcross, Georgia: Institute of Industrial Engineers, Industrial Engineering and Management Press, 1991: Chapter 11, 115-139.

Das, B., and Behara, D.N. "Determination of the Normal Horizontal Working Area: A New Model and Method." *Ergonomics* 1995, 38(4): 734-748.

Das, B., and Grady, R.M. "Industrial Workplace Layout Design: An Application of Engineering Anthropometry." *Ergonomics* 1983a, 26(5): 433-447.

Das, B., and Grady, R.M. "The Normal Working Area in the Horizontal Plane: A Comparative Analysis between Farley's and Squires' Concepts." *Ergonomics* 1983b, 26(5): 449-459.

Das, B., and Sengupta, A.K. "Computer-aided Human Modeling Programs for Workstation Design." *Ergonomics* 1996, 38(9): 1958-1972.

Dreyfus, H. *The Measure of Man—Human Factor in Design*. New York: Whitney Library of Design, 1967.

Eastman Kodak Company. "Ergonomics Design for People at Work." Belmont, CA: Lifetime Learning, 1983.

Farley, R.R. "Some Principles of Methods and Motion Study as Used in Development Work." *General Motors Engineering Journal* 1955, 2: 20-25.

Hertzberg, H.T.E. "Engineering and Anthropometry." *Human Engineering Guide to Equipment Design*. New York: McGraw-Hill, 1972: 15, 1, 3.

Humancad. *MANNEQUIN User Guide*. Melville, NY: Humancad, 1991.

Kern, P., and Bauer, W. "Computer-aided Workplace Design." *Proceedings of the Tenth Congress of the International Ergonomics Association* 1988: 81-83.

Khalil, T.M. "Design Tools and Machines to Fit the Man." *Industrial Engineering* 1972, 4: 32-35.

Konz, S. "Design of Workstations." *Journal of Industrial Engineering* 1967, 18: 413-423.

Konz, S. *Work Design: Industrial Ergonomics*, 4th Edition. Scottsdale, AZ: Publishing Horizons, 1995.

McDaniel, J. "Models for Ergonomic Analysis and Design: COMBIMAN and CREW CHIEF." *Computer Aided Ergonomics: A Researcher's Guide*. London, England: Taylor & Francis, 1990: 138-156.

Porter, J.M. "Man Models and Computer-aided Ergonomics." *Proceedings of the International Conference on Computer-Aided Ergonomics and Safety' 92*. M. Mattila and W. Karwowski, eds. The Netherlands: Elsevier, North Holland, 1992: 13-20.

Rothwell, P.L., and Hickey, D.T. "Three-dimensional Computer Models of Man." *Proceedings of the Human Factors Society, 30th Annual Meeting* 1986: 216-220.

Squires, P. "The Shape of the Normal Work Area." Report No. 275. New London, Connecticut: Navy Department, Medical Research Laboratory, 1956.

Szmrecsanyi, E.J. "CYBERMAN—An Interactive Computer Graphics Manikin for Human Factors Analysis." *Control Data Corporation, Worldwide User Group Conference*, 1982.

Tichauer, E R. "Occupational Biomechanics: The Anatomical Base of Workplace Design." *Medical Centre Rehabilitation Monograph No. 51*. New York: Institute of Rehabilitation Medicine, New York University 1975: 15, 1, 23.

Waterman, D., and Washburn, C.T. "CYBERMAN—A Human Factor Design Tool." Paper presented at the Society of Automotive Engineers Congress and Exposition, Detroit, February 27–March 3. Warrendale, PA: Society of Automotive Engineers, 1978: 15, 1, 26.

CHAPTER 5

A DESIGN AND SELECTION GUIDE FOR HAND HELD TOOLS

F. Aghazadeh
Associate Professor

S.M. Waly
Assistant Professor

Industrial and Manufacturing Systems
Engineering Department
Louisiana State University
Baton Rouge, LA

EVOLUTION OF TOOLS

There is considerable evidence that hand tools have been used for almost a million years. Many of today's basic tools were invented in the prehistoric age and evolved over many thousands of years.

Increased specialization in human activities and the change from agricultural to industrial activities led to the development of specialized tools. It was about 10,000 years ago when specialized tools began to appear (Fraser 1980). These tools aided humans by extending their physical capabilities and improving their control over the environment.

The development of specialized tools was motivated by the need to improve mass production. Changes in tool design have been continually made throughout the history of the human race. However, many tools widely used today, such as hammers, have survived their basic design without any major changes (Mital 1991). Mital (1991) attributed the lack of design changes in many basic hand tools to the absence of motivation for improvements.

TWO CATEGORIES

Freivalds (1987) classified hand tools into two general categories: agricultural/forestry tools and industrial/technological tools either manually-driven or power-driven.

Manually-driven tools include:
- Percussive tools such as axes and hammers;
- Scraping tools such as saws and chisels;
- Drilling and boring tools such as drills and gimlets;
- Holding tools such as pliers, pincers, and tongs;
- Cutting tools such as knives and scissors; and
- Manipulative tools such as screwdrivers and wrenches.

Externally-powered hand tools are normally operated by:
- Electricity, such as power saws, power drills, screwdrivers, hammers, and grinders;
- Internal combustion engine power, such as chain saws;
- Compressed air (pneumatic), such as rotary tools and percussion tools; and
- Explosive-driven tools, such as nail guns and splicers.

PROPER MATCHING

As mentioned earlier, hand held tools are designed mainly to extend human physical capabilities and improve control over the environment. The main objective of ergonomics is to have a proper matching between the worker, hand tool, and tasks performed. A basic understanding of the functional abilities and anatomical limitations of the upper extremities is essential for the design and selection of the proper hand tools. Also, one should be aware of the different types of disorders caused by improper hand tool design.

ANATOMY OF THE UPPER EXTREMITIES

The upper extremities consist of 60 bones in the arms and four bones of the shoulder girdle. These bones include the clavicle, scapula, humerus, radius, ulna, carpal, metacarpals, and phalanges. The muscles that produce motion in the hand, the wrist, and the forearm are summarized in Table 5-1. The upper extremity muscles are connected to the bones by tendons. The tendons, connected to the fingers as well as the various nerves and blood vessels, pass through a narrow channel in the wrist known as the carpal tunnel. There are four major nerves in the forearm and the hand—the ulnar, radial, median, and musculocutaneous that provide control of muscular contraction in the hand and forearm.

Table 5-1. Muscles of the upper extremities

Muscle	Action	Nerve
Flexor pollicis longus	Flexes interphalangeal joint of the thumb. Assists in flexion of metacarpophalangeal and carpometacarpal joints. May assist in flexion of wrist.	Median
Flexor pollicis brevis	Flexes the metacarpophalangeal and carpometacarpal joints of the thumb. Assists in opposition of the thumb toward the little finger. May extend the interphalangeal joint.	Median and ulnar
Opponens pollicis	Flexes and abducts the carpometacarpal joint of the thumb with medial rotation. Places the thumb by flexion into a position to oppose the fingers.	Median
Abductor pollicis longus	Abducts and extends the carpometacarpal joint. Assists in flexing the wrist.	Radial
Abductor pollicis brevis	Abducts the carpometacarpal and metacarpophalangeal joints of the thumb in a ventral direction perpendicular to the palm. May extend the interphalangeal joint of the thumb. May assist in flexion of the metacarpophalangeal joint.	Median
Palmar interosseus	Adducts the thumb, index. Assists in flexion of the metacarpophalangeal joints. Assists in extension of the interphalangeal joints of the same digits.	Ulnar
Dorsal interossei	Abducts the index, middle, and ring fingers from the axial line through the third digit. Assists in the flexion of the metacarpophalangeal joints of the same fingers. May assist in the adduction of the thumb.	Ulnar

Table 5-1. Muscles of the upper extremities (continued)

Muscle	Action	Nerve
Flexor digitorum profundis	Flexes the distal interphalangeal joints of the second through fifth digit. Assists in flexion of the proximal interphalangeal and the metacarpophalangeal joints. Assists in the adduction of the index, ring, and little fingers. Assists in the flexion of the wrist.	Median and ulnar
Flexor digitorum	Flexes the interphalangeal suoerficialis joints of the first, second, third, and fourth fingers. Assists in the flexion of the metacarpophalangeal joints and assists in the flexion of the wrist.	Median
Lumbricales	Extends the interphalangeal joints. Flexes the metacarpophalangeal joints of the fingers.	Median and ulnar
Flexor digiti minimi	Flexes the metacarpophalangeal joint of the fifth phalange. Assists in the opposition of the fifth phalange to the thumb.	Ulnar
Opponens digiti minimi	Helps to cup the palm.	Ulnar
Palmaris brevis	Adjusts skin on ulnar side of palm.	Median
Palmaris longus	Tenses skin of palm. Flexes wrist. May assist in flexion of the elbow and pronation of the forearm.	Median
Flexor carpi ulnaris	Flexes and abducts wrist. May assist in flexion of the elbow.	Ulnar
Flexor carpi radialis	Flexes and abducts wrist. May assist in pronation of forearm and flexion of the elbow.	Median
Abductor digiti minimi	Abducts and assists in flexion of the metacarpophalangeal joint of the fifth phalange.	Ulnar

Table 5-1. Muscles of the upper extremities (continued)

Muscle	Action	Nerve
Extensor digitorum	Extends the metacarpophalangeal joints. Helps to extend the interphalangeal joints of the first through fourth fingers. Assists in the abduction of the first, third, and fourth fingers.	Radial
Extensor digiti minimi	Extends the metacarpophalangeal joint of last digit. Helps in extending the interphalangeal joints of the little phalange. Assists in the abduction of the little finger.	Radial
Extensor indicis	Extends the metacarpophalangeal joint of the index finger. Assists in extending the interphalangeal joints of the first finger. May help in adduction of the first finger.	Radial
Extensor carpi radialis longus	Extends and abducts the wrist. May assist in flexion of the elbow.	Radial
Extensor carpi radialis brevis	Extends the wrist. May help in abduction of the wrist.	Radial
Extensor carpi ulnaris	Extends and adducts the wrist.	Radial
Pronator teres	Pronates the forearm, helps in flexion of the elbow.	Median
Pronator quadratus	Pronates the forearm distal end of the radius.	Median
Adductor pollicis	Adducts the carpometacarpal joint allowing the thumb to move toward the surface of the palm. Adducts and assists in flexion of the metacarpophalangeal joint. Assists in the opposition of the thumb to the little finger.	Ulnar
Abductor pollicis	Abducts and extends the longus carpometacarpal joint of the thumb. Assists in flexing the wrist.	Radial

Table 5-1. Muscles of the upper extremities (*continued*)

Muscle	Action	Nerve
Extensor pollicis brevis	Extends the interphalangeal joint. Assists in the extension of the metacarpophalangeal and the carpometacarpal joints of the thumb. Assists in the abduction and extension of the wrist.	Radial
Supinator	Supinates the forearm.	Radial
Biceps brachii	Flexes the shoulder joint. May assist with abduction of the humerus when laterally rotated. Flexes elbow joint moving the forearm toward the humerus.	Musculotaneous
Brachialis	Flexes the elbow moving the forearm toward the humerus.	Musculotaneous and radial
Brachioradialis	Flexes the elbow. Assists in pronating the forearm to midposition. Assists in supinating the forearm to the midposition.	Radial
Triceps brachii	Extends the elbow. May assist in extension of the shoulder.	Radial
Anconeus	Extends the elbow. May stabilize the ulna during pronation.	Radial

Knowledge of the anthropometric dimensions of the upper extremities is essential in the design and selection of hand tools as well as the overall design of workstations and equipment in general. Hand width ranges from approximately 3.16 in. (80 mm) for a 1st percentile female (Garrett 1969a, 1971) to approximately 3.90 in. (99 mm) for a 99th percentile male (Garrett 1969b, 1971). Kroemer et al. (1986) summarized the body dimensions for U.S. female and male civilians. Table 5-2 provides a summary of the important hand dimensions. The means and standard deviations of ranges of motion of the upper extremities for both males and females are summarized in Table 5-3. Static grip strength of industrial male workers has been estimated to be about 90 pf (400 N) for the fifth percentile. Table 5-4 summarizes the maximum static finger forces (Konz 1990).

INJURIES AND ILLNESSES FROM HAND HELD TOOLS

It has been recognized recently that the design and selection of hand held tools can affect operator performance, efficiency, and productivity. The relationship between tool design and risk of injuries has been well established. It is worth noting that tools were originally designed for occasional use only. However, the industrial revolution brought about rapid and widespread use of all kinds of hand tools. Today, in a large number of jobs, tools may be employed for as many as six hours in an eight-hour shift (Aghazadeh and Mital 1987). Consequently, increasing numbers of workers are experiencing injuries and disabilities due to repetitive work stress. It is now widely known that improper use and design of hand tools may cause injuries and cost billions of dollars in lost work days (Aghazadeh and Mital 1987; Mital and Channaveeraiah 1988; Putz-Anderson 1994).

Improper hand tool design and usage result in a variety of acute and chronic injuries. Acute injuries represent a single incident of exposure, resulting in a cut, a bruise, etc. Illnesses or chronic injuries, on the other hand, are cumulative trauma disorders such as tenosynovitis, carpal tunnel syndrome, tendonitis, ischemia, DeQuervain's disease, Raynaud's syndrome (vibration-induced white finger or dead finger), epicondylitis, and certain forms of arthritis. Workers whose jobs require repetitive hand motions often suffer from cumulative trauma disorders. For definition and detailed description of these types of injuries, readers should refer to Mital (1991).

Table 5-2. Anthropometric measurements of the hands in. (cm)

	5th		50th		95th		Standard deviation	
	Male	Female	Male	Female	Male	Female	Male	Female
Hand length	6.9 (17.60)	6.5 (16.40)	7.5 (19.05)	7.1 (17.95)	8.1 (20.60)	7.8 (19.80)	0.37 (0.93)	0.41 (1.04)
Breadth	3.2 (8.20)	2.8 (7.00)	3.5 (8.88)	3.0 (7.66)	3.9 (9.80)	3.3 (8.40)	0.19 (0.47)	0.16 (0.41)
Circumference	7.8 (19.90)	6.7 (16.90)	8.5 (21.55)	7.2 (18.36)	9.3 (23.50)	7.8 (19.90)	0.43 (1.09)	0.35 (0.89)
Thickness	0.9 (2.40)	1.0 (2.50)	1.1 (2.76)	1.1 (2.77)	1.2 (3.10)	1.2 (3.10)	0.08 (0.21)	0.07 (0.18)
Thumb breadth	0.8 (2.10)	0.7 (1.70)	0.9 (2.29)	0.8 (1.98)	1.0 (2.50)	0.8 (2.10)	0.08 (0.21)	0.05 (0.12)
Thumb length	2.0 (5.10)	1.9 (4.70)	2.3 (5.88)	2.1 (5.36)	2.6 (6.60)	2.4 (6.10)	0.18 (0.46)	0.17 (0.44)
Index breadth	0.7 (1.70)	0.6 (1.40)	0.7 (1.85)	0.6 (1.55)	0.8 (2.00)	0.7 (1.70)	0.05 (0.12)	0.04 (0.10)
Index length	2.7 (6.80)	2.4 (6.10)	3.0 (7.52)	2.7 (6.88)	3.2 (8.20)	3.1 (7.80)	0.18 (0.46)	0.20 (0.52)
Middle breadth	0.7 (1.70)	0.6 (1.40)	0.7 (1.85)	0.6 (1.53)	0.8 (2.00)	0.7 (1.70)	0.05 (0.12)	0.04 (0.09)
Middle length	3.1 (7.80)	2.8 (7.00)	3.4 (8.53)	3.1 (7.77)	3.7 (9.50)	3.4 (8.70)	0.20 (0.51)	0.20 (0.51)
Ring breadth	0.6 (1.60)	0.5 (1.30)	0.7 (1.70)	0.6 (1.42)	0.7 (1.90)	0.6 (1.60)	0.04 (0.11)	0.04 (0.09)
Ring length	2.9 (7.40)	2.6 (6.50)	3.1 (7.99)	2.9 (7.29)	3.5 (8.90)	3.2 (8.20)	0.19 (0.47)	0.21 (0.53)
Little breadth	0.6 (1.40)	0.5 (1.20)	0.6 (1.57)	0.5 (1.32)	0.7 (1.80)	0.6 (1.50)	0.05 (0.12)	0.04 (0.09)
Little length	2.1 (5.40)	1.9 (4.80)	2.4 (6.08)	2.1 (5.44)	2.8 (6.99)	2.4 (6.20)	0.19 (0.47)	0.17 (0.44)

Table 5-3. Ranges of motion of the upper extremities (degrees)

Range of motion	Male		Female	
	Mean	Standard deviation	Mean	Standard deviation
Elbow flexion	142.0	10.0	151.4	7.1
Wrist flexion	90.0	12.0	79.7	15.1
Wrist extension	99.0	13.0	60.6	10.5
Wrist abduction	27.0	9.0	29.7	9.1
Wrist adduction	47.0	7.0	50.4	10.8

Table 5-4. Maximal static force for adult males pounds-force (N)

Finger	Mean	Standard deviation
Thumb versus object	16.10 (71.61)	16.68 (3.75)
Index finger versus object	57.88 (13.01)	12.75 (2.87)
Middle finger versus object	62.78 (14.11)	19.62 (4.41)
Ring finger versus object	49.05 (11.03)	16.68 (3.75)
Little finger versus object	31.39 (7.06)	10.79 (2.43)

It has been estimated that 6% of compensable work injuries in the U.S. are caused by hand tools. The Bureau of Labor Statistics (1994) showed that for all private industries this trend holds true. Over 73,000 injuries involving at least one lost day due to the use of nonpowered hand tools, and 29,000 injuries from powered tools have been reported.

A recent survey revealed that hand tools are the primary source of approximately 9% of all occupational injuries (Aghazadeh and Mital 1985). Hand held tools are used away from work more frequently than any other apparatus. According to the U.S. Consumer Product Safety Commission, 280 deaths and 272,000 injuries were recorded that were associated with the use of home workshop apparatus, tools, and attachments (U.S. Consumer Product Safety Commission 1982). Aghazadeh and Mital (1985) estimated the annual costs associated with hand tool injuries to be in the range of $10 billion in the United States.

A SURVEY

Aghazadeh and Mital (1987) developed a questionnaire that was mailed to various federal and state agencies in the United States to determine the frequency, severity, and annual cost of hand tool-related injuries in industry. It attempted to identify problem areas with regard to tool type, accident type, nature of injury, parts of body affected, industry, and characteristics of the injured worker. Results of the questionnaire showed that hand tool injuries were costly, severe, and frequent. Nonpowered and powered hand tools were responsible for 80% and 20%, respectively, of compensatable hand tool injuries. Higher injury frequency, associated with the nonpowered hand tools, may be due to their frequent use.

The results of this survey indicated that, in the case of nonpowered hand tools, 71.2% of all injuries were caused by "striking by" or "striking against" tools. Overexertion was the second leading cause. "Caught in" or "between" contributed the least. It was noted that knives, hammers, wrenches, and screwdrivers caused the most injuries. They may cause injury by striking by or against the operator. Wrenches and screwdrivers may cause the most overexertion injuries. In the case of powered hand tools, the same types of accidents were responsible for most injuries. "Struck by" and "struck against" types of accidents were caused by saws, drills, hammers, and grinders more than any other powered hand tool.

Parts of the body most injured by nonpowered hand tools were the upper extremities followed by the trunk. The same was true for powered hand tools. In regard to powered hand tools, only, the second most injured part was the lower extremities. For both types of hand tools, about 30% of all body parts affected were fingers. Furthermore, about 56% of all upper extremities affected were fingers. It was concluded that fingers were the most injured part of the body. The head, neck, and eyes were the least injured parts.

INJURY PARAMETERS

The industries where hand tool-related injuries were numerous were manufacturing, mining, construction, and the wholesale and retail groups. Finance, insurance, and real estate groups, where hand tools are seldom used, had the least number of cases. Also contributing to the incidence of hand tool injuries was the service group, comprised of building maintenance workers, automotive and other repair shops, garden services, and health care services—areas where hand tools are used frequently.

Workers between the ages of 18 and 35 years suffered more from hand tool injuries than any other age group. The relative safety of the older workers may be attributed to their experience, conservative behavior, and relatively less demanding jobs. Older and more mature workers know their limits, do not take unnecessary risks, and are usually assigned to less strenuous jobs.

PRINCIPLES OF HAND TOOL DESIGN

Many injuries and illnesses could be prevented if hand tools were designed (or redesigned) to optimize the relationship between the worker and hand tool. Improvements in job design, proper selection of hand tools, and implementation of safe operating procedures can further reduce the incidence of hand tool-related injuries and illnesses. An understanding of anatomical, physiological, kinesiological, anthropometric, mechanical, and psychological considerations is required for the proper selection and design of hand held tools.

In outlining six basic requirements for properly designed hand held tools, Drillis (1963) concluded that the tool:

- Should effectively perform the function for which it was intended;
- Should be properly proportioned to the body dimensions of the operator;
- Should be suitably adjusted to the strength and work capacity of the operator;
- Should not cause premature fatigue;
- Must provide feedback (surface texture, temperature, force, etc.) to the operator; and
- Should be inexpensive and easy to maintain.

The Technical Research Center of Finland (1988) provided guidelines for selecting and evaluating pneumatic screwdrivers and nut runners, based on the review of published scientific information. These guidelines consider a variety of factors.

- Torque—the torque delivered to the workpiece should be approximately 60% of the maximum torque that can be generated by the tool.
- Tool type—straight and angled tools are best for downward screwing; pistol grip type tools (angle of 78° between the shaft and the handle) are best for screwing horizontally.
- The level of noise generated by the tool must not exceed 85 dB for eight hours of exposure.

- The noise vibration level must not exceed 126 dB for eight hours of exposure.
- The tool should be light, well-balanced, proportionately-sized and -shaped.
- The tool should be supported for reaction force if the torque exceeds 53 lbf-in. (6 Nm) for straight type tools, 106 lbf-in. (12 Nm) for pistol type tools, and 443 lbf-in. (50 Nm) for angled-type tools.
- Handle length should be at least 3.93 in. (100 mm). Circular handles should be between 1.10 and 1.50 in. (28 and 38 mm) in diameter. Rectangular handles should be 1.60 × 0.98 in. (40 × 25 mm) in size.

Konz (1990) outlined the following principles for the design and selection of hand tools:

- Use special-purpose tools.
- Design tools to be used by either hand.
- Power with motors rather than with muscles.
- Use a power grip for power and precision grip for precision.
- Make the grip the proper thickness, shape, and length.
- Design the grip surface to be compressible, nonconductive, and smooth.
- Consider the angles of the forearm, grip, and tool.
- Use the appropriate muscle group.

In general, hand tool design guidelines should be based on the principles of biomechanics and ergonomics regardless of the field of application. Some specific guidelines for selecting and designing hand tools are presented.

EFFECTS OF GRIP TYPE, SIZE, AND SHAPE

The shape and size of hand tools have major effects on both performance and stresses on the upper extremities (Aghazadeh et al. 1989). There are steps that can be taken to reduce stresses in the upper extremities and improve performance. Generally, hand tools should be designed to reduce excessive wrist deviation, shoulder abduction, and grip force requirements. Also, they should utilize the proper muscle groups, avoid static loading, and provide finger clearance.

Precision grip involves small muscles, whereas a power grip involves larger muscle groups. It has been estimated that precision grips, on the average, provide only 20% of the strength of a power grip (Swanson et al. 1970). Tools designed primarily for exertion of force, such as a hammer, should use a power grip; and tools designed for manipulation, such as a surgical knife, should use a precision grip.

Grip Size

Pheasant and O'Neill (1975) reported that, for screwdrivers, increasing the diameter (from 0.70 to 1.57 in. [18 to 40 mm]) enables a power grip to generate greater force. Herzberg (1973) reported an increase in grip strength with grip diameter up to 2.6 in. (65 mm) and then a decrease. Ayoub and LoPresti (1971) found a grip diameter of 2 in. (50 mm) produced minimum electromyographic (EMG) activity. Based on the requirements of grip force, EMG, and muscle fatigue, they recommended a grip diameter of 1.5 in. (38 mm). Eastman Kodak Company (1983) recommended a power grip diameter of 1.57 in. (40 mm). Petrofsky et al. (1980) concluded that the optimum hand grip span should be between 1.96 and 2.37 in. (50 and 60 mm). Greenberg and Chaffin (1977) recommended that a power grip diameter be between 1.96 and 3.34 in. (50 and 85 mm) preferably 1.96 in. (50 mm). The maximum span should be 3.93 in. (100 mm) and the minimum span should be 1.96 in. (50 mm). The grip strength is maximized when the span is between 2.95 and 3.14 in. (75 and 80 mm). The maximum force required to close the grip span should be about 19.8 lb (9 kg), which can be produced by 95% of the female population. Replogle (1983) recommended a grip diameter twice the size of the grip span, as it maximized the torques. For powered screwdrivers, Johnson and Childress (1988) recommended a grip diameter of 1.5 in. (38 mm). This recommendation was based on EMG activity that was lower for the 1.5-in. (38-mm) grip diameter compared to the 1.10-in. (28-mm) grip diameter for a screwdriver.

For a precision grip, it was found that increasing a screwdriver handle diameter from 0.32 to 0.64 in. (8 to 16 mm) increased the time to drive a screw from 1.9 seconds to 3.6 seconds. Kao (1976, 1977) recommended the preferred diameter for pens to be 0.52 in. (13 mm). According to Konz (1990), diameters less than 0.24 in. (6 mm) should be avoided for precision grips as they tend to cut into the hand if force is required.

Based on anthropometric data, Konz (1990) recommended a minimum grip length of 4 in. (102 mm) (preferably 5 in. [127 mm] grip length). Eastman Kodak Company (1983) recommended a grip length of 4.8 in. (122 mm). Lindstrom (1973), on the other hand, recommended a handle length of 4.4 in. (112 mm) for men and 4 in. (102 mm) for women. In reality, it is impractical to provide different handle length tools for the same work done by males and females. For external precision grip, the tool shaft must be a least 4 in. (102 mm) in length and must be long enough to be supported at the base of the first finger or thumb. For internal precision grip, the tool should extend past the tender palm, but not so far as to hit the wrist (Konz 1990). In general, the grip length should be selected to avoid excessive compressive forces or pressure on the tender parts of the palm in a way that would not limit the tool head opening.

Grip Shape

The grip shape should maximize the area of contact between the palm and the grip to provide better pressure distribution and reduce the chances of forming pressure ridges or pressure concentration points. This is particularly important for tools that require a power grip. Generally, the tools available in the market have a cylindrical-shape grip. Pheasant and O'Neill (1975) reported that the shape of the handle is not relevant as long as the hand did not slip while around it.

The chances of hand slippage are reduced when a noncircular handle, such as rectangular or triangular, is used. The edges in such grips resist slippage. Mital and Channaveeraiah (1988) reported that the torque exertion capability of individuals with triangular-handle screwdrivers was greater than with cylindrical-handle screwdrivers. For wrenches, torque exertion capability was maximized when cylindrical handles were used. Pheasant and O'Neill (1975) recommended the use of the T-handle for screwdrivers over the straight handle to prevent wrist deviations and increase torque exertion. According to Saran (1973), the T-handle should be slanted (60° angle) to allow the wrist to be straight and should be 1 in. (25 mm) in diameter.

According to Cochran and Riley (1986a), the thrust forces exerted with straight knives are as much as 10% greater with triangular handles as with cylindrical or rectangular handles. A handle perimeter of 4.4 in. (112 mm) is recommended. For knives, a pistol grip is preferable to a straight grip (Armstrong et al. 1982; Karlqvist 1984). The angle of the grip should be 78° from the horizontal (Fraser 1980).

Pheasant and O'Neill (1975) found that screwdrivers with cylindrical and knurled surfaces provided greater torque exertion capability than smooth-surface cylindrical handles. The 2-in. (51-mm) diameter handle with knurled surface maximized torque exertion. According to Mital (1991), grooves and indentations are generally undesirable. The main problems with grooves are that they do not fit people's different anthropometric measurements, creating pressure ridges that may lead to nerve compression and impairment of circulation. Slight and uniform surface indentations, however, are desirable as they allow greater torque exertion capability.

Grip Material

Konz (1990) recommended the use of compressible grip material that dampens vibration and allows better distribution of pressure across the palm/grip contact area. In general, metallic handles should either be avoided or encased in a rubber or plastic sheath. Wu (1975) and Mital (1991) indicated that the grip material should not absorb oil or other liquids and should not permit conduction of heat or electricity. Wood or plastic are desirable handle materials and smooth grip surfaces are desirable. The surface, however, should not be slippery. Sharp indentations or grooves should be avoided as they cut into the palm, as previously indicated.

Guards in front of the grip can prevent injury when the hand slips or when the hand and tool collide against a rigid surface. The guard, in such cases, prevents the hand from slipping and shields the hand against the impact. According to Cochran and Riley (1986b), guards of 0.60-in. (1.52-cm) height provide adequate safety.

EFFECTS OF GLOVES

The use of gloves for safety interferes with an individual's range of motion and grasping abilities. A portion of force generated by muscular contraction is wasted in maintaining the grip that reduces torque production. Swain et al. (1970) and Sudhakar et al. (1988) found that the peak grip strength with rubber and leather gloves was about 15% lower compared to grip strength without gloves. However, no significant differences in muscle activity between gloved and non-gloved conditions were found. Reduction in strength also has been reported by Lyman (1957) and Hertzberg (1973). Gloves also change the effective anthropometric measurements of the hand. According to Damon et al. (1966), the hand thickness increases from 0.2 to 1 in. (5 to 25 mm) and the hand width at the thumb increases from 0.32 to 1.6 in. (8 to 41 mm) when gloves are used. The increased size of the gloved hand should be taken into consideration in the design and selection of hand held tools.

Weidman (1970) studied the influence of different kinds of gloves on manual performance. Performance times decreased by 12.5%, 36%, 45%, and 64% as compared to no gloves, when neoprene, terry cloth, leather, and PVC gloves, respectively, were worn. In some cases, gloves improve speed, enhance friction, and reduce the task strength requirements (Riley et al. 1985; Bradley 1969a, 1969b).

EFFECTS OF WRIST POSITION

It has been well documented that wrist deviations cause various problems, including reducing productivity and grip strength (Tichauer 1976; Terrell and Pursewell 1976). It is recommended to keep the wrist straight to avoid chronic illnesses and loss in productivity. If any bending is required, the tool, rather than the wrist, should bend (Tichauer and Gage 1977). By bending the handles and increasing the length of the upper handle, it is possible to keep the wrist straight and avoid nerve, tissue, and blood vessel compression. Many investigators use bent handles in designing different tools (for example, pliers, soldering irons, hammers) to reduce user fatigue and stress (Yoder et al. 1973; Tichauer 1966; Emanuel et al. 1980; Granada and Konz 1981; Krohn and Konz 1982; Knowlton and Gilbert 1983; Konz and Streets 1984; Konz 1986; Schoenmarklin and Marras 1989a, 1989b). For hammers Konz (1986) concluded that a 10° bend is preferable for subjects. It has been shown that the angle size of bend does not influence performance.

EFFECTS OF TOOL WEIGHT AND MUSCLE GROUP

It has been demonstrated that the tool should not weigh more than 5.06 lb (2.3 kg) to reduce muscle fatigue (Greenberg and Chaffin 1977; Eastman Kodak Company 1983). Johnson and Childress (1988) found that powered screwdrivers weighing 2.4 lb (1.1 kg) or less do not produce significantly different magnitudes of EMG activity.

Power grips are used when exertion of force is required. Larger and stronger forearm muscles, instead of the smaller and weaker finger muscles, should be used for these applications. Trigger strips rather than trigger buttons should be used (Konz 1990). A spring should be used to open the handles because the muscles used for closing the hand are stronger than those that open it (Radonjic and Long 1971).

Effects of Right-handed Versus Left-handed Operators

Tools are designed mostly for right-handed users. Konz (1974) estimated that 10% of the population is left handed. The preferred hand is about 7 to 20% stronger (Shock 1962; Miller 1981), more dexterous (Kellor et al. 1971), and faster (Konz and Warraich 1985). Also, right-hand tools require a different action when used by the left hand (Capener 1956). For instance, the right-handed scissors, when used by a left-handed person, requires a reversal of pressure upon the rings of the scissors. This leads to a tendency to twist it. The shearing surfaces thus lie in the vertical plane instead of the horizontal plane. In many instances, tools designed for the right hand cannot be used by left-handed individuals (Laverson and Meyer 1976). Tools designed for use by either hand can avoid such problems.

SAFETY CONSIDERATIONS

Worker training for proper and safe operation of hand tools is one of the most important factors that should be considered in the use of hand held tools. It has been estimated that nearly 60% of workers who suffer amputation injuries did not receive adequate safety training before working with the tool that caused the injury. A good training program should be implemented to reduce injury incidents.

Specific safe practice rules for using many types of hand tools have been established (U.S. Department of Labor 1981). In general, pinching hazards, sharp corners, and edges should be eliminated. Brake devices on power tools should be installed so the operation can be stopped in case of emergency. Maintenance and periodic inspections are important for safety. Personal protective equipment should be used, when applicable, to avoid injuries. The type of protective equipment depends on the tasks and conditions, which should be analyzed to determine potential hazards. The protective equipment should be selected according to the identified hazard(s).

CONCLUSION

Designers and engineers can apply the basic principles of ergonomics and safety to the design and selection of hand held tools to reduce injury incident rate. Administrative measures can minimize exposure to repetitive tasks, thereby improving productivity and reducing the risk of injury. The following concepts should be considered in the design and/or selection of hand held tools:

1. Reduce excessive wrist deviation and shoulder abduction. It is recommended to keep the wrist straight to avoid chronic illnesses and loss in productivity. If any bending is required, the tool, rather than the wrist, should be bent.

2. Minimize grip force requirements and utilize the proper muscle groups. When exertion of force is required, larger and stronger forearm muscles, instead of the smaller and weaker finger muscles, should be used.
3. A spring should be used to open handles because the muscles used for closing the hand are stronger than those that open it.
4. Avoid static loading. The tool weight should be minimized to reduce muscle fatigue. Use tools weighing less than 5.06 lb (2.3 kg) or support the weight of the tool by mechanical suspension.
5. Tools designed primarily for exertion of force should use a power grip, while those designed for manipulation should use a precision grip.
6. Handle diameter should be 1.6 to 3.4 in. (41 to 86 mm) for power grip and 0.4 to 0.64 in. (10 to 16 mm) for precision grip.
7. Grip length should be at least 4 in. (102 mm) (preferably 5 in. [127 mm] grip length).
8. Grip shape should maximize the area of contact between the palm and the grip.
9. Grooves and indentations are undesirable, however, slight and uniform surface indentations allow greater torque exertion capability.
10. Compressible grip material should be used to dampen vibration and allow better distribution of pressure across the palm/grip contact area.
11. Metallic handles should either be avoided or encased in a rubber or plastic sheath. Grip material should not absorb oil or other liquids and should not permit conduction of heat or electricity. Wood or plastic are desirable handle materials.
12. Guards in front of the grip can prevent injury when the hand slips or when the hand and tool collide against a rigid surface.
13. The increased size of the hand when using gloves should be taken into consideration in the design and selection of hand held tools.
14. A left-handed user is at a disadvantage when using a tool designed for a right-handed individual. Tools should be designed for use by both hands.

15. A good training program should be implemented to reduce injury incidents.
16. Personal protective equipment can be used to avoid injuries.

REFERENCES

Aghazadeh, A., and Mital, A. "Work Related Hand Tool Injuries." *25th American Industrial Hygiene Association Conference.* Las Vegas, Nevada: American Industrial Hygiene Association, 1985.

Aghazadeh, A., and Mital, A. "Injuries Due to Hand Tools: Results of a Questionnaire." *Applied Ergonomics* 1987, 18: 273-278.

Aghazadeh, F., Latif, N.T., and Mital, A. "Guidelines for Preventing Hand Tool Related Accidents and Illnesses." *Ergonomic: SA, I,* 1989: 2-9.

Armstrong, T.J., Foulke, J., Joseph, B., and Goldstein, S. "Investigation of Cumulative Trauma Disorders in a Poultry Processing Plant." *American Industrial Hygiene Association Journal* 1982, 43: 103-116.

Ayoub, M.M., and LoPresti, P. "The Determination of an Optimum-size Cylindrical Handle by Use of Electromyography." *Ergonomics* 1971, 14: 509-518.

Bradley, J.V. "Effect of Gloves on Control Operation Time." *Human Factors* 1969a, 11: 13-20.

Bradley, J.V. "Glove Characteristics Influencing Control Manipulability." *Human Factors* 1969b, 11: 21-35.

Bureau of Labor Statistics. *Occupational Injuries and Illnesses: Counts, Rates, and Characteristics.* Washington, DC: Bureau of Labor Statistics, 1994.

Capener, N. "The Hand in Surgery." *The Journal of Bone and Joint Surgery* 1956, 388: 128-151.

Cochran, D.J., and Riley, M.W. "The Effects of Handle Shape and Size on Exerted Forces." *Human Factors* 1986a, 28: 253-265.

Cochran, D.J., and Riley, M.W. "An Evaluation of Knife Handle Guarding." *Human Factors* 1986b, 28: 295-301.

Damon, A., Stoudt, H.W., and McFarland, R.A. *The Human Body in Equipment Design.* Cambridge, MA: Harvard University Press, 1966.

Drillis, R.J. "Folk Norms and Biomechanics." *Human Factors* 1963, 5: 427-441.

Eastman Kodak Company. *Ergonomic Design for People at Work.* Belmont, CA: Lifetime Learning Publications, 1983.

Emanuel, J.T., Mills, S.J., and Bennett, J.F. "In Search of a Better Handle." *Proceedings of the Symposium on Human Factors and Industrial Design in Consumer Products.* Medford, MA: Tufts University, 1980: 34-40.

Fraser, T. M. "Ergonomics Principles in the Design of Hand Tools." *Occupational Safety and Health Series,* No. 44. Geneva, Switzerland: International Labour Office, 1980.

Freivalds, A. "The Ergonomics of Tools." *International Reviews of Ergonomics* 1987, 1: 43-75.

Garrett, J. "References on Hand Anthropometry for Women." *AMRL Technical Report*, 69-42. Wright Patterson Air Force Base, OH: Army Medical Research Laboratory, 1969a.

Garrett, J. "References on Hand Anthropometry for Men." *AMRL Technical Report*, 69-26. Wright Patterson Air Force Base, OH: Army Medical Research Laboratory, 1969b.

Garret, J. "The Adult Human Hand: Some Anthropometrical and Biomechanical Considerations." *Human Factors* 1971, 13: 117-131.

Granada, J.M., and Konz, S.A. "Evaluation of Bent Hammer Handles." *Proceedings of the Human Factors Society 25th Annual Meeting*. Santa Monica, CA: Human Factors Society, 1981: 322-324.

Greenberg, L., and Chaffin, D.B. *Workers and their Tools*. Midland, MI: Pendell, 1977.

Hertzberg, H. "Engineering Anthropometry." *Human Engineering Guide to Equipment Design*. Washington, DC: U.S. Government Printing Office, 1973.

Johnson, S.L., and Childress, L.J. "Powered Screwdriver Design and Use: Tool, Task, and Operator Effects." *International Journal of Industrial Ergonomics* 1988, 2: 183-191.

Kao, H. "An Analysis of User Preference Toward Handwriting Instruments." *Perceptual and Motor Skills* 1976, 43: 522.

Kao, H. "Ergonomics in Penpoint Design." *Acta Psychologia Taiwanica* 1977, 18: 49-52.

Karlqvist, L. "Cutting Operation at Canning Bench—A Case Study of Hand Tool Design." *Proceedings of the 1984 International Conference on Occupational Ergonomics*. Ontario, Canada: Human Factors Association of Canada, 1984: 452-456.

Kellor, M., Kondrasuk, R., Iverson, I., Frost, J., Silberberg, N., and Hoglund, M. *Hand Strengths and Dexterity Tests*. Manual 721. Minneapolis, MN: Sister Kenny Institute, 1971.

Knowlton, R.G., and Gilbert, J. "Ulnar Deviation and Short-term Strength Reductions as Affected by a Curve-handled Ripping Hammer and a Conventional Claw Hammer." *Ergonomics* 1983, 26: 173-179.

Konz, S. "Design of Hand Tools." *Proceedings of the Human Factors Society 18th Annual Meeting*. Santa Monica, CA: Human Factors Society 1974: 292-300.

Konz, S. "Bent Handle Hammers." *Human Factors* 1986, 28: 317-323.

Konz, S. *Work Design: Industrial Ergonomics,* 3rd Edition. Scottsdale, AZ: Publishing Horizons, Inc., 1990.

Konz, S., and Streets, B. "Bent Hammer Handles Performance and Preference." *Proceedings of the Human Factors Society 28th Annual Meeting*. Santa Monica, CA: Human Factors Society, 1984: 438-440.

Konz, S., and Warraich, N. "Performance Differences Between the Preferred and Nonpreferred Hand When Using Various Tools." *Ergonomics International 1985*. London: Taylor & Francis, 1985.

Kroemer, K.H.E., Kroemer, H.J., and Kroemer-Elbert, K.E. *Engineering Physiology—Physiological Basis of Human Factors/Ergonomics*. New York: Elsevier Science Publishers B.V., 1986.

Krohn, R., and Konz, S. "Bent Hammer Handles." *Proceedings of the Human Factors Society 26th Annual Meeting.* Santa Monica, CA: Human Factors Society, 1982: 413-417.

Laveson, J.I., and Meyer, R.P. "Left Out (Lefties) in Design." *Proceedings of the Human Factors Society 20th Annual Meeting.* Santa Monica, CA: Human Factors Society, 1976: 122-125.

Lindstrom, F.E. "Modern Pliers." Enkoping, Sweden: *BACHO Vertyg,* 1973.

Lyman, J. "The Effects of Equipment Design on Manual Performance." *Production and Functioning of the Hands in Cold Climates.* R. Fisher, ed. Washington, DC: National Academy of Science, National Research Council, 1957: 86-101.

Miller, G.D. *Significance of Dominant Hand Grip Strengths in Hand Tools.* Unpublished M.S. Thesis. University Park, PA: Pennsylvania State University, 1981.

Mital, A. "Hand Tools: Injuries, Illnesses, Design, and Usage." *Workspace, Equipment, and Tool Design.* A. Mital and W. Karwowski, eds. New York: Elsevier Publishers, 1991.

Mital, A., and Channaveeraiah, C. "Peak Volitional Torques for Wrenches and Screwdrivers." *International Journal of Industrial Ergonomics* 1988, 3: 41-64.

Petrofsky, J., Williams, C., Kamen, G., and Lind, A. "The Effect of Handgrip Span on Isometric Exercise Performance." *Ergonomics* 1980, 23: 1,129-1,135.

Pheasant, S., and O'Neill, D. "Performance in Gripping and Turning—A Study in Hand/Handle Effectiveness." *Applied Ergonomics* 1975, 6: 205-208.

Putz-Anderson, V. *Cumulative Trauma Disorders.* London: Taylor and Francis, 1994.

Radonjic, D., and Long, C. "Kinesiology of the Wrist." *American Journal of Physical Medicine* 1971, 50: 57-71.

Replogle, J. "Hand Strength with Cylindrical Handles." *Proceedings of the Human Factors Society 27th Annual Meeting.* Santa Monica, CA: Human Factors Society, 1983: 412-416.

Riley, M., Cochran, D., and Schanbacher, C. "Force Capability Differences Due to Gloves." *Ergonomics* 1985, 28: 441-447.

Saran, C. "Biomechanical Evaluation of T-handles for a Pronation Supination Task." *Journal of Occupational Medicine* 1973, 15: 712-716.

Schoenmarklin, R.W., and Marras, W.S. "Effects of Handle Angle and Work Orientation on Hammering: Part I, Wrist Motion and Hammering Performance." *Human Factors* 1989a, 31: 397-412.

Schoenmarklin, R.W., and Marras, W.S. "Effects of Handle Angle and Work Orientation on Hammering: Part II, Muscle Fatigue and Subjective Preference." *Human Factors* 1989b, 31: 413-420.

Shock, N. "The Physiology of Aging." *Scientific American* 1962, 206: 100-110.

Sudhakar, L.R., Schoenmarklin, R.W., Lavender, S.A., and Marras, W.S. "The Effects of Gloves on Grip Strength and Muscle Activity." *Proceedings of the Human Factors Society 32nd Annual Meeting.* Santa Monica, CA: Human Factors Society, 1988: 647-650.

Swain, A.D., Shelton, G.G., and Rigby, L.V. "Maximum Torque for Small Knobs Operated With and Without Gloves." *Ergonomics* 1970, 13: 201-208.

Swanson, A., Matev, I., and Groot, G. "The Strength of the Hand." *Bulletin of Prosthetics Research*, 1970: 145-153.

Technical Research Center of Finland. *Evaluating and Choosing Pneumatic Screwdrivers and Nut Runners.* Tampere, Finland: Occupational Safety Engineering Laboratory, 1988.

Terrell, R., and Pursewell, J. "The Influence of Forearm and Wrist Orientation on Static Grip Strength as a Design Criterion for Hand Tools." *Proceedings of the Human Factors Society 20th Annual Meeting.* Santa Monica, CA: Human Factors Society, 1976: 28-32.

Tichauer, E.R. "Some Aspects of Stress on Forearm and Hand in Industry." *Journal of Occupational Medicine* 1966, 8: 63-71.

Tichauer, E.R. "Biomechanics Sustains Occupational Safety and Health." *Industrial Engineering* February, 1976: 46-56.

Tichauer, E.R., and Gage, H. "Ergonomic Principles Basic to Hand Tool Design." *American Industrial Hygiene Association Journal* 1977, 38: 622-634.

U.S. Consumer Product Safety Commission. *Annual Report, Parts 1 and 2.* Washington, DC: U.S. Consumer Product Safety Commission, 1982.

U.S. Department of Labor. *OSHA Safety and Health Standards, 29, CFR 1910.* Washington, DC: U.S. Department of Labor, 1981.

Weidman, B. *Effect of Safety Gloves on Simulated Work Tasks.* AD 738981. Springfield, VA: National Technical Information Service, 1970.

Wu, Y. "Material Properties Criteria for Thermal Safety." *Journal of Materials* 1975, 7: 575-579.

Yoder, T.A., Lucas, R.L., and Botzum, C.D. "The Marriage of Human Factors and Safety in Industry." *Human Factors* 1973, 15: 197-205.

CHAPTER 6

COMPUTER-AIDED DESIGN FOR ERGONOMICS AND SAFETY

Markku Aaltonen
Project Manager
Department of Occupational Safety
Finnish Institute of Occupational Health
Vantaa, Finland

Markkus Mattila
Professor, Vice-Rector
Occupational Safety Engineering Department
Tampere University of Technology
Tampere, Finland

GATHERING SAFETY DATA

The use of software for safety and ergonomics in companies is increasing (Canadian Centre for Occupational Health and Safety 1996; Hartung 1993; Ross 1991; Wrench 1990), but it appears that the possibilities of information technology have not yet been utilized sufficiently in safety management and ergonomics (Räsänen et al. 1993).

Almost all respondents in the study of Räsänen et al. (1993) used a computer in their work, and nearly a third used some kind of safety software. The use of computers has been mainly a positive experience. Also communication and data management have improved. Further development of safety information systems was felt to be an important task.

To make effective and appropriate decisions for safety, true and sufficient information about the risks and working conditions is needed. If this information is missing, resources may be directed at irrelevant problems, and the desired results are not achieved. For continuous safety improvements, the effects of implemented safety measures and achievement of desired goals must be realized. If this information is missing, the results are mere guesswork, and the motivational effect of good feedback is lost (Aaltonen 1996).

A workplace is comprised of man/machine systems in which interaction between man and machine in the environment is crucial for safety and health. Safety may be defined as the system's ability to estimate occurrence of unwanted events. The level of safety may be measured by risk, and the quantitative description of safety is a result of risk estimation.

Safety does not occur at random. Rather, it is the definite outcome of decisions made by designers and managers. The design of workplace safety requires expertise and special design methods to be integrated into the design process. We may say that the level of workplace safety is one criterion for designer expertise.

STANDARDIZATION

Ergonomic and safety standardization is one approach closely connected with international integration. In recent years, the number of ergonomic and safety standards has increased rapidly. Standards are necessary to provide quality control and to support legislation and regulations used in establishing an acceptable international market (Parsons 1994).

The European Experience

The importance of workplace safety can be recognized by governmental norms. In recent years, the number of ergonomics and safety standards has increased rapidly. In the European Union, the Council Directive (89/391/EEC) on the introduction of measures to encourage improvements in the safety and health of workers, includes the general principles that the employer shall evaluate the safety and health risks of the workers; and implement measures that avoid risks, evaluate risks, combat risks at source, adopt work to the individual, replace dangerous items with nondangerous or less dangerous ones, and give appropriate instruction to workers.

The European Council (EC) Directive (89/392/EEC) relating to machinery, aims to ensure that machinery taken into service does not endanger human health or safety. The machinery must fulfill the health and safety requirements set in the directive. If a harmonized European standard is prescribed for essential safety requirements, machinery constructed in accordance with this standard shall be presumed to comply with the relevant essential requirements. To certify the conformity of machinery with the provisions of the machinery directive, the manufacturer must draw up an EU declaration of conformity, which consists, where appropriate, of a reference to the harmonized standards. This is why safety and health, as well as ergonomic standards, have become so crucial to the designers of machinery.

European Organization for Standardization (CEN) is working with over 100 safety and ergonomics standards, and the International Organization for Standardization (ISO) also has similar activities. This means a huge amount of ergonomic and safety information and knowledge has to be considered by designers and by design management. There is an urgent need to integrate this information into the design information system (Mattila 1996).

Advanced information technology also provides remarkable possibilities for ergonomic design and safety management (Haijanen et al. 1995; Takala 1992; Aaltonen 1992; Ross 1991; Wrench 1990; Suokas 1990; Kjellén 1987; Successful Accident Prevention 1987). The objective of this chapter is to introduce how computerized information systems can be utilized in ergonomic planning and safety activities.

SAFETY MANAGEMENT INFORMATION SYSTEMS

Safety management requires a comprehensive information system that includes data on incidents and accidents, risks, and safety performance. This means a hierarchical safety measurement system.

The measurement of safety should consist of different methods, depending on which indicator of safety is measured. Methods used to measure safety are different types of accident investigations, risk analyses, and safety audits. Reactive safety outcomes should be measured by changes in insurance premium expenditures or in the number of accidents (Heinrich et al. 1980; Tarrants 1980; Johnson 1988).

Safety responsibilities in a company are connected with the workers' tasks and duties. Therefore, these persons have different needs for safety information, and the safety information systems should be based upon these needs.

Crucial safety information for safe performance consists of material safety data, work and safety instructions, instructions on electrical safety and fire prevention, and maintenance information about the machines and equipment. For safety management decisions, information on risks, accidents, sick leaves, and safety costs are important (Räsänen et al. 1993).

SUPPORTIVE SAFETY INFORMATION

In principle, the need for safety information and the types of safety information systems could be classified into three main categories from the company's point of view (Table 6-1):

1) information available outside a company, 2) information produced within a company, and 3) information needed outside a company.

Table 6-1. Safety information classification

Domain of information need	Detailed domain of information need	Examples of safety software
Supportive safety information available outside a company	Safety legislation, safety agreements between labor unions and employer federations, safety standards, safety literature, educational material, etc.	• Data bases on safety, health, and ergonomics • Bibliographic data bases on safety, health, and ergonomics • Internet services on the topic • Training packages (multimedia applications) • Expert systems for decision support
Safety information produced within a company	• Proactive safety data, for example, material safety data sheets, safety audits and inspections, task observations, hazard screenings, organizational rules and work permits, information on personal protective equipment, etc.	• Chemical safety data sheet information systems • Information systems for work and safety instructions • Audit information systems • Hazard screening information systems
	• Reactive safety data, for example, accidents, near-misses, sickness, fires, losses, costs, etc.	• Accident information systems • Sick leave information systems • Loss control information systems
	• Data on ergonomic design and safety planning, for example, emergency planning, ergonomic design of workplaces, first aid planning, safety analysis, etc.	• CAD software for human modeling • Emergency response information systems • Safety analysis information systems • Job design tools
	• Safety data included in other information systems, for example, in production management systems, maintenance information systems, or communication systems	• Company-specific information systems
The need for safety information outside a company	Administrative safety data which a company should provide for the insurance company or to authorities, for example, labor inspectors	• Accident information system provides specific accident information for labor inspectors and insurance companies • Electronic accident and claim data transfer between companies and insurance companies

Data From Outside a Company

There is a raft of information from outside a company that helps form the basis for an entire safety management policy. Such data include safety legislation, safety agreements between labor unions and employers' federations, safety and ergonomics standards, safety literature and research, and educational material (Takala 1992).

Safety information is available through various media, such as books, journals, and brochures, diskettes, CD-ROM disks, on-line Internet services, or from data banks. Various multimedia solutions can be obtained for safety education. Expert systems for supporting safety decisions are also available.

Data from Within a Company

Information from within a company could be classified as monitoring data on risks and working conditions as well as data on safety planning and execution. Utilizing safety information produced within a company has a huge potential for improving safety, ergonomics, and productivity (Aaltonen 1996; Ross 1991; Saari and Aaltonen 1989; Successful Accident Prevention 1987; Kjellén 1987; Reason 1994).

Data Needed Outside a Company

Typically, a company's administrative safety data need to be provided for the insurance company or for authorities, such as the Occupational Safety and Health Administration (OSHA). Computerized safety information systems can considerably decrease the routine work of preparing information forms.

JOB RISK ASSESSMENT AND PARTICIPATORY SAFETY AND HEALTH

Job and task analysis is needed to support the redesign of problematic tasks in a workplace. Besides the objective risk identification, risk assessment, and risk control by the designer and managers, safety and health procedures must be organized to be effective. Many research reports have shown that participation improves the quality of safety and health of workers in many ways, and also affects the final outcome of development projects (Lund et al. 1993).

Job load and hazard analysis (JLHA) is a participatory method for hazard screening, risk assessment, and safety improvement meant for use by occupational health care or safety staff together with personnel (Mattila 1985). Risks can be grouped into five categories: 1) chemical risks, 2) physical risks, 3) physical workload, 4) accident risks, and 5) mental stress. Identified risks are assessed according to a three-point rating scale (0, 1, 2), where a rating of 2 means the risk is a danger to safety and health, and some preventive measures are needed. Risk identification is performed by the workers and safety and health professionals. The summarized findings are assessed by a team of managers, and proposals for improvements are made.

A computer program has been developed to make data collection and analysis easy and practical, and provides appropriate reports to serve as a background for decisions on risk reduction. It produces: 1) a risk assessment report for jobs analyzed, 2) a report on occupational health, including occupational health care, and 3) a list of recommendations for preventive measures.

The JLHA supports cooperation among the personnel, enhances continuous improvement toward total quality management, and has contributed to reducing workplace risks (Figure 6-1).

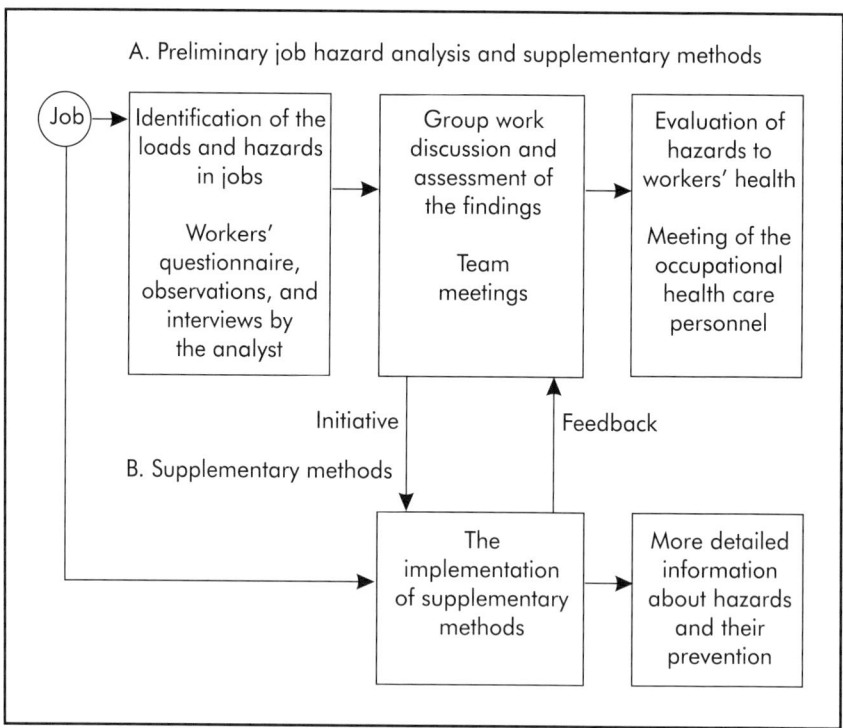

Figure 6-1 Structure of the JLHA method (Mattila 1985).

COMPUTER-AIDED DESIGN FOR ERGONOMICS

Several commercial softwares for ergonomic design and planning recently became available (Canadian Centre for Occupational Health and Safety 1996). These softwares fall into four classes: 1) training, 2) simulation, 3) job design, and 4) integrated information systems.

TRAINING AND INSTRUCTIONAL SOFTWARE PACKAGES

Training and instructional software packages are typically interactive and supportive information systems, either for expert use or for employee motivation and education. The features of these softwares can include, for example, overview information on ergonomics, definitions of ergonomic injuries, examples of risk factors in the workplace, elements of a successful ergonomics program, and various 3D illustrations. Some softwares can be tailored to company-specific needs. For example, established written ergonomics policies and programs can be incorporated into some educational softwares.

SIMULATION SOFTWARES FOR ERGONOMICS

Simulation softwares allow the user to visualize and evaluate the work and the workplace before it is built. The user can simulate the situation of the actual job. These systems are typically compatible with computer-aided design (CAD) systems, and the most advanced systems are called virtual reality (VR) applications.

A Mannequin's Eyes

Simulation software typically allows the user to see the workplace through a mannequin's eyes, and various analyses can be done, such as biomechanical, anthropometric, angular, postural, vision, and reach analyses. The "mannequin" program can represent a variety of the population by modifying anthropometric variables and degrees of freedom for the mannequin.

Simulation makes it easy for the designer to get a realistic overview of the content and form of the object being designed. In this way, CAD simulation and animation provide a tool to support participation in the design process. Potential operators can imagine the system more easily through a simulation model rather than an abstract engineering drawing. Thus, it is a better basis for true participation (Mattila 1996).

Simulation software can be used when analyzing cumulative trauma disorders. It can also be used when evaluating ADA requirements at those companies interested in performing pre-placement screening of new employees.

JOB DESIGN TOOLS FOR ERGONOMICS

Job design tools have been developed for different purposes, such as static strength prediction, estimating energy expenditure rates of manual material handling tasks, video analysis for performing biomechanical and ergonomic evaluations of job tasks, preventing repetitive strain injuries, and analyzing manual lifting jobs.

These softwares are practical tools for solving various ergonomic problems. Usually these systems are based on ergonomic assessment methods or guidelines. For example, analysis of manual lifting jobs is based on NIOSH lifting guidelines.

OWAS Method

Working postures cause significant problems at many workplaces. The Ovako Working Posture Analysis System (OWAS) method has proved to be one of the most successful tools for this analysis. After developing a computer program for the rapid analysis of results, as well as the demonstration reports, OWAS has become a practical tool to support participatory redesign (Mattila et al. 1992). The OWAS computer program gathers data at the workplace and analyzes and reports the results. The OWAS computer program can generate: 1) an overall report on time allocation in different postures and a risk assessment for those jobs analyzed, 2) a ranking of jobs or tasks analyzed according to the amount of poor postures, 3) a list of the most hazardous postures identified, and 4) a list of tasks where the poor postures occurred.

The OWAS analysis may be accompanied by a videotape of the task in question, which makes it easy to determine proposals for new improvements. It has been recommended that the results of the OWAS analysis be handled by a participatory cooperation group where specialists, managers, and operators discuss their findings and make suggestions for improvements. Operators themselves are able to identify poor work situations, which is useful for specifying and promoting needed changes (Figure 6-2).

INTEGRATED SYSTEMS FOR OCCUPATIONAL SAFETY, HEALTH, AND ERGONOMICS

Some multi-task safety and health information systems include modules for ergonomic planning of the workplace. The information on ergonomics is integrated into other safety and health information modules and thus improves its usability. Typically these large information systems run on minicomputers or mainframes.

Computer-aided Design for Ergonomics and Safety

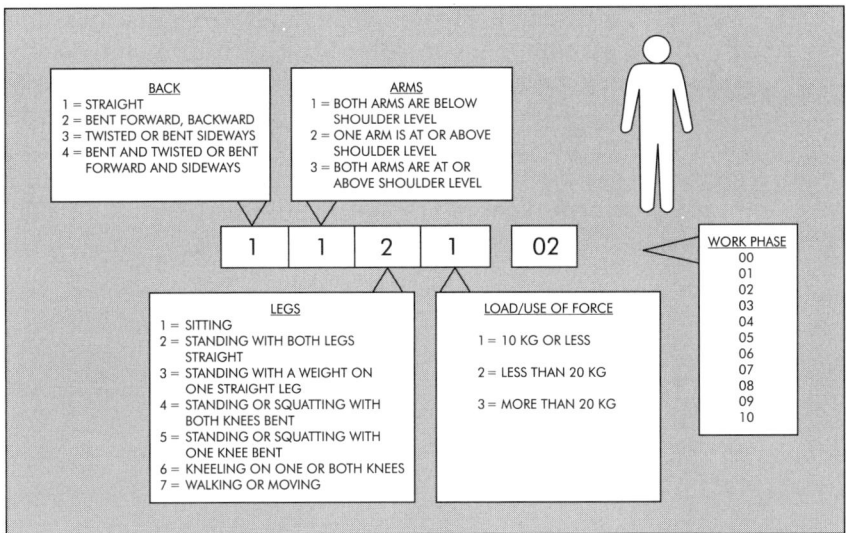

Figure 6-2 Postures in OWAS system.

EXAMPLES OF COMPUTER-AIDED DESIGN FOR ERGONOMICS

Design Support Checklist

One method to help designers evaluate workplace safety is to use safety and health checklists. An ergonomic checklist may contain: (1) physical demands, (2) task visibility, (3) mental demands, (4) machine design, (5) VDT (video display terminal) tasks, (6) safety, (7) ambient environment, and (8) product and process design (Helander 1995).

An example of such a checklist is the Flexible Manufacturing System (FMS) ergonomic and safety checklist (Leppänen et al. 1991). Checklist items are based on essential standards for the ergonomics and safety of flexible manufacturing systems. In this way the checklist is universal and may be used for all kinds of flexible manufacturing systems.

In the software, all ergonomics and safety requirements given for FMS in relevant standards are presented, and these can be checked when the system is designed. With the aid of this software, ergonomics and safety information are integrated into the design process. The designer documents the standards used for the object in question and accepts the liability required by modern legislation.

The checklist is divided into six parts: layout design, mechanical design, electrical design, control design, maintenance design, and installation design of the system. In this way, designers may simultaneously use relevant questions for the specific area under consideration. The questions are formulated in the manner of advice. A positive answer to the question means that the system is safe enough. The structure of the list is shown in Table 6-2. The five questions for mechanical design are presented as examples in Table 6-3.

Table 6-2. Design support checklist structure

Layout design (23 questions)
- Isolation and safeguarding of danger zones;
- Location of controls;
- Location of equipment requiring maintenance; and
- Location and measurements of aisles and service platforms.

Mechanical design (40 questions)
- Location and measurements of electrical equipment;
- Isolation and safeguarding of danger zones;
- Location of equipment requiring maintenance;
- Requirements of structural strength;
- Location of controls;
- Location and measurements of aisles and service platforms; and
- Structure of safety limit switches.

Control design (23 questions)
- Determination of critical components;
- Safeguarding of danger zones;
- Starting of the system;
- Protection of control equipment;
- Information flow in the process; and
- Manual running of the system.

Electrical design (32 questions)
- Control of the system and its components;
- Structure of safety equipment;
- Placement of wiring;
- Housing and labeling of electrical equipment; and
- Safeguarding of danger zones.

Maintenance design (24 questions)
- Safety requirements for maintenance.

Installation design (24 questions)
- Safety requirements for the installation design and installation process.

Table 6-3. Mechanical design questions

1. Are all wires protected against hard mechanical stress, chips, dust, and exposure to liquids?
2. Are the hydraulic pipes protected so that they may not be contacted accidently?
3. Are the engines placed so that they can be easily inspected, serviced, or disassembled, and are the wires easily accessible to disconnect if necessary?
4. If the loading/unloading of the system is done manually, is the working height ergonomically suitable?
5. Has every reachable run-in point, squeeze point, shear point, and catch point in material transportation system(s) been safeguarded?

The checklist is then transferred into an AutoCAD system that allows the user to make his or her own menus in the software. The questions are installed in the ACAD.HLP file, each question having its own cell.

To promote the systematic use of the checklist, seven menus are built into the system. One main menu controls the list, with six submenus, one for each group of questions. When one is chosen, its first question appears on the screen, along with the name and number of the standard on which the question is based. After the first question, the user can continue to the other questions until the list is completed. All AutoCAD commands and the CAD model can be modified interactively while the checklist is being read.

INTERACTIVE ERGONOMIC-ORIENTED PRODUCTION SYSTEM DESIGN

A new ergonomic-oriented information system, ErgoCop™, was developed for workplace designers in a research program called "Good Design Practice" (Lehtelä and Kiiskinen 1993). ErgoCop is intended for use during the process of designing assembly workplaces. Its aim is to provide the designer with all data needed during the design project from the existing information system.

The knowledge of designers and ergonomic experts is collected into the ErgoCop system. Reasonable design solutions for assembly line workplaces are also stored within the system. The program is implemented as a hypermedia application that can be utilized with CAD programs. The information is presented as text, tables, drawings, and photographs. (Lehtelä and Kiiskinen 1993). The ErgoCop program consists of the following:

- Information on ergonomic requirements and design principles for different workplaces (Table 6-4);
- Workplace template cards containing the documents created for each workplace designed;
- Project documentation, including feedback from the operators and safety and health professionals about the workplaces in operation;
- A glossary of photographs and drawings of workplaces and successful ergonomic solutions; and
- A library of components, such as equipment and tools, used for building workplaces. (See Table 6-5.)

Table 6-4. ErgoCop™ workplace information

1. Work content	6. Machines and equipment	11. Other environmental factors
2. Conveyors and material flow	7. Test devices	12. Liquids and other materials used in the process
3. Workplace layout	8. Hand tools	13. Training and information
4. Material handling	9. Workplace materials	14. Ease of maintenance and cleaning
5. Furniture and adjustment	10. Lighting	15. Ease of product
		16. Layout of the assembly line

The basic information in the ErgoCop system is textual. In the text there are so-called "hot words" that function as links to pictures and tables. A workplace card is created for all new workplace designs. The card contains the following items: links to the drawings and sketches; addresses of the workplaces built according to the drawings; a list of components used in building the workplace; information on design decisions and solutions to problems that arose during the design process; and feedback from workers and ergonomic experts when the workplace is tested in real time.

Table 6-5. ErgoCop program components

Block	Contents	Purpose
Ergonomic information	Selection of ergonomic data (in both text and pictures).	Provides the designers with ergonomic knowledge tailored to the company.
Workplace cards	Data from workplace design projects; design decisions, solutions to the problems, and improvements suggested by workers and occupational health personnel.	Designers have a collection of methods and solutions which can be reused in future projects when similar workplaces are designed.
Projects	A glossary containing the layout drawings, timetables, and memos created.	Helps to keep a record of the documents created during the project.
Components	A record of the components used in building workplaces.	Designers easily find the information on frequently used components.
Photographs	A glossary containing photos and drawings of workplaces.	A forum where good ergonomic solutions are presented.

The ErgoCop program is accessible to ergonomic and safety personnel so they can add to the program and give feedback about the designs implemented. In the next design task, the feedback will be considered, poor solutions avoided, and the ergonomic quality of the system improved.

TRANSOM JACK™ HUMAN MODELING

A new generation of computer-aided human modeling systems has been developed in recent years. They offer a large variety of possibilities to analyze and visualize human task behavior in a virtual environment. With the aid of these new virtual reality (VR) tools, it is possible to evaluate ergonomic and safety features of workplaces and machinery already in the early stages of design.

Virtual reality has already been employed in architecture, with virtual offices created and inspected before their designs are committed to reality. In these applications, the expert designer or future user can see the work environment and even move within it.

One of the most advanced human modeling packages is the Transom Jack software. This is a commercial product developed by the

Center for Human Modeling and Simulation at the University of Pennsylvania with the collaboration of NASA.

3D Interactive Environment

Transom Jack provides a 3D interactive environment for controlling articulated figures. It features a detailed human model and includes realistic behavioral controls, anthropometric scaling, task animation and evaluation systems, view analysis, automatic reach and grasp, collision detection and avoidance, and many other useful tools for a wide range of applications. Transom Jack software requires a Silicon Graphics workstation running IRIX 4.x or greater.

Three-dimensional CAD designs of new machines and workstations can be studied with Transom Jack to answer several questions. Are the dimensions suitable for users of different sizes and for those with possible handicaps? Does the worker see all the relevant items in the surrounding space to execute the work safely? What is the strain on the different joints of the body during the work task? Is the worker going to suffer from fatigue after a few iterations (Laitinen et al. 1996)?

Transom Jack uses familiar tools, such as the mouse and pop-up menus that facilitate learning the software. The average designer is able to operate the modeling system after only a few days of training. The software is versatile and advanced. The hardware requirements, the user's expertise in human factors and ergonomics, and the time taken by the user, will increase in proportion to the complexity of the software used during the design process (Haijanen et al. 1995).

Transom Jack software also has interfaces to the electromagnetic movement monitoring system. With motion sensors attached to the worker, the virtual worker simulates his work tasks, and we can carry out various tests, such as strength analysis (Figure 6-3). A data glove and a head-mounted display also can be used. With the head mounted display, you can see what the virtual human figure sees. When you turn your head, the figure turns his head. With data gloves (or in the future a skin-tight body suit with sensors) the worker's movements become the figure's movements. In the future when computing power improves, the virtual human also will be able to hear, feel, and sense the environment (Laitinen et al. 1996).

IGRIP SOFTWARE

The IGRIP program includes 3D models of the most common robots in its library of symbols. This is why the system is especially

Figure 6-3 Testing the usability of a machine with the Transom Jack software.

suitable for the documentation and design of automated production systems. IGRIP also includes the 3D man model, making it possible to analyze the whole man/machine system, the tasks carried out by operators, as well as the man/machine interface. This simulation model even allows for carrying out systematic ergonomic and risk analysis for both the technical production system and the man/machine interface. Kuivanen (1995) studied the reliability and coverage of risks identified through IGRIP simulation models and compared them with results from the same production systems in reality. His results showed that the reliability and the coverage of risk identification are satisfactorily high using this software.

INTEGRATION OF SAFETY AND ERGONOMICS

As previously stated, the EC directive on machine safety (89/392/EEC) and strict liability regulations make it necessary for machine manufacturers to prove that machines and the pertinent documentation meet safety and health requirements. The directive sets out the central requirements and detailed instructions are given in harmonized European standards. Planning a machine in accordance with harmonized safety standards ensures conformity with the requirements specified in the directive.

In practice, the manufacturer: 1) designs and manufactures machinery according to the essential health and safety requirements; 2) prepares the technical documentation and has a type examination made, if necessary; and 3) signs the declaration of conformity and fastens the CE-mark onto the machinery.

The EC directive on machine safety and the harmonized European standards have shown the need to integrate safety and ergonomic design into the design process in a comprehensive and systematic way. The manufacturers and customers need to know that the safety and ergonomics points have been designed sufficiently. Because of this, computer-aided tools are needed.

The aims of a computer-aided design tool are to present information on safety standard directives to designers, document whatever safety measures are taken, and support the designer in producing the document that clarifies compliance with the safety and health legislation (Figure 6-4).

During the machine design process, the CE standard program can be run simultaneously with CAD tools. The program, activated by the progress data base, works in both UNIX and PC environments. Information on safety standards and the EC machine safe-

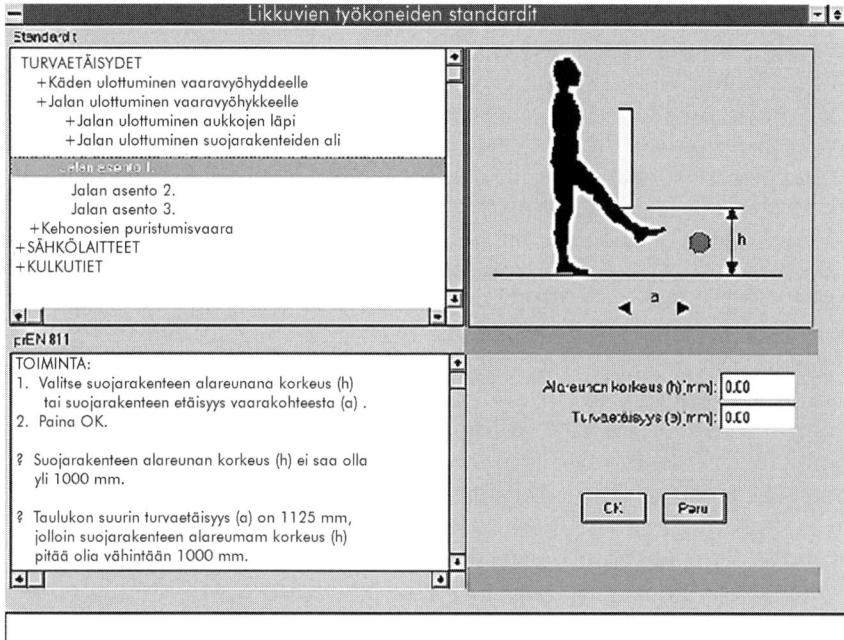

Figure 6-4 User interface of CE-standard software.

ty directive are stored within the data base. The open data base structure also allows the designer to create and add his own design rules and solutions, as well as the company's own standards. In addition, safety documentation can be created by the program. References to the European standards and to the CE machine safety directive are created and added automatically to objects being drawn. The designer can also add his own comments and reasoning to the documentation.

CONCLUSIONS

The safety level is one criterion for assessing the quality of a workplace's design. Design practices should utilize the knowledge and tools available, enabling the integration of safety expertise into the resulting design.

Simulation application is an expanding area in ergonomic planning. The utilization of virtual reality provides flexible tools for the planning of production facilities together with safety considerations (Haijanen et al. 1995).

Applications of artificial intelligence and expert systems have not developed in safety management as rapidly as was anticipated. It might be that the development of an expert system is a laborious task, and the benefits may not be as good as expected. Expert systems are, nevertheless, needed in safety management, and many applications are available to support decisions (Krishnamurthy et al. 1991; Suokas 1990; Lehto and Miller 1987).

The usability of information systems is an important consideration in developing information technology. This applies also to safety information systems. First, the compatibility of organization and information systems should be flexible, so that the safety information systems really support the safety goals of the organization. Computerization and mastering information systems demand more work and training from personnel. It is important to make sure that the implementation of information systems happens smoothly and in compliance with the conditions of the organization. Safety information systems should be integrated operatively into the other information systems of a company (Aaltonen 1992). Safety information flows should be smooth, and the interfaces of software should be user-friendly. Programming techniques should be flexible to provide applications that are easy to tailor to the company's needs.

REFERENCES

Aaltonen, M. "A Consequence and Cost Analysis of Occupational Accidents in the Furniture Industry." Doctoral dissertation. *People and Work,* 6. Helsinki, Finland: Finnish Institute of Occupational Health, 1996.

Aaltonen, M.V.P. "Integration of an Accident Information System into the Computer-aided Production Management Systems in the Chemical Industry." *Computer Applications in Ergonomics, Occupational Safety and Health.* M. Mattila and W. Karwoski, eds. ISSA Amsterdam, The Netherlands: North Holland, 1992: 277-282.

Canadian Centre for Occupational Health and Safety (CCOHS). *OH&S Software Database.* CCINFO CD-ROM Publication. Toronto, Ontario: CCOHS, 1996.

Haijanen, J., Rasa, P.L., Hirvonen, M., Kuusisto, A., and Laitinen, H. "Human Modeling in 3-Dimensional Virtual Environments." *Book of Abstracts of the XVth Congress of the International Society of Biomechanics.* K. Häkkinen, K.L. Leskinen, P.V. Komi, and A. Mero, eds. Jyväskylä, Finland: Gummerus, 1995: 354-355.

Hartung, P. "Selection of Computer Aided Systems for Protection of Health and Safety at Work" (in German with English and French summaries). *Amtliche Mitteilungen.* Düsseldorf, Germany: Bundesanstalt für Arbeitsschutz, 1993.

Heinrich, H.W., Peterson, D., and Roos, N. *Industrial Accident Prevention,* 5th Edition. New York: McGraw-Hill Book Company, 1980.

Helander, M. *A Guide to the Ergonomics of Manufacturing.* London: Taylor & Francis, 1995.

Johnson, S. "Management Accountability for Safety Performance." *Professional Safety* 1988, 33(6): 23-27.

Kjellén, U. "Simulating the Use of a Computerized Injury and Near Accident Information System in Decision Making." *Journal of Occupational Accidents* 1987, 9: 87-105.

Krishnamurthy, K., Chen, J.G., and Fisher, D. "Computer-assisted System for Accident Analysis and Fall Protection in Industrial Construction Industries." *Advances in Industrial Ergonomics and Safety III*. W. Karwowski and J.W. Yates, eds. London: Taylor & Francis, 1991: 895-900.

Kuivanen, R. "Methodology for Simultaneous Robot System Safety Design." *VTT Publications* 219. Espoo, Finland: Technical Research Centre of Finland, 1995.

Laitinen, H., Kuusisto, A., Haijanen, J., Leskinen, T., and Rasa, P.L. "Workplace 2000 Programme: Towards Excellence in Occupational Safety and Health." *Paper presented at the Conference Work in the Information Society*. Helsinki, Finland: Finnish Institute of Occupational Health, 1996.

Lehtelä, J., and Kiiskinen, M. "Information System for Workplace Designers: A Case Study." *Proceedings of Nordiska Ergonomisällskapets årskonferns Ergoomi och Företagsekonomi*. Turku, Finland: Ergonomics Society of Finland, 1993: 157-162.

Lehto, M.R., and Miller, J.M. "Scientific Knowledge Acquisition During the Estension of GSA: An Expert System for Generic Safety Analysis." *International Journal of Industrial Ergonomics* 1987, 2: 61-75.

Leppänen, M., Mattila, M., and Väänänen, M. "Safety Checklist for Flexible Manufacturing Systems." *Reliability and Safety of Processes and Manufacturing Systems*. Y. Malmen and V. Rouhiainen, eds. London: Elsevier Science Publishers, 1991: 183-189.

Lund, R.T., Bishop, A.B., Newman, A.E., and Salzman, H. *Designed to Work: Production Systems and People*. Englewood Cliffs, NJ: Prentice Hall, 1993.

Mattila, M. "Job Load and Hazard Analysis: A Method for the Analysis of Workplace Conditions for Occupational Health Care." *British Journal of Industrial Medicine* 1985, 2: 656-666.

Mattila, M., Vilkki, M., and Tiilikainen, I. "A Computerized OWAS Analysis of Work Postures in the Papermill Industry." *Computer Applications in Ergonomics, Occupational Safety and Health*. M. Mattila and W. Karwowski, eds. Amsterdam, The Netherlands: Elsevier Science Publishers, 1992: 235-239.

Mattila, M. "Design of Safe Workplaces." *Book of Abstracts of the XIV World Congress on Occupational Safety and Health*. Madrid, Spain: ISSA, 1996.

Parsons, K.C. "International Standards: A Sound Basis for the Design of Workplace Environments of the Future." *Proceedings of the 12th Triennal Congress of the International Ergonomics Association, Vol. 5*. Toronto, Canada: International Ergonomics Association, 1994: 168-169.

Räsänen, T., Aaltonen, M., Luukkanen, A., Sartiala, I., and Korjuslommi, E. "Improving the Effectiveness and Profitablity of Occupational Safety and Health with the Aid of the Information Technology" (in Finnish). *Report for the Finnish Work Environment Fund*. Vantaa, Finland: Finnish Institute of Occupational Health, Department of Occupational Safety, 1993.

Reason, J. "A Systems Approach to Organizational Error." *Proceedings of the 12th Triennial Congress of the International Ergonomics Association, Vol. 1.* Toronto, Canada: International Ergonomics Association, 1994: 94-96.

Ross, C.W. Computer Systems for Occupational Safety and Health Management, 2nd Edition. New York: Marcel Dekker, Inc., 1991.

Saari, J., and Aaltonen, M. "Microcomputer-based Accident Database for Companies." *Advances in Industrial Ergonomics and Safety. I.* A. Mital, ed. London: Taylor & Francis, 1989: 345-350.

Successful Accident Prevention. "Recommendations and Ideas Field Tested in the Nordic Countries." *Reviews 12.* Helsinki, Finland: Finnish Institute of Occupational Health, 1987.

Suokas, J., Heino, P., and Karvonen, I. "Expert Systems in Safety Management." *Journal of Occupational Accidents* 1990, 12: 63-78.

Takala, J. "Safety and Health Information Systems: Analysis of Local, National, and Global Methods." Doctoral thesis. *Publications 109.* Tampere, Finland: Tampere University of Technology, 1992.

Tarrants, W.E. *Measurement of Safety Performance.* New York: Garland STPM Press, 1980.

Wrench, C.P. *Data Management for Occupational Health and Safety.* New York: Van Nostrand Reinhold, 1990.

CHAPTER 7

ERGONOMICS TRAINING AND EDUCATION FOR WORKERS AND MANAGERS

Marilyn Joyce
Director
The Joyce Institute/A Unit of Arthur D. Little
Seattle, WA

Training enables an organization to be proactive in the systematic implementation of ergonomics. It maximizes investment in people and technology in ways that can be measured in terms of safety, productivity, and quality.

Training needs to be perceived as only one aspect of a comprehensive ergonomics program (NIOSH 1993). Most important, to be truly effective, training needs to include managers, engineers, safety and health professionals, supervisors, human resources, facilities, and purchasing, as well as hourly workers. This philosophy of training assumes a macroergonomics approach to the implementation of ergonomics. A macroergonomics approach means there is a system that includes both the technical aspects (equipment, tools, environment) and the social aspects (management style, organizational structure, training) (Hendrick 1984).

In his description of the "learning organization," Peter M. Senge states: "Organizations learn only through individuals who learn. Individual learning does not guarantee organizational learning. But without it, no organizational learning occurs" (Senge 1990).

CHARACTERISTICS OF GOOD TRAINING

Characteristics of good training include the nature of the adult learner, the role of the trainer/facilitator, and the environment.

NATURE OF THE ADULT LEARNER

Well-designed training programs are based on assumptions about how people learn. The principles of the adult learning theory, as defined by Malcolm Knowles (Knowles 1984) and updated by him and others are:

1. Adults are motivated to learn when they feel the need or have the interest to know more about something. If they are not ready to learn, they will not learn.
2. Adults learn best when their own experience is an integral part of the learning. Their work and life experiences have defined who they are.
3. Adults learn best when learning to solve a problem or cope with a change in their real-life situations.
4. Adults have a need to be self-directing in the learning process because, as in other aspects of their lives, they want to be responsible for their own decisions.
5. Adults have a wide range of individual differences and, therefore, have varied learning styles, pacing, and timing that must be met.

Training needs to focus on developing habits, which are defined as the "intersection of knowledge, skill, and desire" (Covey 1990). Habits are internalized principles and patterns of behavior that need to be developed if training is to have a measurable impact.

Adults need to be able to integrate new ideas with what they already know if they are to use the information. Also, information that conflicts with what they already know to be true is integrated more slowly because the conflict forces a re-evaluation of old ideas. For example, in many organizations, management has always perceived safety and ergonomics as separate from the core business functions. So, when they are presented with the idea that safety and ergonomics actually contribute to the core business functions and need to be considered in the formative stages of any process, product, or environment change, they often need substantial proof before accepting the idea.

ROLE OF THE TRAINER/FACILITATOR

Many training courses are structured on the traditional model that the trainer is the giver of information and the participants are the receivers. While it is true that the trainer usually has more expertise in the subject matter at hand, it is also true that the learner may know more about the situation in which the training is to be applied. A true trainer is a facilitator, a catalyst, who can draw on the learners' experiences to apply new concepts to real-life situations and who encourages self-directed learning.

Setting the Stage

A facilitator sets the stage by providing a workable physical environment, containing appropriate resources and organizational framework, and by creating an atmosphere where the learner feels free to make mistakes. Perhaps most important, a facilitator is a good listener who provides feedback in a positive way, never embarrassing or ridiculing a learner. Because good trainers understand the principles of adult learning, they follow certain guidelines (Joyce 1989):

1. Provide practice through on-the-job application of the training.
2. Appeal to the learner's senses and learning style by using a variety of instructional techniques: lectures, audiovisual materials, printed matter, demonstrations, role playing, workshops, simulations, software, large and small group discussions, presentations by class members, etc.
3. Reward appropriate responses and behavior, particularly examples where the learner applies the knowledge on the job. Respect the learners and value their input.
4. Keep the learner active.
5. Focus on strategies to help participants change habits. It is important that they believe their actions can make a difference. They must understand why recommendations are made, verbalize their commitment, and demonstrate what they are learning.
6. Help participants understand the change process. They need to know that change of any kind in large organizations is a complex and lengthy process, often taking years. Effective change involves a planned process of communication among the organization's different levels and segments to foster a shared set of values and goals for the change process.

THE ENVIRONMENT

All of us have sat in rooms that are too large or too small, too hot or too cold, too bright or too dark, with distracting noises and odors from the next room. We have found ourselves in uncomfortable seats, straining to see a screen with print too small to read beyond the first row. If you are in a position to organize a training session, address as many of the physical issues as possible. If the space is too large, arrange the tables so that the group occupies only a portion of that space. If the room is too small, limit the number of participants and arrange the room as optimally as possible, removing extraneous items. Contact the facilities person in advance about controlling temperature. Learn where the controls for the lights are and how the window coverings work so that you can adjust for showing items on a screen and for discussion and work group situations. Plan for stand-up rest pauses at least once an hour.

Many organizations underestimate the time and resources required to develop effective in-house training. They often find, after an initial false start, that external resources must be used for at least part of the work.

Cost-justifying the training, based on all available information, is the final step. Considerations such as the expected impact on injury reduction, reduction of lost time, productivity improvements, or reduction of production costs are usually the basis for the cost-justification (Joyce 1994). Even if the training is initiated as part of a corporate policy to train all employees in a wide range of health and safety issues, it is wise to cost-justify ergonomics training to enhance the buy-in of managers.

FIVE PHASES

Why should we train in ergonomics? Whom should we train? What type of training is appropriate? How does ergonomics training relate to and contribute to the core business functions? How do I document the ergonomics changes we are making in the plant? These are the types of questions that many organizations ask as they begin to implement an ergonomics program that includes training.

Successful training involves five phases: planning, development, delivery, measurement of impact, and improvement (Figure 7-1).

PLANNING

The initial step in the planning process involves identifying the need for ergonomics

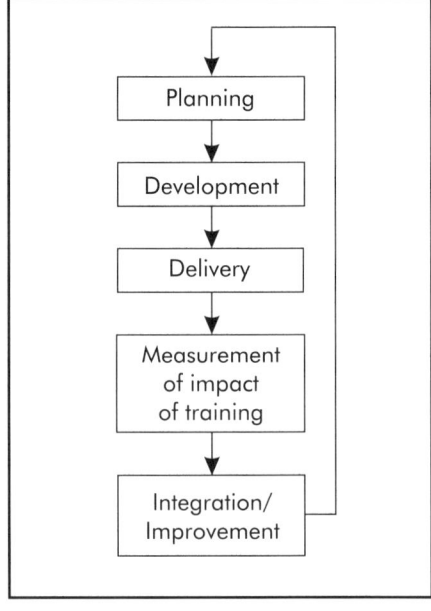

Figure 7-1 Five phases of successful training.

training based on the strategic plan and data collection. The next step is to define the scope in terms of audience, content, goals, and objectives of the training, in collaboration with the department/people to be impacted. The third step is to plan the budget, time frame, and resources for the training. It is at this point that decisions about whether to develop training in-house or to use an external consultant are usually made.

DEVELOPMENT

In the development phase, the first step is to define the responsibilities of those involved in the development, delivery, and evaluation of the training. Often these are shared responsibilities between the ergonomics team members, the training department, and the external consultant. Even if you are using consultants to customize and deliver training, someone internally needs to communicate the goals and objectives to them so that the organization derives more benefits from the training.

The second step involves elaborating on the scope of work outlined in the planning process. Here, objectives are matched with the various participant populations and an outline of the content, training methodology, and materials to be used is established. Part of this step is to define how the impact of the training will be measured over a specific period of time. In the applications section of this chapter is an example of the type of information each population needs for ergonomics development.

The third step involves the detailed development of all the materials: trainer's manual, participant manual, slides, videos, transparencies, learning activities, workshops, and actual experience in the workplace (checklists, work area consultation guidelines).

Generally, the final step of the development stage is a pilot course so that the stakeholders can evaluate all aspects of the training, and so that the training development team or consultant can make changes before proceeding to a major implementation.

If you use an external consultant to provide the training, either through an informal selection process or through a formal request-for-proposal process, consider the criteria outlined in Figure 7-2.

DELIVERY

Successful delivery training requires attention to detail and excellent coordination. In addition to making appropriate arrangements for training rooms, audiovisual equipment, material reproduction and shipment, it is critical to convey the objectives to the participants and their managers.

> **Selection Criteria for an Ergonomics Training Consultant**
>
> 1. Experience of the consultant in providing training in similar industries or type of work (knowledge of the industry alone does not assure good training because the person may not have the ability to communicate the information so that it is understandable).
> 2. Client references, reflecting success over several years.
> 3. Credentials of the organization; check to see how long they have been specialists in ergonomics, whether the ergonomists developing/conducting the training have masters degrees in ergonomics, whether training specialists are involved in the development phases so that the training technologies are sound.
> 4. Ability and willingness of the consultant to customize the training to meet your particular requirements; for example, if you are using the consultant to train using his/her materials, make sure that he/she takes slides and videos of your sites to use in the courses. If you are having the consultant develop proprietary training for your company, without using any of the consultant's materials, be sure that the contract specified that the materials are the property of your organization.
> 5. Documentation of measurable results that training has yielded for other clients.
> 6. Evidence that the organization/ergonomist has the respect of its/his peers (papers, articles, books, etc.).

Figure 7-2 Selection criteria for an ergonomics training consultant.

If possible, schedule a management briefing to outline the objectives and to seek input from the managers regarding the agenda. The briefing will also enable you to motivate managers to support ergonomics efforts. For example, if a group of manufacturing and process engineers is being trained, the engineering manager must be aware of the issues and set the expectation that ergonomics will be considered in modifying or designing workstations and equipment.

The actual facilitation of the training needs to be consistent with the adult learning theory defined earlier in this chapter and the guidelines regarding the use of visuals.

It is critical that training not be simply lecture. For change to occur, the participants must experience a variety of learning activities. The trainer needs to plan those learning activities. Ideally, any videos or slides involved should be of the plant or, at least, of a similar plant, so that participants can relate to familiar situations.

MEASURING THE IMPACT

Often, success measurement ends with the evaluation form that participants complete at the end of the course. This level of evaluation is important because it focuses on the skills and knowledge specified in the training objectives. It provides feedback on the extent to which the training brought attendees to the desired level of proficiency. This type of formative evaluation is conducted to assess the clarity, organization, and comprehensibility of the instruction. Surveys, focus groups, interviews, self-assessment tests, and behavioral demonstrations are common methods of formative evaluation. Information gleaned from these evaluations should be used to refine the training program (NIOSH 1993). However, a second, more meaningful, way is to measure the impact by considering the effect on the person doing the work, the job itself, and the organization (Figure 7-3).

Impact on the participants, their jobs, and the organization can be measured by observing demonstrations in the classroom or on the plant floor, as well as on the job, after the training is completed.

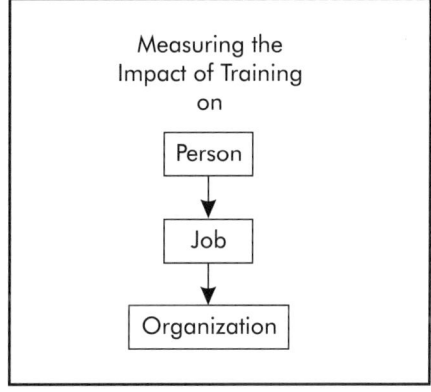

Figure 7-3 Measuring the impact of training.

It is important to build the basis for measuring change into the training. For example, perhaps for every job that must undergo a corrective action, a before-and-after checklist needs to be completed with a short note documenting the results of the change. Health and safety professionals, for example, need to learn a system for documenting trends based on the injury and illness data to track reductions in injuries and their related costs.

Impact on the Person

Engineers being trained can complete an analysis of a videotaped job during a workshop segment of the training; supervisors can complete a checklist while observing a task that an office worker is doing; a safety person can assist a worker in adjusting a workstation; a worker can work with his hands in a neutral position while using a screwdriver; a worker can adjust the height of a chair to improve his or her posture; a product design engineer can eliminate the need for 30% of the fasteners required for an assembly task, based on ergonomics principles.

Impact on the Job

Impact on the job also can be measured by establishing a baseline and tracking individual project results. For example, you might find the cycle time of an electronics assembly operation has been reduced 40% (Figure 7-4).

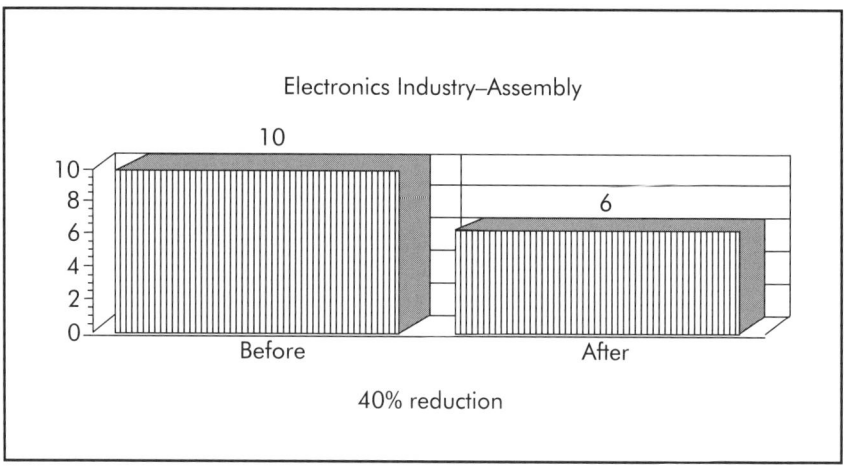

Figure 7-4 Cycle time reduction.

As a result of one ergonomics training program in an office environment, workers improved productivity and quality. Within two weeks, the employees' key strokes per hour were 7,150; operators not in the program took 12 weeks to reach that speed.

One significant measure is to identify the reduction in risk factors (Figure 7-5) with a checklist after changes are made to a job. Example: placing heavy parts near a workstation eliminated a lift-and-carry operation, minimizing the number of lifts from floor level and, consequently, reducing the number of repetitions and force required to complete a task.

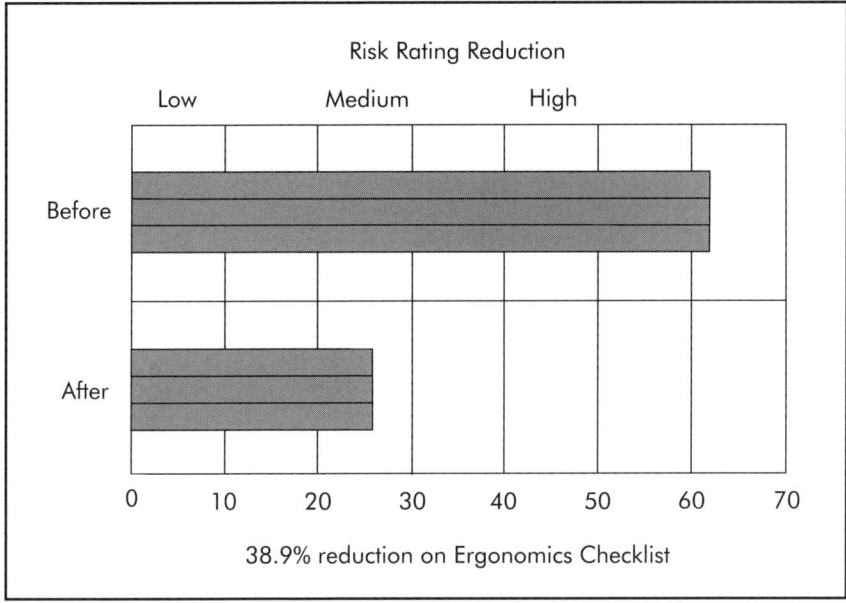

Figure 7-5 The reduction in risk factors after changes are made to a job.

Impact on the Organization

The most meaningful measurement is the impact on the organization. This is the culmination of the training and is the most likely to sustain long-term interest in and support for an ergonomics program. The effect can be measured by relying on company data, such as:

- Reduction in workers' compensation costs or lost days;
- Improvement in productivity data, such as minimizing the time it takes new hires to reach the standard;
- Overall reduction in use of packaging materials; and
- Turnaround time from incoming order to delivery to client.

INTEGRATION AND IMPROVEMENT

The final stage in the training process is to take steps to assure that the effects of the training are integrated into the organization. For example, does the ergonomics team continue to work with Engineering to make sure that changes occur? Does the Purchasing Department follow ergonomics criteria when purchasing tools, furniture, and equipment?

Someone in the organization needs to have ownership of the training process on a sustained basis to assure success. The training needs to be re-evaluated and improved, based on input from participants and changing needs in the organization. For example, after initial training for engineers, more advanced training may be required to take them to a higher level of analysis and problem-solving.

Refresher Training

Research indicates that refresher training needs to be conducted about once a year. "At a minimum, refresher training (both awareness and job/risk specific) should be provided annually to maintain employee motivation, reaffirm organizational commitment, and allow a forum for employee feedback on all factors shown to greatly affect the transfer of training. In addition, targeted training

---

Case Study: Ergonomics Training Leads to Design Change

The Client
A large multinational petrochemical company

The Challenge
A petrochemical company asked for assistance in training its ergonomics task force in the manners used to identify, analyze, and develop solutions for those jobs containing ergonomics stressors. In addition to instructing the individuals regarding identification and quantification of ergonomic stressors, the seminar allowed for the development of design criteria for specific products.

The Result
In a session led by the ergonomist, the group was able to finalize those design criteria which would not only decrease the employee exposure to ergonomic stresses, but also better control the well. The following occurred as a result of a design change:
- The company is now better able to meet the environmental and federal regulations regarding the well's location in relation to bodies of water, reduce possible contamination, and control the well.
- The risk of injury to the employees is reduced by 25% due to the reduction in parts.
- The time required to change the gasket is reduced by 20-50%.

By incorporating a change in the gasket, the client was not only able to annually decrease the amount of time required to change it by 52.1 days/employee, but they were also able to ensure increased compliance with environmental and federal regulations.

should be delivered on an 'as needed' basis when the medical surveillance data or worksite analysis of an existing or modified job indicate a training need" (NIOSH 1993).

JOB-SPECIFIC TRAINING

After having addressed the principles of ergonomics training, it is important to address specific content for various audiences. This section specifies the reasons for training, the types of training that need to be provided, and the audience for each type.

The goal of training and education is to cause changes in behavior at all levels in the organization. The effect of properly designed ergonomics training should be to change the way work is performed and how equipment and processes are designed to reduce the number of ergonomically–related injuries and illnesses. Training also should be geared to enable the organization to conduct worksite analyses as well as develop strategies for hazard prevention and control and medical management. The training must be designed so that information provided is consistent with the roles and responsibilities of each individual. All levels of the organization need to be trained, including corporate and plant management, health and safety professionals, supervisors, facility managers, engineers (both manufacturing and product), and employees. Product design engineers need to be trained so that design criteria related to manufacturing are included in all specifications. If those product specifications are combined with design criteria for workplaces, tools, equipment, and processes, then potential problems can be designed out initially. Timelines need to be established to indicate at what point in the entire process each group needs to be trained.

MANAGERS

The training and briefing for managers needs to occur early in the process to secure commitment. However, it may be wise to wait until after some pilot work has been done and the initial medical safety records review has been completed so that management has some hard data to justify authorization of the program. The training should include:

- Definition of ergonomics;
- Need for ergonomics;
- Cost of injuries nationally and a summary of plant-specific data on injuries/illnesses, quality, and productivity;
- Videos or slides of problem areas at selected sites;

- Benefits of ergonomics; and
- Examples of results after implementing ergonomics at specific sites and companies.

HEALTH AND SAFETY PROFESSIONALS

Training for health and safety professionals, engineers, plant level ergonomics steering team, and facilities managers provides the expertise necessary to initiate a program and plan all its elements. The training should include:

- An overview of ergonomics issues;
- Causes of musculoskeletal injuries/illnesses;
- Prevention strategies, including application of engineering, work practice, and administrative controls;
- Worksite analysis methodologies;
- Risk assessment and prioritization;
- Documentation procedures to measure results and track changes; and
- Guidelines for setting up and implementing an ergonomics plan.

MANUFACTURING ENGINEERS

Three to five days are needed to focus on engineering solutions to problems, simple modifications, and major changes. This enables participants to quantify improvements in existing facilities as well as to work on new facilities. The training should include:

- Benefits of ergonomics;
- Identification and quantification of ergonomic stressors;
- Principles and criteria for design of workstations, tools, etc.;
- Problem-solving methodologies;
- Documentation requirements of safety, productivity, and quality improvements;
- Criteria for designing new workplaces, equipment, and spaces to prevent problems;
- Guidelines for product design manufacturability; and
- Cost justification.

Many typical methods for justifying projects are based on the assumption that injuries will happen if no changes are made or that injuries will be avoided as a result of changes. Other cost justification methods are based on a consideration of probabilities.

The methods allow ergonomics teams and engineers to assess payback, which might be impossible with traditional methods. The approaches used include breakeven point, simple payback period, present worth method, and profit margin. These methods do not excuse the ethical obligation to eliminate known workplace hazards; they simply help prioritize solution alternatives.

SUPERVISORS, TECHNICIANS, ERGONOMICS TEAM LEADERS, LABOR REPRESENTATIVES

This training is critical to the success of an ergonomics program because the supervisors play such a key role as the liaison with employees. Often, they know the jobs very well and are excellent sources of information for identifying problems and suggesting modifications. The training should enable participants to:

- Identify the causes of musculoskeletal injuries/illnesses;
- Assist the health and safety committee in identifying problem jobs (that is, using a checklist);
- Make simple modifications and adjustments to the workplace;
- Apply the principles of ergonomics as they are setting up jobs or developing work procedures;
- Communicate with workers and support design changes; and
- Provide input to the engineers.

EMPLOYEES

The hourly workers need to be involved in the process. However, they should not be trained until the support system (engineers, supervisors, health, and safety personnel) has been established and budgets are in place for making improvements. The training should last about two hours and include such ergonomics and job-specific topics as:

- Information about the body's capabilities and limitations;
- Appropriate positions and movements;
- Appropriate use of tools, equipment, etc.;
- Stretches and massages that can be used to prevent the onset of muscular fatigue and discomfort;
- Information on ergonomics principles so that they can provide valuable input to the health and safety committees and engineers during the solution design process; and
- Information on the early reporting of symptoms of discomfort so the problem can be solved before serious complications occur.

MEDICAL AND HUMAN RESOURCE PROFESSIONALS

Training for medical departments, medical providers, and Human Resource managers is key to returning injured workers to work in a timely manner and to tracking trends in injuries. Medical department training should include:

- Instruction in early recognition, evaluation, treatment, and rehabilitation of ergonomic injuries;
- Principles of ergonomics and epidemiology;
- Characteristics of the jobs the workers are performing;
- Analysis of trends in injury and illness rates;
- Logs and documentation of changes, including ergonomic checklists, before-and-after photos and job summaries;
- Employee symptoms surveys; and
- Policies that focus on the psychological issues.

Ergonomics training is the cornerstone of an effective ergonomics program that contributes to core business functions.

REFERENCES

Covey, S. *The 7 Habits of Highly-Effective People*. New York: Simon & Schuster, 1989.

Hendrick, H., Brown, Jr., O., eds. *Human Factors in Organizational Design and Management*. Amsterdam, The Netherlands: North Holland Publishing Company, 1984.

Joyce, M., and Wallersteiner, U. *Ergonomics: Humanizing the Automated Office*. Cincinnati, OH: South-Western Publishing Co., 1989.

Joyce, M. "Ergonomics Training: The Cornerstone of a Successful Ergonomics Program." *BNAC Communicator*, 1994.

Joyce, M., and Marcotte, A. *Writing and Implementing an Ergonomics Plan*. Seattle, WA: The Joyce Institute/A Unit of Arthur D. Little, 1994.

Knowles, M. *The Adult Learner: A Neglected Species*, 3rd Edition. Houston, TX: Gulf Publishing Co., 1984.

National Institute for Occupational Safety and Health (NIOSH). *Comments on the Occupational Safety and Health Administration Proposed Rule on Ergonomic Safety and Health Management*. Washington, DC: U.S. Department of Health and Human Services, 1993.

Senge, P. *The Fifth Discipline*. New York: Doubleday, 1990.

CHAPTER 8

ASSESSING PHYSICAL WORK LOAD

Veikko Louhevaara
Professor
Regional Institute of
Occupational Health and
University of Kuopio
Kuopio, Finland

Juhani Smolander
Senior Researcher,
Department of Physiology

Tatiana Aminoff
Researcher

Juhani Ilmarinen
Professor

Finnish Institute of
Occupational Health
Vantaa, Finland

PHYSICAL LOAD AT WORK

Physical load at work demands dynamic and static muscle work for the exertion of force. Muscular work in occupational activities can be roughly divided into four groups: heavy dynamic muscle work (mainly moving of own body weight), manual material handling (moving external loads), static work, and repetitive work. Manual material handling and repetitive tasks may be either dynamic or static muscle work in nature or a combination of these two. Static postural work is frequently observed, for example, in office work, the electronics industry, and in repair and maintenance tasks. Repetitive work tasks can be found in the food and wood processing industries, among others.

In many jobs, physical work performed by large muscle groups will remain indispensable in spite of technological developments.

In Finland, it is estimated that in the 1990s, 10-20% of the work force is exposed daily to heavy muscular loads (Louhevaara and Smolander 1993). Heavy muscular exertion is most frequently needed in forestry, agriculture, building, installation, transportation, manual sorting, health, home care, and cleaning. Individuals in special occupations, such as firefighters, police officers, and soldiers also frequently experience heavy muscular exertion.

High-productivity and high-quality work require work ability and good worker health. Work ability, however, cannot be maintained without continuous and appropriate preventive ergonomic, psychosocial, and individual measures carried out by occupational health and safety practitioners. The need for such measures is increasing, particularly among aging workers (Ilmarinen and Louhevaara 1994; World Health Organization 1993). Before selecting and implementing measures for maintaining work ability, it is often necessary to perform reliable field assessments of physical load during actual work.

PHYSIOLOGY OF MUSCULAR WORK

Dynamic Muscle Work

In dynamic work, active skeletal muscles contract and relax rhythmically. Blood flow to the muscles is increased to match the metabolic needs. The increased blood flow is achieved through increased pumping of the heart (cardiac output), decreased blood flow to inactive areas, such as kidneys and liver, and increased number of open blood vessels in the working musculature. Heart rate (HR), blood pressure, and oxygen consumption ($\dot{V}O_2$) increase in linear relation to working intensity of the muscles. Also, pulmonary ventilation is heightened due to deeper breaths and increased breathing frequency. The purpose of activating the whole cardiorespiratory system is to enhance oxygen delivery to the working muscles. $\dot{V}O_2$ measured during dynamic muscle work indicates the intensity of the work. The maximum oxygen consumption ($\dot{V}O_2$max) indicates the person's maximum capacity for aerobic work. $\dot{V}O_2$ values can be translated to energy expenditure (1 liter of $\dot{V}O_2$ per minute equals a 21 kJ/min or 5 kcal/min).

STATIC MUSCLE WORK

In static work, the muscle contraction does not produce visible movement; for example, in a limb. Static work increases the pressure inside the muscle, which, together with the mechanical compression, partly or totally occludes the circulation. The delivery of nutrients and oxygen to the muscle and the removal of metabolic end-products from the muscle are hampered. Thus, in static work, muscles fatigue more easily than in dynamic work.

The most prominent circulatory feature of static work is the rise in blood pressure. HR and cardiac output do not change much. Above a certain intensity, blood pressure increases in direct relation to the intensity and duration of the effort. Also, at the same relative intensity, static work with large muscle groups produces a greater blood pressure response than work with smaller muscles.

In principle, the physiological regulation of ventilation and circulation is similar in static and dynamic work, but the metabolic signals from the muscles are stronger, inducing a different response pattern.

EFFECT OF THE SIZE OF WORKING MUSCLE MASS

The physiological responses to muscular work also depend on the size of active muscle mass. The $\dot{V}O_2$max is higher in exercises requiring large muscle groups than in exercises requiring smaller muscle masses. In young persons, the $\dot{V}O_2$max during arm cranking exercises is about 70% of the $\dot{V}O_2$max during leg exercise. During leg exercise the $\dot{V}O_2$max declines progressively with age, whereas during arm exercise the differences between younger and older persons are smaller (Aminoff et al. 1996). At a given submaximal $\dot{V}O_2$, HR, systolic blood pressure, and pulmonary ventilation are higher during arm work than during leg work or combined arm and leg work (Stenberg et al. 1967).

Standard exercise testing is based on dynamic leg exercise (treadmill, cycle ergometry, or step test), and relative work loads have been expressed as percentages of the individual's $\dot{V}O_2$max (Åstrand 1960; Rutenfranz et al. 1990). The $\dot{V}O_2$max for arm work, however, cannot be estimated from experiments with leg work and vice versa (Asmussen and Hemmingsen 1958). Because the $\dot{V}O_2$max varies greatly between arm and leg work, a work load that is fairly easy for leg work may be quite exhausting for arm work. The assessment of $\dot{V}O_2$max should be done by such a procedure that activates some muscles as during actual work. When the load is related to the maximal capacity of the muscle mass, the differences in physiological strain between the muscle groups become smaller (Lewis et al. 1983; Sargeant and Davies 1973), but are still higher during prolonged arm work than during prolonged leg work (Aminoff et al. 1997).

The recording of HR and the subjective perception of exertion during work is easy, and the values may be used to measure the relative work intensity. The HR is dependent on the fitness of the person and on the size of active muscle mass. HR is higher for unfit persons than for fit persons, and higher during arm work than during leg work at the same *absolute* work load. During exercise at the same *relative* work load, the HR is higher during leg work than during arm work. When the HR is expressed as a percentage of maximal HR for the corresponding active muscle group, the HR response is similar between arm and leg work (Aminoff et al. 1997). When estimating the work load, the recorded HR in the field should be proportioned to the maximal HR for the corresponding active muscle group, and compared with the HR at known work loads in exercises which are similar to actual work.

MUSCULAR OVERLOAD

As previously summarized, the degree of physical strain a worker experiences in muscular work depends on the size of working muscle mass (small, large), type of muscle contractions (static, dynamic), the intensity of contractions, and individual characteristics.

When muscular work load does not exceed a worker's physical capacity, his or her body adapts to the load and recovery is quick when the work is stopped. If the muscular load is too high, fatigue ensues, working capacity is reduced, and recovery slows down. Peak loads or prolonged overload may result in organ damage. On the other hand, muscular work of certain intensity, frequency, and duration may also result in training effects; as diminished muscular demand may cause detraining effects. This model is the expanded stress-strain concept presented by Rohmert (1983) (Figure 8-1).

In general, there is little epidemiological evidence that muscular overload is a risk factor for diseases. However, poor health, disability, subjective overload, and decreased work output is concentrated in physically demanding jobs and is especially true for older workers. Also, many risk factors for work-related musculoskeletal diseases are connected to different aspects of muscular work load. In ergonomics, people have tried to determine acceptable work loads and environments by identifying muscular work loads that are too strenuous. Prevention of chronic effects is the focus of epidemiology, whereas work physiology deals mostly with short-term effects, that is, fatigue during work tasks or work days.

FIELD METHODS FOR ASSESSING PHYSICAL OVERLOAD

The set of field methods of work physiology summarized in Table 8-1 have been commonly used in ergonomics. These methods can be considered basic work physiology and ergonomic methods; their reliability is high when used by professional staff under work site conditions.

Assessing Physical Work Load

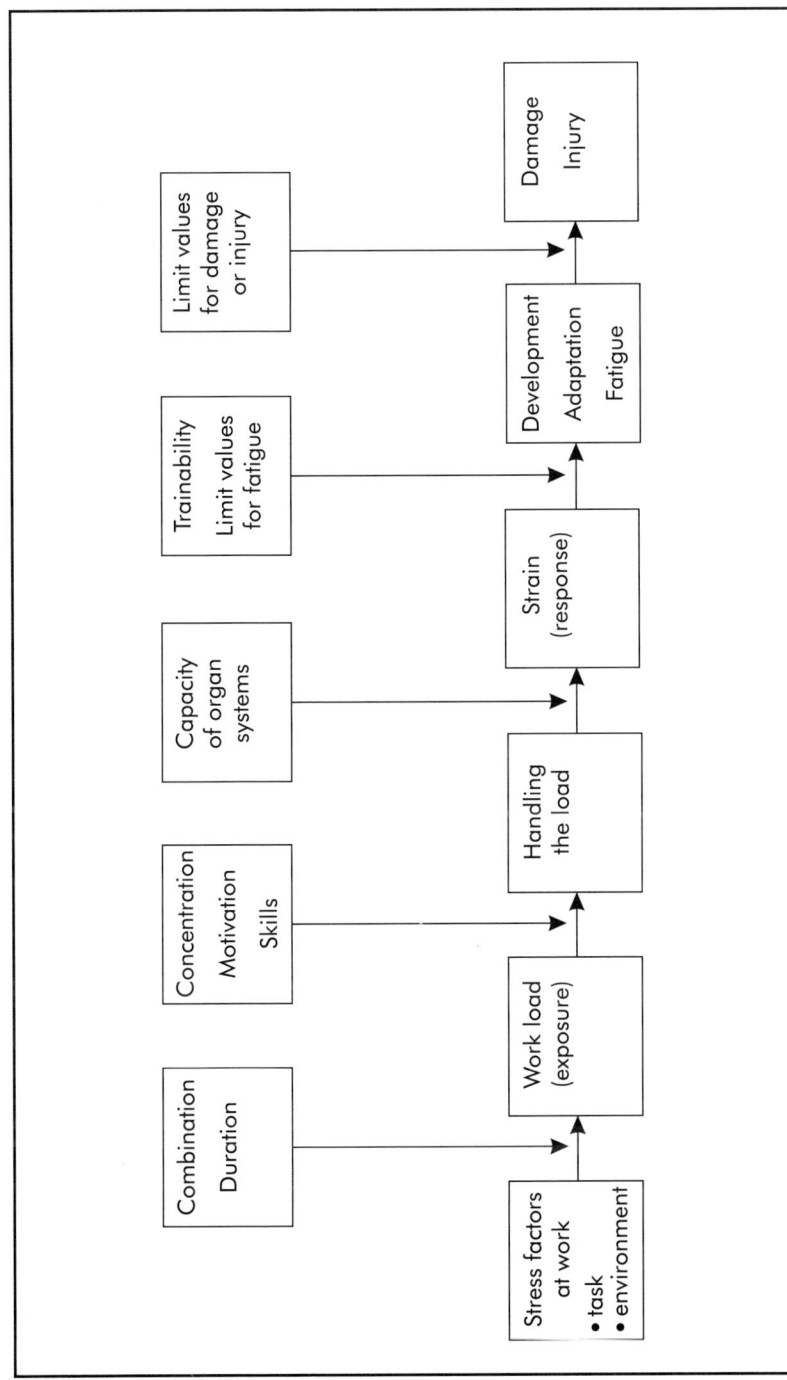

Figure 8-1 The expanded stress-strain model modified from Rohmert (1983).

Table 8-1. Work physiology field methods for the assessment of physical work load at the worksite

Method
1. Ergonomic job analysis (AET) (Rohmert and Landau 1979)
2. Job load and hazard analysis (JOHA) (Mattila 1985)
3. Ergonomic workplace analysis (EAW) (Ahonen et al. 1989)
4. Measurement of oxygen consumption (VO_2) (Louhevaara et al. 1985)
5. Measurement of pulmonary ventilation (V_E or V_I) (Eley et al. 1966)
6. Measurement of heart rate (HR) (Oja et al. 1977)
7. Measurement of systolic blood pressure (SBP) (Theorell et al. 1993)
8. Basic Edholm scale for the estimation of energy expenditure (basic Edholm) (Edholm 1966)
9. Modified Edholm scale for manual material handling (modified Edholm) (Long and Louhevaara 1992)
10. Method for the evaluation of postural load (OWAS method) (Karhu et al. 1977)
11. Rating of perceived exertion (RPE) (Borg 1970)
12. Electromyography (EMG) (Jonsson 1982)

Relevance and feasibility of the methods for work site studies depend mainly on the type of muscular work, aims of the measurements, and the human, technical, economic, and time resources available for research. In association with these methods, basic thermal parameters (air temperature, relative humidity, and air velocity) should be measured or estimated at workplaces. Thermal load may considerably increase physical strain if the environmental conditions markedly deviate from temperate ones, or if the wearing of heavy personal protective equipment is necessary. An unusually slow or rapid work rate may also severely bias the quantification of physical work load.

Physical work load is not directly related to strain responses (Figure 8-1); in fact, HR and the rating of perceived exertion are highly influenced by individual factors. Therefore, the assessment of physical work load based on these responses cannot be established without information about the characteristics and fitness of a worker and environmental conditions. When absolute load parameters are related to a worker's maximal or submaximal physical capacities, individual physical strain is the result.

HEAVY DYNAMIC MUSCLE WORK

Assessing physical work load in dynamic work tasks, such as moving body weight or external loads, has traditionally been based on measurements of $\dot{V}O_2$ (energy expenditure). $\dot{V}O_2$ can be measured with relative ease in the field with portable devices (Douglas-bags, Max Planck respirometer, Oxylog, Cosmed), or it can be estimated from HR recordings, which can be done reliably at the workplace, for example with the SportTester-device. The use of HR in estimating $\dot{V}O_2$ requires that it is individually calibrated against measured $\dot{V}O_2$ in a standard work mode in the laboratory, so that the investigator knows the $\dot{V}O_2$ at a given HR. HR recordings should be considered with caution because they are also affected by many factors, such as physical fitness, environmental temperature, psychological factors, and size of active muscle mass. Thus, HR can lead to overestimates of $\dot{V}O_2$ in the same way as $\dot{V}O_2$ underestimates the total physiological strain by reflecting only energy requirements.

The *relative aerobic strain (RAS)* is defined as the percentage relation of $\dot{V}O_2$ measured on the job to the $\dot{V}O_2$max of the worker measured in the laboratory. Thus, RAS is %$\dot{V}O_2$max. If only HR measurements are available, a close approximation of RAS can be made by calculating percentage HR range (% HR range) with the Karvonen formula (Karvonen et al. 1957), which is:

$$\% \ HR \ range = \frac{\text{Heart rate at work} - \text{heart rate at rest}}{\text{Heart rate maximum} - \text{heart rate at rest}} \times 100$$

HR maximum can be measured in an exercise test or taken from age-specific tables.

According to the Åstrand (Åstrand 1960) classical study, RAS should not exceed 50% during an eight-hour work day. In the experiments at 50% RAS level, body weight decreased, HR did not reach steady state, and subjective discomfort increased during the day. Åstrand recommended a 50% RAS limit for both men and women. Later, she found that construction workers spontaneously chose an average RAS level of 40% (range 25-55%) during a working day. Several more recent studies have indicated that the acceptable RAS is lower than 50%. Most studies recommend a 30-35% RAS level for the entire working day (Rutenfranz et al. 1990).

Originally, the acceptable RAS levels were developed for purely dynamic muscle work, which rarely occurs under real working conditions. It may happen that acceptable RAS levels are not exceeded, for example, in a lifting task, but the local load on the back may greatly exceed acceptable levels.

$\dot{V}O_2$max is usually measured on a bicycle ergometer or treadmill, in which the mechanical efficiency is high (20-25%). When the active muscle mass is smaller or the static component is higher, $\dot{V}O_2$max and mechanical efficiency will be smaller than in exercise with large muscle groups. For example, in sorting postal parcels, the $\dot{V}O_2$max was only 65% of the maximum measured on a bicycle ergometer, and the mechanical efficiency was less than 1% (Louhevaara et al. 1988). When guidelines are based on $\dot{V}O_2$, the test mode in the maximal test should be as close as possible to the real task. This goal, however, is difficult to achieve. Despite its limitations, RAS determination has been widely used in assessing physical strain in different jobs.

Other useful physiological field methods are also available for the quantification of physical stress or strain in heavy dynamic work. Observational techniques can be used in estimating energy expenditure (for example, the Edholm scale). Rating of perceived exertion indicates the subjective accumulation of fatigue. New ambulatory blood pressure monitoring systems allow more detailed analyses of circulatory responses (Table 8-1).

MANUAL MATERIAL HANDLING

Manual material handling includes such work tasks as lifting, carrying, pushing, and pulling of various external loads. Most of the research in this area has focused on lower-back problems in lifting tasks, especially from a biomechanical point of view.

Observation of working postures and use of force (for example, Ovako Working Posture Analysis System [OWAS] method), rating of perceived exertion, and ambulatory blood pressure recordings are also suitable methods for stress and strain assessments in manual material handling. Electromyography can be used to assess local strain responses, such as in arm and back muscles. An RAS level of 21-35% has been recommended for lifting tasks, when the $\dot{V}O_2$ during the task compares to the bicycle ergometer maximum.

Guidelines based on HR are either absolute or related to the resting heart rate. The absolute values for men and women are 90-112 beats/min in continuous manual material handling. These values are about the same as the recommended values for the increase in HR above resting levels, 30-35 beats/min. These guidelines are also valid for heavy dynamic muscle work for young and healthy men and women. However, as mentioned previously, HR data should be treated with caution because it is affected by many factors other than muscle work.

The guidelines for acceptable work load for manual material handling based on biomechanical analysis comprise several factors such as weight of the load, handling frequency, lifting height, distance of the load from the body, and physical characteristics of the person.

In one large-scale field study (Louhevaara et al. 1990), healthy male workers could handle postal parcels weighing 9-11 lb (4-5 kg) during the whole shift without any signs of objective or subjective fatigue. Most of the handling occurred below shoulder level, the average handling frequency was less than eight parcels per minute, and total number of parcels was below 1500 per shift. The mean HR was 101 beats/min and oxygen consumption 1.0 l/min, which corresponded to 31% RAS as related to bicycle maximum.

STATIC MUSCLE WORK

Static muscle work is required primarily when maintaining working postures. The endurance time of static contraction is exponentially dependent on the relative force of contraction. This means, when the static contraction requires 20% of the maximum force, the endurance time is five to seven minutes, and when the relative force is 50%, the endurance time is about one minute (Rohmert 1960).

For the practitioner, fewer field methods are available for the quantification of strain in static work. Some observational methods (for example, OWAS) exist to analyze the proportion of poor working postures, that is, postures deviating from normal middle positions of the main joints. Blood pressure measurement and rating of perceived exertion may be useful, whereas HR is not applicable.

Older studies indicated that no fatigue was developed when the relative force was below 15% of the maximum force (Rohmert 1960). However, more recent studies indicate that the acceptable relative force is specific to the muscle or muscle group, and is less than 8-10% of the maximum static strength (Björkstén and Jonsson 1977). These force limits are, however, difficult to use in practical work situations, because they require laborious electromyographic recordings.

REPETITIVE WORK

Repetitive work with small muscle groups resembles static muscle work according to circulatory and metabolic responses. Typically, in repetitive work muscles contract over 30 times per minute. When the relative force of contraction exceeds 10% of the maximum force, endurance time and muscle force start to decrease. However, there is wide individual variation in endurance times. For example, the endurance time varies between 2-50 minutes, when the muscle contracts 90-220 times per minute at the relative force level of 10-20% (Laurig 1974).

It is very difficult to set any definitive criteria for repetitive work because even very light levels of work such as the use of microcomputer mouse may cause increases in intramuscular pressure, which may lead to swelling of muscle fibers, pain, and reduction in muscle strength. Very few feasible field methods are available for the strain assessment in repetitive work (Table 8-1).

PHYSICAL WORK LOAD ASSESSMENT IN PRACTICE

In practical work situations, it is often impossible to use any measuring instrument, and the analysis must be carried out by observation. In those situations, it is useful to recognize the different types of muscular work (heavy, dynamic muscle work, manual material handling, static work, repetitive work), and the frequency of these activities. For example, the time factor can be classified by:

- None of the time;
- Some of the time;
- Most of the time; and
- All of the time.

Then, the focus of corrective measures can be directed to the most frequent types of muscular load. Also, if needed, more detailed analyses can be employed on the tasks with high occurrence rates.

Age and gender are important factors, especially when heavy dynamic work or manual handling of heavy loads occur. In these tasks, fatigue development and possible injury risk are related to the working capacity ($\dot{V}O_2$max, muscle strength), which is lower among older workers and women.

Epidemiological follow-up studies (Ilmarinen 1992) have shown that for older workers the following high physical demands decrease work ability:

- Static muscular work;
- Use of muscle strength;
- Sudden peak loads;
- Repetitive movements; and
- Simultaneous bent and twisted work postures.

Identifying these demands, and using corrective measures on them, can most likely improve the maintenance of work ability during aging.

PREVENTING PHYSIOLOGICAL OVERLOAD

Relatively little epidemiological evidence exists to show that muscular load is harmful to health. However, work physiological and ergonomic studies indicate that muscular overload results in fatigue (that is, decrease in work capacity), and may reduce productivity and quality of work.

The prevention of muscular overload may be directed to the work content, work environment, and the worker. The load can be adjusted by technical means, which focus on work environment, tools, and working methods. The fastest way to regulate muscular work load is to reorganize the work and, for instance, increase the flexibility of working time individually. This means designing work rest regimens which take into account the work load and the needs and capacities of the individual workers.

When adjusting physical work load, the main principle is that static and repetitive muscle work should be kept at a minimum. Occasional heavy dynamic work phases may be useful for maintaining endurance-type physical fitness. Probably the most useful form of dynamic physical activity, which can be incorporated into a working day, is brisk walking or stair climbing.

Prevention of muscular overload, however, is very difficult if a worker's physical fitness or professional skill is poor. Appropriate training will increase professional skills, and may reduce muscular loads at work. Also, regular physical exercise during work or leisure time will increase the muscular and cardiorespiratory capacities of the worker.

Best results in preventing muscular overload will probably be achieved by the promoting health policies to workers through ergonomics and physical exercise programs at work sites.

REFERENCES

Ahonen, M., Launis, M., and Kuorinka, T. *Ergonomic Workplace Analysis.* Helsinki, Finland: Institute of Occupational Health, Ergonomics Section, 1989.

Aminoff, T., Smolander, J., Korhonen, O., and Louhevaara, V. "Fatigue Development in Arm and Leg Work at Similar Relative Work Loads." *Proceedings of the 13th Triennial Congress of the International Ergonomics Association.* P. Seppälä, T. Luopajärvi, C-H. Nygård, and M. Mattila, eds. Helsinki, Finland: Finnish Institute of Occupational Health, 1997.

Aminoff, T., Smolander, J., Korhonen, O., and Louhevaara, V. "Physical Work Capacity in Dynamic Exercise with Differing Muscle Masses in Healthy Young and Older Men." *European Journal of Applied Physiology and Occupational Physiology* 1996, 73: 180-185.

Asmussen, E., and Hemmingsen, I. "Determination of Maximum Working Capacity at Different Ages in Work With the Legs or With the Arms." *Scandinavian Journal of Clinical and Laboratory Investigation* 1958, 10: 67-71.

Åstrand, I. "Aerobic Work Capacity in Men and Women with Special Reference to Age." *Acta Physiologica Scandinavica* 1960, 49 (suppl 169): 1-92.

Björkstén, M., and Jonsson, B. "Endurance Limits of Force in Long–term Intermittent Static Contractions." *Scandinavian Journal of Work, Environment, and Health* 1977, 3: 23-27.

Borg, G. "Perceived Exertion as an Indicator of Somatic Stress." *Scandinavian Journal of Rehabilitation Medicine* 1970, 2: 92-98.

Edholm, O.G. "The Assessment of Habitual Activity." *Physical Activity in Health and Disease*. K. Evang and K. Lange-Andersen, eds. Oslo, Norway: Universitetforlaget, 1996.

Eley, C., Goldsmith, R., Layman, D., Tan, G.L.E., and Walker, E. "A Respirometer for Use for the Measurement of Oxygen Consumption. 'The Miser,' a Miniature, Indicating and Sampling Electronic Respirometer." *Ergonomics* 1978, 21: 253-264.

Ilmarinen, J. "Job Design for the Aged with Regard to Decline in their Maximal Capacity: Part I–Guidelines for the Practitioner." *International Journal of Industrial Ergonomics* 1992, 10: 53-63.

Ilmarinen, J., and Louhevaara, V. "Productive Aging in Finland." *Aging International* 1994, 21: 34-36.

Jonsson, B. "Measurement and Evaluation of Local Muscular Strain in the Shoulder During Constrained Work." *Journal of Human Ergology* 1982, 11: 73-88.

Karhu, O., Kansi, P., and Kuorinka, I. "Correcting Working Postures in Industry: A Practical Method for Analysis." *Applied Ergonomics* 1977, 8: 199-201.

Karvonen, M., Kentala, K., and Musta, O. "The Effects of Training Heart Rate: A Longitudinal Study." *Ann Med Exper Fenn* 1957, 35: 307-315.

Laurig, W. *Beurteilung Einseitig Dynamischer Muskelarbeit*. W. Berlin, Germany: Beuth, 1974.

Lewis, S.F., Taylor, W.F., Graham, R.M., Pettinger, W.A., Schulte, J.E., and Blomqvist, C.G. "Cardiovascular Responses to Exercise as Functions of Absolute and Relative Work Load." *Journal of Applied Physiology* 1983, 54: 1,314-1,323.

Long, A.F., and Louhevaara, V. "Computerized Collection and Anaysis of Minute-by-minute Physical Activity and Work Phase Data." *Proceedings of the International Conference on Computer-aided Ergonomics and Safety '92—CAES '92*. M. Mattila and W. Karwowski, eds. New York: Elsevier Science, 1992.

Louhevaara, V., Hakola, T., and Ollila, H. "Physical Work and Strain Involved in Manual Sorting of Postal Parcels." *Ergonomics* 1990, 33: 1,115-1,130.

Louhevaara, V., Ilmarinen, J., and Oja, P. "Comparison of Three Field Methods for Measuring Oxygen Consumption." *Ergonomics* 1985, 28: 463-470.

Louhevaara, V., and Smolander, J. "Fyysinen Ylikuormitus Työssä (Physical Overload at Work)." *Työ ja Ihminen* 1993, 7 (suppl.): 17-28.

Louhevaara, V., Teräslinna, P., Piirilä, P., Salmio, S., and Ilmarinen, J. "Physiological Responses During and After Intermittent Sorting of Postal Parcels." *Ergonomics* 1988, 31: 1,165-1,175.

Mattila, M.K. "Job Load and Hazard Analysis: A Method for the Analysis of Workplace Conditions for Occupational Health Care." *British Journal of Industrial Medicine* 1985, 42: 656-666.

Oja, P., Louhevaara, V., and Korhonen, O. "Age and Sex as Determinants of the Relative Aerobic Strain of Nonmotorized Mail Delivery." *Scandinavian Journal of Work, Environment, and Health* 1977, 3: 225-233.

Rohmert, W. "Ermittlung von Erholungspausen für statische Arbeit des Menschen." *Internationale Zeitschrit für angewandte Physiologie Einschliesslich Arbeitsphysiologie* 1960, 18: 123-164.

Rohmert, W. "Formen Menschlicher Arbeit." *Praktische Arbeitsphysiologie*. W. Rohmert and J. Rutenfranz, eds. Stuttgart, Germany: Georg Thieme Verlag, 1983.

Rohmert, W., and Landau, K. *AET—das Arbeitswissenschaftliche Erhebungsverfahren zur Tätigkeitsanalyse (AET—An Ergonomic Job Description Method)*. Bern, Germany: Huber, 1979.

Rutenfranz, J., Ilmarinen, J., Klimmer, F., and Kylian, H. "Work Load and Demanded Physical Performance Capacity Under Different Industrial Working Conditions." *Fitness for Aged, Disabled, and Industrial Worker (International Series on Sport Sciences, Vol. 20)*. M. Kaneko, ed. Champaign, IL: Human Kinetics Books, 1990.

Sargeant, A.J., and Davies, C.T.M. "Perceived Exertion During Rhythmic Exercise Involving Different Muscle Masses." *Journal of Human Ergology* 1973, 2: 3-11.

Stenberg, J., Åstrand, P-O., Ekblom, B., Royce, J., and Saltin, B. "Hemodynamic Responses to Work with Different Muscle Groups, Sitting and Supine." *Journal of Applied Physiology* 1967, 22: 61-70.

Theorell, T., Ahlberg-Hulten, G., Jodko, M., Sigala, F., and de la Torre, B. "Influence of Job Strain and Emotion on Blood Pressure in Female Hospital Personnel During Work Hours." *Scandinavian Journal of Work, Environment, and Health* 1993, 19: 313-318.

World Health Organization. *Aging and Working Capacity* (WHO Tech. Rep. Series 835; Report of a WHO study group). Geneva, Switzerland: World Health Organization, 1993.

CHAPTER 9

STATIC WORK LOAD AND ENDURANCE TIMES

*Nico J. Delleman**
Senior Researcher/Consultant

*Jan Dul***
Head

Innovation Department
NIA TNO
Amsterdam
The Netherlands

MAINTAINING WORKING POSTURES

Many work situations require postures that have to be maintained for an extended period of time, such as machinery operation, assembly, and at a video display unit. A sustained posture (with or without external force exertion) creates a static load, and the muscles involved contract continuously without shortening or lengthening. Static contraction leads to muscle fatigue. For example, approximately one-third of workers in the European Union are involved in painful or tiring postures for more than half the working day (Figure 9-1) (European Foundation for the Improvement of Living and Working Conditions 1997). Pain and fatigue may lead to musculoskeletal diseases, reduced productivity, and deteriorated posture and movement control. The latter can increase the risk of errors and may reduce quality of work while increasing hazardous situations. Sick leave and disability of (skilled) workers directly affect production costs. Static load can be reduced by improving postures through workstation and tool optimization, reducing the holding time of postures, and providing sufficient and properly distributed rest pauses. Variation between and within sitting, standing, and walking is indispensable.

---

*Currently at: TNO Human Factors Research Institute, Soesterberg, The Netherlands.

**Currently at: Erasmus Business Support Centre, Rotterdam, The Netherlands.

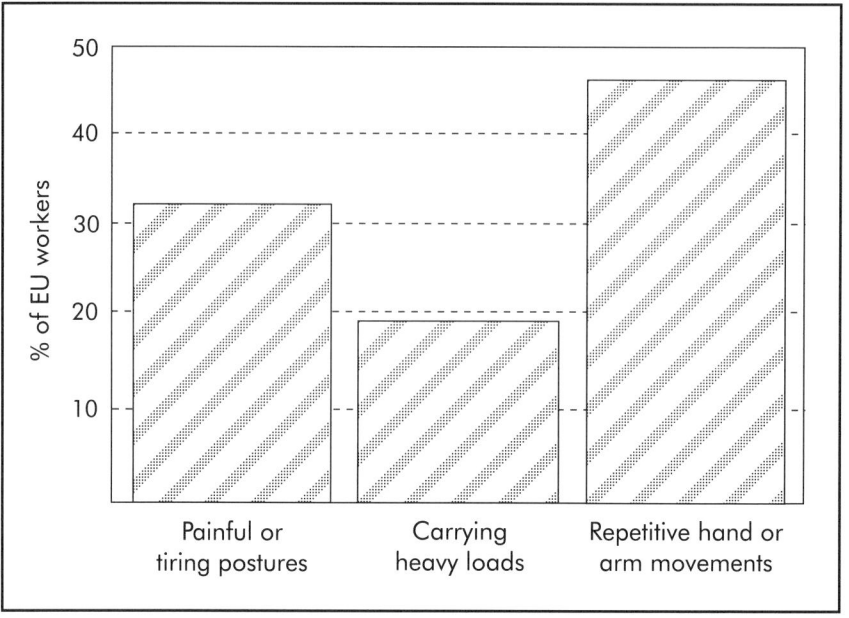

Figure 9-1 Percentage of workers in the European Union involved in painful or tiring postures, carrying heavy loads, or repetitive hand or arm movements for more than half of the working day.

PERCEIVED DISCOMFORT

A suitable measure of the load on muscles, tendons, ligaments, joints, and bones is the discomfort perceived by the worker. One technique for measuring discomfort is the LMD method (Localized Musculoskeletal Discomfort) (Van der Grinten 1991; Van der Grinten and Smitt 1992). In this method, the worker is asked to rate his or her discomfort in 40 regions shown on a diagram of the rear view of a human body, modified after Corlett and Bishop (1976) (Figure 9-2), using a Borg scale (Table 9-1) (Borg 1982). A written or verbal response is given at the beginning and end of a work session. For each body region, the score at the beginning is subtracted from the score at the end. Usually, the resulting scores for several regions are grouped into scores for larger functional units (back, shoulder/arm, etc.), as well as into a whole body score (the sum of the resulting scores on all 40 body regions). The LMD method provides reliable results for comparison of work situations.

Static Work Load and Endurance Times

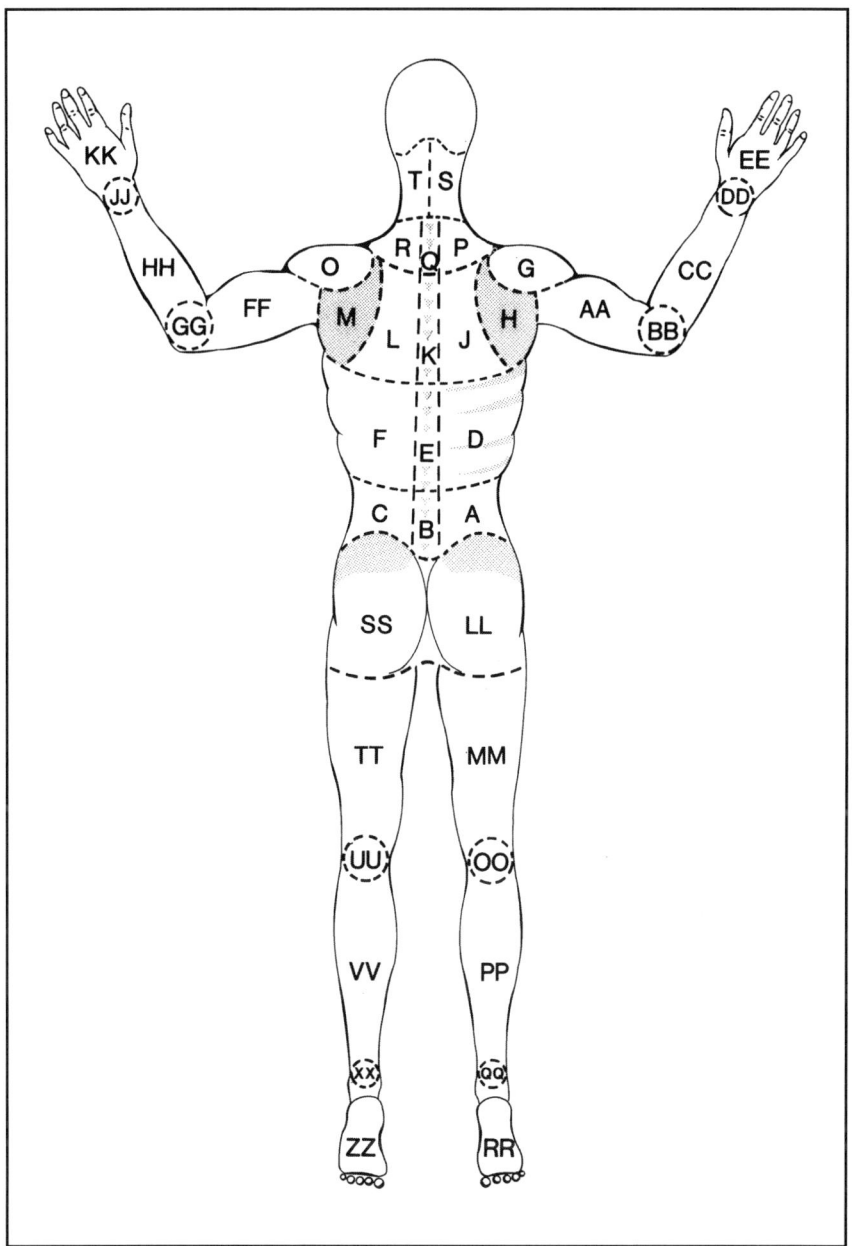

Figure 9-2 LMD method—body diagram.

Table 9-1. The relationship between the discomfort score on the Borg scale, the percentage of the maximum holding time (% MHT), and the remaining endurance capacity (REC)

Discomfort score		% MHT	REC
0	Nothing at all	0	100%
1	Very weak (just noticeable)	10	90%
2	Weak (light)	20	80%
3	Moderate	30	70%
4	Somewhat strong	40	60%
5	Strong (heavy)	50	50%
6		60	40%
7	Very strong	70	30%
8		80	20%
9		90	10%
10	Extremely strong	100	0%

MAXIMUM HOLDING TIME

The maximum holding time (MHT) is the longest duration that a static posture can be maintained continuously from a rested state until maximum discomfort. MHT decreases exponentially as the relative muscle force increases (Figure 9-3) (Rohmert 1960; Björkstén and Jonsson 1977; Sjøgaard 1986). Relative muscle force is the force exerted by the muscle(s) for maintaining a particular posture, expressed as a percentage of the maximum force that can be exerted in the same posture.

MAXIMUM HOLDING TIME VERSUS DISCOMFORT

For groups of subjects, reasonably linear relationships were found between gravitational load and discomfort at a body region (Van der Grinten and Smitt 1992; Boussenna et al. 1982), as well as between discomfort and the percentage of the maximum holding time (%MHT) for a posture (Table 9-1) (Manenica 1986; Meijst et

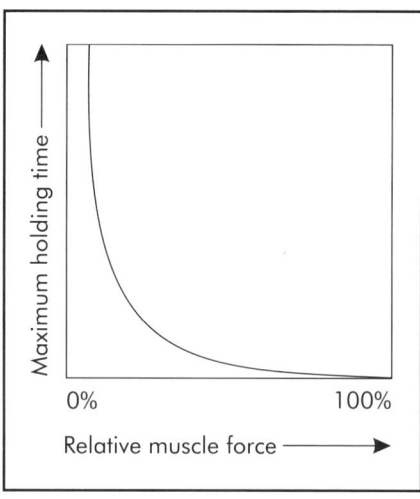

Figure 9-3 The relationship between relative muscle force and maximum holding time.

al. 1995). For example, a discomfort score of 5 after holding a certain posture for 10 minutes means that the MHT of this posture is 20 minutes. The remaining endurance capacity (REC) is defined as 100% − %MHT (Table 9-1).

MAXIMUM ACCEPTABLE LEVEL OF DISCOMFORT

Hagerup and Time (1992) consider a mean discomfort score of between 1 and 3 for a group of workers to be acceptable. Rose and colleagues (1992) found that when subjects were allowed to decide on the duration of static work themselves, they stopped for a pause at approximately 20% of MHT. In ISO 11226, 20% MHT (REC = 80%) is implemented as the maximum acceptable holding time. Because of the linear relationship between %MHT and discomfort, this implies a score of 2 on the Borg scale. The total duration of holding times during intermittent work until reaching a discomfort score of 2 can be considerably longer, depending on the holding time–recovery time regime.

WORKSTATION DESIGN AND ADJUSTMENT

Relatively small workstation adjustments and resulting minor changes of working posture usually have a major effect on work load. Here are two examples, followed by an illustration of the effect of spatial positions of the hands while standing.

SEWING MACHINE OPERATION

Five sewing machine operators (Figure 9-4) worked for 45 minutes at each of four workstation adjustments (Figure 9-5). Pedal positions differed by only 4 in. (10 cm), and desk slopes by 10°. Slight differences were found in the working postures (Figures 9-6 and 9-7). Nevertheless, workers perceived a great improvement while working at workstation adjustment D. The whole body discomfort was reduced, while for the other three adjustments an increase of discomfort was seen (Figure 9-8). Furthermore,

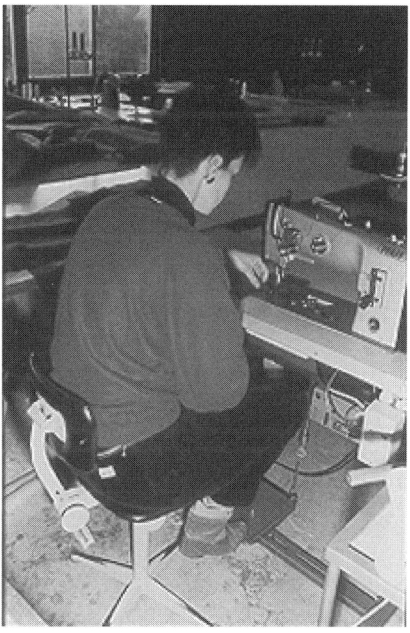

Figure 9-4 Sewing machine operator.

Ergonomics in Manufacturing

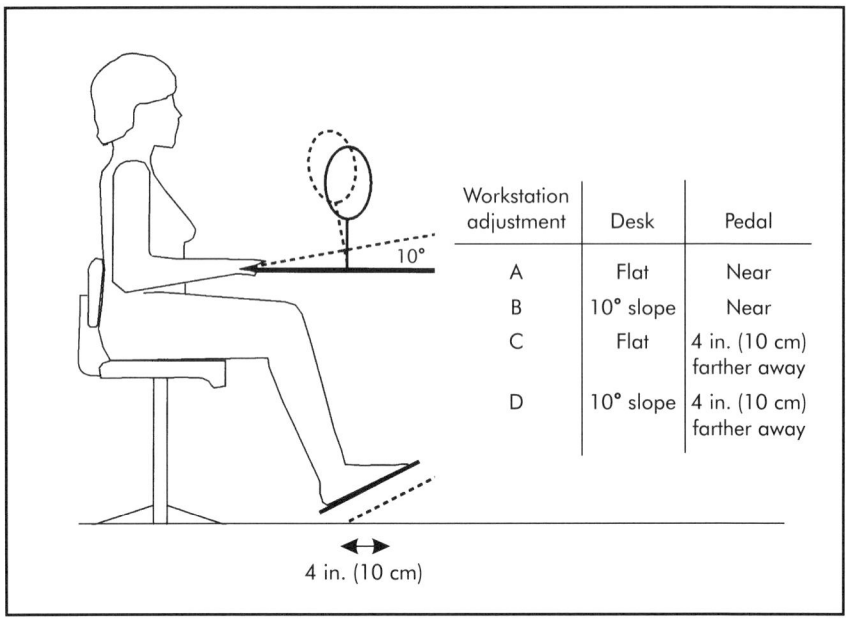

Figure 9-5 Sewing machine operation—the workstation adjustments tested.

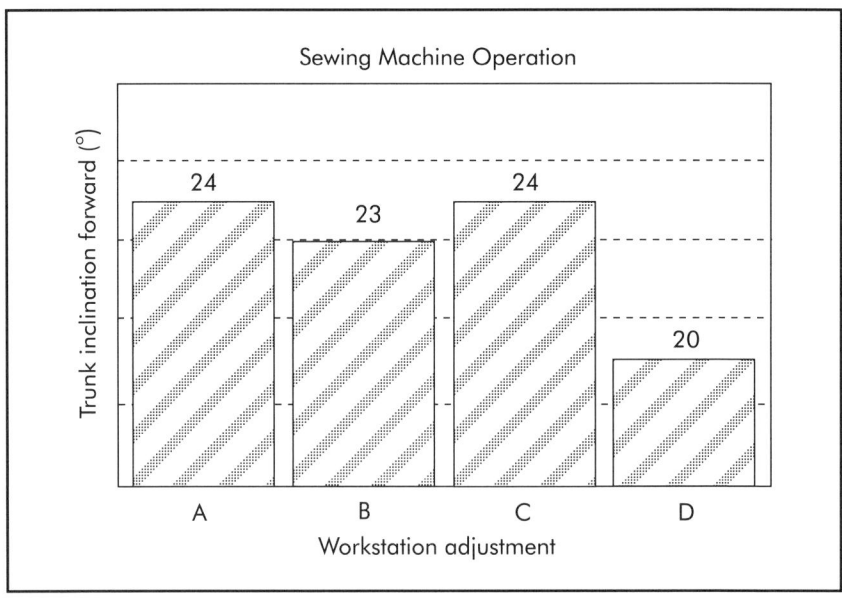

Figure 9-6 Trunk inclination forward for adjustments A-D (refer to Figure 9-5).

140

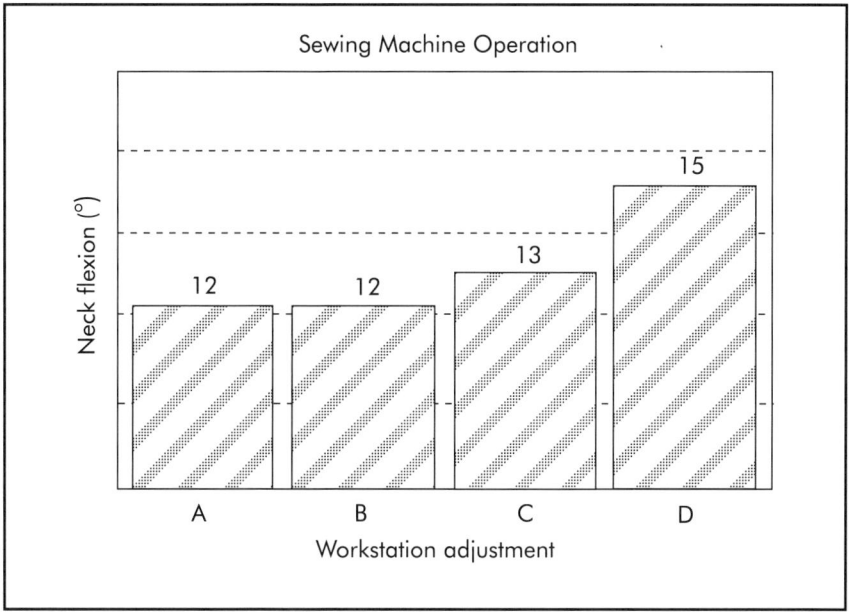

Figure 9-7 Neck flexion for adjustments A-D (refer to Figure 9-5).

the endurance time estimated by the operators was considerably longer (Figure 9-9).

PRESS OPERATION

Eight press operators (Figure 9-10) worked for 25 minutes at each of four reach distances, defined with respect to full arm reach in a reference posture (Figure 9-11). For every 4 in. (10 cm) the reach distance was increased, the trunk inclined forward a little more (Figure 9-12). It turned out that worker perceptions got significantly worse when they stretched slightly beyond full arm reach (Figures 9-13 and 9-14).

HAND POSITIONS

MHT data on 19 different standing postures were taken from the literature (Boussenna et al. 1982; Manenica 1986; Meijst et al. 1995; Corlett and Manenica 1980; Hagberg 1981; Milner 1985; Taksic 1986). All postures were maintained without rest pauses and no external force was exerted. Hand positions were defined as % shoulder height and % arm reach in a reference posture (standing upright). Shoulder height is defined as the vertical distance from the shoulder top (acromion) to the floor. Arm reach is defined as the horizontal distance

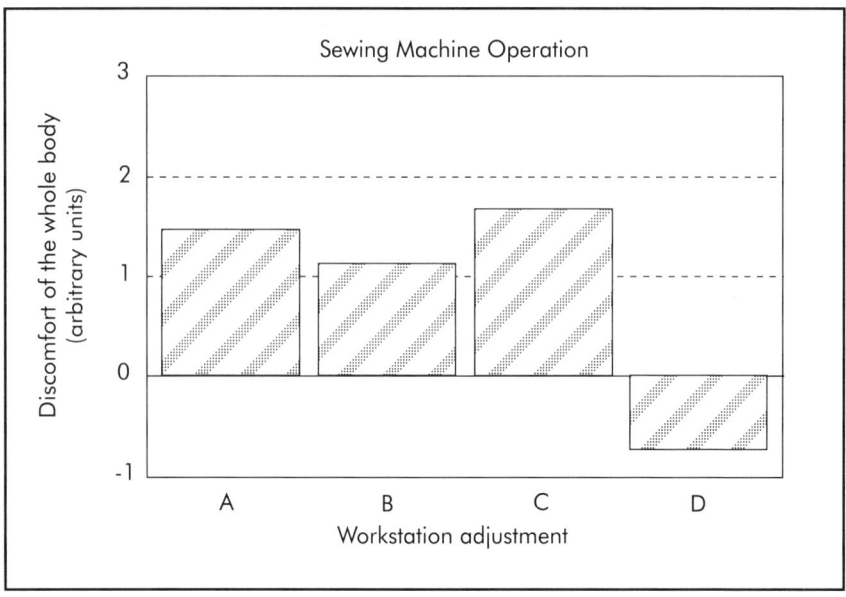

Figure 9-8 Discomfort of the whole body for adjustments A-D (refer to Figure 9-5).

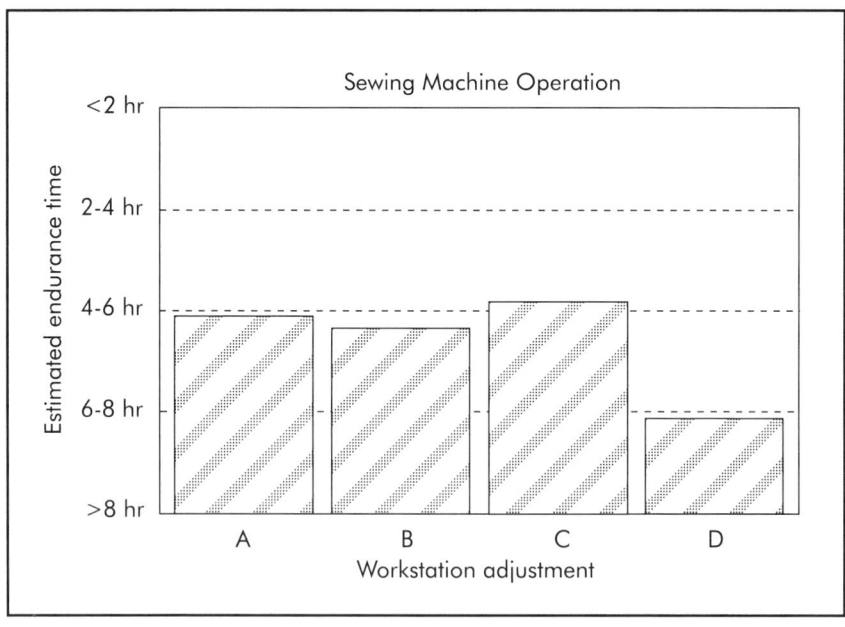

Figure 9-9 Estimated endurance time for adjustments A-D (refer to Figure 9-5).

Static Work Load and Endurance Times

Figure 9-10 Press operator.

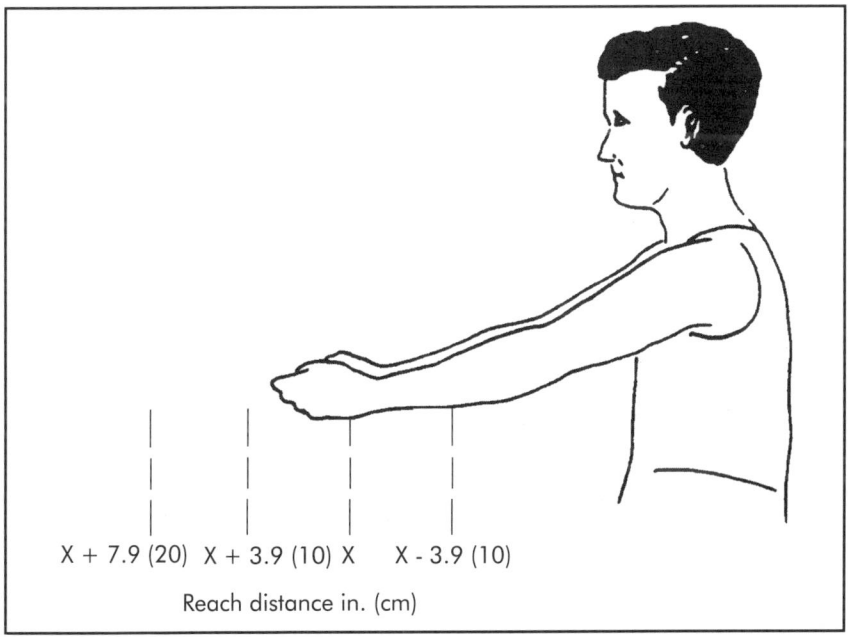

Figure 9-11 Press operation—the reach distances tested (X = full arm reach).

Ergonomics in Manufacturing

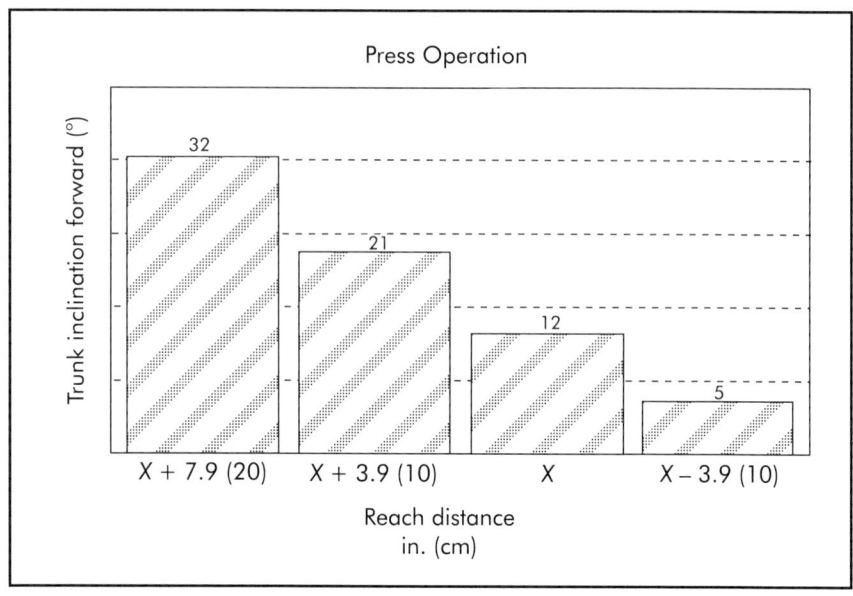

Figure 9-12 Trunk inclination forward for reach distances tested (refer to Figure 9-11).

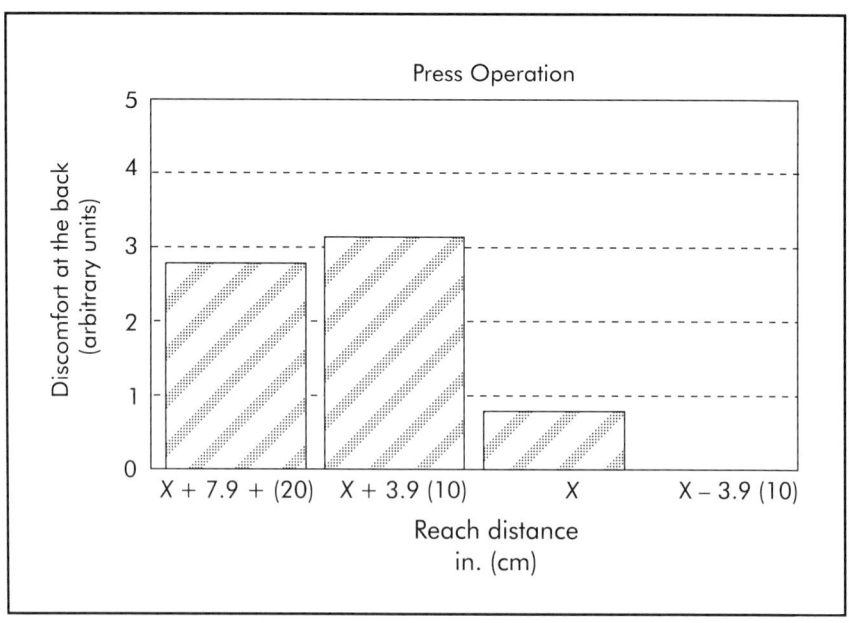

Figure 9-13 Discomfort at the back for reach distances tested (refer to Figure 9-11).

Static Work Load and Endurance Times

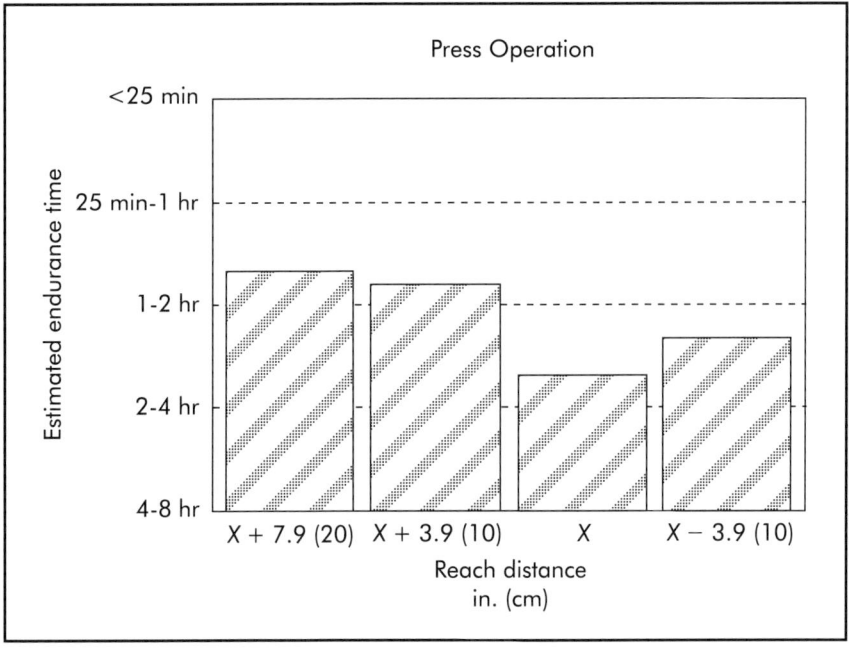

Figure 9-14 Estimated endurance time for reach distances tested (refer to Figure 9-11).

from the knuckles to the wall when standing with the back against the wall and the stretched arms pointing straight forward. The various hand positions (Figure 9-15) were ranked on MHT (Figure 9-16) (Miedema 1992; Dul et al. 1993). Postures with hand positions at or below 50% shoulder height were terminated because of discomfort in the lower back and legs. For postures with hand positions at or below 100% shoulder height, the shoulders and arms are critical. Due to the fact that the data were extracted from various studies (with different subject groups, etc.) some slight inconsistencies may be found in Figure 9-16. However, by grouping the hand positions in terms of relatively low, medium, and high comfort postures, a more reliable result emerges.

STANDARDS

The International Organization for Standardization (ISO) is working on a standard for the evaluation of working postures, ISO 11226. A similar standard, EN 1005-4, is being prepared by the European Committee for Standardization (CEN). The latter standard supports the essential health and safety requirements of the EU-Machinery Directive.

SCOPE

ISO 11226 provides information for all those involved in design or redesign of work, jobs, and products, who are familiar with the basic concepts of ergonomics in general and working postures in particular. It specifies recommended limits for working postures with minimal external force exertion, while taking into account body angles and time aspects. The standard is meant to give reasonable protection to nearly all healthy adults.

CONTENTS

The evaluation procedure used in ISO 11226 considers various body segments and joints independently. The first step considers only the body angles (recommendations are mainly based upon risks for overloading passive body structures, such as ligaments, cartilage, and intervertebral disks). An evaluation may lead to the result "acceptable," "go to step 2," or "not recommended." An evaluation result of "go to step 2" means that the holding time of the posture and the recovery time will also need consideration. Examples are shown in Figures 9-17 and 9-18.

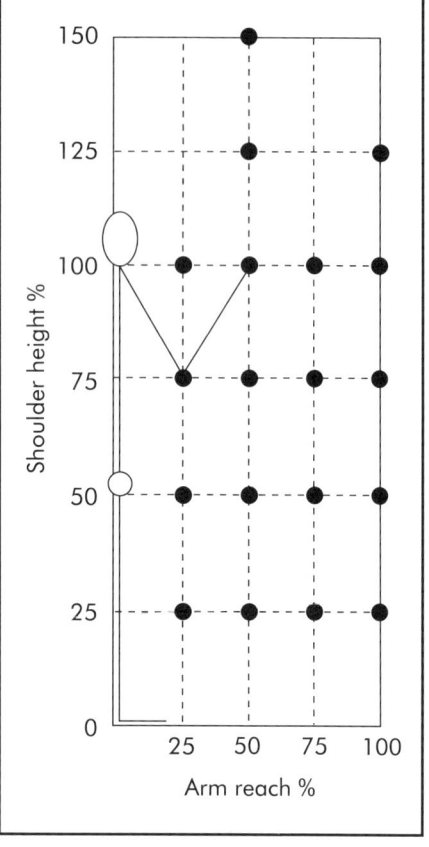

Figure 9-15 The 19 postures found in the literature characterized by the position of the hand(s) in terms of % shoulder height and % arm reach.

WORK-REST MODEL

For ergonomists a mathematical Work-Rest model (WR-model) is available for selecting the most effective holding time/recovery time regimes (Dul et al. 1994). The model predicts the course of

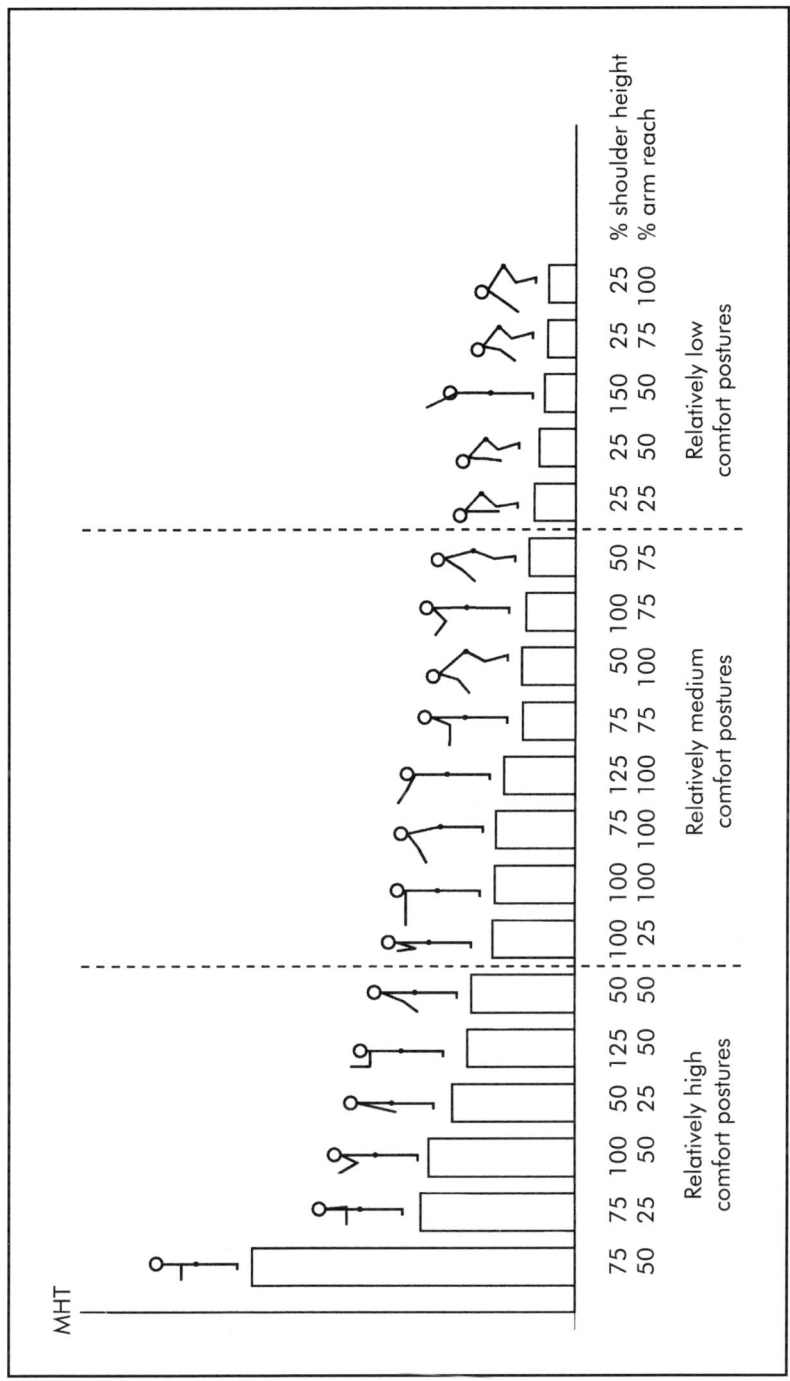

Figure 9-16 Ranking of the 19 postures (refer to Figure 9-15), based on maximum holding time (MHT).

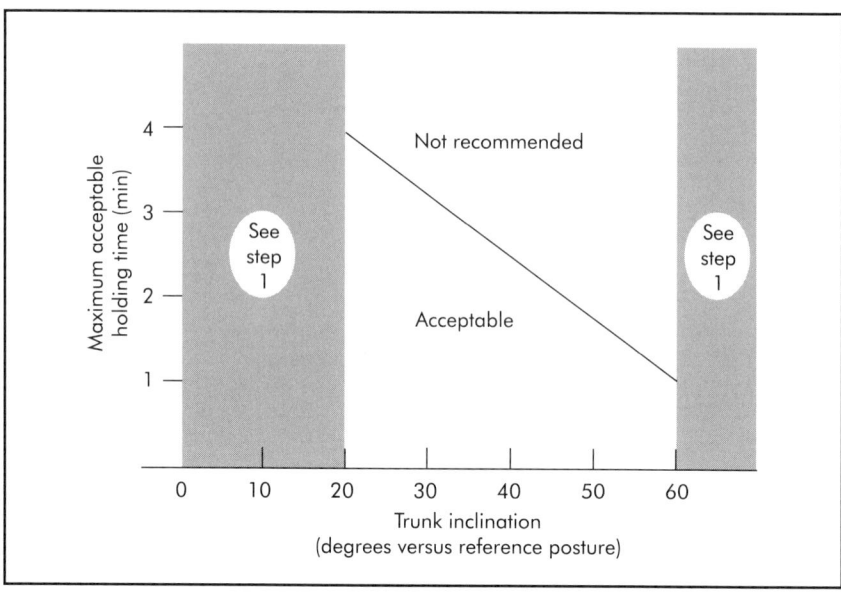

Figure 9-17 ISO 11226—maximum acceptable holding time for trunk inclination (reference posture = trunk upright).

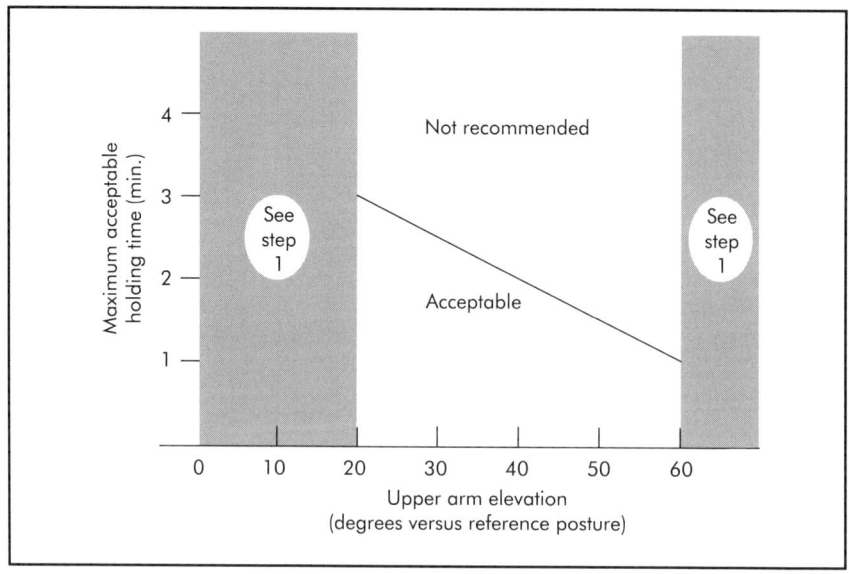

Figure 9-18 ISO 11226—maximum acceptable holding time for upper arm elevation (reference posture = upper arm hanging freely).

muscle fatigue and recovery. The regression equation given by Sjøgaard (1986) was used for the relationship between relative muscle force and MHT. For recovery, the model by Milner (1985) was selected. It is recommended to use the WR-model only for comparison of work situations.

SOFTWARE

A computer program has been developed for analyses with the WR-model. With this software, for example, the remaining endurance capacity (REC) can be calculated for a given combination of relative muscle force and number, duration, and distribution of holding time periods and recovery time periods.

EXAMPLE

The effects of three holding time/recovery time regimes on REC are shown in Figure 9-19. The regimes are equal as far as the same total holding time and total recovery time (in terms of minutes) are concerned. It can be seen that more and shorter holding time periods (more breaks) result in higher REC. In general, it is recommended to consult an expert for evaluating holding time/recovery time regimes. Furthermore, it should be recognized that there are more ways for evaluating holding time/recovery time regimes than shown in the example, for example, based on intervertebral disk or muscle physiology.

REFERENCES

Björkstén, M., and Jonsson, B. "Endurance Limit of Force in Long-term Intermittent Static Contractions." *Scandinavian Journal of Work Environment and Health* 1977, 3: 23-27.

Borg, G. "A Category Scale with Ratio Properties for Intermodal and Interindividual Comparisons." *Psychophysical Judgment and the Process of Perception.* H.G. Geissler and P. Petzold, eds. Amsterdam, The Netherlands: North-Holland, 1982: 25-34.

Boussenna, M., Corlett, E.N., and Pheasant, S.T. "The Relation between Discomfort and Postural Loading at the Joints." *Ergonomics* 1982, 25: 315-322.

Corlett, E.N., and Bishop, R.P. "A Technique for Assessing Postural Discomfort." *Ergonomics* 1976, 19: 175-182.

Corlett, E.N., and Manenica, I. "The Effects and Measurement of Working Postures." *Applied Ergonomics* 1980, 11: 7-16.

Dul, J., Douwes, M., and Miedema, M. "A Guideline for the Prevention of Discomfort of Static Postures." *Advances in Industrial Ergonomics and Safety V.* Proceedings of the Annual International Industrial Ergonomics and Safety Conference, Copenhagen, Denmark, 8-10 June, 1993. R. Nielsen and K. Jørgensen, eds. London and Washington, DC: Taylor & Francis, 1993: 3-5.

Dul, J., Douwes, M., and Smitt, P. "Ergonomic Guidelines for the Prevention of Discomfort of Static Postures can be Based on Endurance Data." *Ergonomics* 1994, 37: 807-815.

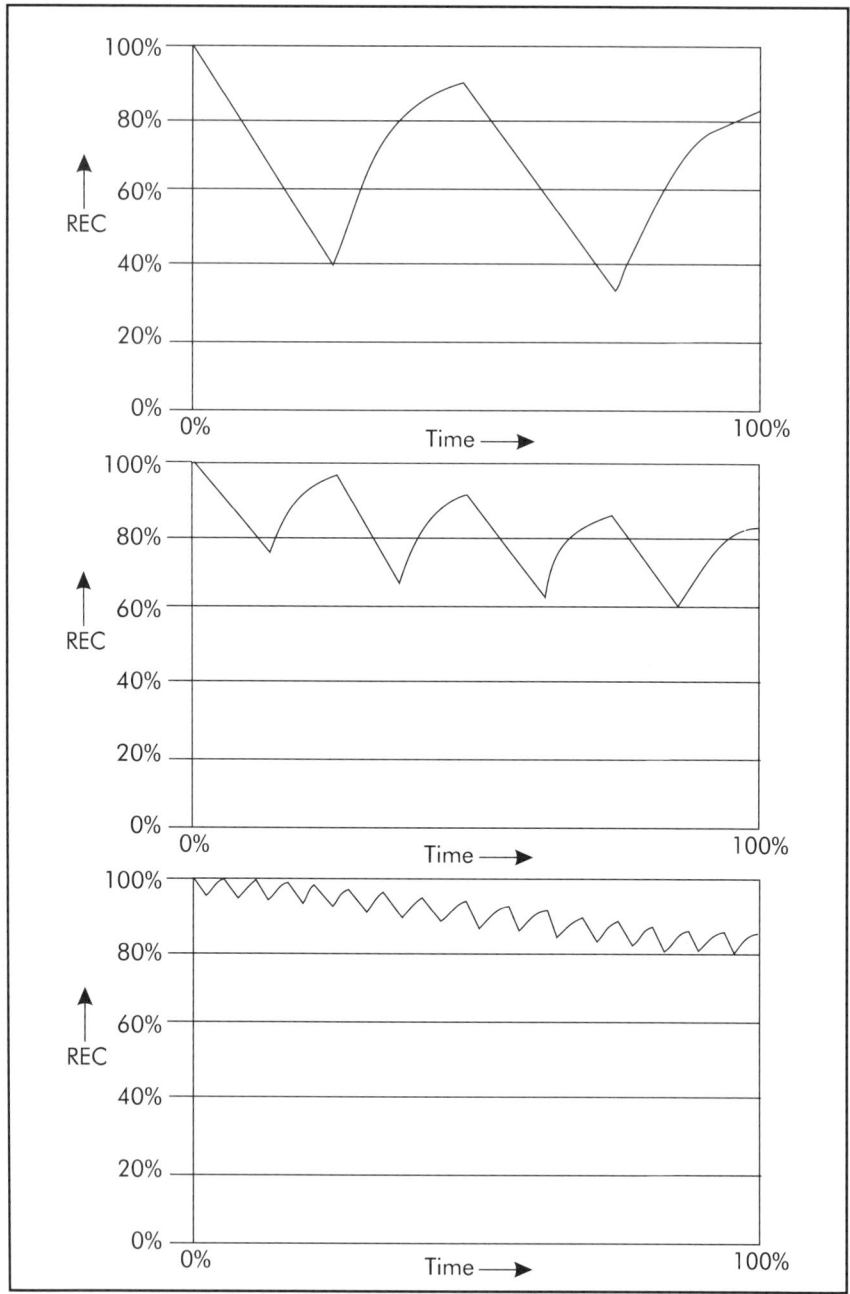

Figure 9-19 The effects of three holding time/recovery time regimes on the remaining endurance capacity (REC).

European Foundation for the Improvement of Living and Working Conditions. *Second European Survey on Working Conditions 1996.* Luxembourg: Office for Official Publications of the European Communities, 1997.

Hagberg, M. "Electromyographic Signs of Shoulder Muscular Fatigue in Two Elevated Arm Positions." *American Journal of Physical Medicine* 1981, 60: 111-121.

Hagerup, A.B., and Time, K. "Felt Load on Shoulder in the Handling of 3 Milking Units with One- and Two-hand-grips in Various Heights." *Proceedings of the International Scientific Conference on Prevention of Work-related Musculoskeletal Disorders-PREMUS,* Stockholm, Sweden, 12-14 May, 1992. M. Hagberg and Å. Kilbom, eds. Stockholm, Sweden: Arbetsmiljöintitutet—National Institute of Occupational Health; 1992, Arbete och Hälsa, 17, 105-107.

Manenica, I. "A Technique for Postural Load Assessment." *The Ergonomics of Working Postures.* E.N. Corlett, J.R. Wilson, and I. Manenica, eds. London and Philadelphia, PA: Taylor & Francis, 1986: 270-277.

Meijst, W.J., Haslegrave, C.M., and Dul, J. *Maximum Holding Times of Static Standing Postures.* Leiden, The Netherlands: TNO Institute of Preventive Health Care, 1995.

Miedema, M. "Static Working Postures." Leiden, The Netherlands: TNO Institute of Preventive Health Care, 1992.

Milner, N. "Modelling Fatigue and Recovery in Static Postural Exercise." Nottingham, United Kingdom: University of Nottingham, 1985.

Rohmert, W. "Ermittlung von Erhohlungspausen für statische Arbeit des Menschen." *Internationale Zeitschrift für angewandte Physiologie einschliesslich Arbeitsphysiologie* 1960, 18: 123-164.

Rose, L., Ericson, M., Glimskär, B., Nordgren, B., and Örtengren, R. "Ergo-index—Development of a Model to Determine Pause Needs After Fatigue and Pain Reactions During Work." Computer Applications in Ergonomics, Occupational Safety and Health, *Proceedings of the International Conference on Computer-aided Ergonomics and Safety '92*—CAES '92, Tampere, Finland, 18-20 May, 1992. M. Mattila and W. Karwowski, eds. Amsterdam, The Netherlands: Elsevier Science Publishers B.V., 1992: 461-468.

Sjøgaard, G. "Intramuscular Changes During Long-term Contraction." *The Ergonomics of Working Postures.* E.N. Corlett, J.R. Wilson, and I. Manenica, eds. London and Philadelphia, PA: Taylor & Francis, 1986: 136-143.

Taksic, V. "Comparison of Some Indices of Postural Load Assessment." *The Ergonomics of Working Postures.* E.N. Corlett, J.R Wilson, and I. Manenica, eds. London and Philadelphia, PA: Taylor & Francis, 1986: 278-282.

Van der Grinten, M.P. "Test/Retest Reliability of a Practical Method for Measuring Body Part Discomfort." *Designing for Everyone.* Y. Queinnec and F. Daniellou, eds. London, New York, and Philadelphia, PA: Taylor & Francis, 1991: 54-56.

Van der Grinten, M.P., and Smitt, P. "Development of a Practical Method for Measuring Body Part Discomfort." *Advances in Industrial Ergonomics and Safety IV,* Proceedings of the Annual International Industrial Ergonomics and Safety Conference, 10-14 June, 1992, Denver, CO. S. Kumar, ed. London and Washington, DC: Taylor & Francis, 1992: 311-318.

CHAPTER 10

WORKER STRENGTH EVALUATION: ERGONOMIC AND LEGAL PERSPECTIVES

Patrick G. Dempsey
Researcher
Liberty Mutual Research Center for Safety and Health
Hopkinton, MA

MUSCULOSKELETAL DISORDERS

Losses associated with musculoskeletal disorders represent a significant problem to employers, employees, and insurers. Manual work, particularly work involving manual materials handling (MMH) tasks, is a primary cause of musculoskeletal disorders in industry. Whether the disorders result from a single overexertion or from repeated microtrauma over a period of time, the direct and indirect costs of the disorders are very high. For example, the mean direct costs of compensable low-back pain cases are approximately $8,300 per claim with medical costs comprising about 32% of the direct costs, and indemnity payments comprising about 68% (Webster and Snook 1994).

Different approaches to the control of musculoskeletal disorders in industry have been developed. With regard to MMH tasks, there are three basic approaches that can be used:

- Ergonomic job design;
- Education and training; and
- Job placement.

ERGONOMIC JOB DESIGN

The approach of ergonomic job design is the most effective method, but it is only partially effective. Supplemental education and training have been widely used in industry, but have provided little evidence of long-term effectiveness to reduce musculoskeletal disorders. Kroemer's (1992) review of the literature concerning training led him to conclude that "the issue of training for prevention of back injuries in MMH is confused, at best."

One method of job placement, medical screening, can be effective for identifying employees who should not perform strenuous work, but it does not necessarily consider the relationship between job demands and worker capacity. The individual's medical history, particularly a history of previous low-back disorders, has been found to be an important predictor of injury propensity. On the other hand, X-rays have not provided effective indication of future injury (Gibson 1987).

Strength and fitness testing has been found the most effective method of job placement (Snook 1987). Strength evaluations reveal mismatches between job demands and capacity of healthy workers. While numerous field studies have examined the effectiveness of strength evaluation, few studies have looked at fitness measures, such as aerobic capacity.

In short, in an effort to reduce MMH-related musculoskeletal disorders in healthy workers, it is recommended to first perform ergonomic job analyses. Jobs that do not accommodate the majority of the population should be redesigned or automated. If jobs cannot be altered to acceptable levels, strength evaluations should be considered as a means of properly matching worker capacity to task demands.

METHODS OF STRENGTH EVALUATION

There are several methods of strength testing to assess a worker's capacity to perform a task. The next sections discuss various methods.

ISOMETRIC TESTING

Isometric (or static) testing is performed by having a subject exert force against a stationary object. Typically, whole-body lifting strength is measured, but it is also possible to isolate specific joints, such as the elbow. Isometric testing requires equipment that offers resistance (typically a platform, chains, and handles) and a force transducer with readout capabilities. Load cells are available with hand-held readout devices. It is possible to feed the output to a computer equipped with an analog-to-digital converter.

Isometric testing is performed by having subjects build up to their maximum voluntary contraction, maintain the exertion for several seconds while data are collected, and then relax. The score is usually the average strength over a 3-5 second period. Use of peak strength across the length of the exertion should be avoided since these measurements can overestimate an individual's strength. Few industrial tasks pose strength demands for such short periods of time.

An advantage of isometric strength testing is that the measurements are easy to conduct and the equipment is typically less expensive than equipment required for other forms of testing. The primary disadvantage is that MMH activities, except holding, are dynamic activities. Dynamic strength tests (isokinetic and isoinertial) have been shown to be better predictors than static measures, particularly for lifting capacity (Dempsey and Ayoub 1996). However, there are more isometric data than dynamic data available for comparative purposes.

ISOKINETIC TESTING

Isokinetic testing refers to testing where the strength testing device holds velocity constant throughout the motion. As with isometric strength, it is possible to test either whole-body strength or the strength of individual joints. In the former case, the worker exerts maximum force against handles that move at a constant vertical velocity. For individual joint testing, the worker is usually restrained to some extent so that a particular joint can be isolated. For instance, the upper leg would be secured for a test of knee strength. Linear (vertical) velocity is held constant when whole-body strength is measured; joint strength is measured while angular velocity is held constant.

The equipment required for isokinetic testing is considerably more expensive than isometric equipment. An isokinetic dynamometer is required as well as equipment connected to the dynamometer. Several systems are available containing all of the necessary components.

ISOINERTIAL TESTING

Isoinertial testing refers to conditions where the mass is constant during a trial. That is, the load the subject lifts, pushes, etc. remains constant. However, the load may be increased across trials, as some isoinertial tests increase the load across trials until the subject cannot successfully complete the task. With respect to lifting, lowering, carrying, pushing, and pulling tasks, isoinertial strength evaluations most resemble the actual tasks.

Most isoinertial equipment is not very expensive, mainly because electronics are not usually required. There are some fairly expensive isoinertial systems. For several types of testing, isoinertial equipment can be less expensive than either isometric or isokinetic testing.

COMPARING STRENGTH EVALUATION METHODS

Several studies have shown that as the ratio of job demands (loads lifted) to isometric strength increases, probability of injury increases (Chaffin and Park 1973; Chaffin 1974; Chaffin et al. 1978; Liles et al. 1984). The studies by Chaffin and his colleagues used raw isometric strength values to determine capacity. Liles et al. (1984) used the isometric strength values with anthropometric measures to predict lifting capacity. The equations predict lifting capacity for different ranges of lift, frequencies, box size, etc. Battié et al. (1989) did not find that isometric strength was related to future low-back injuries, but this research failed to match strength measurements to job demands.

A few field studies have examined the relationship between isokinetic strength and injury probability. Mostardi et al. (1992) did a 2-year study to determine if isokinetic lifting strength, demographic variables, low-back pain, and injury history were correlated with future low-back injuries in a population of 171 nurses. None of the variables correlated with the incidence of pain or injury, leading the authors to conclude that the variables were poor predictors in such a high-risk population. However, the sample size was limited.

A considerable number of studies have looked at the relationship between isokinetic trunk strength and injury. Newton and Waddell (1993) performed an extensive review and analysis of the literature in this area and concluded that there was no scientific evidence verifying tests for predicting injury. These tests only measure the strength of the trunk extensors. The role of trunk strength in predicting MMH capacity is unclear. Furthermore, the results of these tests are not usually related to job demands.

There have been few field studies of isoinertial tests. Snook (1978) reported details on an investigation of 191 low-back injuries associated with MMH. The analysis indicated that 25% of the jobs investigated involved MMH tasks acceptable to less than 75% of the population, based on the psychophysical database reported. Psychophysical limits are determined by having subjects select work loads which are the maximum that they find acceptable (that is, the work loads do not cause undue fatigue, soreness, etc.). Snook (1978) found one half of the low-back injuries resulted from jobs acceptable to less than 75% of the population. These findings led to the conclusion that two out of three low-back injuries could be prevented if jobs were designed to accommodate at least 75% of the population.

Troup et al. (1987) conducted a study using 2,891 subjects selected from various occupations to examine the ability of a psychophysical lifting test to predict future injury. The psychophysical test, based on the rating of acceptable load (RAL), is considerably different from the method developed and used by Snook and his colleagues at the Liberty Mutual Insurance Company. Troup et al. (1987) found RAL to be a poor predictor of the initial episode of low-back pain (LBP), but when the previous history of LBP was known, the test enhanced prediction.

The use of static (isometric) strength for placement in tasks containing dynamic components will likely overestimate an individual's capability to safely perform MMH tasks. For example, Mital and Das (1987) found that isometric strength may overpredict lifting capability, especially for nonindustrial subjects. Thompson et al. (1992) concluded that maximum isometric strength was a poor predictor of peak forces exerted during a dynamic lift. Kumar and Chaffin (1987) found that the relationship between static and dynamic (isokinetic) strength was less predictable as the speed at which dynamic strength was measured increased. Similarly, Rosecrance et al. (1991) compared maximum isometric strength to maximum weight that subjects with a history of low-back disability were able to lift. The average correlation between static and dynamic measurements was 0.51, with the isometric strength values being greater than the dynamic ones.

Thus, it appears that maximum isometric strength is most relevant to predicting the ability to exert isometric forces. Although dynamic strength is a better predictor of handling capacity, field studies utilizing dynamic measures have been rare.

SELECTING A STRENGTH EVALUATION METHOD

The strength test selected should be matched to the tasks for which it is intended. For example, if one is using strength evaluations to determine lifting capacity, then the test should resemble lifting as closely as possible. The test also should be predictive of lifting capacity. With respect to lifting in particular, dynamic strength tests (isokinetic and isoinertial) are better predictors of lifting capacity than static tests (Dempsey and Ayoub 1996).

Overall, worker strength evaluation tests should be examined from the perspective of safety and scientific integrity. Chaffin's (1982) criteria for such tests are: the test is safe to administer; the test gives reliable, quantitative values; the test is related to specific job requirements; the test is practical; and the test predicts risk of future injury. When deciding on a technique to use, or selecting from tests provided by external sources, each of these criterion should be examined carefully.

The reliability and safety of a test should be determined prior to employing the test in the field. Almost all the isometric, isokinetic, and isoinertial tests discussed earlier are safe and reliable. These factors are not usually a problem for most types of tests. Determination of safety and reliability is usually performed in laboratory studies.

The need to have a test related to specific job requirements provides legal implications. From the standpoint of loss prevention, a test that is not predictive of future injury is not useful. Scientific evidence supporting a specific test should be examined very carefully. Some indication of the integrity of the results could be found if the results have been published in the peer-reviewed scientific literature. However, some consulting firms do not publish their results due to the proprietary nature.

LEGAL IMPLICATIONS

THE AMERICANS WITH DISABILITIES ACT

In 1990, the Americans with Disabilities Act (ADA), formally Public Law 101-336, became law. The specific regulations that enforce the employment provisions of this law are contained in Code 29 of Federal Regulations Part 1630—Regulations to Implement the Equal Employment Provisions of the Americans with Disabilities Act. Title 1 (Employment) of this legislation extends the Civil Rights Act to include "qualified individuals with disabilities" from being discriminated against. Employers with 15 or more employees are required to comply with the ADA.

Relevant Definitions

Because of the litigation potential from pre-employment and/or pre-placement strength evaluations due to the ADA, correct definitions must be applied when determining the legality of a given test. The following definitions are contained in CFR 29, Part 1630.

A *disability* is a physical or mental impairment that substantially limits one or more of the major life activities of an individual, a record of such an impairment, or being regarded as having such an impairment. *Major life activities* are functions such as performing manual tasks, walking, seeing, hearing, etc.

Substantially limits means that an individual is unable to perform a major life activity that the average person in the general population can perform, or is significantly restricted as to the condition, manner, or duration in which an individual can perform a particular major life activity.

This includes restricted ability to perform either a class of jobs or a broad range of jobs in various classes compared to the average person. Inability to perform a single job does not constitute a substantially limited ability to perform the major life activity of working (EEOC 1992).

A *qualified individual with a disability* is an individual with a disability who satisfies the requisite skill, experience, education, and other job-related requirements of the employment position that he or she holds or desires, and who, with or without reasonable accommodation, can perform the essential functions of such a position.

A *reasonable accommodation* involves making existing facilities accessible and usable by individuals with disabilities. This may include modifications or adjustment of the work environment, job restructuring, modifying work schedules, etc. It may be necessary to include the qualified individual with a disability in determining the reasonable accommodation. Financial issues and the impact of the accommodation on the facility, such as the ability of other individuals to perform their jobs with the accommodation in place, may be taken into consideration when determining whether or not an accommodation is reasonable.

A reasonable accommodation is an accommodation that does not cause *undue hardship*, which is defined as significant difficulty or expense incurred by a covered entity. Determination of undue hardship requires a consideration of factors such as financial resources of the organization, number of locations of the entity, number of employees, and impact of the accommodation on the facility's operation, including impact on the ability of other employees to perform their duties (EEOC 1992).

Perhaps the most important factor in determining the appropriateness of a screening test is *essential function*. A job function may be considered essential if:

- The position exists to perform that function;
- A limited number of employees can perform the function; and/or
- The person in the position is hired to perform the function because of his or her expertise.

Determining whether or not a function is essential includes, but is not limited to, the employer's judgment, written job descriptions, amount of time spent performing the function, consequences of not requiring the person holding the position to perform the function, and collective bargaining agreements (EEOC 1992). As will be later illustrated with examples, infrequent tasks are less likely to be essential than are tasks performed throughout the day.

Implications of the ADA for the Use of Tests of Individual Capability

Perhaps the most important part of a pre-employment or placement exam is a written job description that explicitly describes the essential functions of a job before it is advertised or candidates are interviewed. Overall, accurate job descriptions can be very useful to the ergonomist.

Developing a job description requires knowledge of what functions employees in the position perform and whether removing a function would fundamentally change the job. Caution must be taken when including specific functions, such as "lifting 50 pounds," in a job description if a reasonable accommodation can remove the function from the job, since this could lead to liability if an accommodation is not considered (Pimentel et al. 1993). Worker strength evaluations should only be used for essential functions.

The next step requires determining which essential functions can be performed by a qualified individual with a disability, with or without reasonable accommodation. In some cases, this will require consultation with the individual to isolate his or her physical capabilities, identify potential accommodations, and assess their effectiveness (EEOC 1992). If a reasonable accommodation allows a qualified individual with a disability to perform the job function, the pre-placement test may be unnecessary.

Is it Job-related?

The criteria for establishing qualification standards, such as tests of physical capacity, is that they be both job-related and consistent with business necessity. To be considered job-related, a test must be a legitimate measure of the specific job. It cannot be a test for a class of jobs. There are also generic concerns for employee selection procedures contained in CFR 29, Part 1607—Uniform Guidelines on Employee Selection Procedures. Briefly, these guidelines require that a test must be predictive of, or significantly correlated with, elements of job performance (criterion-related validity); representative of important aspects of performance on the job (content validity); or measure the degree to which candidates have identifiable characteristics important in successfully performing the job (construct validity). These requirements are independent of the ADA, but are nonetheless important.

Once the actual interview/hiring process begins, certain procedures must be followed. Individuals may not be asked to describe their disability or any information about it. They can, however, be asked to describe or demonstrate how they will perform specific job functions if such a demonstration is required of everyone applying for the job. If the applicant has a known disability (for example, applicant has one arm), he or she may be asked to provide the demonstration, even if everyone applying for the job is not required to do so (EEOC 1992). The ADA requires that the test be performed in a manner not requiring use of the impaired skill, unless the test is specifically designed to measure that skill.

Physical tests can be given at any point during the application or employment process, provided that the test is given to all applicants and follows the guidelines above (CFR 29, Ch. XIV, Section 1630.14[a]). Although pre-employment medical examinations are prohibited by the ADA, physical capacity tests are allowed, if given to all applicants, as they are not considered a medical examination (EEOC 1992).

Because the ADA is so complex and intricate, a case-by-case analysis is required, sometimes at the level of the individual. For this reason, Johns et al. (1994) recommends that decisions be made by a multidisciplinary team, including legal counsel.

The issue of job-relatedness is easily justified if the test has been shown to correlate with the ability to perform MMH tasks, or if test performance has been related to injury probability through epidemiological studies. The issue of business necessity requires that the test be related to the ability to perform a specific function. For example, a specific function would be lifting a 12 in. × 12 in. × 12 in. (30 cm × 30 cm × 30 cm) box weighing 20 lb (9 kg) from the ground to a height of 40 in. (102 cm), three times per minute. However, just lifting 20 lb (9 kg) is not considered specific enough. These issues should be addressed by validation studies before the strength evaluation technique is used in the field.

EXAMPLES

An example for which strength evaluation tests are not particularly appropriate is infrequent MMH tasks. Consider a job where a fork lift operator must occasionally replace a 70-lb (32 kg) battery. Although the ability to lift 70 lb (32 kg) is a job-related standard, a reasonable accommodation, such as having another employee perform the task, is justified (EEOC 1992). However, if this task is required every 15 minutes, then it may be difficult to have another employee perform the task.

Another example of a reasonable accommodation would be a case where medical documentation indicates that a person can lift a 50-

lb (23 kg) sack to waist height, but cannot carry it to the storage room. In this case, a dolly would be a reasonable accommodation (EEOC 1992), as pushing a dolly is much easier than carrying a load. Likewise, this would be an inexpensive solution that decreases task demands for all workers. These two examples represent cases where tests of physical capacity violate the ADA.

A test of physical capacity is legal under the ADA in those situations where the position exists to perform MMH tasks, as is common in warehousing and shipping/receiving departments. Additionally, many production jobs require frequent MMH, such as loading and unloading stock. Typically, removing the need for MMH requires some sort of automation, which often extends beyond a reasonable accommodation and would be considered an undue hardship. In such cases, tests of physical capacity are allowed under the ADA.

SUMMARY AND CONCLUSIONS

Few experts would disagree that the probability of MMH-related injuries is based on the discrepancy between task demands and an individual's physical capacity to perform that task. The concept of ensuring that task demands do not exceed human capabilities is one of the fundamental principles of ergonomics. Figure 10-1 presents a summary of the role of worker strength evaluations in ergonomic job analyses.

Ergonomic changes or automation should be used to design jobs that are within the capacity of the population (preferably at least 90%). Strength evaluations can be used to assess whether or not the majority of the population is accommodated. When ergonomic job design changes are not sufficient to result in task demands that are within the capabilities of the majority of the workers performing a job, worker strength evaluations can effectively supplement the ergonomic changes. However, these evaluations should only be considered after ergonomic job design changes have taken place.

When selecting a strength evaluation method, there are many factors to consider. The first should be the legality of testing workers for the particular task or set of tasks. The next step is to select a technique from the various tests relevant for predicting the capacity required for the tasks in question. It should follow that the more accurately a given testing protocol predicts an individual's MMH capacity in a given situation, the more accurate the estimate of injury risk. There are numerous factors that need to be consid-

Worker Strength Evaluation: Ergonomic and Legal Perspectives

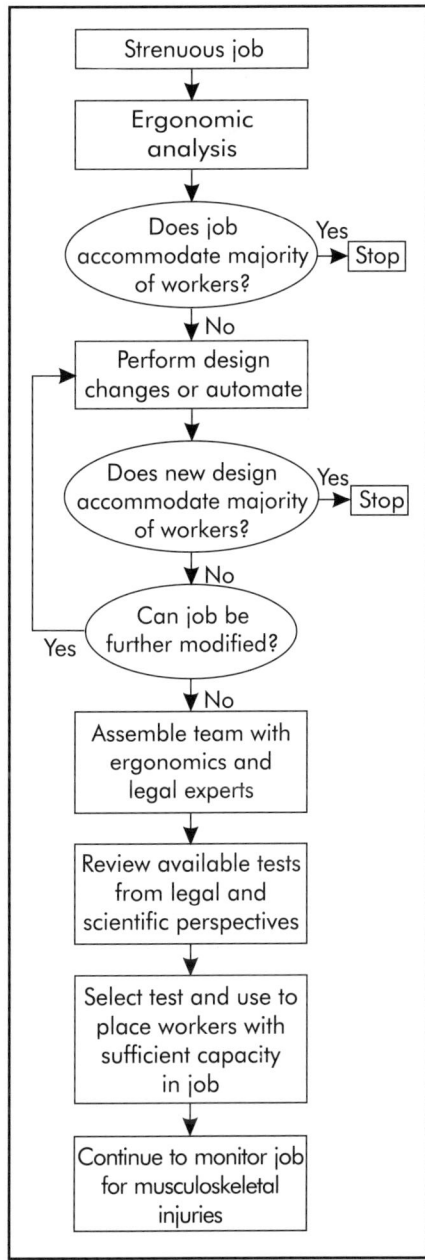

Figure 10-1 Summary of the role of worker strength evaluations in the job (re)design process.

ered in the selection process, such as reliability, validity, safety, and correlation of test results with future injury prediction.

Due to the complexity of worker strength evaluations, it is likely that legal and/or ergonomic experts would be consulted during the process. This will ensure that the methods selected are satisfactory from the standpoint of regulatory requirements and scientific integrity.

When used correctly, worker strength evaluations can be an effective supplemental tool to prevent musculoskeletal disorders associated with manual work. A final note is that the strength evaluations should only be considered one component of the overall program to reduce overexertion injuries. The comprehensive program should include other factors, such as ergonomics, injury treatment, enlightened management, and cooperative unions (Snook 1987).

REFERENCES

Battié, M.C., Bigos, S.J., Fisher, L., Hansson, T.H., Jones, M.E., and Wortley, M.D. "Isometric Lifting Strength as a Predictor of Industrial Back Pain Reports." *Spine* 1989, 14(8): 851-856.

Chaffin, D.B. "Functional Assessment and Heavy Physical Labor." M.H. Alderman and M.J. Hanley, eds. *Clinical Medicine for the Occupational Physician*. New York: Marcel Dekker, Inc., 1982.

Chaffin, D.B. "Human Strength Capability and Low-back Pain." *Journal of Occupational Medicine* 1974, 16(4): 248-254.

Chaffin, D.B., Herrin, G.D., and Keyserling, W.M. "Pre-employment Strength Testing." *Journal of Occupational Medicine* 1978, 20(6): 403-408.

Chaffin, D.B., and Park, K.S. "A Longitudinal Study of Low-back Pain as Associated with Occupational Weight Lifting Factors." *American Industrial Hygiene Association Journal* 1973, 34(12): 513-525.

Dempsey, P.G., and Ayoub, M.M. "The Role of Power in Predicting Lifting Capacity." *Proceedings of the Human Factors and Ergonomics Society 40th Annual Meeting.* Santa Monica, CA: Human Factors and Ergonomics Society, 1996: 609-613.

Equal Employment Opportunity Commission (EEOC). *Technical Assistance Manual on the Employment Provisions (Title 1) of the Americans with Disabilities Act.* Washington, DC: EEOC, 1992.

Gibson, E.S. "The Value of Preplacement Screening Radiography of the Low Back." *Spine: State of the Art Reviews* 1987, 2(1): 91-107.

Johns, R.E., Bloswick, D.S., Elegante, J.M., and Colledge, A.L. "Chronic, Recurrent Low-back Pain: A Methodology for Analyzing Fitness for Duty and Managing Risk under the Americans with Disabilities Act." *Journal of Occupational Medicine* 1994, 36(5): 537-547.

Kroemer, K.H.E. "Personnel Training for Safer Material Handling." *Ergonomics* 1992, 35(9): 1,119-1,134.

Kumar, S., and Chaffin, D.B. "Static and Dynamic Lifting Strength of Young Males." B. Jonsson, ed. *Biomechanics X-A*. Champaign, IL: Human Kinetic Publishers, 1987: 25-28.

Liles, D.H., Deivanayagam, S., Ayoub, M.M., and Mahajan, P. "A Job Severity Index for the Evaluation and Control of Lifting Injury." *Human Factors* 1984, 26(6): 683-693.

Mital, A., and Das, B. "Human Strengths and Occupational Safety." *Clinical Biomechanics* 1987, 2: 97-106.

Mostardi, R.A., Noe, D.A., Kovacik, M.W., and Porterfield, J.A. "Isokinetic Lifting Strength and Occupational Injury: A Prospective Study." *Spine* 1992, 17(2): 189-193.

Newton, M., and Waddell, G. "Trunk Strength Testing with Iso-machines Part 1: Review of a Decade of Scientific Evidence." *Spine* 1993, 18: 801-811.

Pimentel, R.K., Bell, C.G., and Lolito, M.J. *The Job Placement—ADA Connection.* Chatsworth, CA: Milt Wright and Associates, Inc., 1993.

Rosecrance, J.C., Cook, T.M., and Golden, N.S. "A Comparison of Isometric Strength and Dynamic Lifting Capacity in Men with Work-related Low-back Injuries." *Journal of Occupational Rehabilitation* 1991, 1(3): 197-205.

Snook, S.H. "Approaches to Preplacement Testing and Selection of Workers." *Ergonomics* 1987, 30(2): 241-247.

Snook, S.H. "The Design of Manual Handling Tasks." *Ergonomics* 1978, 21(12): 963-985.

Snook, S.H., Campanelli, R.A., and Hart, J.W. "A Study of Three Preventive Approaches to Low-back Injury." *Journal of Occupational Medicine* 1978, 20(7): 478-481.

Thompson, D.D., Chaffin, D.B., Hughes, R.E., and Evans, O. "The Relationship of Isometric Strength to Peak Dynamic Hand Forces During Submaximal Weight Lifting." *International Journal of Industrial Ergonomics* 1992, 9: 15-23.

Troup, J.D.G., Foreman, T.K., Baxter, C.E., and Brown, D. "The Perception of Back Pain and the Role of Psychophysical Tests of Lifting Capacity." *Spine* 1987, 12(7): 645-657.

Webster, B.S., and Snook, S.H. "The Cost of 1989 Workers' Compensation Low-back Pain Claims." *Spine* 1994, 19(10): 1,111-1,116.

CHAPTER 11

METHODS FOR EVALUATING POSTURAL WORK LOAD

W. Monroe Keyserling
Professor
Industrial and Operations Engineering
The University of Michigan
Ann Arbor, MI

WORK POSTURE

Work posture in the workplace is determined by the interaction of many factors, including workstation layout (heights of conveyors, reach distances to access pallets and storage bins), equipment design (positions of machine controls, location of visual displays), and work methods (sequence of work tasks, tool selection, work technique). In addition, body-size characteristics of a worker interact with all the workplace factors to determine specific postures used to perform a job (Keyserling 1990).

To the greatest extent possible, jobs should be designed to accommodate a *neutral* posture. For a standing worker, this means that the trunk and neck should be nearly vertical, with minimal twisting or bending (forward, backward, or sideways). Furthermore, both arms should hang comfortably down from the shoulders, roughly parallel to the trunk.

Awkward postures occur when there is a mismatch between a worker's body size and the job requirements. Many are caused by excessive reach requirements. For example, bending into bins or reaching to the side of the body to place or retrieve parts bends or twists the trunk; reaching overhead to high shelves or conveyors, or reaching in front of the body to activate machine-operating controls, elevates the shoulders. If awkward postures are assumed repetitively or for prolonged periods, increased rates of fatigue, discomfort, and/or injury may occur, resulting in reduced productivity and higher costs.

HEALTH EFFECTS OF AWKWARD POSTURE

Due to the weight of the head, any forward, backward, or sideways neck bending results in biomechanical strain on muscles and connective tissue in the neck (Kumar and Scaife 1979; Harms-Ringdahl et al. 1986). Industry-based studies of workers show that excessive neck flexion (forward bending) produces fatigue and musculoskeletal disorders in the neck and shoulder region. Complaints of pain and fatigue are positively related to the magnitude of the flexion angle, the duration of time spent with the neck flexed, and number of neck flexions per hour (Kilbom et al. 1986). At maximal forward flexion, pain occurs in as little as 15 minutes (Harms-Ringdahl and Ekholm 1986). Neck extension (backward bending) and neck twisting/sideways bending also have been associated with elevated rates of fatigue and disorders in both laboratory and industrial studies (Kilbom et al. 1986; Van Wely 1970; Harms-Ringdahl et al. 1986; Tola et al. 1988).

Upper Arms

Laboratory and industry-based studies of shoulder posture show that prolonged elevation of the upper arms causes extreme levels of muscle fatigue and discomfort in the neck/shoulder region. Cases of posture-related tendonitis have been documented on jobs where overhead or forward reaches require workers to raise the elbow above the level of the mid-torso (Kilbom et al. 1986; Hagberg 1982; Hagberg 1984; Keyserling et al. 1993). Shoulder extension (moving the upper arm backward to retrieve an object behind the body) also has been shown to cause injury to the neck/shoulder region (Feldman et al. 1983).

Trunk

Trunk postures that deviate from the neutral upright and forward-facing position cause fatigue and contribute to occupational back pain. Laboratory studies show that forward bending, sideways bending, and twisting the spine increase stresses on the spinal muscles and intervertebral discs (Schultz et al. 1982). In a recent study of automobile assembly workers, non-neutral postures, such as forward bending more than 20° from the vertical and twisting more than 20° from a forward-facing position, were significant factors in work-related back pain (Punnett et al. 1991). Prolonged sitting also has been related to increased cases of back pain in professional drivers (Kelsey and Hochberg 1988).

Prolonged awkward postures of the lower body may cause pain, discomfort, and injury to the legs and feet, including compression injuries to nerves. Studies of workers from a variety of industries have shown that prolonged or repeated use of a foot pedal while standing, or prolonged kneeling, and/or squatting contributes to elevated rates of discomfort and injury in the lower body (Feldman et al. 1983; Corlett and Bishop 1976).

PRODUCTIVITY EFFECTS OF AWKWARD POSTURE

Awkward work postures contribute to fatigue, resulting in a decrease in productivity. Table 11-1 presents supplemental rest allowances recommended by the International Labor Organization for jobs that require the use of awkward work postures (International Labor Organization 1964). The figures in this table show the recommended rest time, expressed as a fraction of the active work time, for a worker to recover from the effects of an awkward posture. For example, a worker who stands all day in a neutral posture should be given a rest allowance of 0.11. During an 8-hour work shift, this person would work 432.5 minutes and would need 47.5 minutes of rest. If the job requires the use of a bent trunk posture, the rest allowance would increase to 0.13, resulting in 425 minutes of work and 55 minutes of rest. If the job requires extreme reaches or stretches, the rest allowance increases to 0.18, resulting in 407 minutes of work and 73 minutes of rest. It is clear from this example that increased rest requirements decrease productivity for jobs that require the use of awkward postures.

Table 11-1. Rest allowances suggested by the International Labor Organization

Allowance	Rest/recovery time*
Base (personal needs and basic fatigue)	.09
Additional for awkward posture	
Standing upright	.02
Standing with bent trunk	.04
Standing with extreme reaches/stretching	.09

*Expressed as a fraction of work time.

Awkward work postures may also slow down basic work motions and may add nonvalue-added time to production activities. For example, the time required to perform the basic hand motion of moving a tool or part increases by 15% when the upper arm performs work in an overhead reach posture (60° above the horizontal) compared to performing the same task with the shoulder in a neutral posture. Furthermore, the corresponding time required to precisely position a tool or part increases by 26% in the overhead reach position compared to the neutral posture (Wiker 1986). The simple act of bending the trunk to reach with the hands to a location below the knees requires 2.2 seconds for each lowering and subsequent raising of the body (Zandin 1990). Ignoring the previously discussed effects of fatigue, this "body motion" time further decreases productivity in jobs where workers must repeatedly reach to low locations, such as the bottom of bins and pallets during material handling activities.

POSTURE ANALYSIS METHODS

Posture analysis techniques can be classified into two major categories: 1) exposure assessment methods, and 2) root cause analysis methods.

Exposure assessment methods determine whether or not a worker uses postures likely to have adverse effects on health and productivity. They generally measure the presence and time duration of awkward postures. Often, benchmarks are provided so that the user can determine if a posture problem exists on a job. Most exposure assessment techniques are designed to minimize the time required to collect and analyze data. In achieving this time efficiency, these methods may not provide sufficient information to determine the underlying causes of awkward posture.

Root cause analysis methods are used to understand how workplace attributes and work tasks cause awkward postures. In addition to measuring the presence and time duration of awkward postures, these methods also involve detailed documentation of workstation layout and work methods. Because these techniques are quite time consuming, they are frequently limited to situations where awkward postures are known to exist and a commitment has already been made to provide the resources for job redesign.

EXPOSURE ASSESSMENT METHODS

Posture Checklists

A posture checklist is an exposure assessment tool that can be used by persons with relatively little formal training in either posture analysis or ergonomics. Posture checklists are used in situations when the primary goal is to quickly analyze a large number of jobs to identify situations where work posture may cause excessive fatigue and injury.

Checklists are used often in conjunction with facility-wide ergonomic surveys (evaluating all jobs at a work site) when time to analyze each job is limited to only a few minutes. In this sense, the checklist serves as a preliminary screening tool that classifies a job as either "acceptable" or "requiring further study."

Because checklists do not provide precise quantitative measures of exposure or identify root causes of awkward posture, follow-up analyses must be performed on all "further study" jobs. Despite this limitation, checklists remain useful, particularly when an ergonomics program is first introduced at a work site. In this situation, the level of ergonomics expertise within the plant is quite limited and there is a pressing need to screen a large number of jobs in a small amount of time to establish priorities for subsequent actions.

Figure 11-1 presents a checklist for evaluating posture of the lower body, trunk, neck, and shoulders. This checklist was developed and used as part of a joint labor/management ergonomics program in a large U.S. manufacturer of cars and trucks (Keyserling et al. 1993; Keyserling et al. 1992). Because this program stressed floor-level participatory ergonomics, the checklist was designed to be used by line supervisors and production workers with little formal training in ergonomics. Users observed the job (typically for 15-20 minutes) to determine if certain awkward postures were used (see specific checklist questions in Figure 11-1). If a listed awkward posture was never used, a check mark was placed in the "never" column following the question. If an awkward posture was used, it was necessary to judge its duration as either less than or greater than one-third of the work cycle, and to place a mark in the appropriate column.

The response to each question in Figure 11-1 produced a qualitative estimate of risk, using the following categories:

1. *Acceptable*: exposure to the posture for the indicated duration presented insignificant risk of posture-related fatigue or injury to workers.

2. *Moderate risk*: exposure for the indicated duration presented possible risk of posture-related fatigue or injury to some workers.

3. *Significant risk*: exposure for the indicated duration produced posture-related fatigue and was likely to cause posture-related injury to many workers.

Results of the checklist evaluation were used by site-based ergonomics teams as one of several criteria to select priority jobs for further analysis and ergonomic intervention. Other criteria used for prioritization included: corroborating evidence on employee health records of excessive posture-related illness or injury; em-

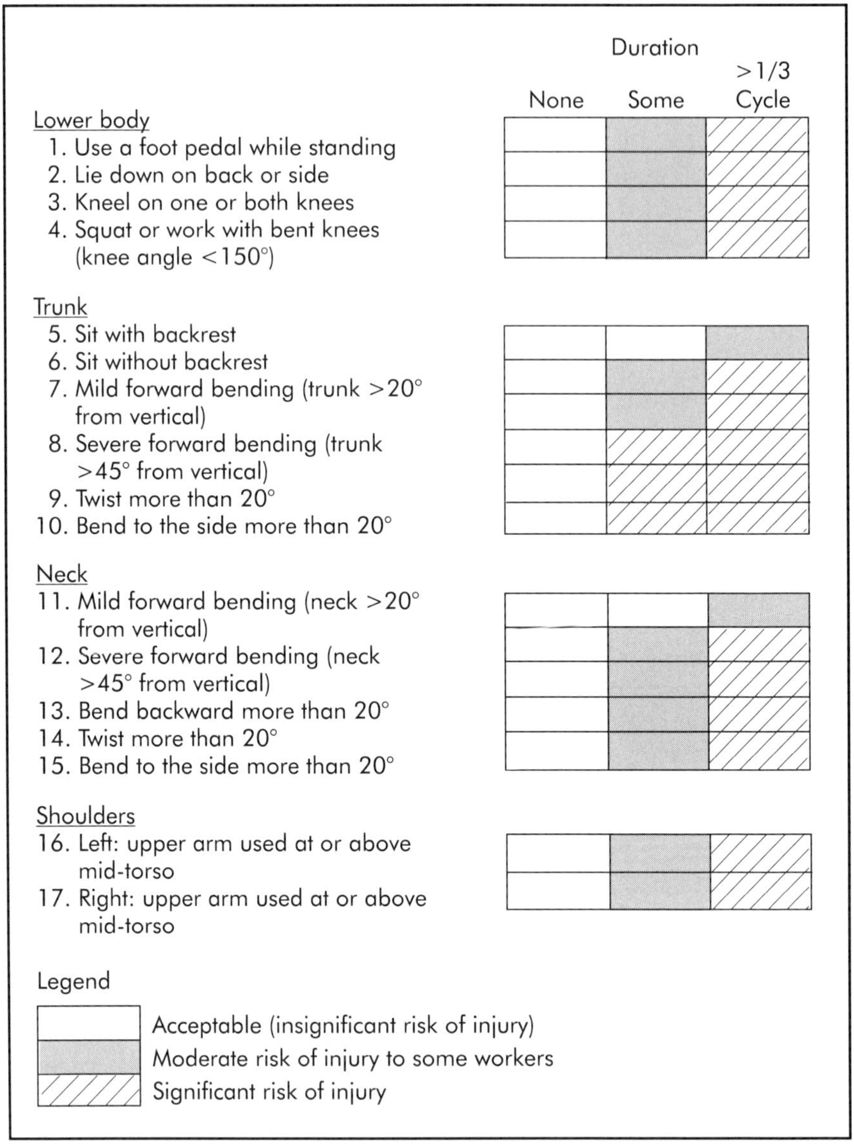

	Duration		
	None	Some	>1/3 Cycle

Lower body
1. Use a foot pedal while standing
2. Lie down on back or side
3. Kneel on one or both knees
4. Squat or work with bent knees (knee angle <150°)

Trunk
5. Sit with backrest
6. Sit without backrest
7. Mild forward bending (trunk >20° from vertical)
8. Severe forward bending (trunk >45° from vertical)
9. Twist more than 20°
10. Bend to the side more than 20°

Neck
11. Mild forward bending (neck >20° from vertical)
12. Severe forward bending (neck >45° from vertical)
13. Bend backward more than 20°
14. Twist more than 20°
15. Bend to the side more than 20°

Shoulders
16. Left: upper arm used at or above mid-torso
17. Right: upper arm used at or above mid-torso

Legend
- ☐ Acceptable (insignificant risk of injury)
- ▨ Moderate risk of injury to some workers
- ▥ Significant risk of injury

Figure 11-1 A checklist for assessing exposure to awkward work postures (Keyserling et al. 1993; Keyserling et al. 1992).

ployee/supervisor complaints of excessive fatigue or discomfort while performing the job; the number of employees who would benefit from ergonomic enhancements to the job; the time required to design and install job improvements; and the expected benefits and costs associated with implementing job changes (Keyserling et al. 1993; Keyserling et al. 1992).

The checklist questions in Figure 11-1 were used to evaluate 335 jobs at four work sites (an engine manufacturing plant, a metal stamping plant, and two warehouses). A subset of these jobs was also evaluated by ergonomic experts using a method that will be described later. Major findings included the following (Keyserling et al. 1993; Keyserling et al. 1992):

- Checklist results were generally in agreement with those of expert ergonomists, however the checklist tended to overestimate the seriousness of certain exposures.
- Users considered the checklist to be a helpful and easy-to-use job analysis tool.
- At some sites, the checklist was used to identify root causes of postural stress.

Whenever a question produced a mark in either the "marginal risk" or "significant risk" column, supplemental notes were taken to document workstation features and/or work methods that caused the awkward posture. Results of the checklist analysis were also used to select jobs for ergonomic improvements.

Statistical Sampling

Statistical sampling is an exposure assessment tool that provides unbiased quantitative information on the amount of time (usually expressed as a percentage of the work shift) spent in neutral versus awkward postures. The procedure is based on work sampling, a work measurement tool used by industrial engineers since the early 1900s (Niebel 1993). To perform a statistical sampling study, it is first necessary to designate posture categories for the body segments of interest. Figure 11-2 presents a system for classifying the lower body, trunk, shoulders, and neck into designated posture categories. This system, called the Ovako Working Posture Analysis System (OWAS), was developed during the mid-1970s in Finland by Ovako Oy, a private steel company, and the Finnish National Institute for Occupational Health (Karhu et al. 1977). The OWAS system also provides benchmarks for interpreting the results of a sampling study. Four action categories are defined, representing increasing levels of urgency for implementing job improvements:

1. *Acceptable*: the duration of exposure to the posture is not harmful and no corrective actions are necessary.

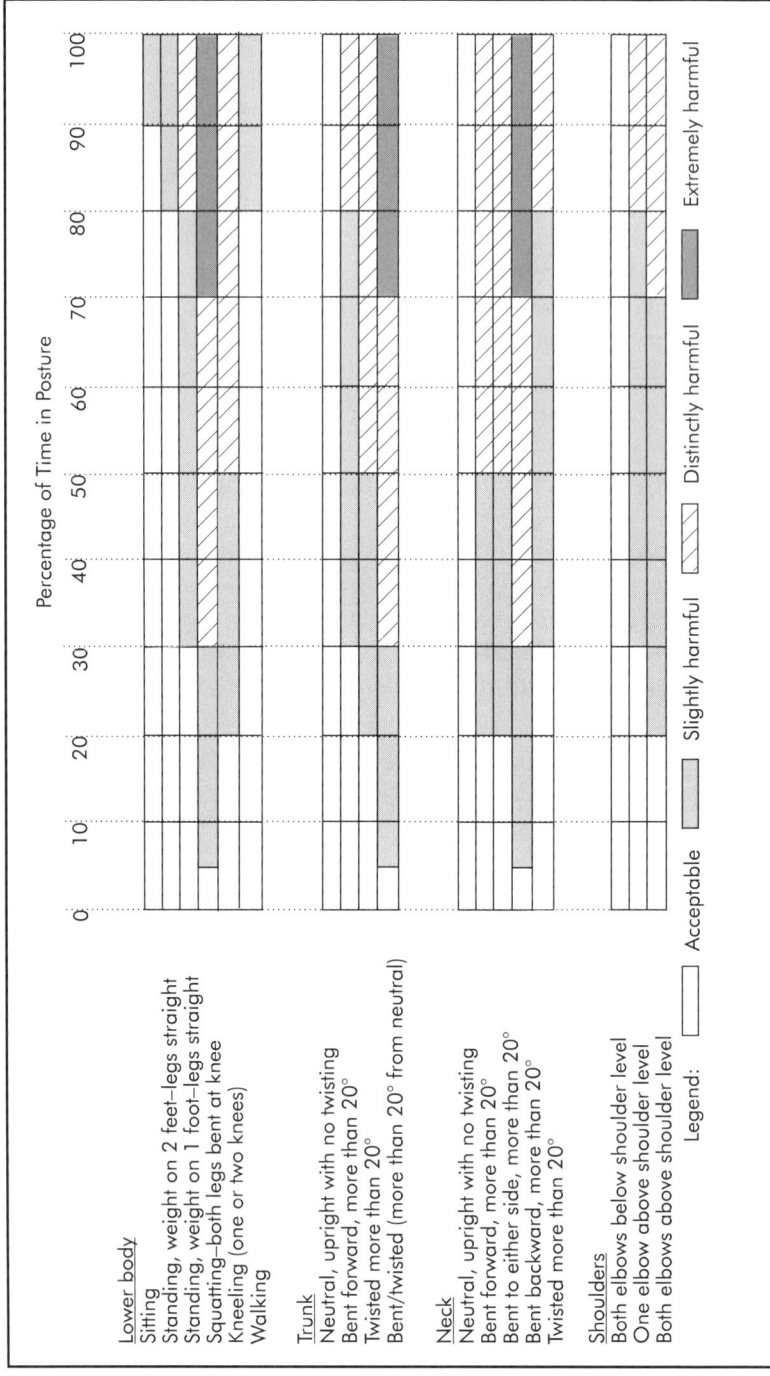

Figure 11-2 OWAS method for evaluating stressfulness of selected work postures (Karhu et al. 1977; Kilbom et al. 1985).

2. *Slightly harmful*: the duration of exposure is sufficiently long to cause fatigue or discomfort. Corrective actions should be taken in the near future.
3. *Distinctly harmful*: the duration of exposure is sufficiently long to cause extreme fatigue, discomfort, and/or injury. Corrective actions should be taken as soon as possible.
4. *Extremely harmful*: immediate action is needed to correct an unacceptable level of postural stress.

Benchmarks for determining action categories based on the percentage of time spent in various postures using the OWAS classification system are included in Figure 11-2 (Mattila et al. 1993).

Data collection in a statistical sampling study involves taking a large number (typically 500-1,000) of randomly-scheduled, instantaneous observations of work posture. When using OWAS, each observation involves classifying and recording the position of four body locations: the lower body (six categories), the trunk (four categories), the neck (five categories), and the shoulder (three categories). Once all the observations have been taken, calculations are performed at each body location to determine the percentage of time spent in each designated posture using the ratio:

$$\frac{\text{number of observations of the specific work posture} \times 100}{\text{total number of observations in the study}}$$

The observed percentages are then compared to the benchmarks in Figure 11-2 to determine whether corrective action is needed.

Example: A statistical sampling study (750 random observations) of a metal stamping operation was performed using the OWAS system. In this job, a seated worker manually loaded small blanks to an open-back, inclined stamping press and then activated two-handed palm buttons to initiate the press cycle. Approximately 5,000 units were produced during a 9-hour work shift. Unprocessed blanks were obtained from a feed hopper located to the left of the worker. The worker periodically stood up to reposition parts in the feed hopper. A small amount of walking was required to talk with supervisors and perform preventive inspections and maintenance.

The results of the OWAS analysis are summarized in Table 11-2. The analysis of lower body posture shows that this job is acceptable; over 80% of the job can be performed in a seated position with an occasional shift to standing or walking. Forward bending of the trunk occurs during 54% of the work shift. This posture, rated as "slightly harmful," occurs during two activities: 1) leaning

Table 11-2. Results of an OWAS posture analysis for a stamping press operation

Posture	Count	%	OWAS class
Lower body			
Sitting	622	83	Acceptable
Standing, weight on two feet, legs straight	41	5	Acceptable
Standing, weight on one foot, legs straight	23	3	Acceptable
Squatting, both legs bent at knee	0	0	Acceptable
Kneeling (one or two knees)	0	0	Acceptable
Walking	64	9	Acceptable
Trunk			
Neutral	297	40	Acceptable
Bent forward, more than 20°	407	54	Slightly harmful
Twisted more than 20°	0	0	Acceptable
Bent and twisted	46	6	Acceptable
Neck			
Neutral	346	46	Acceptable
Bent forward, more than 20°	339	45	Slightly harmful
Bent to either side, more than 20°	0	0	Acceptable
Bent backward, more than 20°	0	0	Acceptable
Twisted more than 20°	65	9	Acceptable
Shoulders			
Both elbows below shoulder	702	94	Acceptable
One elbow above shoulder	48	6	Acceptable
Both elbows above shoulder	0	0	Acceptable

forward while seated to load parts into the press, and 2) leaning forward while standing to arrange parts in the feed hopper. Forward bending of the neck occurs for 45% of the shift. This slightly harmful posture also occurs during two activities: 1) visually inspecting the position of the blank on the die prior to initiating the press cycle, and 2) arranging parts in the feed hopper. Shoulder posture is acceptable, with only occasional elevation of the upper arm when arranging parts in the hopper.

Statistical sampling methods, such as OWAS, are quite easy to learn. While study design requires a good knowledge of statistics to assure that sample sizes are adequate to provide desired levels of confidence and precision, data collection can be performed by a technician and does not require an extensive background in either ergonomics or statistics. Compared to checklist methods, statistical sampling provides a considerably more precise measurement of the time spent in awkward work postures. Data collection requires more time, however, due to the large number of required observations. Like checklists, sampling methods do not identify root causes of awkward posture, and follow-up analyses are required to implement effective job changes.

ROOT CAUSE ANALYSIS METHODS

To develop effective interventions for reducing postural stress on a job, it is first necessary to understand how factors such as a worker's body size, workstation layout, equipment, tools, and work methods interact to influence working posture. To do this, one must measure and record working posture (often at multiple joints) and work activities on a common, continuous time scale. This is a complex data collection activity, for which the pencil and paper methods previously described are no longer feasible. Instead, videotape is used to create a permanent recording of the job that can be played and replayed (in slow motion, if needed) by the job analyst. Personal computers are also used to assist the analyst in recording posture and work activities, and to generate reports.

Video-based, computer-aided systems were developed during the mid-1980s at the Swedish National Institute for Occupational Health (Kilbom et al. 1985) and at the University of Michigan (Keyserling 1990; Keyserling 1986) to analyze the working posture of the lower body, trunk, shoulders, and neck during repetitive assembly line jobs. Because the computer simplifies data recording activities and reduces analysis time, the system can be used to quickly analyze postural stresses in the workplace.

Similarity to Time Study

The video computer-aided system for observing and recording work posture is developed from the time study methods used by industrial engineers. Time study and posture analysis are conceptually similar; the goal of the time study is to measure the amount of time required to complete a work element, while the goal of posture analysis is to measure the amount of time the worker spends in certain postures (Niebel 1993; Keyserling 1986). To make posture analysis feasible, it is first necessary to define standard postures. Along with a menu of standard postures, time study methods can be used to measure the time spent in each posture.

Posture Classification

The posture classification system presented in Figure 11-3 defines a menu of standard postures for the lower extremities, trunk, shoulders, and neck. This menu was developed at the University of Michigan by observing videotapes of a variety of work activities in several manufacturing facilities to establish a taxonomy of common work postures, and by reviewing the literature to identify specific work postures associated with the development of fatigue and/or musculoskeletal disorders (Keyserling 1990). For each joint of interest, the neutral posture was considered to be a position that placed relatively low levels of stress on the musculoskeletal system and that could be maintained for prolonged periods without unusual fatigue or discomfort. Non-neutral postures were associated with increased levels of muscular activity and/or strain on connective tissues. Finally, posture categories were selected to facilitate the need for the analyst to make rapid classification decisions while viewing a videotape.

When using the system, it is necessary to independently describe the posture of the trunk, neck, left shoulder, right shoulder, and lower extremities using these standard positions. Trunk posture is classified with respect to an absolute neutral posture (vertical orientation with no twisting). Neck and shoulder postures are classified relative to the position of the trunk. Lower extremity posture is classified based on activities and/or the method of supporting body weight.

Equipment and Procedures

The first step in the analysis is to obtain a continuous videotape of the job, using any portable video camera and recorder. A color system is recommended, however, because it increases the detail and quality of the playback.

When making the tape, it is essential that the camera angle be chosen so that all joints of interest (trunk, neck, shoulders, and lower extremities) are not obstructed. In some cases, it is advisable to use more than one camera angle to record the job. For highly repetitive work, such as an assembly line job with a cycle time of less than two minutes, several work cycles should be recorded continuously from each camera angle. This assures that all major work activities will be documented. A dimensioned workplace sketch, showing the location of the operator, equipment, tools, material handling devices, etc., should be made at the time of videotaping. This sketch is used in conjunction with the results of the posture analysis to evaluate the workstation layout.

The second step of the analysis process is to develop a sequential description of major tasks, or work elements, required to perform a complete work cycle. (Note: A work cycle is the sequence of

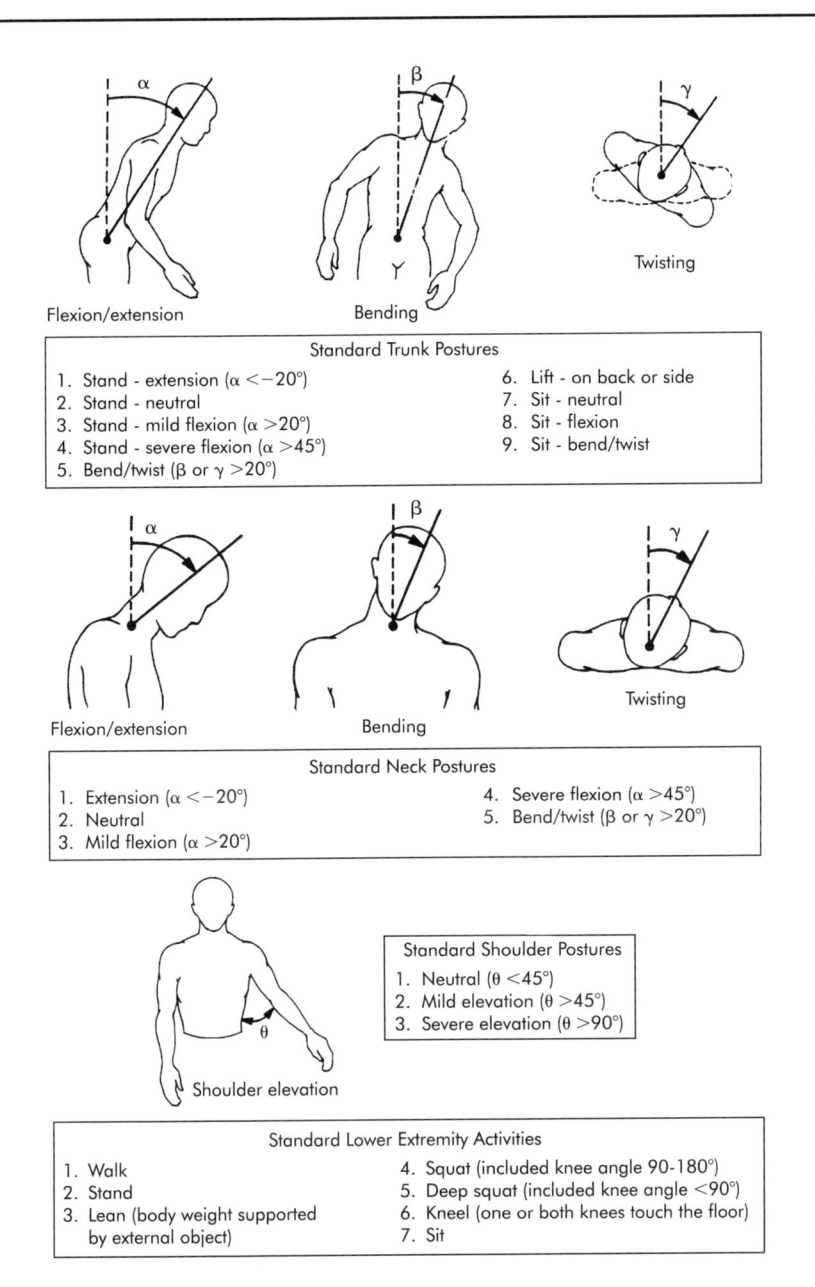

Figure 11-3 Standard categories for classifying posture at the trunk, neck, shoulders, and lower extremities in a video computer-aided posture analysis system (Keyserling 1990).

elements performed by a worker to produce one unit of product.) After the elements have been identified and listed sequentially, a stopwatch or other timing device is used to measure the duration of each element and its end point relative to the start of the work cycle. This procedure establishes a common time scale so that work postures can be associated with specific work activities.

The third step is to collect the postural data. To keep all analyses on a common time scale, it is necessary to analyze the same sequence of videotape that was used to develop the task description. A personal computer is used to assist the analyst in collecting and recording time and posture data. The computer keyboard is used to enter data while the tape is played, and the computer's internal clock is used to measure time. To do this, each of the standard postures in Figure 11-3 is assigned a key.

Whenever the worker changes posture, the analyst hits a key corresponding to the new posture. The value of the new posture and the time of the posture change (measured by the internal clock relative to the start of the work cycle) are stored for subsequent analysis and archiving. Because the computer performs all required timekeeping and clerical functions, the analyst can devote uninterrupted attention to the videotape. This is an essential feature due to the very short time (sometimes less than 0.5 seconds) between posture changes on highly dynamic jobs.

To eliminate the need for observing multiple joints simultaneously, the tape is played one time for each joint of interest (lower extremities, trunk, left shoulder, right shoulder, and neck). The joints can be analyzed in any order since each analysis is performed on a common time scale, with zero corresponding to the start of the work cycle.

System Reports

Following the final data entry, the computer generates a posture profile for each joint. These profiles provide basic descriptive statistics for postural activity (the total time spent in each standard posture during the work cycle, the minimum and maximum times spent in each standard posture, the number of times the posture was entered, etc.). Posture profiles also may be used to evaluate the effectiveness of changes in workstation layout and work methods.

The system also has the capability to generate a graph that shows posture changes over time for any joint of interest. For the purposes of job redesign, it is useful to consider postural activity and task activities on a common time scale. This graph can be used in conjunction with the sequential task description to identify specific work activities associated with awkward postures (Keyserling 1990; Keyserling 1986).

A CASE STUDY

This case study (Keyserling et al. 1993) is taken from a parts distribution center where bulk shipments of automotive replacement parts are unitized (wrapped or boxed into individual containers) prior to warehouse storage. The case study job involves a routine maintenance procedure on a unitizing machine that wraps small, lightweight parts in an envelope formed from a roll of plastic film. When the roll of film is depleted or when a different width of film is required for a new batch of parts, the machine operator must perform a roll change. Depending on production requirements, the roll change activity is performed between two and six times per shift. Because of the posture stresses on this job, the previously described posture analysis system was used.

PRE-CHANGE ANALYSIS

Workstation Layout and Awkward Trunk Posture

Initially, the depleted roll of plastic film was positioned on a spindle on the wrapping machine approximately 15 in. (38 cm) above the floor as shown in Figure 11-4. This location required the worker to bend the trunk forward more than 45° from the neutral upright position when changing the film and when adjusting the machine to accommodate the new roll. Furthermore, incoming rolls of plastic film (weighing 47.4 lb [21.5 kg]) were stored in boxes on pallets, as shown in Figure 11-5. Note that the boxes at the bottom

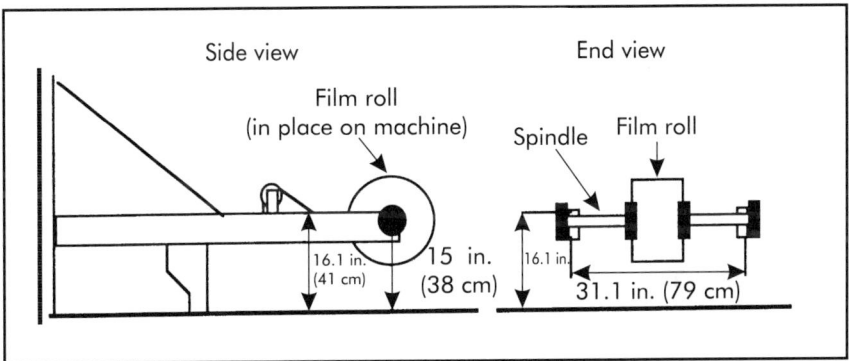

Figure 11-4 Location of film roll on wrapping machine prior to ergonomic changes.

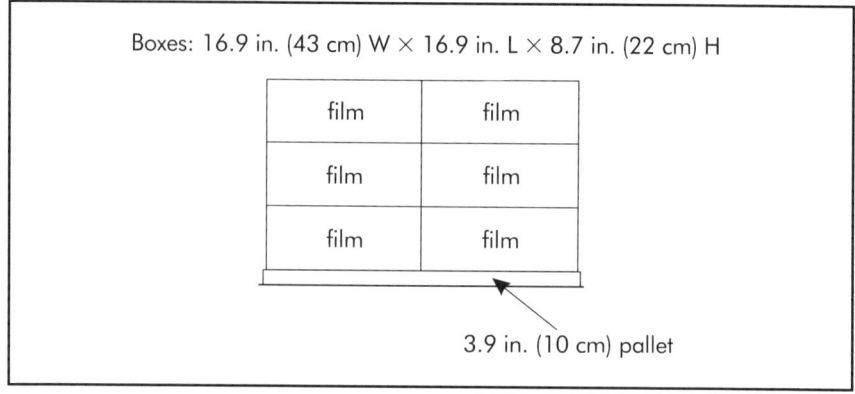

Figure 11-5 Pallet for storing film rolls prior to ergonomic changes.

of the pallet were only 3.9 in. (10 cm) above the floor, requiring the worker to bend forward when obtaining or replacing rolls at this level.

Work Elements

Prior to changes to improve working posture, a typical roll change involved the following tasks:

1. Cut film, remove old roll from spindle, and carry it to storage location.
2. Walk to film storage pallet, open box containing new roll.
3. Remove new roll and place on floor.
4. Roll new roll across the floor to the wrapping machine (distance up to 13 ft [4.0 m]).
5. Insert spindle into new roll and lift assembly into machine.
6. Secure new roll in feed mechanism and splice leader to remnant of old roll.

Posture Analysis

Due to the low height of plastic roll on the wrapping machine and the low storage height of boxes on the pallet, elements #1, 3, 4, 5, and 6 were performed with the worker's trunk bent forward. As shown in Table 11-3, 87% of the work cycle was performed with the trunk in severe flexion (forward bending more than 45°). In addition, mild forward flexion of both shoulders was required when performing some of the low reaches.

Table 11-3. Results of posture analyses on case study job (Keyserling et al. 1993)

Observed postures	% of cycle		Time in posture (seconds)	
	Pre-change	Post-change	Pre-change	Post-change
Trunk				
Stand, neutral	13	72	9.5	30.6
Stand, mild flexion	0	23	0.0	8.4
Stand, severe flexion	87	2	70.5	3.1
Obscured view	0	3	0.0	7.0
Neck				
Neutral	75	58	17.0	22.4
Mild flexion	24	38	5.7	9.6
Severe flexion	0	<1	0.0	0.7
Twisting	3	<1	1.4	1.5
Obscured view	0	3	0.0	7.0
Left shoulder				
Neutral	31	88	10.9	113.0
Mild elevation	68	10	23.8	16.0
Severe elevation	<1	0	0.3	0.0
Obscured view	0	3	0.0	6.8
Right shoulder				
Neutral	21	89	10.7	46.6
Mild elevation	79	8	51.8	4.6
Severe elevation	<1	0	0.3	0.0
Obscured view	0	3	0.0	6.9

Note: Lower body posture was not considered during this case study.

ERGONOMIC CHANGES

Three significant changes were implemented to reduce ergonomic stresses on this job:

1. The location of the film roll on the machine was raised from 15 in. (38 cm) to 31.1 in. (79 cm). See Figure 11-6.

2. The pallet used for storing film rolls was positioned on a 22.8-in. (58-cm) "backsaver" rack. See Figure 11-7.

3. A lift cart was provided for raising, lowering, and transporting rolls. Not only did this eliminate the need to stoop while rolling the roll across the floor from the storage pallet to the

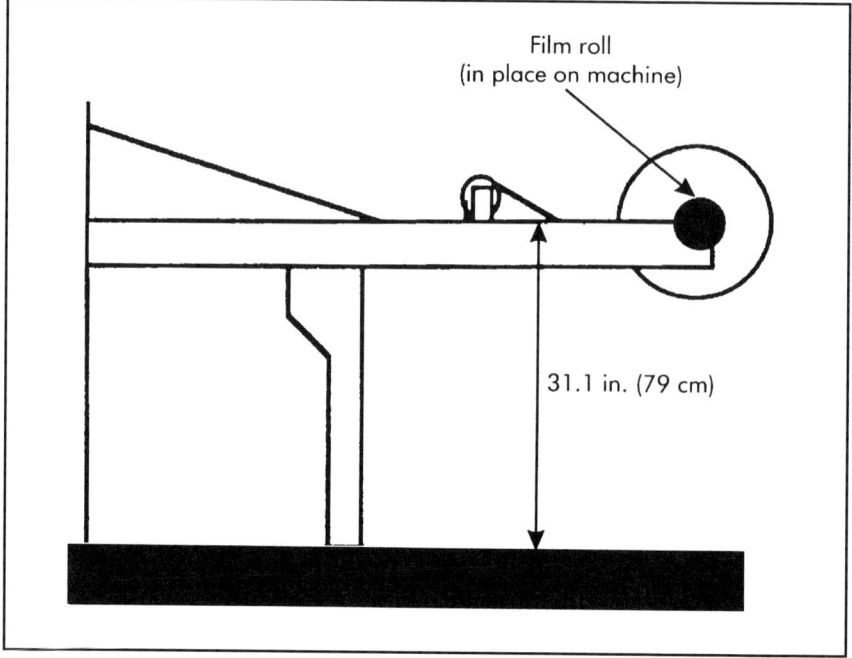

Figure 11-6 Location of film roll on wrapping machine after ergonomic changes. Raising spindle location reduced awkward trunk posture during roll change operations.

machine, but it also eliminated the need to manually lift and lower a heavy load—66 lb (30 kg), including spindle.

POST-CHANGE ANALYSIS

Workstation Layout and Work Elements

The effort of the worker to reach to low locations was ameliorated by raising the height of the film roll on the machine and placing storage pallets on racks. Furthermore, with the addition of the lift cart, the following sequence of tasks was used to perform a roll change:

1. Cut film and use lift cart to remove old roll and spindle from machine.
2. Use cart to transfer used roll to storage location; place used roll on storage pallet.
3. Push cart to pallet holding boxes of new rolls.
4. Remove new roll from box and place on cart.

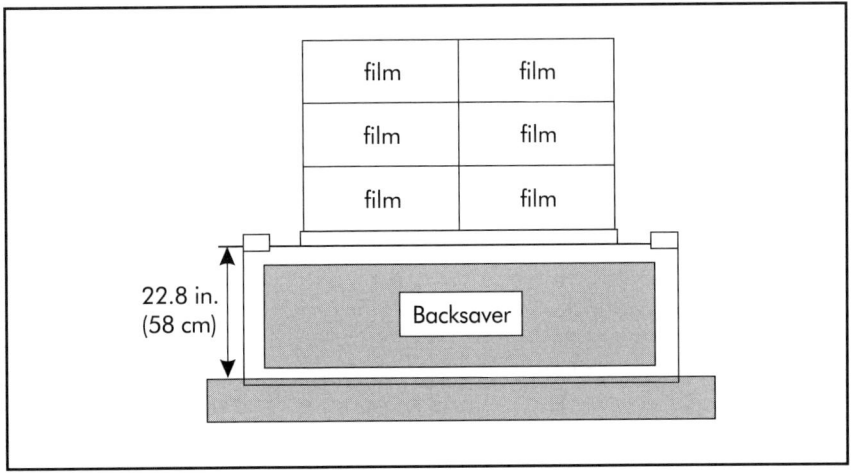

Figure 11-7 Location of film roll on machine after ergonomic changes. Placement of the pallet on a "backsaver" rack raises rolls off the floor, reducing awkward trunk posture.

5. Push cart to wrapping machine.
6. Insert spindle in new roll, use cart to lower assembly into machine.
7. Secure new roll in feed mechanism and splice leader to remnant of old roll.

Posture Analysis

As a result of changes to workstation layout and equipment, it was no longer necessary for the worker to position the trunk in severe flexion. As shown in Table 11-3, the new configuration allowed the worker to maintain a neutral posture for 72% of the work cycle. Mild trunk flexion was required during 23% of the cycle and a twisted/bent posture was required for only 2% of the cycle. The time spent with the left and right shoulders in an elevated posture was also decreased as a result of reduced reach requirements.

Case Review

A posture analysis of this job revealed that roll changing required prolonged usage of awkward trunk postures. This information was used to develop changes in workstation layout and equipment to reduce the duration and severity of awkward trunk postures. Furthermore, the new lift cart significantly reduced the amount of lifting and carrying required to perform a roll change. Both of these changes reduced worker exposure to risk factors associated with occupational low-back pain. There was also a slight reduction in awkward shoulder postures as a result of the layout changes.

SUMMARY

Awkward working posture occurs when there is a mismatch between a worker's body size and job requirements. When awkward postures are used repetitively or for extended periods, undesired outcomes may occur, including fatigue, discomfort, injury, and reduced productivity.

Ergonomists have developed a number of job analysis methodologies to evaluate awkward postures. Some of these methods (for example, posture checklists and OWAS) are easy to use and do not require special equipment. These methods can be used as part of a rapid screening program to identify jobs where postural demands are excessive. When the rapid screening methods identify a job with excessive postural stress, it is necessary to perform a detailed analysis to determine the root causes of awkward posture. This analysis provides the essential information for developing practical and effective workplace interventions for controlling awkward posture.

REFERENCES

Corlett, E.N., and Bishop, R.P. "A Technique for Assessing Postural Discomfort." *Ergonomics* 1976, 19: 175-182.

Feldman, R., Goldman, R., and Keyserling, W. "Peripheral Nerve Entrapment Syndromes and Ergonomic Factors." *American Journal of Industrial Medicine* 1983, 4: 661-681.

Hagberg, M. "Local Shoulder Muscular Strain—Symptoms and Disorders." *Journal of Human Ergology* 1982, 11: 99-108.

Hagberg, M. "Occupational Musculoskeletal Stress and Disorders of the Neck and Shoulder: A Review of Possible Pathophysiology." *International Archives of Occupational and Environmental Health* 1984, 53: 269-278.

Harms-Ringdahl, K., and Ekholm, J. "Intensity and Character of Pain and Muscular Activity Levels Elicited by Maintained Extreme Flexion Position of the Lower-cervical-upper-thoracic Spine." *Scandinavian Journal of Rehabilitation Medicine* 1986, 18: 117-126.

Harms-Ringdahl, K., Ekholm, J., Schuldt, K., Nemeth, G., and Arborelius, U.P. "Load Moments and Myoelectric Activity when the Cervical Spine is Held in Full Flexion and Extension." *Ergonomics* 1986, 29: 1,539-1,552.

International Labor Organization (ILO). *Introduction to Work Study.* Geneva, Switzerland: ILO, 1964.

Karhu, O., Kansi, P., and Kuorinka, I. "Correcting Working Postures in Industry: A Practical Method for Analysis." *Applied Ergonomics* 1977, 8: 199-201.

Kelsey, J., and Hochberg, M. "Epidemiology of Chronic Musculoskeletal Disorders." *Annual Review of Public Health* 1988, 9: 379-401.

Keyserling, W.M. "A Computer-aided System to Evaluate Postural Stress in the Workplace." *American Industrial Hygiene Association Journal* 1986, 47: 641-649.

Keyserling, W.M. "Computer-aided Posture Analysis of the Trunk, Neck, Shoulders, and Lower Extremities." *Computer-aided Ergonomics*. W. Karwowski, A.M. Genaidy, and S.S. Asfour, eds. London: Taylor and Francis, 1990.

Keyserling, W.M., Brouwer, M., and Silverstein, B.A. "A Checklist for Evaluating Ergonomic Risk Factors Resulting from Awkward Postures of the Legs, Trunk, and Neck." *International Journal of Industrial Ergonomics* 1992, 9: 283-301.

Keyserling, W.M., Brouwer, M.L., and Silverstein, B.A. "The Effectiveness of a Joint Labor-Management Program in Controlling Awkward Postures of the Trunk, Neck, and Shoulders: Results of a Field Study." *International Journal of Industrial Ergonomics* 1993, 11: 51-65.

Keyserling, W.M., Stetson, D.S., Silverstein, B.A., and Brouwer, M.L. "A Checklist for Evaluating Ergonomic Risk Factors Associated with Upper Extremity Cumulative Trauma Disorders." *Ergonomics* 1993, 36: 807-831.

Kilbom, A., Persson, J., and Jonsson, B.G. "Disorders of the Cervicobrachial Region among Female Workers in the Electronics Industry." *International Journal of Industrial Ergonomics* 1986, 1: 37-47.

Kilbom, A., Persson, J., and Jonsson, B.G. "Risk Factors for Work-related Disorders of the Neck and Shoulder—with Special Emphasis on Working Postures and Movements." *Proceedings: International Symposium of the Ergonomics of Working Posture*. London: Taylor and Francis, 1985.

Kumar, S., and Scaife, W.G.S. "A Precision Task: Posture and Strain." *Journal of Safety Research* 1979, 11: 28-36.

Mattila, M., Karwowski, W., and Vilkki, M. "Analysis of Working Postures in Hammering Tasks on Building Construction Sites: The Computerized OWAS Method." *Applied Ergonomics* 1993, 24: 405-412.

Niebel, B.W. *Motion and Time Study*, 9th Edition. Homewood, IL: Richard D. Irwin, Inc., 1993.

Punnett, L., Fine, L.J., Keyserling, W.M., Herrin, G.D., and Chaffin, D.B. "Back Disorders and Non-neutral Trunk Postures of Automobile Assembly Workers." *Scandinavian Journal of Work, Environment, and Health* 1991, 17: 337-346.

Schultz, A., Andersson, G., Ortengren, R., Nachemson, A., and Haderspeck, K. "Loads on the Lumbar Spine: Validation of a Biomechanical Analysis: Measurements of Intradiscal Pressures and Myoelectric Signals." *Journal of Bone and Joint Surgery* 1982, 64A: 713-720.

Tola, S., Riihimaki, H., Videman, T., Viikari-Juntura, E., and Hanninen, K. "Neck and Shoulder Symptoms among Men in Machine Operating, Dynamic Physical Work and Sedentary Work." *Scandinavian Journal of Work, Environment, and Health* 1988, 14(5): 299-305.

Van Wely, P. "Design and Disease." *Applied Ergonomics* 1970, 1: 262-269.

Wiker, S.F. *Effects of Hand Location Upon Movement Time and Fatigue*. Ph.D. Dissertation. Ann Arbor, MI: Department of Industrial and Operations Engineering, University of Michigan, 1986.

Zandin, K.B. *MOST Work Measurement Systems*, 2nd Edition. New York: Marcel Dekker, Inc., 1990.

CHAPTER 12

MANUAL MATERIAL HANDLING: DESIGN DATA BASES

Christin L. Shoaf, Doctoral Student
Ashraf M. Genaidy, Associate Professor
Mechanical and Industrial and
Nuclear Engineering Department
University of Cincinnati
Cincinnati, OH

Handling objects manually is a major source of lower back injuries in the workplace (Troup and Edwards 1985; Centers for Disease Control 1983; Bureau of Labor Statistics 1982; Snook 1988). Designers, as well as practitioners, of manual material handling have relied on job design as a means to reduce the frequency and severity of back injuries (Snook 1978). Job design involves tailoring job demands to worker ability. For job design to succeed in industry, information predicting the manual material handling capabilities of workers is necessary. There are mathematical models that allow such predictions to be made. The models presented in this chapter can be used to calculate a recommended weight or force for handling activities for a given set of input conditions—lifting, lowering, pushing, pulling, or carrying activities.

Safe manual handling limits should be based on integration of biomechanical, physiological, and psychophysical criteria. Previous attempts have concentrated on establishing such limits for manual lifting (Waters et al. 1993), and consider task parameters such as frequency and distance traveled. Furthermore, since the practice of industrial manual handling operations consists of the previously mentioned activities, it is quite obvious that there is an urgent need to establish safe limits for all types of manual handling activities.

MODELS

STRUCTURE

The mathematical models presented in this chapter are patterned after the empirical equation proposed by Drury and Pfeil (1975) for lifting activities. Their model was described by a base weight (the maximum weight that can be lifted under ideal conditions). It is multiplied by a series of factors representing effects of the task variables. Optimum conditions of a factor are represented by a value of one, whereas non-ideal conditions are scaled to yield factors of less than one. The model assumes that all factors are interacting in a multiplicative way. The lifting equations formulated by the National Institute for Occupational Safety and Health (NIOSH) (Waters et al. 1993) are based on a similar approach.

The models calculate a recommended load for a given task by using a base weight, the maximum weight or force that can be handled under ideal conditions, multiplied by a series of factors that represent the effects of the task and personal variables (Hidalgo et al. 1997; Shoaf et al. 1997). For all factors except body weight, optimum conditions are represented by a value of one, whereas non-ideal conditions are scaled to yield factors of less than one. Based on the models' output, resulting load capacities can be predicted for workers of both genders and for any percentage of the worker population performing all types of manual material handling activities.

Each mathematical model was built in a three stage process. Initially, these models were based on the psychophysical data established by Snook and Ciriello (1991). Base weights and discounting factor multiplier curves were generated for both male and female populations. Then, the biomechanical approach (Tichauer 1973; Tichauer 1978) was used to modify the recommended base weights. Finally, the physiological approach was used to refine the frequency multiplier curves.

Lifting

The structure of the lifting model (Hidalgo et al. 1997) is:

$$LC = W_B \times H \times V \times D \times F \times AG \times BW \times TD \times T \times C \times HS$$

where:

LC = lifting capacity (kg)
W_B = maximum load acceptable to a specified percentage of the worker population (kg) (see Table 12-1)
H = multiplier for horizontal distance away from the body with respect to the midpoint between the ankles (cm) (see Table 12-2)
V = multiplier for the vertical distance of lift (cm) (see Table 12-3)

Table 12-1. Base weights for lifting and lowering

Population %	Male lb (kg)	Female lb (kg)
5	86.2 (39.1)	50.3 (22.8)
10	82.7 (37.5)	48.9 (22.2)
15	79.4 (36.0)	47.4 (21.5)
20	76.1 (34.5)	46.1 (20.9)
25	72.8 (33.0)	44.8 (20.3)
30	69.4 (31.5)	43.4 (19.7)
35	66.1 (30.0)	42.1 (19.1)
40	62.8 (28.5)	40.8 (18.5)
45	59.5 (27.0)	39.5 (17.9)
50	56.2 (25.5)	38.1 (17.3)
55	52.9 (24.0)	36.8 (16.7)
60	49.6 (22.5)	35.5 (16.1)
65	46.3 (21.0)	34.2 (15.5)
70	42.8 (19.4)	32.8 (14.9)
75	39.5 (17.9)	31.5 (14.3)
80	36.2 (16.4)	30.2 (13.7)
85	32.8 (14.9)	28.9 (13.1)
90	29.5 (13.4)	27.6 (12.5)
95	26.2 (11.9)	26.2 (11.9)

Table 12-2. Horizontal distance multiplier for lifting

Distance in. (cm)	Male	Female
9.8 (25)	1.00	1.00
11.8 (30)	0.87	0.84
13.8 (35)	0.79	0.73
15.7 (40)	0.73	0.66
17.7 (45)	0.70	0.64
19.7 (50)	0.68	0.65
21.7 (55)	0.63	0.61
23.6 (60)	0.52	0.50
24.8 (63)	0.43	0.41

Table 12-3. Vertical travel distance multiplier for lifting

Distance in. (cm)	Male	Female
9.8 (25)	1.00	1.00
11.8 (30)	0.97	0.99
13.8 (35)	0.94	0.97
15.7 (40)	0.91	0.96
17.7 (45)	0.88	0.95
19.7 (50)	0.86	0.94
21.7 (55)	0.84	0.92
23.6 (60)	0.83	0.89
25.6 (65)	0.81	0.87
27.6 (70)	0.80	0.84
29.5 (75)	0.79	0.82
31.5 (80)	0.78	0.80
33.5 (85)	0.77	0.79
35.4 (90)	0.76	0.78
37.4 (95)	0.75	0.77
39.4 (100)	0.73	0.77
41.3 (105)	0.72	0.76
43.3 (110)	0.71	0.75
45.3 (115)	0.69	0.75
47.2 (120)	0.68	0.75
49.2 (125)	0.67	0.74
51.2 (130)	0.66	0.74
53.1 (135)	0.64	0.74
55.1 (140)	0.63	0.74
57.1 (145)	0.61	0.74
59.1 (150)	0.59	0.74
61.0 (155)	0.57	0.73
63.0 (160)	0.55	0.73
65.0 (165)	0.52	0.73
66.9 (170)	0.49	0.72

D = multiplier for the vertical distance of the hands between the origin and the destination of lift (cm) (see Table 12-4)
F = multiplier for the frequency of lift (times/minute) (see Table 12-5)

Table 12-4. Vertical distance multiplier for lifting

Distance in. (cm)	Male	Female
0 (0)	0.62	0.74
2.0 (5)	0.64	0.76
3.9 (10)	0.67	0.77
5.9 (15)	0.69	0.79
7.9 (20)	0.72	0.81
9.8 (25)	0.75	0.83
11.8 (30)	0.77	0.84
13.8 (35)	0.80	0.86
15.7 (40)	0.82	0.88
17.7 (45)	0.85	0.90
19.7 (50)	0.87	0.91
21.7 (55)	0.90	0.93
23.6 (60)	0.92	0.95
25.6 (65)	0.95	0.97
27.6 (70)	0.98	0.98
29.5 (75)	1.00	1.00
31.5 (80)	0.99	0.99
33.5 (85)	0.98	0.98
35.4 (90)	0.97	0.96
37.4 (95)	0.96	0.95
39.4 (100)	0.94	0.93
41.3 (105)	0.93	0.90
43.3 (110)	0.92	0.88
45.3 (115)	0.91	0.86
47.2 (120)	0.89	0.83
49.2 (125)	0.88	0.81
51.2 (130)	0.87	0.78
53.1 (135)	0.84	0.76
55.1 (140)	0.83	0.72
57.1 (145)	0.82	0.70
59.1 (150)	0.80	0.67
61.0 (155)	0.79	0.65
63.0 (160)	0.77	0.62
65.0 (165)	0.76	0.60
66.9 (170)	0.73	0.56
68.9 (175)	0.71	0.54

Table 12-5. Frequency multiplier for lifting

Frequency Times/min	Male	Female
0.20	1.00	1.00
0.50	0.99	0.99
1.00	0.95	0.91
2.00	0.89	0.87
3.00	0.83	0.84
4.00	0.78	0.80
5.00	0.73	0.77
6.00	0.69	0.74
7.00	0.65	0.70
8.00	0.62	0.68
9.00	0.59	0.66
10.00	0.56	0.65
11.00	0.54	0.64
12.00	0.52	0.63
13.00	0.50	0.63
14.00	0.49	0.62
15.00	0.47	0.61
16.00	0.46	0.60

AG = multiplier for age group (see Table 12-6)

BW = multiplier for body weight (see Table 12-7)

TD = multiplier for task duration (hours) (see Table 12-8)

T = multiplier for trunk twisting angle (°) (see Table 12-9)

C = multiplier for coupling factor (see Table 12-10)

HS = multiplier for heat stress (° C, wet bulb globe temperature) (see Table 12-11).

Lowering

The structure of the lowering model (Shoaf et al. 1997) is:

$$LOC = W_B \times H \times V \times F \times AG \times BW \times TD$$

Manual Material Handling: Design Data Bases

Table 12-6. Age multiplier

Age	Male	Female
20	1.00	1.00
25	0.91	0.95
30	0.88	0.90
35	0.88	0.87
40	0.86	0.82
45	0.78	0.79
50	0.69	0.72
55	0.62	0.64
60	0.59	0.49

Table 12-8. Task duration multiplier

Duration (hr)	Multiplier
0	1.00
1	1.00
2	0.77
3	0.67
4	0.60
5	0.58
6	0.54
7	0.50
8	0.45

Table 12-7. Body weight multiplier

Weight lb (kg)	Male	Female
88.2 (40)	0.70	1.00
99.2 (45)	0.70	1.00
110.2 (50)	0.70	1.00
121.3 (55)	0.70	1.00
132.3 (60)	0.70	1.00
143.3 (65)	0.80	1.20
154.3 (70)	1.00	1.40
165.3 (75)	1.20	1.68
176.4 (80)	1.30	1.85
187.4 (85)	1.41	1.98
198.4 (90)	1.45	2.05
209.4 (95)	1.45	2.05
220.5 (100)	1.45	2.05

Table 12-9. Twisting multiplier for lifting

Twisting angle (°)	Multiplier
0	1.00
10	0.98
20	0.95
30	0.93
40	0.90
50	0.88
60	0.86
70	0.83
80	0.81
90	0.78
100	0.76
110	0.74
120	0.71
130	0.69

Table 12-10. Coupling multiplier for lifting

Coupling	Multiplier
Good and comfortable handles/firm holds to initiate the lift	1.000
Poor quality handles/limited or slippery hold	0.925
No handles/holds to initiate the lift	0.850

Table 12-11. Heat stress multiplier for lifting

Heat stress °F (°C)	Multiplier
66-81 (19-27)	1.00
82 (28)	0.98
84 (29)	0.95
86 (30)	0.93
88 (31)	0.90
90 (32)	0.88
91 (33)	0.86
93 (34)	0.83
95 (35)	0.81
97 (36)	0.78
99 (37)	0.76
100 (38)	0.74
102 (39)	0.71
104 (40)	0.69

where:

LOC = lowering capacity (kg)

W_B = maximum load acceptable to a specified percentage of the worker population (kg) and is also a function of level of lowering (where FK = knuckle to floor, KS = shoulder to knuckle, SR = arm reach to shoulder) (see Table 12-1)

H = multiplier for the horizontal distance away from the body with

Table 12-12. Horizontal distance multiplier for lowering

Distance in. (cm)	Male V <29.5 in. (75 cm)	Female V <29.5 in. (75 cm)	Male V >29.5 in. (75 cm)	Female V >29.5 in. (75 cm)
15.4 (39)	0.785	0.642	0.753	0.810
16.1 (41)	0.760	0.620	0.739	0.799
16.9 (43)	0.738	0.609	0.736	0.796
17.7 (45)	0.719	0.606	0.740	0.800
18.5 (47)	0.701	0.605	0.748	0.806
19.3 (49)	0.682	0.604	0.754	0.811
20.1 (51)	0.662	0.597	0.755	0.812
20.9 (53)	0.639	0.581	0.746	0.804
21.7 (55)	0.611	0.555	0.717	0.775
22.4 (57)	0.577	0.523	0.665	0.718
23.2 (59)	0.539	0.484	0.595	0.638
24.0 (61)	0.497	0.439	0.510	0.541
24.8 (63)	0.453	0.387	0.414	0.433

respect to the midpoint between the ankles (cm) (see Table 12-12)

V = multiplier for the vertical distance of lowering (cm) (see Table 12-13)

F = multiplier for the frequency of lowering (times/minute) (see Table 12-14)

AG = multiplier for age group (see Table 12-6)

BW = multiplier for body weight (see Table 12-7)
TD = multiplier for task duration (hours) (see Table 12-8)

Table 12-13. Vertical distance multiplier for lowering

Distance in. (cm)	Male V < 29.5 in. (75 cm)	Female V < 29.5 in. (75 cm)	Male V > 29.5 in. (75 cm)	Female V > 29.5 in. (75 cm)
9.8 (25)	1.000	1.000	1.000	1.000
13.8 (35)	0.945	0.985	0.939	0.942
17.7 (45)	0.904	0.961	0.885	0.895
21.7 (55)	0.887	0.925	0.839	0.865
25.6 (65)	0.877	0.873	0.799	0.840
29.5 (75)	0.866	0.827	0.773	0.815
33.5 (85)	0.855	0.801	0.760	0.789
37.4 (95)	0.844	0.779	0.749	0.763
41.3 (105)	0.830	0.748	0.737	0.737
45.3 (115)	0.807	0.713	0.726	0.712
49.2 (125)	0.777	0.675	0.714	0.688
53.1 (135)	0.745	0.638	0.703	0.663
57.1 (145)	0.715	0.604	0.692	0.638
61.0 (155)	0.691	0.574	0.678	0.609
65.0 (165)	0.669	0.549	0.656	0.560
68.9 (175)	0.649	0.529	0.629	0.500

Table 12-14. Frequency multiplier for lowering

Frequency Times/min	Male V <29.5 in. (75 cm)	Female V <29.5 in. (75 cm)	Male V >29.5 in. (75 cm)	Female V >29.5 in. (75 cm)
0.2	1.000	1.000	1.000	1.000
0.5	0.925	0.947	0.972	0.889
1.0	0.825	0.842	0.833	0.778
2.0	0.738	0.789	0.800	0.748
3.0	0.686	0.762	0.784	0.730
4.0	0.657	0.743	0.762	0.723
5.0	0.634	0.720	0.716	0.724
6.0	0.612	0.696	0.665	0.727
7.0	0.596	0.680	0.631	0.718
8.0	0.582	0.670	0.606	0.698
9.0	0.568	0.661	0.583	0.676
10.0	0.555	0.652	0.562	0.652
11.0	0.541	0.643	0.544	0.630
12.0	0.525	0.632	0.528	0.611
13.0	0.508	0.619	0.512	0.596
14.0	0.490	0.606	0.497	0.581
15.0	0.470	0.593	0.484	0.567
16.0	0.450	0.579	0.472	0.556

Pushing

For pushing and pulling, two different forces are included in each activity. The initial force is the force required to place an object in motion and is a function of the subject's acceleration. The sustained force is the force required to keep an object in motion. The structure of the pushing model (Shoaf et al. 1997) is:

$$PC = F_B \times V \times T \times F \times AG \times BW \times TD$$

where:

PC = pushing capacity (kg)
F_B = maximum force acceptable to a specified percentage of the worker population (kg) (see Table 12-15)
V = multiplier for the vertical distance from floor to hands (cm) (see Table 12-16)
T = multiplier for the traveled distance (m) (see Table 12-17)
F = multiplier for the frequency of push (times/minute) (see Table 12-18)
AG = multiplier for age group (see Table 12-6)

Table 12-15. Base forces for pushing

Population %	Male Initial lb (kg)	Female Initial lb (kg)	Male Sustained lb (kg)	Female Sustained lb (kg)
5	146.6 (66.5)	87.3 (39.6)	97.4 (44.2)	69.4 (31.5)
10	137.1 (62.2)	82.2 (37.3)	91.5 (41.5)	64.8 (29.4)
15	131.2 (59.5)	79.1 (35.9)	87.5 (39.7)	61.7 (28.0)
20	127.4 (57.8)	76.0 (34.5)	84.2 (38.2)	59.3 (26.9)
25	122.8 (55.7)	74.7 (33.9)	81.6 (37.0)	57.1 (25.9)
30	118.8 (53.9)	72.8 (33.0)	79.1 (35.9)	55.1 (25.0)
35	115.0 (52.2)	71.0 (32.2)	76.9 (34.9)	53.4 (24.2)
40	112.0 (50.8)	69.2 (31.4)	74.7 (33.9)	51.8 (23.5)
45	108.7 (49.3)	67.5 (30.6)	72.5 (32.9)	50.0 (22.7)
50	105.8 (48.0)	66.1 (30.0)	70.5 (32.0)	48.5 (22.0)
55	102.5 (46.5)	64.8 (29.4)	68.6 (31.1)	47.0 (21.3)
60	101.0 (45.8)	63.0 (28.6)	66.4 (30.1)	45.2 (20.5)
65	96.8 (43.9)	61.5 (27.9)	64.2 (29.1)	43.7 (19.8)
70	93.0 (42.2)	59.5 (27.0)	61.9 (28.1)	41.9 (19.0)
75	89.0 (40.4)	58.0 (26.3)	59.5 (27.0)	39.9 (18.1)
80	85.5 (38.8)	56.7 (25.7)	56.9 (25.8)	37.7 (17.1)
85	80.0 (36.3)	53.4 (24.2)	53.6 (24.3)	35.3 (16.0)
90	75.0 (34.0)	50.7 (23.0)	49.6 (22.5)	32.2 (14.6)
95	65.9 (29.9)	45.9 (20.8)	43.7 (19.8)	27.6 (12.5)

Table 12-16. Vertical height multiplier for pushing

Height in. (cm)	Male Initial	Female	Male Sustained	Female
35.4 (90)	0.988	0.971	0.989	0.983
37.4 (95)	0.996	0.984	0.995	0.992
39.4 (100)	1.000	0.993	0.999	0.998
41.3 (105)	0.999	0.998	1.000	1.000
43.3 (110)	0.993	1.000	0.999	0.999
45.3 (115)	0.982	0.998	0.996	0.994
47.2 (120)	0.966	0.992	0.990	0.985
49.2 (125)	0.945	0.982	0.983	0.973
51.2 (130)	0.920	0.969	0.972	0.958
53.1 (135)	0.889	0.952	0.960	0.939
55.1 (140)	0.854	0.931	0.945	0.917

Table 12-17. Travel distance multiplier for pushing

Distance ft (m)	Male Initial	Female	Male Sustained	Female
65.6 (20)	0.732	0.741	0.597	0.637
82.0 (25)	0.6667	0.719	0.552	0.583
98.4 (30)	0.614	0.710	0.511	0.537
114.8 (35)	0.577	0.708	0.474	0.520
131.2 (40)	0.548	0.713	0.440	0.534
147.6 (45)	0.523	0.711	0.409	0.536
164.0 (50)	0.499	0.695	0.383	0.504
180.4 (55)	0.476	0.671	0.360	0.455
196.8 (60)	0.455	0.638	0.341	0.338
213.3 (65)	0.438	0.597	0.326	0.305

Table 12-18. Frequency multiplier for pushing

Frequency Times/min	Male Initial	Female	Male Sustained	Female
0.002	1.000	1.000	1.000	1.000
0.016	0.901	0.956	0.894	0.877
0.030	0.854	0.933	0.844	0.818
0.100	0.843	0.919	0.830	0.795
0.200	0.833	0.900	0.813	0.773
0.500	0.813	0.800	0.719	0.727
1.000	0.792	0.767	0.688	0.682
4.000	0.542	0.667	0.438	0.545
6.000	0.557	0.600	0.203	0.455

BW = multiplier for body weight (see Table 12-7)
TD = multiplier for task duration (hours) (see Table 12-8)

Pulling

The structure of the pulling model (Shoaf et al. 1997) is:

$$PLC = F_B \times V \times T \times F \times AG \times BW \times TD$$

where:

PLC = pulling capacity (kg)
F_B = maximum force acceptable to a specified percentage of the worker population (kg) (see Table 12-19)
V = multiplier for the vertical distance from floor to hands (cm) (see Table 12-20)
T = multiplier for the traveled distance (m) (see Table 12-21)
F = multiplier for the frequency of pull (times/minute) (see Table 12-22)
AG = multiplier for age group (see Table 12-6)

Table 12-19. Base forces for pulling

Population %	Male Initial lb (kg)	Female Initial lb (kg)	Male Sustained lb (kg)	Female Sustained lb (kg)
5	137.8 (62.5)	83.8 (38.0)	98.1 (44.5)	67.7 (30.7)
10	131.0 (59.4)	79.6 (36.1)	91.9 (41.7)	63.5 (28.8)
15	125.9 (57.1)	76.7 (34.8)	88.0 (39.9)	60.6 (27.5)
20	123.0 (55.8)	74.7 (33.9)	84.7 (38.4)	58.4 (26.5)
25	119.0 (54.0)	72.1 (32.7)	81.8 (37.1)	56.4 (25.6)
30	115.3 (52.3)	70.5 (32.0)	79.4 (36.0)	54.7 (24.8)
35	113.8 (51.6)	69.0 (31.3)	76.9 (34.9)	52.9 (24.0)
40	110.9 (50.3)	67.2 (30.5)	74.7 (33.9)	51.4 (23.3)
45	108.5 (49.2)	65.7 (29.8)	72.8 (33.0)	50.0 (22.7)
50	105.8 (48.0)	64.2 (29.1)	70.5 (32.0)	48.5 (22.0)
55	103.4 (46.9)	62.6 (28.4)	68.3 (31.0)	47.0 (21.3)
60	101.4 (46.0)	59.3 (26.9)	66.1 (30.0)	45.6 (20.7)
65	98.8 (44.8)	57.3 (26.0)	64.2 (29.1)	44.1 (20.0)
70	95.9 (43.5)	57.5 (26.1)	61.7 (28.0)	42.3 (19.2)
75	93.0 (42.2)	56.2 (25.5)	59.3 (26.9)	40.6 (18.4)
80	89.7 (40.7)	53.6 (24.3)	56.4 (25.6)	38.6 (17.5)
85	85.8 (38.9)	52.2 (23.7)	53.1 (24.1)	36.4 (16.5)
90	81.1 (36.8)	48.5 (22.0)	49.2 (22.3)	33.5 (15.2)
95	74.7 (33.9)	44.3 (20.1)	43.0 (19.5)	29.3 (13.3)

Table 12-20. Vertical height multiplier for pulling

Height in. (cm)	Male Initial	Female Initial	Male Sustained	Female Sustained
23.6 (60)	1.000	1.000	1.000	1.000
25.6 (65)	0.983	0.993	0.993	0.995
27.6 (70)	0.966	0.987	0.984	0.990
29.5 (75)	0.947	0.981	0.974	0.985
31.5 (80)	0.928	0.975	0.962	0.979
33.5 (85)	0.908	0.969	0.949	0.973
35.4 (90)	0.887	0.964	0.935	0.967
37.4 (95)	0.865	0.958	0.919	0.960
39.4 (100)	0.842	0.953	0.901	0.953
41.3 (105)	0.818	0.949	0.882	0.945
43.3 (110)	0.794	0.944	0.862	0.937
45.3 (115)	0.768	0.940	0.840	0.929
47.2 (120)	0.742	0.936	0.817	0.920
49.2 (125)	0.715	0.932	0.792	0.911
51.2 (130)	0.687	0.929	0.765	0.902
53.1 (135)	0.658	0.926	0.738	0.892
55.1 (140)	0.628	0.922	0.708	0.882

Table 12-21. Travel distance multiplier for pulling

Distance ft (m)	Male Initial	Female Initial	Male Sustained	Female Sustained
3.3 (1)	1.000	1.000	1.000	1.000
16.4 (5)	0.930	0.950	0.831	0.972
32.8 (10)	0.878	0.856	0.743	0.877
49.2 (15)	0.845	0.752	0.697	0.750
65.6 (20)	0.785	0.739	0.631	0.696
82.0 (25)	0.717	0.726	0.562	0.655
98.4 (30)	0.657	0.713	0.514	0.625
114.8 (35)	0.614	0.700	0.490	0.604
131.2 (40)	0.577	0.687	0.466	0.587
147.6 (45)	0.547	0.674	0.442	0.565
164.0 (50)	0.524	0.657	0.418	0.532
180.4 (55)	0.505	0.631	0.394	0.492
196.9 (60)	0.491	0.600	0.370	0.446
213.3 (65)	0.485	0.568	0.347	0.393

BW = multiplier for body weight (see Table 12-7)
TD = multiplier for task duration (hours) (see Table 12-8)

Table 12-22. Frequency multiplier for pulling

Frequency Times/min	Male	Female	Male	Female
	Initial		Sustained	
0.002	1.000	1.000	1.000	1.000
0.016	0.898	0.958	0.909	0.864
0.030	0.851	0.938	0.865	0.800
0.100	0.842	0.924	0.852	0.783
0.200	0.830	0.906	0.838	0.760
0.500	0.787	0.813	0.730	0.680
1.000	0.766	0.781	0.703	0.640
4.000	0.700	0.738	0.598	0.620
6.000	0.663	0.696	0.539	0.568

Carrying

The structure of the carrying model (Shoaf et al. 1997) is:

$$CC = W_B \times V \times T \times F \times AG \times BW \times TD$$

where:

CC = carrying capacity (kg)
W_B = maximum weight acceptable to a specified percentage of the worker population (kg) (see Table 12-23)
V = multiplier for the vertical distance from floor to hands (cm) (see Table 12-24)
T = multiplier for the traveled distance (m) (see Table 12-25)
F = multiplier for the frequency of carry (times/minute) (see Table 12-26)
AG = multiplier for age group (see Table 12-6)
BW = multiplier for body weight (see Table 12-7)
TD = multiplier for task duration (hours) (see Table 12-8)

EXAMPLES

EXAMPLE 1

What is the maximum amount of weight an average (50th percentile) female worker, age 30, weighing 120 lb (54 kg) could carry at a height of 31.5 in. (80 cm) for 32.8 ft (10 m), two times per hour for one hour twice per day?

Look up the multipliers required for the carrying capacity equation found in the carrying model. Using Table 12-23, Base weights for carrying, for a 50th percentile worker, W_B = 35.9 lb (16.3 kg); using Table 12-24, Vertical height multiplier for carrying, for a ver-

Manual Material Handling: Design Data Bases

Table 12-23. Base weights for carrying

Population %	Male	Female
	lb (kg)	
5	81.1 (36.8)	47.2 (21.4)
10	78.0 (35.4)	45.9 (20.8)
15	73.9 (33.5)	44.5 (20.2)
20	71.9 (32.6)	43.4 (19.7)
25	68.6 (31.1)	42.1 (19.1)
30	65.5 (29.7)	40.8 (18.5)
35	62.4 (28.3)	39.7 (18.0)
40	59.3 (26.9)	38.4 (17.4)
45	56.2 (25.5)	37.0 (16.8)
50	52.9 (24.0)	35.9 (16.3)
55	49.8 (22.6)	34.6 (15.7)
60	46.7 (21.2)	33.3 (15.1)
65	43.7 (19.8)	32.2 (14.6)
70	40.3 (18.3)	30.9 (14.0)
75	37.3 (16.9)	29.5 (13.4)
80	34.2 (15.5)	28.4 (12.9)
85	31.1 (14.1)	27.1 (12.3)
90	27.8 (12.6)	25.8 (11.7)
95	24.7 (11.2)	24.7 (11.2)

Table 12-24. Vertical height multiplier for carrying

Height in. (cm)	Male	Female
29.5 (75)	1.000	1.000
31.5 (80)	0.976	0.978
33.5 (85)	0.953	0.957
35.4 (90)	0.929	0.935
37.4 (95)	0.905	0.913
39.4 (100)	0.882	0.892
41.3 (105)	0.858	0.870
43.3 (110)	0.834	0.848

Table 12-25. Travel distance multiplier for carrying

Distance ft (m)	Male	Female
6.6 (2)	1.000	1.000
13.1 (4)	0.871	0.910
19.7 (6)	0.811	0.860
26.2 (8)	0.819	0.849
32.8 (10)	0.894	0.878

Table 12-26. Frequency multiplier for carrying

Frequency Times/min	Male	Female
0.002	1.000	1.000
0.016	0.904	0.842
0.030	0.854	0.769
0.100	0.796	0.768
0.200	0.750	0.769
0.500	0.688	0.731
1.000	0.667	0.731
4.000	0.521	0.577
6.000	0.417	0.300

tical height of 31.5 in. (80 cm), $V = 0.978$; using Table 12-25, Travel distance multiplier for carrying, for a traveled distance of 32.8 ft (10 m), $T = 0.878$; using Table 12-26, Frequency multiplier for carrying, for a frequency of two times per hour (.033 times per minute) $F = 0.769$; using Table 12-6, Age multiplier, for a 30-year-old female worker, $AG = 0.90$; using Table 12-7, Body weight multiplier, for a 120-lb (54.5-kg) female worker, $BW = 1.00$; and using Table 12-8, Task duration multiplier, for a task duration of one hour, $TD = 1.00$.

Therefore, carrying capacity (CC) is:

$$CC = W_B \times V \times T \times F \times AG \times BW \times TD$$

$CC = 16.3 \text{ kg} \times 0.978 \times 0.878 \times 0.769 \times 0.90 \times 1.00 \times 1.00$
$CC = 9.7 \text{ kg}$

The above specified worker could be expected to carry 21 lb (9.7 kg) for the described task.

EXAMPLE 2

A worker is required to lift 51-lb (23-kg) bags of an asphalt mixture. It is July and very hot. The worker is male, 25 years old, weighs 198.4 lb (90 kg) and is in excellent physical condition (5th percentile). Horizontal distance for this task is equal to 19.7 in. (50 cm); vertical travel distance = 35.4 in. (90 cm), and vertical distance = 29.5 in. (75 cm). Twisting angle is 10°. Is it reasonable to expect this worker to be able to perform this task 10 times per hour for a two-hour time period?

Look up the multipliers required for the lifting capacity equation found in the lifting model. Using Table 12-1, Base weights for lifting and lowering, for a 5th percentile worker, W_B = 39.1 kg; using Table 12-2, Horizontal distance multiplier for lifting, for a distance of 19.7 in. (50 cm), H = 0.68; using Table 12-3, Vertical travel distance multiplier for lifting, for a distance of 35.4 in. (90 cm), V = 0.76; using Table 12-4, Vertical distance multiplier for lifting, for a distance of 29.5 in. (75 cm), D = 1.00; using Table 12-5, Frequency multiplier for lifting, for a frequency of 10 times per hour (0.17 times per minute), F = 1.00; referring back to Table 12-6, Age multiplier, for a 25 year old male worker, AG = 0.91; referring back to Table 12-7, Body weight multiplier, for a 198.4-lb (90-kg) male worker, BW = 1.45; referring back to Table 12-8, Task duration multiplier, for a task duration of two hours, TD = 0.77; using Table 12-9, Twisting multiplier for lifting, for an angle of 10°, T = 0.98; using Table 12-10, Coupling multiplier for lifting, for a limited hold, C = 0.925; and using Table 12-11, Heat stress multiplier for lifting, for a hot July day assume 95° F (35° C) HS = 0.81.

Therefore, lifting capacity *(LC)* is:

$LC = W_B \times H \times V \times D \times F \times AG \times BW \times TD \times T \times C \times HS$

$LC = 39.1 \text{ kg} \times 0.68 \times 0.76 \times 1.00 \times 1.00 \times 0.91 \times 1.45 \times 0.77$
$\quad \times 0.98 \times 0.925 \times 0.81$
$LC = 15 \text{ kg}$

Therefore, because the predicted lifting capacity is 33 lb (15 kg) for the described task, it is not reasonable for this worker to lift 51 lb (23 kg). The task should be redesigned to meet the worker's capacity.

REDUCING EXPOSURE TO MANUAL MATERIAL HANDLING HAZARDS

The most effective means of protecting workers from manual material handling injury is to redesign job tasks as well as the work environment so that exposure to manual handling hazards is minimized. Worker education can then be utilized to maintain a less hazardous environment. Figure 12-1 provides a summary of recommended methods (Grandjean 1988; NIOSH 1994) for reducing the exposure to manual material handling hazards.

As manual material handling activities are a major source of overexertion injuries in the United States (NIOSH 1981), the minimization and prevention of these injuries represent a significant challenge for practitioners of occupational safety and health.

Task redesign	• Avoid twisting or rotating movements • Use gravity to move load when possible • Use mechanical aids, such as handles, trolleys, hoists, or slides • Reduce load weight to lowest possible level • Handle load close to body, between knuckle and shoulder height
Work environment	• Provide frequent activity breaks • Consider job rotation
Worker education	• Train workers to identify and report hazards • Train workers in proper handling technique

Figure 12-1 Summary of methods to reduce exposure to manual material handling hazards.

REFERENCES

Bureau of Labor Statistics. *Back Injuries Associated with Lifting.* Bulletin #2144. Washington: Bureau of Labor Statistics, 1982.

Centers for Disease Control. "Musculoskeletal Injuries." *Morbidity and Mortality Weekly Report* 1983, 32(14).

Drury, C.G., and Pfeil, R.E. "A Task-based Model of Manual Lifting Performance." *International Journal of Production Research* 1975, 13: 137-148.

Grandjean, E. *Fitting the Task to the Man.* London: Taylor & Francis, 1988.

Hidalgo, J., Genaidy, A.M., Karwowski, W., Christensen, D.M., Huston, R., and Stambough, J. "A Comprehensive Lifting Model Beyond the NIOSH Lifting Equation." *Ergonomics* 1997, 40: 916-927.

National Institute for Occupational Safety and Health (NIOSH). *Work Practices Guide for Manual Lifting,* DHHS (NIOSH) Publication No. 81-122. Cincinnati, OH: NIOSH, 1981.

NIOSH. *Back Belts—Do They Prevent Injury?* DHHS (NIOSH) Publication No. 94-127. Cincinnati, OH: NIOSH, 1994.

Shoaf, C., Genaidy, A., Karwowski, W., Waters, T., and Christensen, D. "Comprehensive Manual Handling Limits for Lowering, Pushing, Pulling, and Carrying." *Ergonomics* 1997, 40: 1,183-1,200.

Snook, S.H. "The Design of Manual Materials Handling." *Ergonomics* 1978, 21: 963-985.

Snook, S.H. "The Costs of Back Pain in Industry." *Occupational Medicine: State of the Art Reviews* 1990, 3(1): 1-5.

Snook, S., and Ciriello, V.M. "The Design of Manual Handling Tasks: Revised Tables of Maximum Acceptable Weights and Forces." *Ergonomics* 1991, 34: 1,197-1,213.

Tichauer, E. *The Biomechanical Basis of Ergonomics.* New York: John Wiley & Sons, 1978.

Tichauer, E. "The Industrial Environment—Its Evaluation and Control." U.S. Department of Health and Human Services. Cincinnati, OH: NIOSH, 1973.

Troup, J.D.G, and Edwards, F.C. *Manual Handling and Lifting.* London: Her Majesty's Stationary Office, 1985.

Waters, T.R., Putz-Anderson, V., Garg, A., and Fine, L.J. "Revised NIOSH Equation for the Design and Evaluation of Manual Lifting Tasks." *Ergonomics* 1993, 36: 749-776.

CHAPTER 13

ASSESSMENT OF MANUAL LIFTING—THE NIOSH APPROACH

Thomas R. Waters
Research Physiologist

Vern Putz-Anderson
Ergonomist

Applied Psychology and Ergonomics Branch
National Institute for Occupational Safety and Health
Cincinnati, Ohio

THE BODY OF WORK

Historically, the National Institute for Occupational Safety and Health (NIOSH) has recognized and addressed the problem of work-related back injuries and published the *Work Practices Guide for Manual Lifting* (*WPG*) in 1981 (NIOSH 1981). The *WPG* contains a summary of the lifting-related literature up to 1981, analytical procedures, a lifting equation for calculating a recommended weight for specified two-handed, symmetrical lifting tasks, and an approach for controlling the hazards of low-back injury from manual lifting. The approach to hazard control was coupled with the *action limit* (AL), a term that denoted the recommended weight derived from the lifting equation.

In 1985, NIOSH convened an ad hoc committee of experts who reviewed the current literature on lifting, including the NIOSH *WPG*. The literature review was summarized in a document containing updated information on the physiological, biomechanical, psychophysical, and epidemiological aspects of manual lifting (NIOSH 1991). Based on the results of the literature review, the ad hoc committee recommended criteria for defining the lifting capacity of healthy workers. The committee used the criteria to formulate the revised lifting equation (Waters et al. 1994). Subsequently, NIOSH staff developed documentation for the equation and played a prominent role in recommending methods for interpreting the results of the lifting equation. The revised lifting equation reflects new findings and provides methods for evaluating asymmetrical lifting tasks and lifts of objects with less than optimal couplings between the object and the worker's hands. The revised lifting equation also provides guidelines for a more diverse range of lifting tasks than the earlier equation (NIOSH 1981).

The rationale and criteria for developing the revised NIOSH lifting equation are provided in a journal article by Waters et al. (1993). The article provides a better understanding of the data and decisions made in formulating the revised equation. It offers an explanation on selecting the biomechanical, physiological, and psychophysical criteria, and describes the derivation of each component of the revised lifting equation. However, for those individuals primarily concerned with applying the revised lifting equation, this chapter provides a more complete description of the method and limitations.

Although the revised lifting equation has not been fully validated, the recommended weight limits derived from the revised equation are consistent with, or lower than, those generally reported in the literature. Moreover, the proper application of the revised equation is more likely to protect healthy workers for a wider variety of lifting tasks than methods that rely solely on a single task factor or criterion.

Finally, it should be stressed that the NIOSH lifting equation is only one tool in a comprehensive effort to prevent work-related low-back pain and disability. Examples of other approaches are described elsewhere (ASPH/NIOSH 1986). Moreover, lifting is only one cause of work-related low-back pain and disability. Other causes that have been hypothesized or established as risk factors include whole body vibration, static postures, prolonged sitting, and direct trauma to the back. Psychosocial factors, appropriate medical treatment, and job demands also may be particularly important in influencing the transition of acute low-back pain to chronic disabling pain (see Chapter 12, "Manual Material Handling").

RECOMMENDED WEIGHT LIMIT (RWL)

The principal product of the revised NIOSH lifting equation is the *recommended weight limit* (*RWL*). The *RWL* is defined for a specific set of task conditions as the weight of the load that nearly all healthy workers could perform over a substantial period of time (up to eight hours) without an increased risk of developing lifting-related low-back pain (LBP). "Healthy workers" are those who are free of adverse health conditions that would increase their risk of musculoskeletal injury.

The concept behind the revised NIOSH lifting equation is to start with a recommended weight that is considered safe for an "ideal" lift (load constant equal to 51 lb [23 kg]) and then reduce the weight as the task becomes more stressful (the task-related factors become less favorable). The precise formulation of the revised lifting equation for calculating the *RWL* is based on a multiplicative model that provides a weighting (multiplier) for each of six task variables:

1. Horizontal distance of the load from the worker (*H*) (Table 13-1).
2. Vertical height of the lift (*V*) (Table 13-2).
3. Vertical displacement during the lift (*D*) (Table 13-3).
4. Angle of asymmetry (*A*) (Table 13-4).
5. Frequency (*F*) and duration of lifting (Table 13-5).
6. Quality of the hand-to-object coupling (*C*) (Tables 13-6 and 13-7).

Table 13-1. Horizontal multiplier

Horizontal distance in.	Horizontal multiplier	Horizontal distance cm	Horizontal multiplier
≤10	1.00	≤25	1.00
11	0.91	28	0.89
12	0.83	30	0.83
13	0.77	32	0.78
14	0.71	34	0.74
15	0.67	36	0.69
16	0.63	38	0.66
17	0.59	40	0.63
18	0.56	42	0.60
19	0.53	44	0.57
20	0.50	46	0.54
21	0.48	48	0.52
22	0.46	50	0.50
23	0.44	52	0.48
24	0.42	54	0.46
25	0.40	56	0.45
>25	0.00	58	0.43
		60	0.42
		63	0.40
		>63	0.00

Table 13-2. Vertical multiplier

Vertical distance in.	Vertical multiplier	Vertical distance cm	Vertical multiplier
0	0.78	0	0.78
5	0.81	10	0.81
10	0.85	20	0.84
15	0.89	30	0.87
20	0.93	40	0.90
25	0.96	50	0.93
30	1.00	60	0.96
35	0.96	70	0.99
40	0.93	80	0.99
45	0.89	90	0.96
50	0.85	100	0.93
55	0.81	110	0.90
60	0.78	120	0.87
65	0.74	130	0.84
70	0.70	140	0.81
>70	0.00	150	0.78
		160	0.75
		170	0.72
		175	0.70
		>175	0.00

Table 13-3. Distance multiplier

Distance in.	Distance multiplier	Distance cm	Distance multiplier
≤10	1.00	≤25	1.00
15	0.94	40	0.93
20	0.91	55	0.90
25	0.89	70	0.88
30	0.88	85	0.87
35	0.87	100	0.87
40	0.87	115	0.86
45	0.86	130	0.86
50	0.86	145	0.85
55	0.85	160	0.85
60	0.85	175	0.85
70	0.85	>175	0.00
>70	0.00		

Table 13-4. Asymmetric multiplier

Angle of asymmetry A (°)	Asymmetric multiplier
0	1.000
15	0.95
30	0.90
45	0.86
60	0.81
75	0.76
90	0.71
105	0.66
120	0.62
135	0.57
>135	0.00

The weightings are expressed as coefficients that serve to decrease the load constant, which represents the maximum recommended load weight to be lifted under ideal conditions. For example, as the horizontal distance between the load and the worker increases from 10 in. (25 cm), the recommended weight limit for that task would be reduced from the ideal starting weight.

Table 13-5. Frequency multiplier (FM)

Frequency (F)* lifts/min	Work duration					
	≤1 hr		>1 but ≤2 hr		>2 but ≤8 hr	
	V** < 30	V ≥ 30	V < 30	V ≥ 30	V < 30	V ≥ 30
≤0.2	1.00	1.00	0.95	0.95	0.85	0.85
0.5	0.97	0.97	0.92	0.92	0.81	0.81
1.0	0.94	0.94	0.88	0.88	0.75	0.75
2.0	0.91	0.91	0.84	0.84	0.65	0.65
3.0	0.88	0.88	0.79	0.79	0.55	0.55
4.0	0.84	0.84	0.72	0.72	0.45	0.45
5.0	0.80	0.80	0.60	0.60	0.35	0.35
6.0	0.75	0.75	0.50	0.50	0.27	0.27
7.0	0.70	0.70	0.42	0.42	0.22	0.22
8.0	0.60	0.60	0.35	0.35	0.18	0.18
9.0	0.52	0.52	0.30	0.30	0.00	0.15
10.0	0.45	0.45	0.26	0.26	0.00	0.13
11.0	0.41	0.41	0.00	0.23	0.00	0.00
12.0	0.37	0.37	0.00	0.21	0.00	0.00
13.0	0.00	0.34	0.00	0.00	0.00	0.00
14.0	0.00	0.31	0.00	0.00	0.00	0.00
15.0	0.00	0.28	0.00	0.00	0.00	0.00
>15.0	0.00	0.00	0.00	0.00	0.00	0.00

* For lifting < once per 5 min, set F = 0.2 lifts/min
** Values of V are in inches

Table 13-6. Hand-to-container coupling classification

Good	Fair	Poor
1. For containers of optimal design, such as some boxes, crates, etc., a "good" hand-to-object coupling would be defined as handles or handhold cutouts of optimal design (see notes 1 to 3 below).	1. For containers of optimal design, a "fair" hand-to-object coupling would be defined as handles or handhold cutouts of less than optimal design (see notes 1 to 4 below).	1. Containers of less than optimal design or loose parts or irregular objects that are bulky, hard to handle, or have sharp edges (see note 5 below).
2. For loose parts or irregular objects, which are not usually containerized, such as castings, stock, and supply materials, a "good" hand-to-object coupling would be defined as a comfortable grip in which the hand can be easily wrapped around the object (see note 6 below).	2. For containers of optimal design with no handles or handhold cutouts or for loose parts or irregular objects, a "fair" hand-to-object coupling is defined as a grip in which the hand can be flexed about 90° (see note 4 below).	2. Lifting nonrigid bags (i.e., bags that sag in the middle).

- An optimal handle design has 0.75 to 1.5 in. (1.9 to 3.8 cm) diameter, $\geq$ 4.5 in. (11.4 cm) length, 2 in. (5 cm) clearance, cylindrical shape, and a smooth, nonslip surface.
- An optimal handhold cutout has these approximate characteristics: $\geq$ 1.5 in. (3.8 cm) height, 4.5 in. (11.4 cm) length, semi-oval shape, $\geq$ 2 in. (5 cm) clearance, smooth nonslip surface, and $\geq$ 0.25 in. (0.64 cm) container thickness (e.g., double thickness cardboard).
- An optimal container design has $\leq$ 16 in. (41 cm) frontal length, $\leq$ 12 in. (30 cm) height, and a smooth nonslip surface.
- A worker should be capable of clamping the fingers at nearly 90° under the container, such as required when lifting a cardboard box from the floor.
- A container is considered less than optimal if it has a frontal length > 16 in. (40 cm), height > 12 in. (30 cm), rough or slippery surfaces, sharp edges, asymmetric center of mass, unstable contents, or requires the use of gloves.
- A worker should be able to comfortably wrap the hand around the object without causing excessive wrist deviations or awkward postures, and the grip should not require excessive force.

Table 13-7. Coupling multiplier

Coupling type	Coupling multiplier	
	V < 29.5 in. (75 cm)	V ≥ 29.5 in. (75 cm)
Good	1.00	1.00
Fair	0.95	1.00
Poor	0.90	0.90

The RWL is defined as

$$RWL = LC \times HM \times VM \times DM \times AM \times FM \times CM$$

where:

		U.S. Customary	Metric
LC	= load constant	= 51 lb	23 kg
HM	= horizontal multiplier	= (10/H)	(25/H)
VM	= vertical multiplier	= $1 - (.0075\|V-30\|)$	$1 - (.003\|V-75\|)$
DM	= distance multiplier	= $0.82 + (1.8/D)$	$0.82 + (4.5/D)$
AM	= asymmetric multiplier	= $1 - (0.0032A)$	$1 - (0.0032A)$
FM	= frequency multiplier	= from Table 13-5	from Table 13-5
CM	= coupling multiplier	= from Table 13-7	from Table 13-7

The term *task variables* refers to the task-related measurements used as input data for the formula (H, V, D, A, F, and C), whereas, the term *multipliers* refers to the reduction coefficients in the equation (HM, VM, DM, AM, FM, and CM).

MEASUREMENT REQUIREMENTS

The following list briefly describes the measurements required to use the revised NIOSH lifting equation. Details for each variable are presented later in this chapter.

H = Horizontal location of the hands from the midpoint between the inner ankle bones. Measure at the origin and destination of the lift (in. or cm).

V = Vertical location of the hands from the floor. Measure at the origin and destination of the lift (in. or cm).

D = Vertical travel distance between the origin and the destination of the lift (in. or cm).

A = Angle of asymmetry or angular displacement of the load from the worker's sagittal plane. Measure at the origin and destination of the lift (°).

F = Average frequency rate of lifting, measured in lifts per minute. Duration is defined to be less than 1 hour; 1-2 hours; or 2-8 hours, assuming appropriate recovery allowances (see Table 13-5).

C = Quality of hand-to-object coupling (interface between the worker and the load lifted). The quality of the coupling is classified as good, fair, or poor depending upon the type and location of the coupling, the physical characteristics of the load, and the vertical height of the lift.

LIFTING INDEX (LI)

The *lifting index* (*LI*) offers a relative estimate of the level of physical stress associated with a particular manual-lifting task. The estimate of the level of physical stress is defined by the relationship between the weight of the load lifted and the recommended weight limit. *LI* is defined by the equation

$$LI = \frac{L}{RWL}$$

where:

LI = lifting index
L = load weight of the object lifted (lb or kg)
RWL = recommended weight limit

MISCELLANEOUS TERMS

Lifting task. The act of manually grasping an object of definable size and mass with two hands and vertically moving the object without mechanical assistance.

Load weight (*L*). Weight of the object to be lifted, in pounds or kilograms, including the container.

Horizontal location (*H*). Distance of the hands away from the midpoint between the ankles, in inches or centimeters (measure at the origin and destination of lift, see Figure 13-1).

Vertical location (*V*). Distance of the hands above the floor, in inches or centimeters (measure at the origin and destination of lift, see Figure 13-1).

Vertical travel distance (*D*). Absolute value of the difference between the vertical heights at the destination and origin of the lift, in inches or centimeters.

Angle of asymmetry (*A*). Angular measure of how far the *object* is displaced from the front (midsagittal plane) of the worker's body at the beginning or end of the lift, in degrees (measure at the origin and destination of lift, see Figure 13-2). The asymmetry angle is defined by the location of the load relative to the worker's midsagittal plane, as defined by the neutral body posture, rather than the position of the feet or extent of body twist.

Assessment of Manual Lifting—The NIOSH Approach

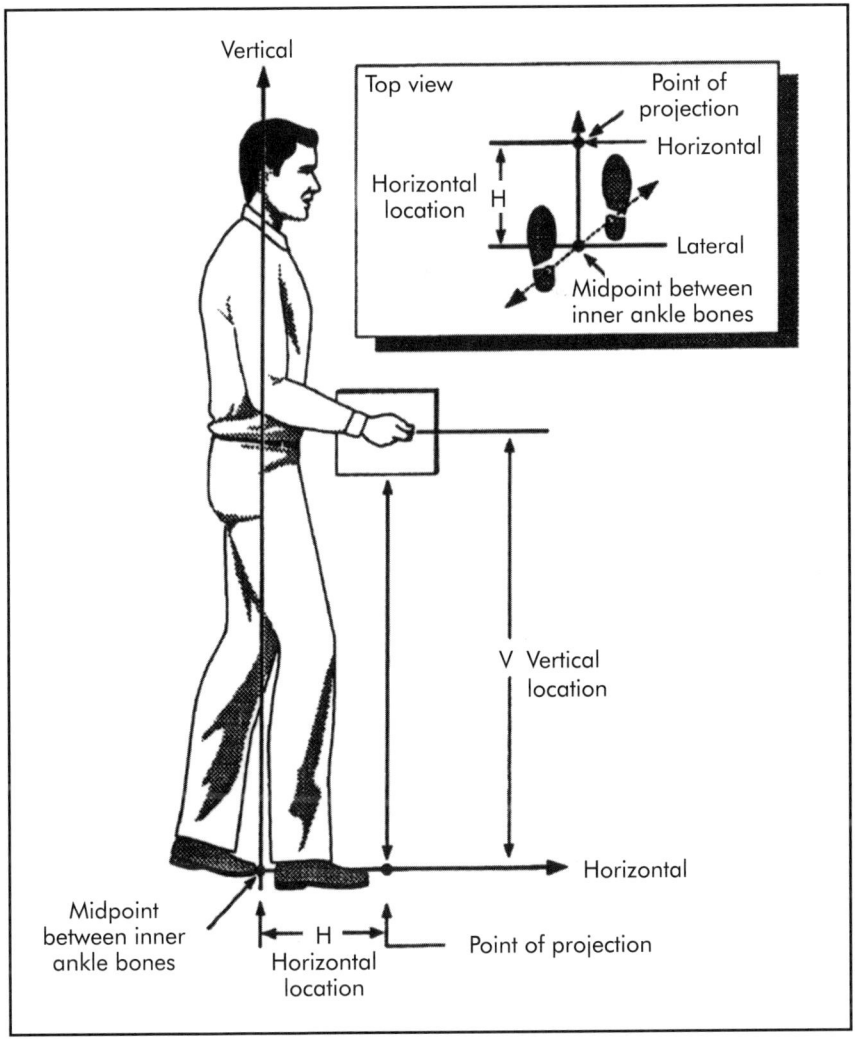

Figure 13-1 Graphic representation of hand location.

Neutral body position. Position of the body where the hands are directly in front of it and there is minimal twisting of the legs, torso, or shoulders.

Frequency of lifting (F). Average number of lifts per minute over a 15-minute period.

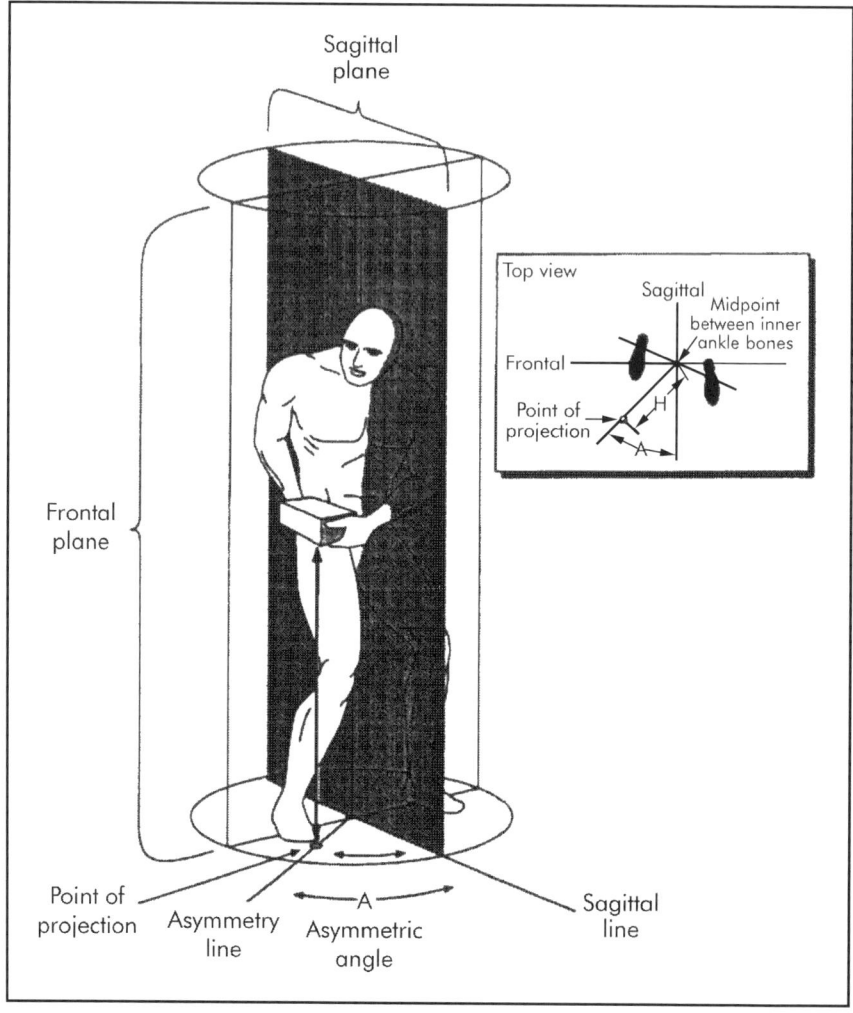

Figure 13-2 Graphic representation of angle of asymmetry (A).

Duration of lifting. Three-tiered classification of lifting duration specified by the distribution of work time and recovery time (work pattern). Duration is classified as either short (1 hour), moderate (1-2 hours), or long (2-8 hours), depending on the work pattern.

Coupling classification. Classification of the quality of the hand-to-object coupling (handle, cut-out, or grip). Coupling quality is classified as good, fair, or poor.

Significant control. A condition requiring precision placement of the load at the destination of the lift. This is usually the case when the worker has to re-grasp the load near the destination of the lift, the worker has to momentarily hold the object at the destination, or when the worker has to carefully position or guide the load at the destination.

EQUATION LIMITATIONS

The lifting equation is a tool for assessing the physical stress of two-handed manual lifting tasks. As with any tool, its application is limited to those conditions for which it was designed. The lifting equation was designed to meet specific lifting-related criteria that encompass biomechanical, physiological, and psychophysical assumptions and data used to develop the equation. To the extent that a given lifting task accurately reflects these underlying conditions and criteria, this lifting equation may be appropriately applied.

The following list identifies a set of work conditions where the application of the lifting equation could either underestimate or overestimate the extent of physical stress associated with a particular work-related activity. Each of the following task limitations also highlights topics in need of further research to extend the application of the lifting equation to a greater range of real-world lifting tasks.

The revised NIOSH lifting equation does not apply if the operator performs lifting or lowering:

- With one hand;
- For over 8 hours;
- While seated or kneeling;
- In a restricted work space;
- Of objects considered unstable;
- While carrying, pushing, or pulling;
- With wheelbarrows or shovels;
- With high-speed motion (faster than 30 in./second [76 cm/second]);
- With unreasonable foot/floor coupling (< 0.4 coefficient of friction between the sole and the floor); or
- In an unfavorable environment (temperature significantly outside the 66-79° F [19-26° C] range; relative humidity outside the 35-50% range).

HORIZONTAL COMPONENT

Definition and Measurement

Horizontal location (H) is measured from the midpoint of the line joining the inner ankle bones to a point projected on the floor directly below the midpoint of the hand grasps (load center), as defined by the large middle knuckle of the hand (Figure 13-1). Typically, the worker's feet are not aligned with the midsagittal plane, as shown in Figure 13-1, but may be rotated inward or outward. If this is the case, then the midsagittal plane is defined by the worker's neutral body posture as defined above. If significant control is required at the destination (precision placement), then H should be measured at both the origin and destination of the lift.

Horizontal distance (H) should be measured. In those situations where the H value cannot be measured, then H may be approximated from the following equations:

<u>U.S. Customary (all distances in inches)</u>

$$H = 8 + W/2 \text{ for } V \geq 10 \text{ in.}$$
$$H = 10 + W/2 \text{ for } V < 10 \text{ in.}$$

<u>Metric (all distances in centimeters)</u>

$$H = 20 + W/2 \text{ for } V \geq 25 \text{ cm}$$
$$H = 25 + W/2 \text{ for } V < 25 \text{ cm}$$

where:

H = horizontal distance
W = width of the container in the sagittal plane
V = vertical location of the hands from the floor

Horizontal Restrictions

If the horizontal distance is less than 10 in. (25 cm), then H is set to 10 in. (25 cm). Although objects can be carried or held closer than 10 in. (25 cm) from the ankles, most objects that are closer cannot be lifted without encountering interference from the abdomen or hyperextending the shoulders. Although 25 in. (64 cm) was chosen as the maximum value for H, it is probably too great a distance for shorter workers, particularly when lifting asymmetrically. Furthermore, objects at a distance of more than 25 in. (64 cm) from the ankles normally cannot be lifted vertically without some loss of balance.

Horizontal Multiplier

The horizontal multiplier (*HM*) is 10/H for H measured in inches and 25/H for H measured in centimeters. If H is less than or equal to 10 in. (25 cm), the multiplier is 1.0. *HM* decreases with an increase in H value. The multiplier for H is reduced to 0.4 when H is 25 in. (64 cm). If H is greater than 25 in. (64 cm), then $HM = 0$. The *HM* value can be computed directly or determined from Table 13-1.

VERTICAL COMPONENT

Definition and Measurement

Vertical location (*V*) is defined as the vertical height of the hands above the floor. *V* is measured vertically from the floor to the midpoint between the hand grasps, as defined by the large middle knuckle. The coordinate system was illustrated in Figure 13-1.

Vertical Restrictions

Vertical location (*V*) is limited by the floor surface and upper limit of vertical reach for lifting (70 in. [178 cm]). The vertical location should be measured at the origin and destination of the lift.

Vertical Multiplier

To determine the vertical multiplier (*VM*), the absolute value or deviation of *V* from an optimum height of 30 in. (76 cm) is calculated. A height of 30 in. (76 cm) above floor level is considered "knuckle height" for a worker of average height (66 in. [168 cm]). The vertical multiplier (*VM*) is $(1 - (0.0075|V-30|))$ for *V* measured in inches, and *VM* is $(1 - (0.003|V - 75|))$, for *V* measured in centimeters.

When *V* is at 30 in. (76 cm), the vertical multiplier (*VM*) is 1.0. The value of *VM* decreases linearly with an increase or decrease in height from this position. At floor level, $VM = 0.78$, and at 70 in. (178 cm) height, $VM = 0.7$. If *V* is greater than 70 in. (178 cm), then $VM = 0$. The *VM* value can be computed directly or determined from Table 13-2.

DISTANCE COMPONENT

Definition and Measurement

The distance variable (*D*) is defined as the vertical travel distance of the hands between the origin and destination of the lift. For lifting, *D* can be computed by subtracting the vertical location (*V*) at the origin of the lift from the corresponding *V* at the destination of the lift (*D* is equal to *V* at the destination minus *V* at the origin). For a lowering task, *D* is equal to *V* at the origin minus *V* at the destination.

Distance Restrictions

The distance variable (D) is assumed to be at least 10 in. (25 cm) and no greater than 70 in. (178 cm). If the vertical travel distance is less than 10 in. (25 cm), then D should be set to the minimum distance of 10 in. (25 cm).

Distance Multiplier

The distance multiplier (DM) is $(0.82 + (1.8/D))$ for D measured in inches and $(0.82 + (4.5/D))$ for D measured in centimeters. For D less than 10 in. (25 cm), D is assumed to be 10 in. (25 cm), and DM is 1.0. The distance multiplier, therefore, decreases gradually with an increase in travel distance. The $DM = 1.0$ when D is set at 10 in. (25 cm); $DM = 0.85$ when D is 70 in. (178 cm). Thus, DM ranges from 1.0-0.85 as the D varies from 0-70 in. (0-178 cm). The DM value can be computed directly or determined from Table 13-3.

ASYMMETRY COMPONENT

Definition and Measurement

Asymmetry refers to a lift that begins or ends outside the mid-sagittal plane (see Figure 13-2). In general, asymmetric lifting should be avoided. If it cannot be avoided, however, the recommended weight limits are significantly less than those used for symmetrical lifting.*

An asymmetric lift may be required under the following task or workplace conditions:

- The origin and destination of the lift are oriented at an angle to each other;
- The lifting motion is across the body, such as in swinging bags or boxes from one location to another;
- The lifting is done to maintain body balance in obstructed workplaces, on rough terrain, or on littered floors; and
- Productivity standards require reduced time per lift.

The asymmetric angle (A), depicted in Figure 13-2, is operationally defined as the angle between the asymmetry line and the mid-

---

It may not always be clear if asymmetry is an intrinsic element of the task or just a personal characteristic of the worker's lifting style. Regardless of the reason for the asymmetry, any observed asymmetric lifting should be considered an intrinsic element of the job design and should be considered in the assessment and subsequent redesign. Moreover, the design of the task should not rely on worker compliance, but rather the design should discourage or eliminate the need for asymmetric lifting.

sagittal line. The *asymmetry line* is defined as the line that joins the midpoint between the inner ankle bones and the point projected on the floor directly below the midpoint of the hand grasps, as defined by the large middle knuckle. The *sagittal line* is defined as the line passing through the midpoint between the inner ankle bones and lying in the midsagittal plane, as defined by the neutral body position (hands directly in front of the body, with no twisting at the legs, torso, or shoulders). Note: the asymmetry angle is not defined by foot position or the angle of torso twist, but by the location of the load relative to the worker's midsagittal plane.

In many cases of asymmetric lifting, the worker will pivot or use a step turn to complete the lift. Because this may vary significantly between workers and between lifts, we have assumed that no pivoting or stepping occurs. Although this assumption may overestimate the reduction in acceptable load weight, it will provide the greatest protection for the worker.

The asymmetry angle (A) must always be measured at the origin of the lift. If significant control is required at the destination, however, then angle A should be measured at both the origin and the destination of the lift.

Asymmetry Restrictions

Angle A is limited to a range of 0-135°. If $A > 135°$, then AM is set to equal zero, which results in a RWL of zero, or no load.

Asymmetric Multiplier

The asymmetric multiplier (AM) is $1 - (0.0032A)$. AM has a maximum value of 1.0 when the load is lifted directly in front of the body and decreases linearly as the angle of asymmetry (A) increases. The range is from a value of 0.57 at 135° of asymmetry to a value of 1.0 at 0° of asymmetry (symmetric lift). If A is greater than 135°, then $AM = 0$, and the $RWL = 0.0$. The AM value can be computed directly or determined from Table 13-4.

FREQUENCY COMPONENT

Definition and Measurement

The frequency multiplier is defined by 1) number of lifts per minute (frequency), 2) amount of time engaged in the lifting activity (duration), and 3) vertical height of the lift from the floor. Lifting frequency (F) refers to the average number of lifts made per minute, as measured over a 15-minute period. Because of the potential variation in work patterns, analysts may have difficulty obtaining an accurate or representative 15-minute work sample for computing F. If significant variation exists in the frequency of lifting over the course of the day, analysts should employ standard work sampling techniques to obtain a representative work sample for determining the number of lifts per minute. For those jobs where the frequency varies from session to session, each session should be analyzed separately, but the overall work pattern must still be considered. For more information, most standard industrial engineering or ergonomics texts provide guidance for establishing a representative job sampling strategy (Eastman Kodak 1986).

Lifting Duration

Lifting duration is classified into three categories based on the pattern of continuous work time and recovery time (light work) periods. A continuous *work time* (WT) period is defined as a period of uninterrupted work. *Recovery time* (RT) is defined as the duration of light work activity following a period of continuous lifting. Examples of light work include activities such as sitting at a desk or table, monitoring operations, and light assembly work. The three categories are short duration, moderate duration, and long duration.

Short duration. Short duration lifting tasks are those that have a work duration of one hour or less, followed by a recovery time equal to 1.2 times the work time (at least a 1.2 recovery-time to work-time ratio [RT/WT]). For example, to be classified as short-duration, a 45-minute lifting job must be followed by at least a 54-minute recovery period prior to initiating a subsequent lifting session. If the required recovery time is not met for a job of one hour or less and a subsequent lifting session is required, then the total lifting time must be combined to correctly determine the duration category. Moreover, if the recovery period does not meet the time requirement, it is disregarded for purposes of determining the appropriate duration category.

As another example, assume a worker lifts continuously for 30 minutes, then performs a light work task for 10 minutes, and then lifts for an additional 45-minute period. In this case, the recovery time between lifting sessions (10 minutes) is less than 1.2 times the initial 30-minute work time (36 minutes). Thus, the two work times (30 minutes and 45 minutes) must be added together to determine the duration. Since the total work time (75 minutes) exceeds one hour, the job is classified as moderate duration. On the other hand, if the recovery period between lifting sessions was increased to 36 minutes, then the short duration category would apply, resulting in a larger *FM* value.

Moderate duration. Moderate duration lifting tasks are those that have a duration of more than 1 hour, but not more than 2 hours, followed by a recovery period of at least 0.3 times the work time (at least a 0.3 recovery time to work time ratio [RT/WT]).

For example, a worker who continuously lifts for 2 hours would need at least a 36-minute recovery period before initiating a subsequent lifting session. If the recovery time requirement is not met and a subsequent lifting session is required, then the total work time must be added together. If the total work time exceeds 2 hours, then the job must be classified as a long duration lifting task.

Long duration. Long duration lifting tasks are those that have a duration of between 2-8 hours, with standard industrial rest allowances (morning, lunch, and afternoon rest breaks). Note: no weight limits are provided for more than 8 hours of work.

The difference in the required RT/WT ratio for the short (<1 hour) duration category, which is 1.2, and the moderate (1-2-hour) duration category, which is 0.3, is due to the difference in the magnitudes of the frequency multiplier values associated with each of the duration categories. Since the moderate category provides a larger reduction in the RWL than the short category, there is less need for a recovery period between sessions than for the short duration category. In other words, the short duration category would result in higher weight limits than the moderate duration category, so larger recovery periods would be needed.

SPECIAL FREQUENCY ADJUSTMENT PROCEDURE

A special procedure has been developed for determining the appropriate lifting frequency (F) for certain repetitive lifting tasks in which workers do not lift continuously during the 15-minute sampling period. This occurs when the work pattern is such that the worker lifts repetitively for a short time and then performs light work for a short time before starting another cycle. For work patterns such as this, F may be determined as follows, as long as the actual lifting frequency does not exceed 15 lifts per minute:

1. Compute the total number of lifts performed for the 15-minute period (lift rate times work time).
2. Divide the total number of lifts by 15.
3. Use the resulting value as the frequency (F) to determine the frequency multiplier (FM) from Table 13-5.

For example, if the work pattern for a job consists of a series of cyclical sessions requiring 8 minutes of lifting followed by 7 minutes of light work, and the lifting rate during the work sessions is 10 lifts per minute, then the frequency rate (F) that is used to determine the frequency multiplier for this job is equal to (10 × 8)/15, or 5.33

lifts per minute. If the worker lifted continuously for more than 15 minutes, however, then the actual lifting frequency (10 lifts per minute) would be used.

When using this special procedure, the duration category is based on the magnitude of the recovery periods *between* work sessions, not *within* work sessions. In other words, if the work pattern is intermittent and the special procedure applies, then the intermittent recovery periods that occur during the 15-minute sampling period are *not* considered as recovery periods for purposes of determining the duration category. For example, if the work pattern for a manual lifting job was composed of repetitive cycles consisting of 1 minute of continuous lifting at a rate of 10 lifts per minute, followed by 2 minutes of recovery, the correct procedure would be to adjust the frequency according to the special procedure (F = (10 lifts per minute $\times$ 5 minutes)/15 minutes = 50/15 = 3.3 lifts per minute). The 2-minute recovery periods would not count toward the RT/WT ratio, however, and additional recovery periods would have to be provided as described above.

Frequency Restrictions

Lifting frequency (F) for repetitive lifting may range from 0.2 lifts per minute to a maximum frequency that is dependent on the vertical location of the object (V) and the duration of lifting (Table 13-5). Lifting above the maximum frequency results in a RWL of 0.0 (except for the special case of discontinuous lifting previously discussed, where the maximum frequency is 15 lifts per minute).

Frequency Multiplier

The FM value depends upon the average number of lifts per minute (F), the vertical location (V) of the hands at the origin, and the duration of continuous lifting. For lifting tasks with a frequency less than 0.2 lifts per minute, set the frequency equal to 0.2 lifts per minute. Otherwise, the FM is determined from Table 13-5.

COUPLING COMPONENT

Definition and Measurement

The nature of the hand-to-object coupling or gripping method can affect not only the maximum force a worker can or must exert on an object, but also the vertical location of the hands during the lift. A "good" coupling will reduce the maximum grasp forces required and increase the acceptable weight for lifting, while a "poor" coupling will generally require higher maximum grasp forces and decrease the acceptable weight for lifting.

The effectiveness of the coupling is not static, but may vary with the distance of the object from the ground, so that a good coupling could become a poor coupling during a single lift. The entire range of the lift should be considered when classifying hand-to-object couplings, with classification based on overall effectiveness. The analyst must classify the coupling as good, fair, or poor. The three categories are defined in Table 13-6. If there is any doubt about classifying a particular coupling design, the more stressful classification should be selected.

The decision tree shown in Figure 13-3 may be helpful in classifying the hand-to-object coupling.

Limitations

There are no limitations for classifying the coupling, but both hands must be observed when assessing the coupling. If one hand is predominately used to lift the load and the fingers are flexed at 90° under the load, then the coupling should be rated as fair, regardless of the position of the other hand.

Coupling Multiplier

Based on the coupling classification and vertical location of the lift, the coupling multiplier (CM) is determined from Table 13-7.

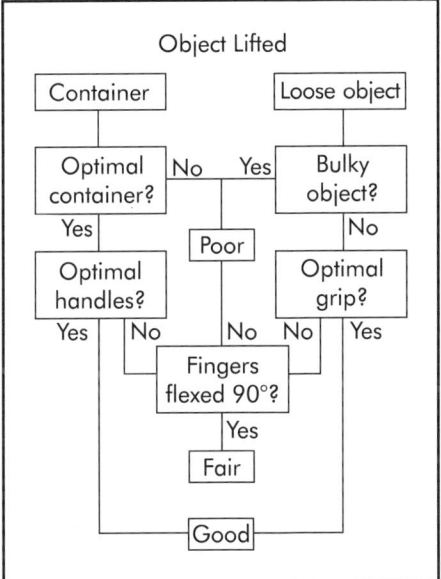

Figure 13-3 Decision tree for coupling quality.

PROCEDURES

Prior to data collection, the analyst must decide if the job should be analyzed as a single- or multi-task manual lifting job and if significant control is required at the lift's destination. This is necessary because the procedures differ according to the type of analysis required.

A manual lifting job may be analyzed as a single task if the task variables do not differ from task to task, or if only one task is of interest (single most stressful task). This may be the case if one of the tasks clearly has a dominant effect on strength demands, localized muscle fatigue, or whole-body fatigue. On the other hand, if the task variables differ significantly between tasks, it may be more appropriate to analyze a job as a multi-task manual lifting job. A multi-task analysis is more difficult to perform than a single-task analysis because additional data and computations are required. The multi-task approach, however, will provide more detailed information about specific strength and physiological demands.

For many lifting jobs, it may be acceptable to use either the single- or multi-task approach. The single-task analysis should be used when possible, but when a job consists of more than one task and detailed information is needed to specify engineering modifications, then the multi-task approach is a reasonable method of assessing the overall physical demands. The multi-task procedure is more complicated than the single-task procedure, and requires a greater understanding of assessment terminology and mathematical concepts. Therefore, the decision to use the single- or multi-task approach should be based on: the need for detailed information about all facets of the multi-task lifting job; the need for accuracy and completeness of data in assessing the physiological demands of the task; and the analyst's level of understanding of the assessment procedures.

The decision about control at the destination is important because the physical demands on the worker may be greater at the destination of the lift than at the origin, especially when significant control is required. When significant control is required at the destination, for example, the physical stress is increased because the load will have to be accelerated upward to slow down its descent. This acceleration may be as great as the acceleration at the origin of the lift and may create high loads on the spine. Therefore, if significant control is required, then the RWL and LI should be determined at both locations, and the lower of the two values will specify the overall level of physical demand.

To perform a lifting analysis using the revised lifting equation, two steps are undertaken: data are collected at the worksite, and the recommended weight limit and lifting index values are computed using the single- or multi-task analysis procedure.

STEP 1: COLLECT DATA

The relevant task variables must be carefully measured and clearly recorded in a concise format. As mentioned previously, these variables include the horizontal location of the hands (H), vertical location of the hands (V), vertical displacement (D), asymmetric angle (A), lifting frequency (F), and coupling quality (C). Job analysis work sheets, as shown in Figure 13-4 for single-task jobs, or Figure 13-5 for multi-task jobs, provide a simple form for recording the task variables and the data needed to calculate the *RWL* and the *LI* values. A thorough job analysis is required to identify and catalog each independent lifting task in the worker's complete job. For multi-task jobs, data must be collected for each task.

STEP 2: SINGLE- AND MULTI-TASK PROCEDURES

Single-task Procedure

For a single-task analysis, step 2 consists of computing the *RWL* and the *LI*. This is accomplished as follows.

Calculate the *RWL* at the origin for each lift. For lifting tasks requiring significant control at the destination, calculate the *RWL* at both the origin and the destination of the lift. This procedure is required if the worker has to re-grasp the load near the destination of the lift, the worker has to momentarily hold the object at the destination, or the worker has to position or guide the load at the destination. The purpose of calculating the *RWL* at both the origin and destination of the lift is to identify the lift's most stressful location. Therefore, the lowest *RWL* value at the origin or destination should be used to compute the *LI* for the task, as this value would represent the limiting set of conditions.

The assessment is completed on the single-task work sheet by determining the *LI* for the task of interest. This is accomplished by comparing the actual weight of the load (*L*) lifted with the *RWL* value obtained from the lifting equation.

Multi-task Procedure

For a multi-task analysis, step two has three parts:

1. Compute the frequency-independent recommended weight limit (*FIRWL*) and single-task recommended weight limit (*STRWL*) for each task.

2. Compute the frequency-independent lifting index (*FILI*) and single-task lifting index (*STLI*) for each task.

3. Compute the composite lifting index (*CLI*) for the overall job.

Frequency-independent recommended weight limit (*FIRWL*). Compute the *FIRWL* value for each task by using the respective task variables and setting the frequency multiplier (*FM*) to a value of 1.0. The *FIRWL* for each task reflects the compressive force and muscle strength demands for a single repetition of that task. If significant control is required at the destination for any individual

Ergonomics in Manufacturing

Figure 13-4 Single-task job analysis work sheet.

Assessment of Manual Lifting—The NIOSH Approach

Multi-task Job Analysis Work Sheet

Department _____   Job description _____
Job title _____
Analyst's name _____
Date _____

Step 1. Measure and record task variables

Task no.	Object weight (lb)		Hand location (in.)				Vertical distance (in.)	Asymmetric angle (°)		Frequency rate		Duration (hr)	Coupling
	L (avg)	L (max)	Origin		Destination		D	Origin	Destination	Lifts/min			C
			H	V	H	V		A	A	F			

Step 2. Compute multipliers and FIRWL, STRWL, FILI, and STLI for each task

Task no.	LC	×	HM	×	VM	×	DM	×	AM	×	CM	FIRWL	×	FM	STRWL	FILI = L/FIRWL	STLI = L/STRWL	New task number
	51																	
	51																	
	51																	
	51																	
	51																	

Step 3. Compute the composite lifting index for the job (after renumbering tasks)

CLI = $STLI_1$ + $\Delta FILI_2$ + $\Delta FILI_3$ + $\Delta FILI_4$ + $\Delta FILI_5$

$FILI_2(1/FM_{1,2} - 1/FM_1)$ $FILI_3(1/FM_{1,2,3} - 1/FM_{1,2})$ $FILI_4(1/FM_{1,2,3,4} - 1/FM_{1,2,3})$ $FILI_5(1/FM_{1,2,3,4,5} - 1/FM_{1,2,3,4})$

CLI = _____

Figure 13-5 Multi-task job analysis work sheet.

task, the *FIRWL* must be computed at both the origin and the destination of the lift, as previously described for single-task analysis.

Single-task recommended weight limit (*STRWL*). Compute the *STRWL* for each task by multiplying its *FIRWL* by its appropriate *FM* value. The *STRWL* for a task reflects the overall demands of that task, assuming it was the only task being performed. Note: this value does not reflect the overall demands of the task when the other tasks are considered. Nevertheless, it is helpful in determining the extent of excessive physical stress for an individual task.

Frequency-independent lifting index *(FILI)*. The *FILI* is computed for each task by dividing the maximum load weight (L) for that task by its respective *FIRWL*. The maximum weight is used to compute the *FILI* because the maximum weight determines the maximum biomechanical loads to which the body will be exposed, regardless of the frequency of occurrence. Thus, the *FILI* can identify individual tasks with potential strength problems for infrequent lifts. If any of the *FILI* values exceed a value of 1.0, then job design changes may be needed to decrease the strength demands.

Single-task lifting index (*STLI*). The *STLI* is computed for each task by dividing the average load weight (L) for that task by the respective *STRWL*. The average weight is used to compute the *STLI* because it provides a better representation of the metabolic demands, which are distributed across the tasks, rather than dependent on individual tasks. The *STLI* can be used to identify individual tasks with excessive physical demands (tasks that would result in fatigue). The *STLI* values do not indicate the relative stress of the individual tasks in the context of the whole job, but they can be used to prioritize the individual tasks according to the magnitude of their physical stress. Thus, if any of the *STLI* values exceed a value of 1.0, then ergonomic changes may be needed to decrease the overall physical demands of the task. Note: it may be possible to have a job in which all individual tasks have a *STLI* less than 1.0 and yet is physically demanding due to the combined demands of the tasks. In cases where the *FILI* exceeds the *STLI* for any task, the maximum weights may represent a significant problem, making careful evaluation necessary.

Composite lifting index *(CLI)*. The assessment is completed on the multi-task work sheet by determining the composite lifting index (*CLI*) for the overall job. The *CLI* is computed as follows:

1. The tasks are renumbered in order of decreasing physical stress, from the task with the greatest *STLI* down to the task

with the smallest *STLI*. The tasks are renumbered in this way so that the more difficult tasks are considered first.

2. The *CLI* for the job is then computed according to the following formula:

$$CLI = STLI_1 + \Sigma \Delta LI$$

where:

$$\Sigma \Delta LI = FILI_2 \times ((1 \div FM_{1,2}) - (1 \div FM_1))) + FILI_3 \times ((1 \div FM_{1,2,3}) - (1 \div FM_{1,2}))) + FILI_4 \times ((1 \div FM_{1,2,3,4}) - (1 \div FM_{1,2,3}))) + FILI_n \times ((1 \div FM_{1,2,3,4,\ldots,n}) - (1 \div FM_{1,2,3,\ldots,(n-1)})))$$

Note: the numbers in the subscripts refer to the new task numbers; and the *FM* values are determined from Table 13-5, based on the sum of the frequencies for the tasks listed in the subscripts.

An Example

The following example is provided to demonstrate this step of the multi-task procedure. Assume that an analysis of a typical three-task job provided the results shown in Table 13-8.

Table 13-8. Computations from multi-task example

Task #	Load (L) weight lb (kg)	Task frequency (F)	FIRWL	FM	STRWL	FILI	STLI	New task #
1	66.1 (30)	1	20	0.94	18.8	1.50	1.6	1
2	44.1 (20)	2	20	0.91	18.2	1.00	1.1	2
3	22.0 (10)	4	15	0.84	12.6	0.67	0.8	3

To compute the composite lifting index (*CLI*) for this job, the tasks are renumbered in order of decreasing physical stress, beginning with the task with the greatest *STLI*. In this case, as shown in Table 13-8, the task numbers do not change. Next, the *CLI* is computed according to the formula previously shown. The task with the greatest *CLI* is Task 1 (*STLI* = 1.6). The sum of the frequencies for Tasks 1 and 2 is 1 + 2 or 3, and the sum of the frequencies for Tasks 1, 2, and 3 is 1 + 2 + 4, or 7. Then, from Table 13-5, FM_1 is 0.94, $FM_{1,2}$ is 0.88, and $FM_{1,2,3}$ is 0.70. Finally, the *CLI* = 1.6 +

1.0 (1/0.88 − 1/0.94) + 0.67(1/0.70 − 1/0.88) = 1.6 + .07 + .20 = 1.9.
Note: the FM values were based on the sum of the frequencies for the subscripts, the vertical height, and the duration of lifting.

APPLYING THE EQUATIONS

USING THE RWL AND LI TO GUIDE ERGONOMIC DESIGN

The recommended weight limit (RWL) and lifting index (LI) can be used to guide ergonomic design in several ways.

1. The individual multipliers can be used to identify specific job-related problems. The relative magnitude of each multiplier indicates the relative contribution of each task factor (horizontal, vertical, frequency).

2. The RWL can be used to guide the redesign of existing manual lifting jobs or to design new manual lifting jobs. For example, if the task variables are fixed, then the maximum weight of the load could be selected so the RWL is not exceeded; if the weight is fixed, then the task variables could be optimized so that the weight is not exceeded by the RWL.

3. The LI can be used to estimate the relative magnitude of physical stress for a task or job. The greater the LI, the smaller the fraction of workers capable of safely sustaining the level of activity. Thus, two or more job designs could be compared.

4. The LI can be used to prioritize ergonomic redesign. For example, a series of suspected hazardous jobs could be rank-ordered according to the LI, and a control strategy could be developed (jobs with lifting indices above 1.0 or higher would benefit the most from redesign).

RATIONALE AND LIMITATIONS FOR LI

The NIOSH RWL and LI equations are based on the concept that the risk of lifting-related low-back pain increases as the demands of the lifting task increase. In other words, as the magnitude of the LI increases, the level of risk for a given worker increases, and a greater percentage of the work force is likely to be at risk for developing lifting-related low-back pain. The shape of the risk function, however, is not known. Without additional data showing the relationship between low-back pain and the LI, it is impossible to predict the magnitude of the risk for an individual or the exact percent of the work population who would be at an elevated risk for low-back pain.

To gain a better understanding of the rationale for the development of the *RWL* and *LI*, consult Waters et al. (1993), which provides a discussion of the criteria underlying the lifting equation and of the individual multipliers, and identifies both the assumptions and uncertainties in the scientific studies that associate manual lifting and low-back injuries.

JOB-RELATED INTERVENTION STRATEGY

The lifting index may be used to identify potentially hazardous lifting jobs or to compare the relative severity of two jobs for the purpose of evaluating and redesigning them. From the NIOSH perspective, it is likely that lifting tasks with a $LI > 1.0$ pose an increased risk for lifting-related low-back pain for some fraction of the work force (Waters et al. 1993). Hence, lifting jobs should be designed to achieve a *LI* of 1.0 or less.

Some experts believe, however, that worker selection criteria may be used to identify workers who can perform potentially stressful lifting tasks (lifting tasks that would exceed a *LI* of 1.0) without significantly increasing their risk of work-related injury above the baseline level (Chaffin and Andersson 1984). Those who endorse the use of selection criteria believe that the criteria must be based on research studies, empirical observations, or theoretical considerations that include job-related strength testing and/or aerobic capacity testing. Even these experts agree, however, that many workers will be at a significant risk of a work-related injury when performing highly stressful lifting tasks (lifting tasks that would exceed a *LI* of 3.0). Also, "informal" or "natural" selection of workers may occur in many jobs that require repetitive lifting tasks. According to some experts, this may result in a unique work force that may be able to work above a lifting index of 1.0, at least in theory, without substantially increasing their risk of low-back injuries above the baseline rate of injury.

EXAMPLE PROBLEMS

Two sample problems are provided to demonstrate the proper application of the lifting equation and procedures. The procedures offer a method for determining the level of physical stress associated with a specific set of lifting conditions and assist in identifying the contribution of each job-related factor. The examples can help guide the development of an ergonomic redesign strategy. Specifically, for each example, a job description, job analysis, hazard assessment, redesign suggestion, illustration, and completed work sheet are provided.

A series of general design/redesign suggestions for each job-related risk factor are provided in Table 13-9. These suggestions can be used to develop a practical ergonomic design/redesign strategy.

Table 13-9. General design/redesign suggestions

If $HM < 1.0$	Bring the load closer to the worker by removing any horizontal barriers or reducing the size of the object. Lifts near the floor should be avoided; if unavoidable, the object should fit easily between the legs.
If $VM < 1.0$	Raise/lower the origin/destination of the lift. Avoid lifting near the floor or above the shoulders.
If $DM < 1.0$	Reduce the vertical distance between the origin and the destination of the lift.
If $AM < 1.0$	Move the origin and destination of the lift closer together to reduce the angle of twist, or move the origin and destination further apart to force the worker to turn the feet and step, rather than twist the body.
If $FM < 1.0$	Reduce the lifting frequency rate, reduce the lifting duration, or provide longer recovery periods (i.e., light work period).
If $CM < 1.0$	Improve the hand-to-object coupling by providing optimal containers with handles or handhold cutouts, or improve the handholds for irregular objects.
If the RWL at the destination $<$ at the origin	Eliminate the need for significant control of the object at the destination by redesigning the job or modifying the container/object characteristics. (See requirements for significant control in text.)

Loading Supply Rolls, Example 1

Job description. With both hands directly in front of the body, a worker lifts the core of a 35-lb (16-kg) roll of paper from a cart, then shifts the roll in the hands and holds it by the sides to position it on a machine, as shown in Figure 13-6. Significant control of the roll is required at the destination of the lift. Also, the worker must crouch at the destination of the lift to support the roll in front of the body, but does not have to twist.

Job analysis. The task variable data are measured and recorded on the job analysis work sheet (Figure 13-7). The vertical location of the hands is 27 in. (69 cm) at the origin, and 10 in. (25 cm) at the

Assessment of Manual Lifting—The NIOSH Approach

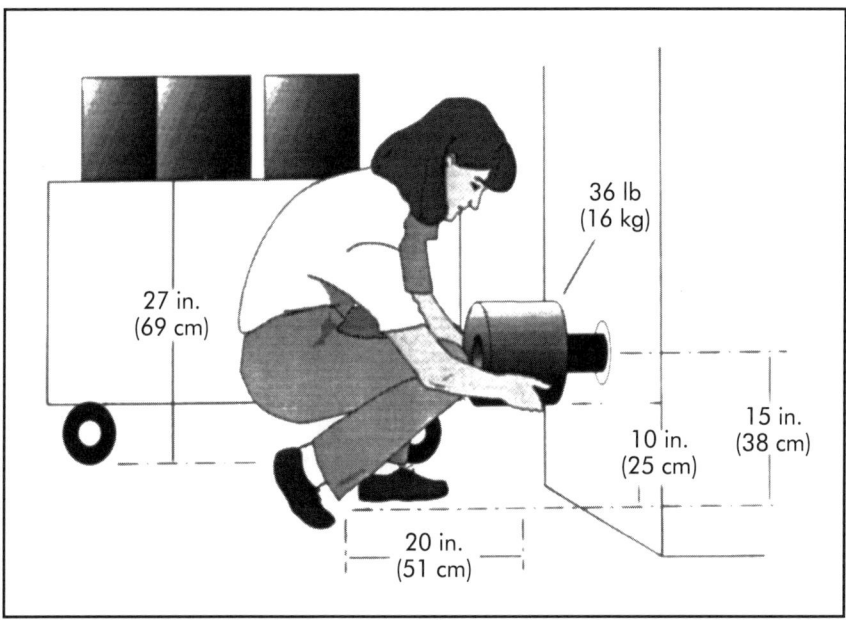

Figure 13-6 Loading supply rolls.

destination. The horizontal location of the hands is 15 in. (38 cm) at the origin and 20 in. (51 cm) at the destination. The asymmetric angle is 0° at both the origin and the destination, and the frequency is four lifts per shift (less than 0.2 lifts per minute for less than 1 hour—(see Table 13-5).

Using Table 13-6, the coupling is classified as poor because the worker must reposition the hands at the destination of the lift and he or she cannot flex the fingers to the desired 90° angle (hook grip). No asymmetric lifting is involved ($A = 0$), and significant control of the object is required at the destination of the lift. Thus, the RWL should be computed at both the origin and the destination of the lift. The multipliers are computed from the lifting equation or determined from the multiplier tables (Tables 13-1 to 13-5, and Table 13-7). As shown in Figure 13-7, RWL = 28.0 lb (13 kg) at the origin and RWL = 18.1 lb (8 kg) at the destination for this activity.

Hazard assessment. The weight to be lifted (35 lb [16 kg]) is greater than the RWL at both the origin and destination of the lift (28.0 lb [13 kg] and 18.1 lb [8 kg], respectively). At the origin, LI = 35 lb [16 kg]/28.0 lb [13 kg] = 1.3; at the destination, LI = 35 lb [16 kg]/18.1 lb [8 kg] = 1.9. These values indicate that this job

Job Analysis Work Sheet

Department: Shipping
Job title: Packager
Analyst's name: _____
Date: _____

Job description: Loading paper supply rolls

Example 1

Step 1. Measure and record task variables

Object weight (lb)		Hand location (in.)				Vertical distance (in.)	Asymmetric angle (°)		Frequency rate	Duration	Object coupling
		Origin		Destination			Origin	Destination	Lifts/min	(hr)	
L (avg)	L (max)	H	V	H	V	D	A	A	F		C
35	35	15	27	20	10	17	0	0	<.2	<1	Poor

Step 2. Determine the multipliers and compute the RWLs

$RWL = LC \times HM \times VM \times DM \times AM \times FM \times CM$

Origin: $RWL = \boxed{51} \times \boxed{.67} \times \boxed{.98} \times \boxed{.93} \times \boxed{1.0} \times \boxed{1.0} \times \boxed{.90} = \boxed{28.0 \text{ lb}}$

Destination: $RWL = \boxed{51} \times \boxed{.50} \times \boxed{.85} \times \boxed{.93} \times \boxed{1.0} \times \boxed{1.0} \times \boxed{.90} = \boxed{18.1 \text{ lb}}$

Step 3. Compute the lifting index

Origin: Lifting index = $\dfrac{\text{Object weight (L)}}{RWL} = \dfrac{35}{28.0} = \boxed{1.3}$

Destination: Lifting index = $\dfrac{\text{Object weight (L)}}{RWL} = \dfrac{35}{18.1} = \boxed{1.9}$

Figure 13-7 Job analysis work sheet on which task variable data is measured and recorded.

is somewhat stressful at the origin, but more stressful at the destination of the lift.

Redesign suggestions. The first choice for reducing the risk of injury for workers performing this task would be to adapt the cart so that the paper rolls could be easily pushed into position on the machine, without manually lifting them.

If the cart cannot be modified, then the results of the equation may be used to suggest task modifications. The work sheet displayed in Figure 13-7 indicates that the multipliers with the smallest magnitude (those providing the greatest penalties) are 0.50 for the *HM* at the destination, 0.67 for the *HM* at the origin, 0.85 for the *VM* at the destination, and 0.90 for the *CM* value. Using Table 13-9, the following job modifications are suggested:

1. Bring the load closer to the worker by making the roll smaller so that it can be lifted from between the worker's legs. This will decrease the *H* value, which in turn will increase the *HM* value.
2. Raise the height of the destination to increase the *VM*.
3. Improve the coupling to increase the *CM*.

If the size of the roll cannot be reduced, then the vertical height (*V*) at the destination should be increased. Figure 13-8 shows that if *V* were increased to about 30 in. (76 cm), then *VM* would be increased from 0.85 to 1.0; the *H* value would be decreased from 20 in. to 15 in. (51 cm to 38 cm), which would increase *HM* from 0.50 to 0.67; and *DM* would be increased from 0.93 to 1.0. As shown in Figure 13-8, the final *RWL* would be increased from 18.1 lb to 30.8 lb (8 kg to 14 kg), and the *LI* at the destination would decrease from 1.9 to 1.1.

In some cases, redesign may not be feasible. In these cases, the use of a mechanical lift may be more suitable. As an interim control strategy, two or more workers may be assigned to lift the supply roll.

Comments. The horizontal distance (*H*) is a significant factor that may be difficult to reduce, because the size of the paper rolls may be fixed. Moreover, redesign of the machine may not be practical. Therefore, elimination of the manual lifting component of the job may be more appropriate than job redesign.

Dish-washing Machine Unloading, Example 2

Job description. A worker manually lifts trays of clean dishes from a conveyor at the end of a dish-washing machine and loads them on a cart as shown in Figure 13-9. The trays are filled with assorted dishes (glasses, plates, bowls) and silverware. The job takes

Job Analysis Work Sheet

Department __Shipping__
Job title __Packager__
Analyst's name _____
Date _____

Job description
__Loading paper supply rolls__
__Modified example 1__

Step 1. Measure and record task variables

Object weight (lb)		Hand location (in.)				Vertical distance (in.)	Asymmetric angle (°)		Frequency rate	Duration	Object coupling
		Origin		Destination			Origin	Destination	Lifts/min	(hr)	
L (avg)	L (max)	H	V	H	V	D	A	A	F		C
35	35	15	27	15	30	3	0	0	<.2	<1	Poor

Step 2. Determine the multipliers and compute the RWLs

$$RWL = LC \times HM \times VM \times DM \times AM \times FM \times CM$$

Origin $RWL = 51 \times .67 \times .98 \times 1.0 \times 1.0 \times 1.0 \times .90 = 30.1 \text{ lb}$

Destination $RWL = 51 \times .67 \times 1.0 \times 1.0 \times 1.0 \times 1.0 \times .90 = 30.8 \text{ lb}$

Step 3. Compute the lifting index

Origin   Lifting index = $\dfrac{\text{Object weight (L)}}{\text{RWL}} = \dfrac{35}{30.1} = 1.2$

Destination Lifting index = $\dfrac{\text{Object weight (L)}}{\text{RWL}} = \dfrac{35}{30.8} = 1.1$

Figure 13-8 Modified job analysis work sheet.

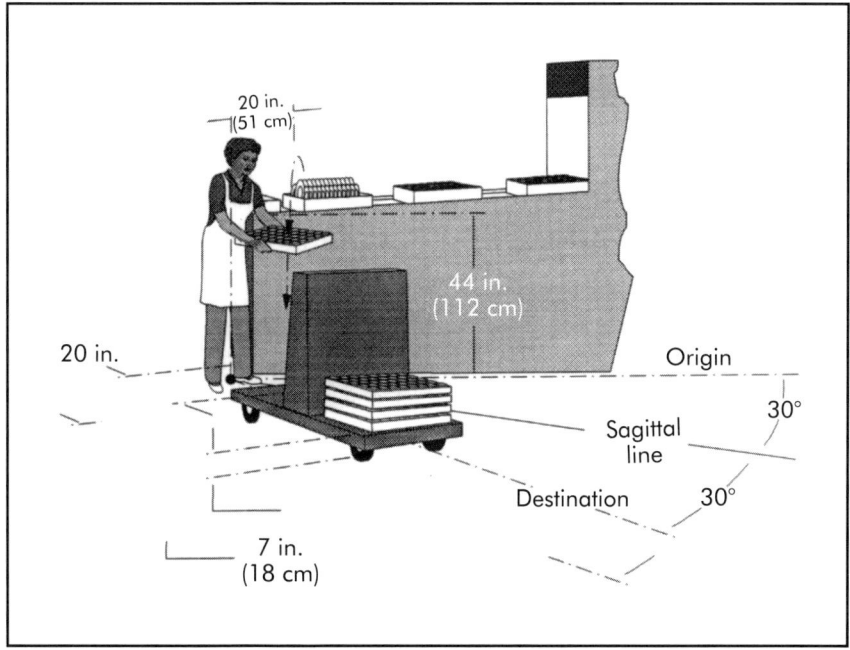

Figure 13-9 Dishwashing machine unloading.

between 45 minutes and 1 hour to complete, and the lifting frequency rate averages five lifts per minute. Workers usually twist their body to one side to lift the trays (asymmetric lift) and then rotate their body to the other side to lower the trays to the cart in one smooth continuous motion. The asymmetric angle (A) varies between workers and within workers, but there is usually equal asymmetry to either side. During the lift, the worker may take a step toward the cart. The trays have well-designed handhold cutouts and are made of lightweight materials.

Job analysis. The task variable data are measured and recorded on the job analysis work sheet (Figure 13-10). At the origin of the lift, the horizontal distance (H) is 20 in. (51 cm), the vertical distance (V) is 44 in. (112 cm), and the angle of asymmetry (A) is 30°. At the destination of the lift, H is 20 in. (51 cm), V is 7 in. (18 cm), and A is 30°. The trays normally weigh between 5 lb (2.3 kg) and 20 lb (9.1 kg), but for this example, assume that all of the trays weigh 20 lb (9.1 kg).

Using Table 13-6, the coupling is classified as "good." Significant control is required at the destination of the lift. Using Table 13-5, the

Ergonomics in Manufacturing

Job Analysis Work Sheet

Department: Food service
Job title: Cafeteria worker
Analyst's name: _____
Date: _____

Job description: Unloading a dish-washing machine

Example 2

Step 1. Measure and record task variables

Object weight (lb)		Hand location (in.)				Vertical distance (in.)	Asymmetric angle (°)		Frequency rate	Duration	Object coupling
		Origin		Destination			Origin	Destination	Lifts/min	(hr)	
L (avg)	L (max)	H	V	H	V	D	A	A	F		C
20	20	20	44	20	7	37	30	30	5	<1	Good

Step 2. Determine the multipliers and compute the RWLs

$$RWL = LC \times HM \times VM \times DM \times AM \times FM \times CM$$

Origin: RWL = $51 \times .50 \times .90 \times .87 \times .90 \times .80 \times 1.0$ = 14.4 lb

Destination: RWL = $51 \times .50 \times .83 \times .87 \times .90 \times .80 \times 1.0$ = 13.3 lb

Step 3. Compute the lifting index

Origin: Lifting index = $\dfrac{\text{Object weight (L)}}{RWL} = \dfrac{20}{14.4}$ = 1.4

Destination: Lifting index = $\dfrac{\text{Object weight (L)}}{RWL} = \dfrac{20}{13.3}$ = 1.5

Figure 13-10 Job analysis work sheet for cafeteria worker.

FM is determined to be 0.80. As shown in Figure 13-10, the *RWL* is 14.4 lb (6.5 kg) at the origin and 13.3 lb (6 kg) at the destination.

Hazard assessment. The weight to be lifted (20 lb [9.1 kg]) is greater than the *RWL* at either the origin or destination of the lift (14.4 lb [6.5 kg] and 13.3 lb [6 kg], respectively). The *LI* at the origin is 20/14.4 = 1.4 and the *LI* at the destination is 1.5. These results indicate that this lifting task would be stressful for some workers.

Redesign suggestions. The work sheet shows that the smallest multipliers (the greatest penalties) are 0.50 for the *HM*, 0.80 for the *FM*, 0.83 for the *VM*, and 0.90 for the *AM*. Using Table 13-9, the following job modifications are suggested:

- Bring the load closer to the worker to increase *HM*;
- Reduce the lifting frequency rate to increase *FM*;
- Raise the destination of the lift to increase *VM*; and
- Reduce the angle of twist to increase *AM* by either moving the origin and destination closer together or moving them further apart.

Since the horizontal distance (*H*) is dependent on the width of the tray in the sagittal plane, this variable can only be reduced by using smaller trays. Both the *DM* and *VM*, however, can be increased by lowering the height of the origin and increasing the height of the destination. For example, if the height at both the origin and destination is 30 in. (76 cm), then *VM* and *DM* are 1.0, as shown in the modified work sheet (Figure 13-11). Moreover, if the cart is moved so that reaching to the side is eliminated, the *AM* can be increased from 0.90 to 1.00. As shown in Figure 13-11, with these redesign suggestions, the *RWL* can be increased from 13.3-20.4 lb (6-9.3 kg), and the *LI* values are reduced to 1.0.

Comments. This analysis was based on a 1-hour work session. If a subsequent work session begins before the appropriate recovery period has elapsed (1.2 hours), then the 1-2-hour or 2-8-hour category would be used to compute the *FM* value.

REFERENCES

Association of Schools of Public Health (ASPH)/National Institute for Occupational Safety and Health (NIOSH). *Proposed National Strategies for the Prevention of Leading Work-related Diseases and Injuries: Part 1.* Cinncinnati, OH: Association of Schools of Public Health under a cooperative agreement with the National Institute for Occupational Safety and Health, 1986.

Ayoub, M.M., and Mital, A. *Manual Materials Handling.* London: Taylor & Francis, 1989.

Chaffin, D.B., and Andersson, G.B.J. *Occupational Biomechanics.* New York: John Wiley and Sons, 1984.

Job Analysis Work Sheet

Department: Food service
Job title: Cafeteria worker
Analyst's name: _____
Date: _____

Job description: Unloading a dish-washing machine

Modified example 2

Step 1. Measure and record task variables

Object weight (lb)		Hand location (in.)				Vertical distance (in.)	Asymmetric angle (°)		Frequency rate Lifts/min	Duration (hr)	Object coupling
		Origin		Destination		D	Origin	Destination	F		C
L (avg)	L (max)	H	V	H	V		A	A			
20	20	20	30	20	30	0	0	0	5	<1	Good

Step 2. Determine the multipliers and compute the RWLs

$$RWL = LC \times HM \times VM \times DM \times AM \times FM \times CM$$

Origin: RWL = 51 × .50 × 1.0 × 1.0 × 1.0 × .80 × 1.0 = 20.4 lb

Destination: RWL = 51 × .50 × 1.0 × 1.0 × 1.0 × .80 × 1.0 = 20.4 lb

Step 3. Compute the lifting index

Origin: Lifting index = Object weight (L) / RWL = 20 / 20.4 = 1.0

Destination: Lifting index = Object weight (L) / RWL = 20 / 20.4 = 1.0

Figure 13-11 Modified job analysis work sheet showing the decreased LI values.

Eastman Kodak. *Ergonomic Design for People at Work,* Vol. 2. New York: Van Nostrand Reinhold, 1986.

NIOSH. *Scientific Support Documentation for the Revised 1991 NIOSH Lifting Equation: Technical Contract Reports, May 8, 1991.* Available from the National Technical Information Service (NTIS) (No. PB-91-226274). Cincinnati, OH: U.S. Department of Health and Human Services, National Institute for Occupational Safety and Health, 1991.

NIOSH. *Work Practices Guide for Manual Lifting.* NIOSH Technical Report No. 81-122. Cincinnati, OH: U.S. Department of Health and Human Services, National Institute for Occupational Safety and Health, 1981.

Waters, T.R., Putz-Anderson, V., and Garg, A. *Applications Manual for the Revised NIOSH Lifting Equation.* Technical Report DHHS (NIOSH) Pub. No. 94-110. Available from the National Technical Information Service (No. PB-94-176930). Cincinnati, OH: U.S. Department of Health and Human Services, National Institute for Occupational Safety and Health, 1994.

Waters, T.R., Putz-Anderson, V., Garg, A., and Fine, L.J. "Revised NIOSH Equation for the Design and Evaluation of Manual Lifting Tasks." *Ergonomics* 1993, 36(7): 749-776.

CHAPTER 14

PERSPECTIVE ON LIFTING BELTS FOR MATERIAL HANDLING

Malgorzata J. Rys
Assistant Professor
Department of Industrial and Manufacturing Systems Engineering
Kansas State University
Manhattan, KS

Luis Rene Contreras
Professor
Department of Industrial and Manufacturing Engineering
Institute of Engineering and Technology
Autonomous University of Ciudad Juarez
Ciudad Juarez, Chihuahua, Mexico

Work-related back injuries exact a heavy toll on industry and the service sector. In fact, back injuries and low-back pains, representing approximately 30% of industry's total injuries, rank second only to the common cold as the leading cause of work time lost in the United States. In 1991, back injuries and low-back pain were related to 31% of all workers' compensation claims in the United States (National Safety Council 1992). The term *back injury* refers to all back disorders, injuries, or pain.

Even though back trouble is rarely life threatening, the cost of such a prevalent problem is enormous. Marras et al. (1993) report that in the United States, back injuries account for up to 40% of compensation costs. The average cost of a back injury runs $7,400, according to data gathered on a large number of low-back injury cases (Mahone 1994). Howell (1992) estimates that in many supermarket companies, low-back injuries account for 25-35% of time lost from work and for 50-60% of workers' compensation costs. Back injuries and low-back pain could be devastating for industry; not only do industries pay workmens' compensation, but they also spend billions of dollars on tests, treatments, claims, lawsuit awards, settlements, and surgeries (Tomecek 1992). Reducing back injuries and their associated costs are concerns for many companies.

PREVENTION ATTEMPTS

Over the years, industry, business, and government have tried many avenues of relief for these costs. Traditionally, three approaches—personnel training, personnel selection, and job design—have been tested to make jobs as safe and efficient as possible. The first two approaches fit the person to the job, the third approach fits the job to the person. The three strategies are highly related because both job demands and related human capabilities must be known to match them. According to Kroemer (1992), all three strategies should be successful if properly applied. Additionally, many companies have been involved in the use of an assortment of preventive devices, with back support belts representing one of the most frequently used.

PERSONNEL TRAINING

Training is expected to reduce back injuries by developing specific material handling skills, promoting further awareness and self-responsibility, and improving specific physical characteristics (physical training). However, experts disagree about the value of training. On the pro-training side, several publications (Chaffin et al. 1978; Liles 1985; Klaber et al. 1986) have reported that training programs helped reduce the incidence and severity of back injuries. However, Brown (1975) and Yu et al. (1984) concluded that for over four decades, the educational approach has been unsuccessful in significantly reducing incidents of back injury.

There is also the unexpected finding that stronger persons were found to be injured more frequently than weaker colleagues (Battié et al. 1989).

PERSONNEL SELECTION

The second strategy for controlling back injuries involves the use of employee placement procedures. In an attempt to select a worker population whose capacities match or exceed the demand of a given job, numerous employee screening or placement programs have been, and are being, used (Ayoub and Mital 1989). Some of these techniques are: back X-ray films, strength testing, medical examinations, psychological tests, job simulators, and rating methods. If taken as sole measures of a worker's capacity, the previously mentioned screening procedures are not adequate. They can be useful only in some aspects of pre-employment screening.

Finally, these procedures must be carefully used due to possible legal implications under the Americans with Disabilities Act.

JOB DESIGN

The best hope of eliminating or reducing back injuries is through job design and implementation of engineering controls. Eliminating or reducing hazardous tasks, decreasing job demands, and minimizing body movements are three goals of job design. A strong advantage of this approach is that its results are permanent, as opposed to the temporary effects of personnel selection and training.

EXTERNAL SUPPORT DEVICES

While all three strategies have been useful in preventing back injuries (National Institute for Occupational Safety and Health [NIOSH] 1994), these options are not always feasible (Udo et al. 1992). Thus, to alleviate back pain and prevent back injuries, companies have begun to use back support belts as an additional strategy (Tomecek 1992).

Back belts are used by production and service workers, but there are no conclusive results about their effectiveness. There are many contradictory opinions (Klaber et al. 1986; Imker 1994; Lund and Rambo 1994) in magazines, newspapers, and conferences, as well as in a few research studies (Grew and Deane 1982; Harman et al. 1989; Hilgen et al. 1992; McCoy et al. 1988; McGill et al. 1990; Mitchell et al. 1994; Reddell et al. 1992; Udo et al. 1993; Walsh and Schwartz 1990). Recently, the U.S. Government (NIOSH 1994) concluded that based on a review of scientific literature, there was no scientific evidence indicating that belts prevent back problems. Furthermore, if the belts work, it must be established whether they work in the same way for different MMH activities, and if they work the same for males and females.

BACK SUPPORT BELTS

Back support belts are known as lifting belts, weight-lifting devices, supports, aids, abdominal support belts, and pelvic belts. The term *back belt* is applied to therapeutic devices, such as lumbosacral orthoses, back or spinal braces, supports, and corsets. Back support belts, regardless of the name, are external support devices tightened around the lower-back and abdomen or pelvis.

There are three basic types of back support belts: orthotic devices prescribed by a physician to help treat low-back pain of injured people; belts used for weight lifters to prevent back injuries while lifting heavy weights; and industrial back support devices used to reduce injury in the workplace.

Orthotic Devices

Medically-prescribed orthotic devices can be grouped into two major categories: corrective, and supportive and immobilizing (Walters and Morris 1970). Corrective braces are used in treating such disorders as scoliosis and kyphosis, while supportive braces and corsets are prescribed for the relief of low-back pain.

There are many types and constructions of the medically-prescribed lumbar spinal supports (Grew and Deane 1982). They can be made of fabric, leather, or nylon. These supports can be rigid, wide, or narrow. Some have plastic removable inserts, while others do not. They all usually cover from the sacrum or pelvic region to the thoracic region.

Million et al. (1981) describes the lumbar corset used in their experiment as a wide wrap-over body belt that comes in various sizes, fastened by a belt. In the back of the corset there was a large pocket to hold a plastic insert. Kumar and Godfrey (1986) cite the Camp lumbosacral corset and the Harris brace as two examples of the medically-prescribed back supports.

The Camp lumbosacral corset extends from the mid-thoracic to the sacroiliac region. It is firmly bonded in the back and reinforced around with pliable semirigid uprights to ensure better pressure distribution. The device has self-adjusting laces on the side to apply pressure on the abdomen by means of three pull straps.

The Harris brace is a low-back orthosis with a pelvic and thoracic band and four upright bars connecting the two bands. Two of the uprights are paraspinal in position and the other two are lateral. In front, two canvas flaps covering the abdomen can be tightened by four pull straps to provide abdominal pressure.

New spinal braces are frequently designed and old ones modified (Kumar and Godfrey 1986). Generally, the alternations have been to improve the fit, incorporate new materials, and sometimes to include additional mechanical forces.

WEIGHT-LIFTING BELTS

Weight-lifting belts are normally made of leather, but there are some new nonleather designs. The are two basic types of belts. One is wide in the center (4 or 6 in. [10 or 15 cm]), but tapered 2.4 or 2.8 in. (6 or 7 cm) at both ends near the buckle (see Figure 14-1). The other belt has the same width (4-in. [10-cm] wide) for its entire length (see Figure 14-2).

INDUSTRIAL BACK SUPPORT BELTS

Industrial back support belts come in a variety of styles, materials, sizes, and colors. This type of belt usually is designed to be lightweight and provide long-lasting back and abdominal support. Made of breathable fabric or nylon and many other synthetic materials, the belts are designed to be worn outside the employee's clothing.

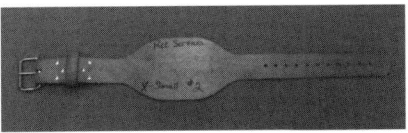

Figure 14-1 Leather weight-lifting belt.

There are three basic ways in which the belts give support: some have flexible steel stays, some are of thick semirigid foam, and some are air filled. Some belts provide an additional support to the lumbosacral region with a padded insert.

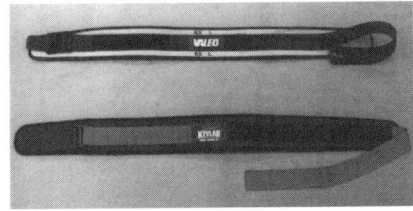

Figure 14-2 Weight-lifting belts by Valeo and Kevlar.

Some belts have detachable suspenders. Suspenders allow the worker to loosen the belt without having to take it off. This comes in handy because workers can always have their belts on them. A possible problem with suspenders is that they can become caught in machinery or other objects.

The industrial back support belts come in different widths. The most common widths are 4 in. (10 cm) and 6 in. (15 cm) wide. But some belts are available in 5-, 8-, and 9-in. (12.5-, 20-, and 22.5-cm) widths.

The belts come in a wide range of waist sizes, from extra small (XS) to extra extra large (XXL). Depending on the manufacturer, the waist sizes are grouped in six categories: XS (22-26 in. [56-66 cm]), S (26-30 in. [66-76 cm]), M (30-34 in. [76-86 cm]), L (34-38 in. [86-97 cm]), XL (38-42 in. [97-107 cm]), and XXL (42-46 in. [107-117 cm]). Also, there are some one-size-fits-all models available.

Most of the industrial back supports belts feature Velcro® closures for adjustable tension and some are available in washable and waterproof materials.

The OK-1 Model

The OK-1 Back Support Belt™ model SS-4 has a continuous 4-in. (10-cm) width (Figure 14-3). Overall thickness is 0.2-in. (0.5 cm). This belt comes with a permanently attached lumbar pad measuring 0.8 in. (2 cm). The length of the belt is adjustable by overlapping the inside layer of self-gripping fasteners. It also has an outside strap, designed to lock the belt in place without tightening it. It is made of synthetic material and represents the type of support belt characterized by a semirigid construction and uniform width.

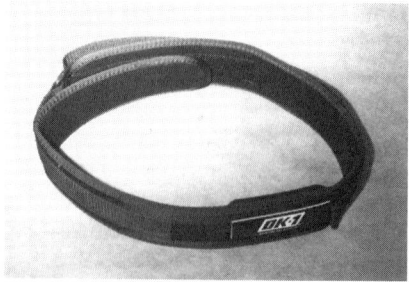

Figure 14-3 The OK-1 Back Support Belt™ Model SS4.

Japanese Belt

The Japanese belt represents a flexible type of back support designed in two styles: male belt and female belt. Figure 14-4 shows a picture of both belts. The female belt (Figure 14-4a) has a maximum width of 6 in. (15 cm) in the center, tapering to 4 in. (10 cm) at both ends; the upper edge has more curvature than the male model (to adjust to female hip shape). The male belt (Figure 14-4b) has a maximum width of 6.5 in. (17 cm) in the center, tapering to 4 in. (10 cm) at both ends. The belt is made of different sections of rubber and a synthetic mesh. The circumference size of the belt is adjustable by two straps that loop through buckles on the left side and double back to Velcro secured to the right side. A characteristic of this belt is that it is designed to be held around the pelvic region.

ProFlex™

The standard model of the ProFlex™ back support belt, is adjustable in circumference with Velcro fastening and measures 8 in. (20 cm) at the center, tapering to 4 in. (10 cm) at the fastener (see Figure 14-5). It is made of a synthetic type of material (spandex) and is tightened by stretching two elastic bands to provide proper tension. A characteristic of this belt is the suspenders attached to it. This belt also represents a flexible/elastic type of back support.

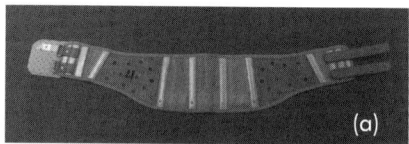

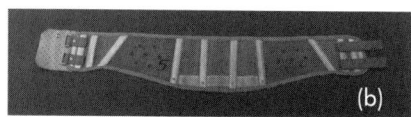

Figure 14-4 The Japanese belt designed for females (a) and males (b).

UNKNOWN FACTORS

Some belt manufacturers claim that belts remind workers to use proper body mechanics when lifting, bending, and carrying. Others claim the belts (combined with proper lifting, bending, and carrying techniques) help reduce both new and chronic injuries. But, with such a wide range of back support belt characteristics, the problem is which of these characteristics makes a back support belt effective in preventing or reducing back injuries. For example, since there is no difference in male and female belts (except for the Japanese belt), it is not known if females get the same support as males. Also, it is not known which type of belt support is the best or which width is optimal.

Figure 14-5 The ProFlex™ back support belt.

Another problem is that an optimum fit has not been established. Workers customize the fit to suit their personal comfort. However, everyone has a different sense of fit, so some workers may be benefiting more than others. According to Tomecek (1992), the only thing certain is that the back support belt is increasing in popularity and its manufacturers are reporting a dramatic increase in sales.

RATIONALE FOR USING BACK SUPPORT BELTS

Walters and Morris (1970) report that the spinal supports prescribed for relieving low-back pain are presumed to achieve their effect by providing all of the following:

- Decreasing movement at the intervertebral joint;
- Shifting a portion of the upper body's load from the spine to the rest of the trunk by producing abdominal compression;
- Decreasing lumbar lordosis, therefore, mechanical stresses, on the spine;
- Providing a placebo (psychological) effect; and
- Providing sufficient support in the lumbar region, allowing relaxation of the trunk muscles.

According to Grew and Deane (1982), some reasons why spinal supports are prescribed in terms of their therapeutic effects could be that they limit movement, alter intracavity pressures, modify muscle actions, and warm skin (the increase in the blood flow helps relieve back pain).

Reddell et al. (1992) state that the most cited reason for the use of back support belts is that their use increases the intra-abdominal pressure (IAP), which reduces compressive forces around the lumbar spine.

McGill (1990) suggests that belts may possibly support shear loading on the spine, resulting from the effect of gravity acting on the hand-held load and mass of the upper body when the trunk is flexed. He also cites a claim that back support belts may reduce back injuries by reminding workers to lift properly.

Udo (1993) reports that the basis for back support belts is that the raising of IAP reduces the load on the lumbar vertebra by 30%. The rationale that Udo gives is that when the abdomen is pressed, the abdominal cavity is shaped like a rugby ball, which supports the load as a pillow.

EFFECT OF BACK SUPPORT BELTS

While the back support belt manufacturers provide impressive injury reduction numbers to support the use of belts (21% reduction in injury rate when using the Champion Ergonomics Work Belt™ [Morris and Del Vecchio 1992]), the scientific community disagrees on whether back support devices are an effective means of reducing or preventing low-back injuries.

De Ruiter (1990) believes that back supports are not a complete solution, but are part of an overall approach that includes education, training, and warm-ups. In his opinion, belts contribute to employee satisfaction and productivity by reducing the risk of lower-back pain.

Hilgen et al. (1991) suggest that fitting workers with abdominal belts should include formal training to eliminate any mental enhancement that may be given by an abdominal belt in manual material handling tasks. They concluded that abdominal belts appeared to be beneficial in stooped lifting tasks.

According to Fograscher (1991), back braces, or lifting belts, are not endorsed by Ohio's Division of Safety and Hygiene as tools for preventing back injuries. He states that back belts in no way eliminate a worker's exposure to hazards and offer no long-term prevention of low-back disorders.

Perry (1992) mentioned that lifting belts can be used as an additional support during heavy or extended physical exertion to prevent fatigue and the resulting spinal stress. He concluded that the best back support in the world will not supplant an exercise program to keep workers in good physical shape.

Congleton et al. (1993), based their research on six separate studies (McCoy et al. 1988; Sherwood 1988; Amendola 1989; Horsford 1989; Wilson 1989; Reddell et al. 1992) on lifting belts conducted at Texas A&M University, which concluded that back support belts have not proven to be an effective piece of personal protective equipment and are not an effective means of increasing lifting capacity in any controlled study.

The research on back belts has covered studies using medically-prescribed spinal supports, weight-lifting belts, and industrial back support belts. The research ranges from epidemiological studies to many types of physical studies (psychophysical, biomechanical, physiological, or a combination of these). This chapter concentrates on epidemiological studies, and includes a brief summary of physical research. A full discussion of physical research is provided in the appendix to this chapter.

EPIDEMIOLOGICAL RESEARCH

Walsh and Schwartz (1990) investigated the effect of multimodel intervention ("back school" training and back support belts) in the prevention of back injury, and evaluated the potential adverse side effects of using a lumbosacral corset in the workplace. The subjects were 90 healthy male warehouse workers. The authors concluded that the study supported the concept of using education and prophylactic bracing to prevent back injury and reduce lost time. However, the use of a lumbosacral orthoses only during working hours (8 hours) appeared to have a negative effect on abdominal muscle strength. Berry (1991) concluded that the Schwartz and Walsh findings could be contaminated by Hawthorne and placebo effects.

Holmstrom and Moritz (1992) studied the effects of lumbar belts on trunk muscle strength and endurance using construction workers. After 2 months of daily use, a lumbar spinal support did not influence the trunk extensor strength or endurance, but the trunk flexor strength increased significantly.

The study by Reddell et al. (1992) indicates that individuals who wear a lifting belt, then discontinued its use, have a higher injury rate than other groups. As subjects, the study used 642 baggage handlers, working for a major airline company. They concluded that when employees discontinue using lifting belts, an exercise program must be implemented to strengthen the abdominal and back muscles. They recommended that close attention be given to injuries occurring shortly after the belt is discontinued, with particular emphasis on off-the-job injuries.

Udo et al. (1992, 1993) investigated the effect of a belt on the incidence of low-back pain in an exchange rolling mill task, a rice-carrying task, and a crane task on male workers who had experienced low-back pain. The belt used in the experiments was attached at pelvic position. The test of the belt on 33 rolling mill workers for 6 months established that the belt did not damage the paravertebral muscle (expressed in terms of reduction of muscular strength). The authors concluded that wearing a back support belt is useful in reducing the lumbar load and healing the lumbago of workers who were engaged in heavy MMH (rice-carrying task) and "unneutral" postural tasks (crane operation). Among the subjects who were graded as suffering from low-back pain, the percentages that showed an improvement in the kinetic pain score were 50%, 63%, and 80% for the exchanging milling task, the rice-carrying task, and the crane operation, respectively. Among the rice-carrying workers, who always were at risk of suffering acute lumbar sprain, the belt prevented the incidence of lumbar sprain.

Mitchell et al. (1994) administered a survey to 1,316 workers who perform lifting activities at an Air Force base to examine the effectiveness of back support belts in reducing back injuries and the cost-effectiveness of their use. Results showed that workers wearing back belts had significantly fewer lost days than those not wearing belts. The best predictors of whether a worker was going to suffer low-back injury were the amount of time spent performing lifts and the history of injury. Training programs seemed effective in preventing many of the back problems. However, the expense of injury while wearing a belt was found to be higher than if the worker was not wearing a belt.

Thompson et al. (1994) conducted a study of hospital workers to investigate the influence of wearing back belts on employee job attitudes and the experience of back pain. There were 41 subjects in the belt groups and 19 in the control group. Attitudes were found to be significantly improved as a result of the back belt program. Employees not only perceived belts as helping them avoid injury, but they reported a decrease of low-back pain.

PHYSICAL RESEARCH

Based on the extensive review of physical studies (see the appendix of this chapter for details), including psychophysical, biomechanical, physiological, or a combination of these, it was concluded that:

1. No studies provided conclusive evidence that actual trunk muscle forces, predicted spinal compression, or sheer forces were significantly reduced by wearing a back support belt.
2. Some studies concluded that back support belts reduce spinal loading under certain conditions.
3. A small number of studies suggest that back support belts may reduce the range of spinal motion for a person wearing a belt while lifting.
4. Based on physiological studies, the use of back support belts can put a temporary strain on the cardiovascular system.
5. Biomechanical studies suggest that long-term use of back support belts may decrease abdominal muscle tone, increasing the chances of back injury if the user discontinues use of the belt.
6. There are some concerns that a back support belt may alter a worker's perception of capacity to lift heavy loads.

SUMMARY

Industry today is looking for new methods of dealing with the increasing cost of—and lost work days due to—lower-back injuries. Industrial back support belts are remedies that more employers are utilizing to protect workers. Companies choose to use the belts to demonstrate, quickly and inexpensively, their wish to prevent back injuries.

Studies indicate that back support belts have potential disadvantages as well as advantages. While they seem to reduce lifting stress, they may lead to a false sense of security, so that injury occurs when they are not being worn. There also are comfort problems with some belts.

Most studies were conducted in controlled environments (campus laboratories) and have used mainly college-age (typically 24 years), weight-lifting men for short periods of time (less than 6 months, and often less than 6 hours).

More long-term studies in uncontrolled environments (in the workplace) need to be conducted.

Summarizing from NIOSH (1994):

1. There is insufficient data indicating that back belts reduce back stress during manual lifting.
2. There is insufficient scientific evidence that back belts reduce back injuries.
3. There is insufficient data relating the back injuries of healthy workers to the discontinued use of back belts.
4. Back belts may produce a temporary strain on the cardiovascular system.
5. The effectiveness of using belts to lessen the risk of back injury among uninjured workers remains unproven. The report did not recommend the use of back belts to prevent injuries among uninjured workers and did not consider back belts to be personal protective equipment.
6. Back belts do not mitigate the hazards posed by repeated lifting, pushing, pulling, twisting, or bending.

In the meantime, employers should be looking for ways to eliminate or reduce back injuries through ergonomic valuation and implementation of engineering control. More scientific research is needed before any conclusions can be drawn about the positive, negative, or long-term effects of lifting belts. NIOSH has started to write guidelines for epidemiological and biological studies. A nationwide discount retailer has been selected for the epidemiological study. Unfortunately, these studies will not be completed for several years.

For now, without objective and definitive back belt research, employers should remain cautious about using back belts to prevent back injuries on the job or to reduce the risk of injury.

APPENDIX

Walters and Morris (1970) found that wearing a spinal brace has little or no effect upon muscle activities when the subjects were standing, at rest, or walking.

According to Ahlgren and Hansen (1978), chronic lumbago is the most common ailment for which a corset is prescribed. This form of treatment is sometimes suggested because it provides support, with or without the relief of pain.

Million et al. (1981) suggested that a lumbar corset relieves symptoms by restricting spinal motion, rather than acting as an abdominal binder.

Grew and Deane (1982) confirmed that spinal supports influence the movement, intra-abdominal pressure (IAP), and skin temperature (directly under the support) of the wearer. The IAP was 0.28 in. (7 mm) Hg higher on average, and the skin temperature rose almost 34° F (2° C). The authors also suggested that, over the period of treatment, patients become accustomed to the orthosis, and subconsciously adopt it as part of the spinal support mechanism. They affirmed the importance of using other treatments, such as exercise regimes, along with a spinal support, especially when patients cease to wear the corset.

The mechanical function of medically-prescribed lumbar spine orthoses was studied by Nachemson et al. (1983). Only four subjects (one male and three females) without previous back injury were used in the study. The results indicated that wearing lumbar spine orthoses can significantly unload the trunk in some situations, but has no effect in others. Lumbar spine compression was reduced by about one-third during trunk flexion. The back muscle activity was reduced in some experiments by one-third with a back support. In others, it increased by the same amount. None of the back supports raised the intragastric pressure significantly. No one orthosis was found superior in mechanical effectiveness in the tasks studied.

Hemborg et al. (1985) studied the immediate effect of two types of back belts on the activity of the oblique abdominal muscles and the erector spinae muscle, and on the intra-abdominal and intrathoracic pressure during lifting. They used 20 subjects from the construction industry with chronic low-back pain and 10 well-trained weight lifters. They showed that the two belts did not interfere with activation of the oblique abdominal muscles and the erector spinae muscle. Both belts made the intra-abdominal pressure rise moderately at all times, while the intrathoracic pressure was only slightly increased during some types of lifts.

Kumar and Godfrey (1986) demonstrated that spinal braces can increase the IAP. They also found that there was no significant difference in the IAP generated by six different braces.

Lantz and Shultz (1986) studied the effects of wearing three commonly-prescribed back braces and corsets on restrictions of gross body motions and trunk muscle myoelectric activity. They used five male university students with no history of significant back pain. All three orthoses restricted some trunk motions (flexion, extension, lateral bending, twisting). Based on the results of the first study, the authors concluded that all three back supports would likely reduce loads placed on the lumbar spine.

The results of a second study showed that none of the orthoses were consistently effective in reducing measured myoelectric activity and, in some cases, signal levels increased when orthoses were worn. The authors concluded that lumbar orthoses are mechanically effective only sometimes, often are not effective, and sometimes are counterproductive. They said that even when lumbar orthosis wearing is effective, the load reduction is not dramatic.

Alaranta and Hurri (1988) indicated that the application of a corset is a reasonable complement in conservative treatment for patients with chronic low-back pain. They decided that trunk muscles may be weakened by physical inactivity, not by wearing a corset. Hence, recommendations of proper trunk exercises should be a rule, rather than an exception, when prescribing a lumbar support.

McCoy et al. (1988) evaluated the role of lifting belts in manual lifting. The 12 healthy male college participants reported that the belts increased the perceived maximum acceptable work load compared with the use of no belt (from 248.5 lb ft/minute [34.4 kg m/minute] to at least 280 lb ft/minute [38.8 kg m/minute]).

In a study using 12 college females, Sherwood (1988), evaluated the effects of lifting with three types of belts—leather, Air Belt™, and CompVest™—and compared it to a control group using no belt. A psychophysical lifting task and subjective surveys administered to the participants regarding their belt preference and body part discomfort were used as criteria. Results showed that there were no significant differences between groups wearing belts and the control group when raising the maximum acceptable weight of lift. Body part discomfort, with the exception of the buttocks, also was nonsignificant when comparing belts versus control. However, for the buttocks, the CompVest produced significantly more body part discomfort. The subjective survey of the belts found no significant differences in comfort or support provided by the belts. Subjects preferred the no-belt condition and the CompVest slightly more than the other two conditions (leather belt and Air Belt).

Amendola (1989) assessed the utility of belts for manual lifting. He used the Air Belt, the CompVest, and a combination of both. The experiment used four methods of data collection: biomechanical, psychophysical, subjective survey (rating and ranking), and body part discomfort. His results yielded no significant difference between the belt conditions and a control without the belt at maximum acceptable lift weight. No significant differences were found among belts in alleviating the compressive force in the lower-back. For body part discomfort and preference, the ranking indicated no clear favorite among the belts.

Horsford (1989) investigated the effectiveness of back support belts on 12 healthy college males in a lifting experiment at 50% and 75% of maximum voluntary dynamic lifting capability. He used the same criteria as Wilson (1989). In the surveys, the participants responded favorably to the belt treatment; in the ranking they preferred the belt treatment over no belt. However, he concluded that the belts tested did not aid individuals involved in lifting tasks.

Hunter et al. (1989) reported the effects of a weight training belt on blood pressure during exercise. They used five healthy males and one healthy female. The exercises used were aerobic bicycle, on-arm bench press exercise, and isometric dead-lift exercise. The results of this study showed that blood pressure is affected by the use of a weight-lifting belt during both rest and exercise. They warned that individuals with compromised cardiovascular systems are probably at greater risk when undertaking exercise with back support. They recommended back belts not be cinched tightly while performing aerobic exercise.

Rovere et al. (1989) reported that the lumbosacral corset is used to limit motion and provide external support in subjects with back injuries resulting from athletic activities.

Wilson (1989) evaluated back support belts on 12 healthy college females. The four criteria were biomechanical, psychophysical, subjective survey (rating and ranking), and body part discomfort. She concluded that, at 50% and 75% of voluntary maximum lift, an analysis of the shear and compressive forces of the L5/S 1 disc revealed no significant difference between the use of belts and a control situation without a belt. The body part discomfort survey and the subjective comfort and support questionnaires yielded no difference among belts. The conclusion of this study was that back support devices cannot be recommended as an aid in lifting.

Harman et al. (1989) and Lander et al. (1990) analyzed the effect of a weight-lifting belt during performance of dead-lift and squat exercises, respectively. Results confirmed that a weight-lifting belt can aid in supporting the trunk by increasing IAP (0.6 in. [15 mm] Hg and 0.7 in. [18 mm] Hg higher for the peak IAP, respectively). Lander et al. (1992) studied the effect of a weight-lifting belt during multiple repetitions of the squat exercise. They suggested that the use of a weight-lifting belt aids in supporting the trunk due to the clearly increased IAP (approximately 11% from the first to last repetition) during the squat exercise.

McGill et al. (1990) studied the effect of an abdominal belt on trunk muscle activity and IAP during squat lifts. They concluded that the muscle activity and IAP results of their study made it difficult to justify the prescription of abdominal belts to workers. Belts do not appear to contribute to support of the loaded lumbar spine, based on the erector spinae activity. However, the authors do not exclude possible benefits from the belt, such as restricting the amount of forward flexion or axial twist of the spine, forcing a worker to pivot by moving the feet while working, or reminding the lifter to maintain proper posture when the task at hand is physically dangerous.

Later, McGill et al. (1994) studied the effect of breath holding, using a weight-lifting belt, on trunk stiffness. They used 37 college students without history of back pain. They found that both breath holding and leather weight-lifting belts appeared to stiffen the torso while subjects performed lateral bending and axial rotation. However, the rigid belts did not consistently restrict forward bending or straightening up.

Woodhouse et al. (1990) concluded that no statistically significant isokinetic changes existed relative to improving functional lifting capacity, between subjects wearing various types of lumbar/sacral supports and a control condition. Their study used 10 well-conditioned male athletes. The results suggested that there were no functional lifting qualities (increased lifting capacity) attributed to utilizing some back supports while performing maximal lifting tasks.

The effect of a weight-lifting belt on spinal shrinkage was studied by Bourne and Reilly (1991). Spinal loading during weight lifting is reflected in changes of stature (shrinkage). The authors concluded that wearing a weight-lifting belt tends to induce less absolute spinal shrinkage (0.144 in. [3.7 mm] without the belt, and 0.116 in. [2.9 mm] with the belt) and causes significantly less discomfort compared to lifting without a belt. They supported the hypothesis that a weight-lifting belt helps stabilize the trunk.

Penrose et al. (1991) found that the use of an inflatable lumbar corset may contribute significantly to regaining muscular strength and flexibility lost through an injury in the lower-back and at the same time it may lessen the pain.

Axelsson et al. (1992) studied the stabilizing effect of back supports on the intervertebral mobility of the lower lumbar region. They used seven patients who had back surgery. The patients used the lumbar supports for 5 months after surgery. The authors concluded that neither of the two types of lumbar support had any stabilizing effect on the intervertebral mobility of the lower lumbar spine. They also suggested that any kind of lumbar support can be used during lumbar fusion healing as long as the support reminds and helps the patient to keep the trunk straight.

Hamonet and Meziere (1993) investigated the activity level of the abdominal muscles (rectus abdominis and transversus abdominus), with and without a flexible back belt, on 12 subjects who had been suffering from lumbar pain for more than 4 years. Based on 75% of the cases, they concluded that wearing a flexible back belt does not affect the activity level in normal positions and ordinary daily movements. This, in their opinion, is an argument against the theory that wearing a back belt causes muscle atrophy.

Lavender and Kenyeri (1994) compared the maximum acceptable weight of lift with and without the use of an elastic type of back belt. They used 16 subjects, 11 males and five females. They concluded that the use of lifting belts does not appear to offer a significant biomechanical or motivational advantage to the user when handling loads considered acceptable in repetitive material handling tasks. More than half of the subjects found the belts uncomfortable. Based on this, the authors predict poor compliance with organizational demands.

Lavender et al. (1994) studied the effect of lifting belts on trunk motions, using eight male and eight female nursing personnel and students. As dependent measures, they used torso kinematic data in the sagittal, frontal, and transverse planes. They found that the belt reduced lateral trunk motion if foot movement was restricted. Also, the combination of belt, asymmetry, and foot movement conditions affected the most extreme benign and twisting postures. These results lead to the recommendation that ergonomists should focus on the design of workplace layouts that encourage foot movement, rather than providing a lifting belt.

Reyna et al. (1995) found that lumbar belts neither augment isolated lumbar muscle strength nor increase dynamic lifting capacity. They studied lifting performance in 22 healthy untrained subjects, with and without a commercially-available back belt. They pointed out that the belt still may have other benefits, but it does not strengthen the spine.

REFERENCES

Ahlgren, S.A., and Hansen, T. "The Use of Lumbosacral Corsets Prescribed for Low-back Pain." *Prosthetics and Orthotics International* 1978, 2: 101-104.

Alarantra, H., and Hurri, H. "Compliance and Subjective Relief by Corset Treatment in Chronic Low-back Pain." *Scandinavian Journal of Rehabilitation Medicine* 1988, 20: 133-136.

Amendola, A.A. "An Investigation of the Effects of External Supports on Manual Lifting." Ph.D. dissertation. College Station, TX: Texas A&M University, 1989.

Axelsson, P., Johnsson, R., and Stomqvist, B. "Effect of Lumbar Orthosis on Intervertebral Mobility: A Roentgen Sterophotogrammetric Analysis." *Spine* 1992, 17(6): 678-681.

Ayoub, M.M., and Mital, A. *Manual Materials Handling*. London: Taylor & Francis, 1989.

Battie, M.C., Bigos, S.J., Fischer, L., Hansson, T.H., Jones, M.E., and Wortley, M.D. "Isometric Lifting Strength as a Predictor of Industrial Back Pain Reports." *Spine* 1989, 14: 851-856.

Berry, J.F. "Letters to the Editor." *American Journal of Physical Medicine & Rehabilitation* 1991, 70: 111-112.

Bourne, N.D., and Reilly, T. "Effect of a Weight-lifting Belt on Spinal Shrinkage." *British Journal of Sports Medicine* 1991, 25(44): 209-212.

Brown, J.R. "Factors Contributing to the Development of Low-back Pain in Industrial Workers." *American Industrial Hygiene Association Journal* 1975, 36: 26-31.

Chaffin, D.B., Herrin, G.D., and Keyserling, W.M. "Pre-employment Strength Testing: an Updated Position." *Journal of Occupational Medicine* 1978, 20: 403-408.

Congleton, J.J., Amendola, A.A., Horsford, W.H., McCoy, M.A., Reddell, C.R., Sherwood, C.A., and Wilson, K.F. "Brief Summary of Lifting Belt Research Conducted at Texas A&M University." W.S. Marras, W. Karwowski, J.L. Smith, and L. Pacholski, eds. *The Ergonomics of Manual Work*. London: Taylor and Francis, 1993: 139-140.

De Ruiter, M. "Back to Basics with ProFlex." *Ergonomics* 1990, 33(3): 383-385.

Fograscher, J. "A Sound Approach to Ergonomics." *Ohio Monitor,* December 1991.

Grew, N.D., and Deane, G. "The Physical Effect of Lumbar Spinal Supports." *Prosthetics and Orthotics International* 1982, 6: 79-87.

Hamonet, C.L., and Meziere, C. "Comparative Study of the Activity Level of the Abdominal Muscles With and Without a Flexible Lumbar Belt in Back Pain." *Rhumatologie* 1993, 45(7): 165-170.

Harman, E.A., Rosenstein, R.M., Frykman, P.N., and Nigro, G.A. "Effects of a Belt on Intra-abdominal Pressure During Weight-lifting." *Medicine and Science in Sports and Exercise* 1989, 21(2): 186-190.

Hemborg, B., Mortiz, U., Holmstrom, E., and Akesson, I. "Lumbar Spinal Support and Weight-lifter's Belts: Effect on Intra-abdominal and Intra-thoracic Pressure and Trunk Muscle Activity During Lifting." *Journal of Manual Medicine* 1985, 1: 86-92.

Hilgen, T.H., Smith, L.A., and Lander, J.E. "The Minimum Abdominal Belt-aided Lifting Weight." *Advances in Industrial Ergonomics and Safety III.* W. Karwowski and J.W. Yates, eds. London: Taylor and Francis, 1991: 271-224.

Holmstrom, E., and Mortiz, U. "Effects of Lumbar Belts on Trunk Muscle Strength and Endurance: A Follow-up Study of Construction Workers." *Journal of Spinal Disorders* 1992, 5(3): 260-266.

Horsford, W.H. "An Ergonomic Evaluation of Back Support Devices Used as Lifting Aids by Males." Master's thesis. College Station, TX: Texas A&M University, 1989.

Howell, M.E. "Back Support Devices Pro's & Con's." *FMI Ergonomics and Workplace Safety Conference.* Cambridge, MA: FMI, Sept. 14, 1992.

Hunter, G.R., McGuirk, J., Mitrano, N., Pearman, P., Thomas, B., and Arrington, R. "The Effects of a Weight Training Belt on Blood Pressure During Exercise." *Journal of Applied Sport Science Research* 1989, 3(1): 13-18.

Imker, F.W. "The Back Support Myth." *Ergonomics in Design*, April 1994: 8-12.

Klaber, J.A., Chase, S.M., Portek, I., and Ennis, J.R. "A Control Perspective Study to Evaluate the Effectiveness of Back School and the Relief of Chronic Low-back Pain." *Spine* 1986, 11: 120-123.

Kroemer, K.H.E. "Personnel Training for Safer Material Handling." *Ergonomics* 1992, 35(9): 1,119-1,134.

Kumar, S., and Godfrey, C.M. "Spinal Braces and Abdominal Support." *Trends in Ergonomics/Human Factors III.* North Holland, The Netherlands: Elsevier Science Publishers B.V., 1986: 717-726.

Lander, J.E., Hundley, J.R., and Simonton, R.L. "The Effectiveness of Weight Belts During Multiple Repetitions of the Squat Exercise." *Medicine and Science in Sports and Exercise* 1992, 22(1): 117-126.

Lander, J.E., Simonton, R.L., and Giacobbe, J.K.F. "The Effectiveness of Weight Belts During the Squat Exercise." *Medicine and Science in Sports and Exercise* 1990, 22(1): 117-126.

Lantz, S.A., and Schultz, A.B. "Lumbar Spine Orthosis Wearing I: Restriction of Gross Body Motions." "Lumbar Spine Orthosis Wearing II: Effect on Trunk Muscle Myoelectric Activity." *Spine* 1986, 11(8): 834-842.

Lavender, S.A., and Kenyeri, R. "Lifting Belts: A Psychophysical Analysis." *Ergonomics* 1995, 38(9): 1,723-1,729.

Lavender, S.A., Thomas, J.S., Chang, D., and Andersson, G.B.J. "The Effect of Lifting Belts on Trunk Motions." F. Aghazadeh, ed. *Advances in Industrial Ergonomics and Safety VI.* London: Taylor and Francis, 1994: 667-670.

Liles, D.H. "Using NIOSH Lift Guide Decreases Risk of Back Injuries." *Occupational Health and Safety* 1985, 54(12): 57-60.

Lund, K., and Rambo, J. "Weighing the Evidence: Controversy Over Back Belts Raises Questions about Workplace Use." *Occupational Health and Safety* 1994, 63(11): 33-35.

Mahone, D.B. "Manual Materials Handling: Stop Guessing and Design." *Industrial Engineering* 1994, 26(3): 29-31.

Marras, W.S., Karwowski, W., Smith, J.L., and Pacholski, L., eds. *The Ergonomics of Manual Work.* London: Taylor and Francis, 1993.

McCoy, M.A., Congleton, J.J., and Johnston, W.L. "The Role of Lifting Belts in Manual Lifting." *International Journal of Industrial Ergonomics* 1988, 2: 259-266.

McGill, S.M., Norman, R.W., and Sharratt, M.T. "The Effect of an Abdominal Belt on Trunk Muscle Activity and Intra-abdominal Pressure During Squat Lifts." *Ergonomics* 1990, 33(2): 147-160.

McGill, S.M., Seguin, J., and Bennett, G. "Passive Stiffness of the Lumbar Torso in Flexion, Extension, Lateral Bending, and Axial Rotation: Effect of Belt Wearing and Breath Holding." *Spine* 1994, 19(6): 696-704.

Million, R., Haavik Nilsen, K., Jayson, M.I.V., and Backer, R.D. "Evaluation of Low-back Pain and Assessment of Lumbar Corsets with and without Back Supports." *Annals of the Rheumatic Diseases* 1981, 40: 449-454.

Mitchell, L.V., Lawler, F.H., Bowen, D., Mote, W., Asundi, P., and Purswell, J. "Effectiveness and Cost-effectiveness of Employer-issued Back Belts in Areas of High Risk for Back Injury." *Journal of Occupational Medicine* 1994, 36: 90-94.

Morris, T.L., and Del Vecchio, D.C. "Evaluation of the Champion Ergonomics Work Belt: Interim Report." Dallas, TX: *Advanced Ergonomics, Inc.*, 1992.

Nachemson, A., Schultz, A., and Anderson, G.B.J. "Mechanical Effectiveness Studies of Lumbar Spine Orthoses." *Scandinavian Journal of Rehabilitation Medicine* (Supplement) 1983, 9: 139-149.

National Institute for Occupational Safety and Health (NIOSH). *Workplace Use of Back Belts.* DHHS (NIOSH), No. 94-122. Cincinnati, OH: NIOSH, 1994.

National Safety Council. *Accident Facts l992 Edition.* Itasca, IL: National Safety Council, 1992.

Penrose, K.W., Chook, K., and Stump, J.L. "Acute and Chronic Effects of Pneumatic Lumbar Support on Muscular Strength, Flexibility, and Functional Impairment Index." *Sports Training, Medicine, and Rehabilitation* 1991, 2: 121-129.

Perry, G.F. "Occupational Medicine Forum." *Journal of Occupational Medicine* 1992, 34 (7): 679-680.

Reddell, C.R., Congleton, J.J., Huchingson, R.D., and Montgomery, J.F. "An Evaluation of a Weight-lifting Belt and Back Injury Prevention Training Class for Airline Baggage Handlers." *Applied Ergonomics* 1992, 23(5): 319-329.

Reyna, J.R., Leggett, S.H., Kenney, K., Holmes, B., and Mooney, V. "The Effect of Lumbar Belts on Isolated Lumbar Muscle: Strength and Dynamic Capacity." *Spine* 1995, 20(1): 68-73.

Rovere, G.D., Curl, W.W., and Browninig, D.G. "Bracing and Taping in an Office Sports Medicine Practice." *Clinics in Sports Medicine* 1989, 8(3): 497-515.

Sherwood, C.A. "An Evaluation of Back Support Devices for Females Involved in Lifting Tasks." Master's thesis. College Station, TX: Texas A&M University, 1988.

Thompson, L., Pati, A.B., Davidson, H., and Hirsh, D. "Attitudes and Back Belts in the Workplace." *Work* 1994, 4(1): 22-27.

Tomecek, S. "Shop Smart to Help Your Workers' Backs." *Safety and Health,* November 1992: 30-38.

Udo, H., Seo, A., Koda, S., Kurumatani, N., Dejima, M., Hisashige, A., Fujimura, T., Mutuura, Y., Matumura, K., and Iki, M. "The Effect of a Preventive Belt on the Incidence of Low-back Pain (Part II): Investigation in Rice-carrying Work." *Journal Science of Labour* 1992, 68(10): 503-520.

Udo, H., Tanida, H., Isawa, S., Umino, H., Inoue, Y., Hirano, S., and Iyadomi, Y. "The Effect of a Preventive Belt on the Incidence of Low-back Pain, the First Report: Discussion on the Changing Rolling Roller Work." *Journal of Labour Hygiene in Iron and Steel Industry* 1992, 39(2): 1-6.

Udo, H., Yoshinaga, F., Tanida, H., Umino, H., and Yoshioka, M. "The Effect of a Preventive Belt on the Incidence of Low-back Pain (Part III): Investigation in Crane Work." *Journal Science of Labour* 1993, 69(1): 10-21.

Walsh, N.E., and Schwartz, R.K. "The Influence of Prophylactic Orthoses on Abdominal Strength and Low-back Injury in the Workplace." *American Journal of Physical Medicine & Rehabilitation* 1990, 69: 245-250.

Walters, R.L., and Morris, J.M. "Effect of Spinal Support on the Electrical Activity of Muscles of the Trunk." *The Journal of Bone and Joint Surgery* 1970, 52-1(1): 51-60.

Wilson, K.F. "An Evaluation of Lower-back Support Devices for Females." Master's thesis. College Station, TX: Texas A&M University, 1989.

Woodhouse, M.L., Heinen, J.R.K., Shall, L., and Bragg, K. "Selected Isokinetic Lifting Parameters of Adult Male Athletes Utilizing Lumbar/Sacral Supports." *Journal of Orthopedics and Sports Physical Therapy* 1990, 11(11): 467-473.

Yu, T., Roht, L.H., Wise, R.A., Kilian, D.J., and Weir, F.W. "Low-back Pain in Industry: An Old Problem Revisited." *Journal of Occupational Medicine* 1984, 26: 517-525.

CHAPTER 15

TRAINING AND EDUCATION IN BACK INJURY PREVENTION

Glenda L. Key
President
KEY Method
Minneapolis, MN

When it comes to back injury, the plant manager's major responsibility is to be educated in prevention (Figure 15-1). There are four points at which the plant manager makes the biggest impact on preventing back injuries—prior to the hire, after the hire and prior to the injury, after the injury, and when the injured employee returns to work.

IMPACT POINTS

PRIOR TO THE HIRE

Hiring the right person minimizes the possibility of costly, fraudulent, or inflated workers' compensation claims. What are the legal and economic consequences if an injury occurs after you hire a person without knowing his or her physical capabilities for a job with unknown physical requirements? Having quality information about potential hires is critical.

Job placement assessments and job analysis are two pieces of information that need to be linked in an organized, standardized manner in the hiring process. After integrating this strategy with their hiring process, one of the world's largest trucking firms has recorded no new incidents of injury.

AFTER THE HIRE

Prior to the Injury

Trade publications are flooded with programs and tools available to the plant manager and staff for preventing injury. Some education, fitness, and exercise programs have been effective. Options are numerous, but any that pull employees off work are costly. It is difficult, however, to sort the pertinent from the unsuitable.

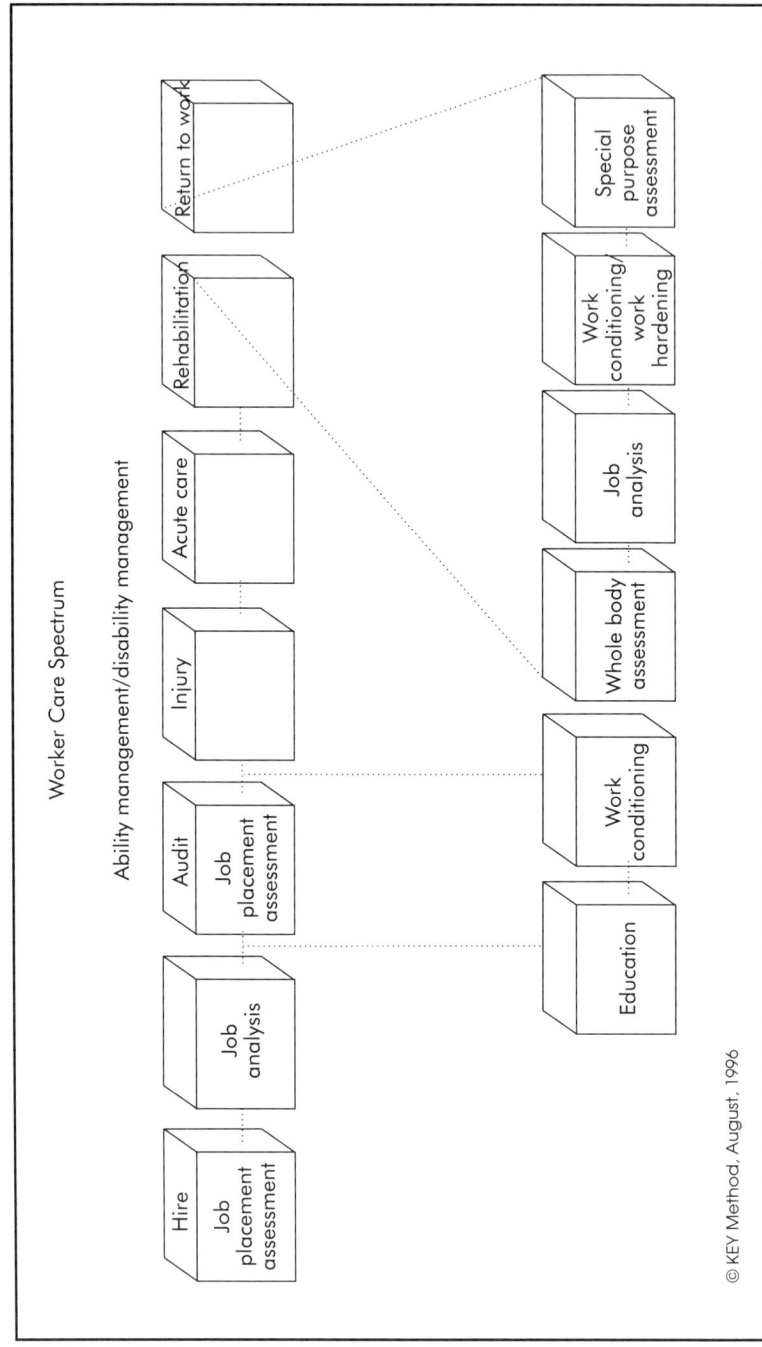

Figure 15-1 Overview of the ability/disability management techniques a manager should know to make sound judgments about trying to avoid back injuries and managing the problem if such an injury does occur.

Outcomes are the most important measurements of an injury prevention program. Successful results ensure that the risks and investments have a positive effect on the bottom line.

After the Injury

What are the tools for expediting an injured employee's return to work? The system has now come full circle to lay the control of this process in the hands of the employer. This is a very difficult and frustrating task for a manager. It is his or her responsibility to keep costs down by trying to avoid injury and expediting resolution of any injury cases that do occur. How do you know the returning employee is going to be able to do the job? Are there ways to identify the individuals who are trying to abuse the system by malingering? What assurances exist to predict that once the worker is readmitted to employment, there will be no re-injuries?

The critical information needed for decision making is found in a functional capacity assessment, which identifies an injured worker's return to work capabilities. There is great possibility for legal action if individuals return to work at a level that results in re-injury or if they are prevented from returning to work because of an underestimation of ability, thereby limiting their income. The case becomes exponentially complicated if the individual perceives action taken or not taken as the result of gender, racial, age, or disability discrimination.

Are there treatment programs nearby that cover the specific function and work needs of the employee and employer? This is important for the acute stage, as well as the longer-term treatment of a back injury.

After the Injured Employee Returns to Work

To maintain the capability level of rehabilitated workers, it is imperative to follow all recommendations from the treatment programs or functional assessment. It is important to not only follow the guidelines, but also to pay close attention to the participation level or motivation determinant. If only numbers of capability are provided, without a specific participation determination, you will not know whether the person is giving less than full capability, or overdoing it and working beyond what is actually safe.

PREVENTION PROGRAMS

Some theorize that recurrence of back injuries can be prevented with proper job placement and injury prevention programs. Several of these programs are discussed in this section.

AUDIT ASSESSMENT

The audit assessment is a tool that can be used on a periodic basis to track the physical capability of workers. In addition to being very important for an organization that has an aging work force with little turnover, this assessment is especially appropriate for heavy labor or emergency action jobs. The audit assessment is generally a repeat of the job placement assessment. It may or may not have added components. Using the same, or many of the same assessment components allows the worker and the manager to see changes easily, minimizing discrepancy in interpreting the results or matching the information. All parties can immediately see the changes, then compare them to the job description to confirm the employee is still meeting minimum requirements of the job.

Other applications for an audit assessment include:

1. Rotation of jobs. Rotation in and of itself may not be a successful strategy unless there is assurance that each individual meets the minimum requirements for the physical demands of the job.

2. Transfer to other jobs. It is important to know that individuals have the physical capabilities to perform the job prior to moving them. This becomes even more important when more than one employee applies for a transfer to the same job at the same time.

BACK BELTS

The issue of back belts is covered in another chapter; it is mentioned briefly here to recognize it as an option in a back injury prevention program. If literature that represents case and experience outcomes were to demonstrate no significant change in injury rates or cost savings, back belts would fade quickly away—as did the use of X-rays for pre-employment screening. There are as many stories that support non-use as there are that support the use of back belts. It would benefit the reader to learn more about the circumstances that made applications successful or unsuccessful. Identify the likenesses and circumstances and make an educated decision to include it or not.

EDUCATION

Although an education program in back injury prevention is imperative because of changing technology, inconsistent findings in literature, a workers' compensation system that needs improvement, and a continuing shortage of time make it an often complicated and challenging endeavor. In addition to reducing injuries, the goals and objectives of education programs include minimizing injuries, returning workers to work more quickly, preventing productivity loss, decreasing employee lost work days, and avoiding litigation.

An education program offers opportunities for all, from all levels of management and all levels of workers, to have an impact and make a difference. The difference can be felt physically, medically, and economically at all levels. This chapter does not cover specific content of an education program, but assists in understanding the organization and general content issues.

General implementation guidelines critical in bringing about successful results include:

1. All management and hourly employees must participate (Melnik 1995; Saunders and Saunders 1995; Saunders et al. 1995).
2. Accept that this is a process, not a program (Melnik 1995). There is a beginning point, but no end.
3. Education is the underlying force that moves all successful programs forward.
4. Interaction is required. A lecture and listen format will not bring about the desired outcomes.
5. Suppliers, insurance carriers, and medical providers must participate in the process.

The material handling content is fairly standard across the programs. Delivery of the information is not. The delivery needs to be captivating and immediately applicable.

Figure 15-2 shows the results of Saunders' et al. (1995) study of the impact of employee back injuries on lost work days in a rural nursing home. Back injuries were reduced from 705 in 1990 to only 39 in 1991, following intervention. This represents a decrease of 666 days, a 95% reduction in back injury lost work days (Saunders et al. 1995).

The fear of education producing an injury epidemic is a legitimate concern. If information is provided that has minimal verbiage on how to procure the injury, minimal medical jargon, and a focus on taking responsibility for one's own health through prevention techniques, this will not occur.

A 40% decrease in lost work days was reported by Melton in a study of eight different industries. "Although an increase in the reports of lower back pain was noted, those reporting showed an 86% reduction in lost time days" (Saunders and Saunders 1995).

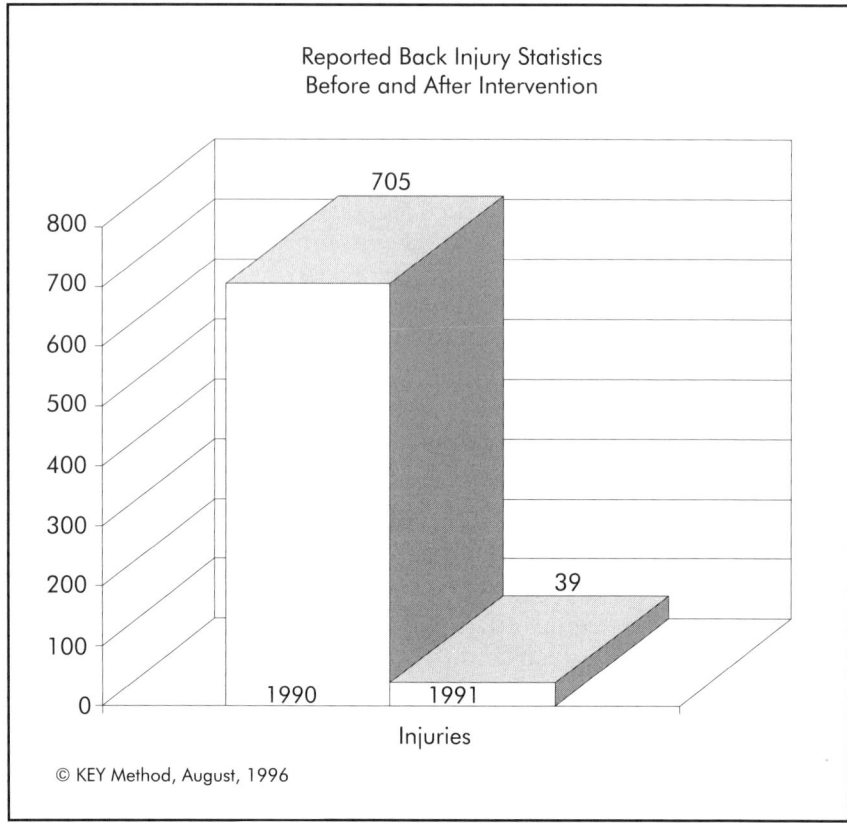

Figure 15-2 Chart shows dramatic decline in employee back injuries in just one year due to interventions by the employer in a rural nursing home.

As with all programs, it is important to call on references to confirm the legitimacy of the consultant's claim. An expected cost-to-benefit ratio should be written up between the buyer and the seller of the program. Outcomes can then be tracked and reported against it.

ERGONOMICS

Ergonomics is covered in greater detail throughout this book. It is of critical importance to acknowledge ergonomics as one of the areas of impact for preventing back injuries.

Manual lifting may save the cost of an automated piece of equipment. But the money spent on medical care, workers' compensation premiums, and lost work days might very likely pay for the purchase of a newer technology, which could be responsible for decreasing costs in the future, while increasing productivity (Aders 1996).

EXERCISE

On-the-job exercise programs can be an important component of a back injury prevention program. Many of the issues relating to establishing an exercise program are covered in the section on fitness. The goals of a well-organized program, with high participation, are to control fatigue, increase strength, improve flexibility, build endurance, stimulate blood flow, reduce stress, and offer reverse positions for the body.

Frequency and Structure

Stretching and other fatigue control exercises have the greatest results when performed at intervals or points in time that are easy for employees to remember. These include just prior to starting work and at close of day, immediately after break or lunch, and after completing a work cycle, such as after loading each pallet. Saunders supports the philosophy that in companies with 200 or more employees, there should always be someone stretching, or in some form of exercise, at any point in time (Saunders et al. 1995).

Exercise programs have typically been structured around group activities. Exercise programs should be based on data gathered during the hiring process, especially through job placement assessment. Some companies offer audit assessments on an annual basis. This provides the company and the worker with data for participation decisions in fitness or exercise programs.

FITNESS

A well-run, well-promoted employee fitness program is a sound corporate investment in health care. Fortune 500 companies comprise a major share of the companies with physical fitness programs. This is in contrast to only 75 in 1973. Xerox, IBM, Control Data, General Foods, General Electric, Johnson & Johnson, Kimberly Clark, and Prudential Insurance are all offering a wellness program (Volski 1995). Fitness programs range in size and content. Activities most commonly included in a fitness program include:

- Low impact exercise (minimal joint compression);
- Stretching (increases and improves flexibility);
- Back education for lifting activities (prevents injury);
- Aerobic exercise (cardiovascular);
- Strength and endurance exercises (with or without fitness equipment);

- Walking or running routine (cardiovascular and endurance building); and
- Activities and education for weight loss and control.

Although a fitness or wellness program can be structured in many ways, three common options are (Volski 1995):

1. Company-operated, in-house. The location of the fitness center is on-site and the individuals operating and instructing are employees of the plant.
2. Company-sponsored, in-house. The location is on-site and the individuals operating and instructing are contracted from elsewhere.
3. Company reimbursed. The location is off-site owned by others and employees are either reimbursed for membership or membership is paid by the employer to the facility.

Goals and Benefits

Goals of the program should be clearly defined and articulated in short-term and long-term classifications. Short-term categories would include issues such as attendance, exercise increments, and frequency of participation in specific programs. Long-term goals would include categories such as:

1. Decreased disability days. Significant reduction in disability days was demonstrated by employers when comparing corporate fitness program participation. The Association of Quality Clubs compiled and reviewed the results of several studies, summarized in Figure 15-3 (Volski 1995).
2. Decreased health care costs. General Electric reduced health care costs by 38% in 18 months, attributing the changes to the inauguration of a fitness program (Figure 15-4) (Volski 1995).
3. Reduced employee turnover rates. British Columbia Hydroelectric found that employees who participated in a company fitness program left the company less frequently, as compared with the average turnover rate (Figure 15-5) (Benefits of Employee Health Programs 1991).
4. Improved morale and productivity. Figure 15-6 demonstrates that morale improved and productivity increased for the fitness participants at Saatchi & Saatchi Advertising (Volski 1995). Participants in an exercise program at NASA demonstrated improved work performance. Nonexerciser's efficien-

cy decreased during the final 2 hours of the day by 50%, whereas exercise adherents worked at full efficiency all day (Volski 1995).

FUNCTIONAL CAPACITY ASSESSMENTS

Each year nearly 500,000 U.S. workers are unable to resume their jobs for long periods of time as the result of injury. Workers' compensation costs have tripled since 1980, reaching $70 billion per year. Knowing a worker's functional capabilities is one of the most valuable pieces of information for injury prevention a professional can have in the workplace today. The Functional Capacity Assessment (FCA) is a process that determines an individual's physical functional capabilities through measuring, recording, and analyzing data gathered during a standardized testing procedure. The capabilities assessed should include three categories:

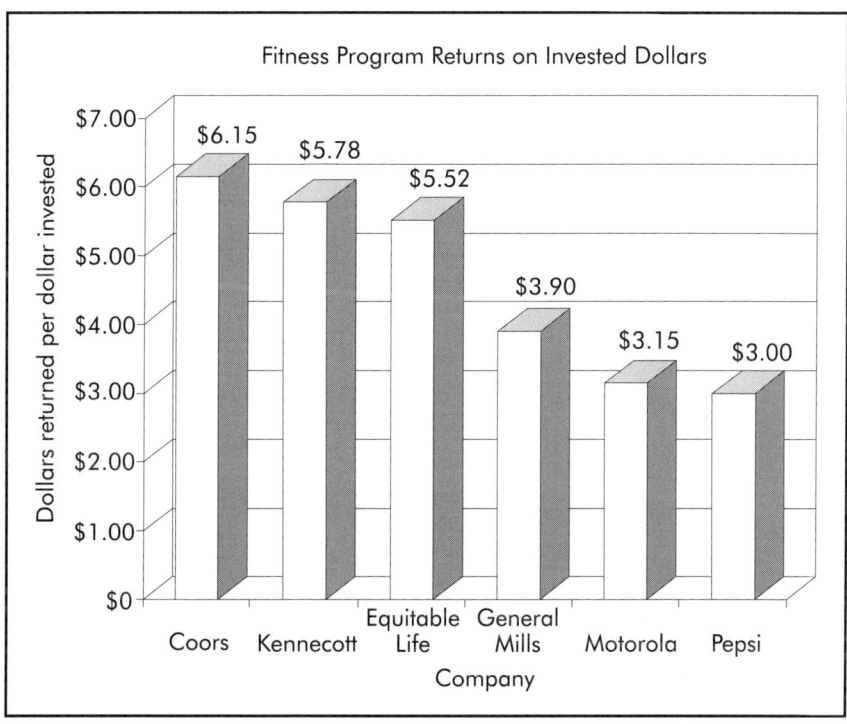

Figure 15-3 Fitness program returns on dollars invested by various employers.

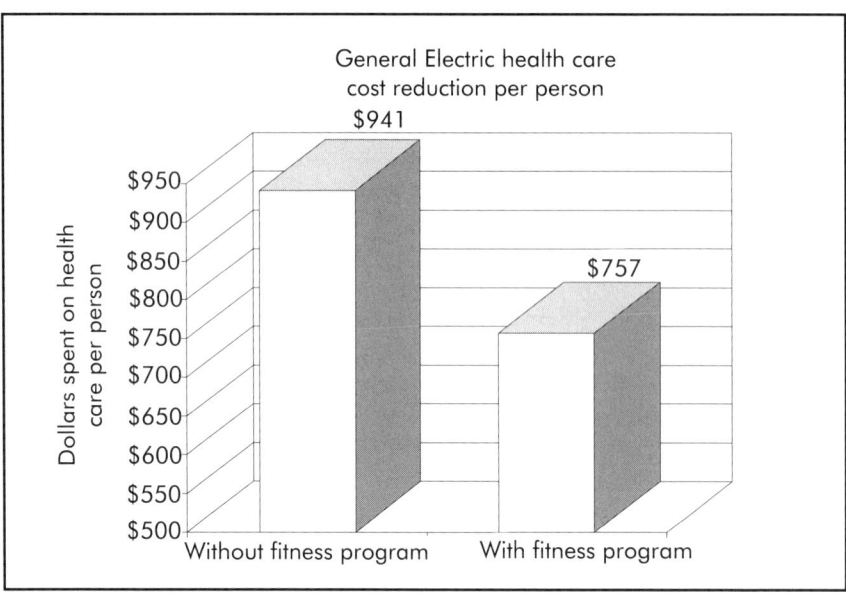

Figure 15-4 After inaugurating a fitness program for employees, General Electric reduced its health care costs by 38%.

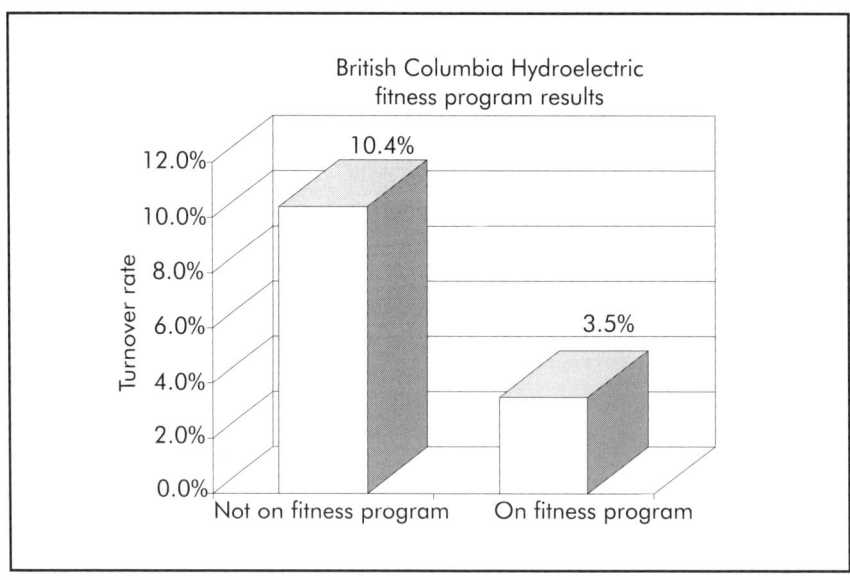

Figure 15-5 Employees participating in a fitness program at British Columbia Hydroelectric averaged a 3.5% turnover rate, compared with the overall average of 10.4%.

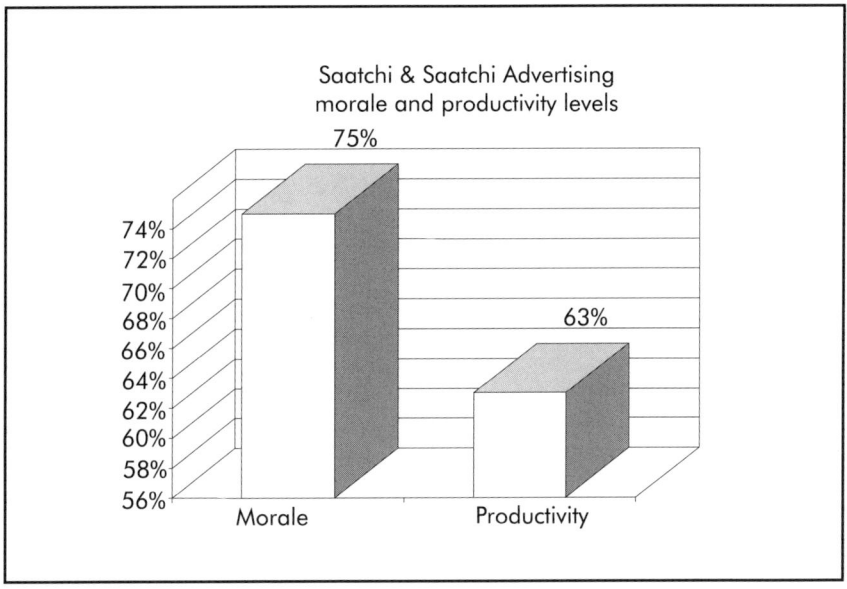

Figure 15-6. Many people claim that an exercise program can lead to increased morale and productivity. This was the experience at Saatchi and Saatchi Advertising. An exercise program there led to a 75% rise in morale and a 63% increase in productivity.

1. Weighted capabilities includes analyzing an individual's material handling capabilities, such as lifting at standard heights, carrying, pushing, and pulling. Tolerances of posture should be included within each of the components. A detailed list of components is found in Table 15-1.
2. Tolerance and endurance assessment should also provide formulas used to identify specific tolerances to the workday itself. A detailed list of these components is found in Table 15-2.
3. Validity of participation determinants should be evaluated through statistical methods of process and decision science (Grossman 1983).

Determinants for validity of participation have been developed by one vendor of FCA equipment and protocols (Key 1984). The statistical validation is based upon algorithms using principles of science and medicine (Grossman 1983). As an example, certain vital signs, such as heart rate, respond predictively as the individual approaches and reaches exertion level (Worker Data Bank 1994, 1995, 1996; Astrand and Ryhming 1954). The delineations of per-

Table 15-1. Weighted capabilities

Lifting desk to chair
Lifting chair to floor
Lifting above shoulder
Carrying
Pulling
Pushing

Table 15-2. Tolerance and endurance

Balancing
Bending
Cervical mobility
Circuit board tolerance
Climbing
Crawling
Crouching
Fastener board tolerance
Fine manipulation
Firm grasping
Grip strength
Keyboard tolerance
Kneeling
Reaching
Repetitive foot motion
Simple grasping
Sitting
Squatting
Standing
Stooping
Tool station work tolerance
Walking
Workday tolerances

formance levels are (Key 1984; Worker Data Bank 1994, 1995, 1996; Gilliand 1986; Key 1995):

- Valid participation—the individual participated with full effort;
- Invalid participation—the individual intentionally provided less than full effort;
- Conditionally valid participation—the results reflect the individual's perception of his or her capability even though he or she has demonstrated less than full capability and can physically do more;
- Conditionally invalid participation—the individual has demonstrated beyond what would be considered full, safe levels for extended work periods.

FCA PRINCIPLES

Principles to look for in selecting an FCA include:

1. It must identify the validity of individual participation.
2. The methodology must be consistent from tester to tester and test to test.
3. The equipment must be standardized and the same procedure followed with each assessment.
4. The assessment administrator must be thoroughly trained and objective.
5. The assessment needs to recognize the psychology of the personality as well as the kinesiology of the activities (Schmidt et al. 1989). There is a growing body of literature supporting the theory that there is a relationship between low-back pain

and an individual's psychosociological factors as tested by the Minnesota Multiphasic Personality Inventory (MMPI) (Bigos et al. 1986). The MMPI also has been able to predict when poor response to surgery or conservative care would be the outcome (Schmidt et al. 1989; Block et al. 1996), and to predict the occurrence of job-related low-back pain (Bigos et al. 1986).

FCA REPORTING AND OUTCOMES

Figure 15-7 is a sample report comparing results of the assessment with the physical demands of the job. The decision maker can make an informed, unbiased, nondiscriminating, defensible decision relating to the return-to-work of an injured employee. The decision process is made easier through the use of a visual display of the results, as in Figure 15-8, comparing results with data bank norms, job requirements, and norms of uninjured individuals with similar profiles. For one to rely on recommendations, the predictive ability needs to be demonstrated based on a track record of return-to-work without re-injury. Through outcome surveys and studies, FCAs are becoming increasingly efficient and are able to demonstrate predictive capabilities of safe working levels and resultant reduced risk of injury (Personnel Decisions, Inc. 1994). Key (1995) demonstrates a 99% success rate of return-to-work without recurrence of injury.

The primary outcomes that one should be looking for include decreased re-injury rates, elapse of time from date of injury to date of return-to-work, incidents and costs of litigation, and time spent by management. The State of Colorado studied the impact of FCAs on shortening the amount of time required for vocational evaluation of injured workers. Waite (1987) found that including the KEY Method FCA resulted in a median of 18 fewer days of rehabilitation, which could save the State of Colorado over $200,000 annually.

FCA STANDARDIZATION

When looking for an assessment, standardization and validity determinants are the primary elements of defense against the occurrence of litigation. One of the standardized systems maintains a complete data bank of all assessments performed (Worker Data Bank 1994, 1995, 1996). The consistency of assessment results, including validity, across the regions of the U.S. was analyzed and is represented in Table 15-3. Using two-way analysis by ranks, no differences were found (Aitken 1996; Portney and Watkins 1993). Figure 15-9 demonstrates the consistency of the results of the assessment when administered in three different countries: the United States, Canada, and Australia (Worker Data Bank 1994, 1995, 1996).

Essential Functions and Physical Job Requirements Overview

Name: __John E. Doe__ Date of assessment: __03/14/94__
Job title: __Materials Packager__ DOT Code: __923__
Validity determination: __Valid__ SIC code: __4763__
Job information source: __Risk Safety Manager__
Referral source: __Jane E. Smith, CRC__

Activity	Client capabilities				Physical job requirements			
Workday	8 hours				7 hours			
Sit–(EF)	7 to 8 hours at 60-minute durations, regular breaks				7 hours at 60-minute durations, regular breaks			
Stand–(EF)	5 hours at 45-minute durations, regular breaks				5 hours at 60-minute durations, regular breaks			
Walk–(EF)	2 to 3 hours at frequent, moderate distances				2 hours at frequent, long distances			
Activity	N	Occ	Freq	Cont	N/EF	Occ	Freq	Cont
Bend/stoop		X			N			
Squat			X		EF	X		
Stairs			X		EF		X	
Crawl		X			N			
Crouch		X			EF	X		
Kneel			X		EF	X		
Balance		X			EF	X		
Above shoulder–right								
Above shoulder–left								
Above shoulder–bilat.		43.4	34.6			50	25	
Desk/chair–right					EF			
Desk/chair–left					EF			
Desk/chair–bilat.		52.8	25.8		EF	50	25	
Chair/floor–right								
Chair/floor–left								
Chair/floor–bilat.		52.8	34.6			50	25	
Push		106.3	86.3		EF	100	50	
Pull		89.1	54.1		EF	100	50	
Carry–right		17.0	17.0		EF	20	10	
Carry–left		17.0	17.0		EF	20	10	
Foot–right			X		N			
Foot–left			X		N			
Hand–simp. grasp right		X			EF	X		
Hand–simp. grasp left			X		EF	X		
Hand–firm grasp right		X			EF	X		
Hand–firm grasp left			X		EF	X		
Hand–fine manip. right			X		N			
Hand–fine manip. left		X			N			
Head/neck–static		X			N			
Head/neck–flexion			X		N			
Head/neck–rotation			X		N			

Occ = 0 to 2.5 hours (1-33%)
Freq = 2.5 to 5.5 hours (34-66%)
Cont = Over 5.5 hours (67-100%)

N = Not at all
EF = Essential function
All weights listed are in pounds

© KEY Functional Assessments, Inc., 1985

Figure 15-7 This form helps determine if an injured employee is ready to resume work, based on the demands of the job and other factors.

Training and Education in Back Injury Prevention

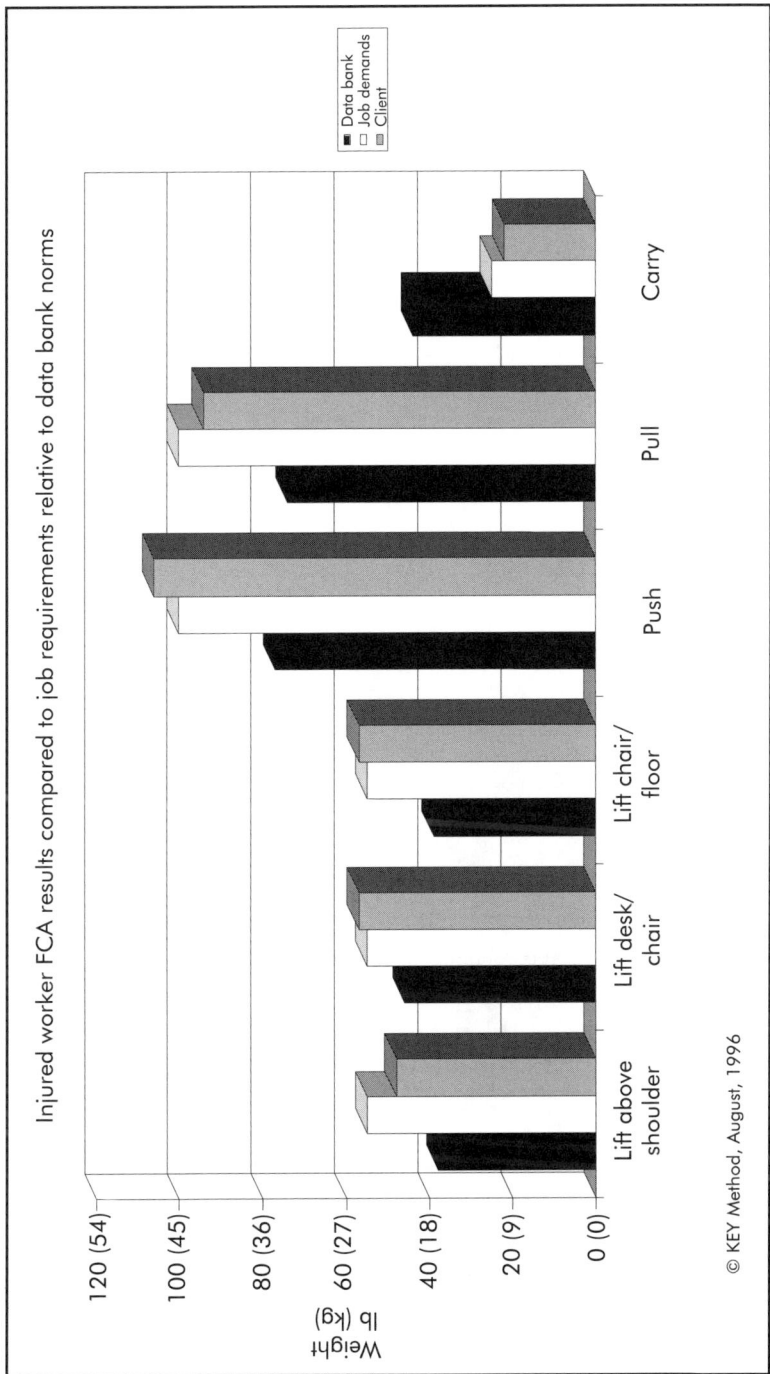

Figure 15-8 The decision on whether an injured employee should return to work is crystallized by this type of chart that compares results from data bank norms, job requirements, and norms of uninjured workers.

Table 15-3. Validity in each territory

	West %	Central %	Northeast %	Southeast %
Valid	78	78	78.5	76
Invalid	5	3	5	5
Conditionally valid	15	17	15	17
Conditionally invalid	1	2	1.5	1

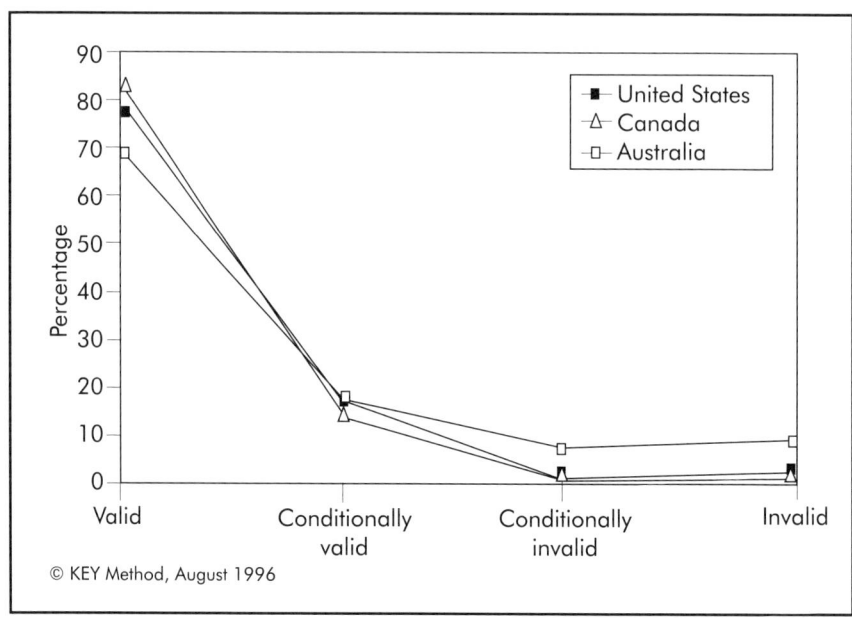

Figure 15-9 Consistency of assessment results in three countries.

Approximately 80% of the population will be afflicted with low-back pain at some point during their lives (Wheeler and Hanley 1995). Proactive strategies for return-to-work should be reflected in a prevention focus.

FUNCTIONAL THERAPY

The entire economic structure of a company is affected when a worker is injured. Injured individuals should receive immediate, aggressive, and function-specific care designed and documented with a focus on return-to-work. This type of intervention is standard procedure when working with athletes.

Among the ways to manage the system is to require that all medical providers support the philosophy of back injury prevention and also support a treatment concept of function/dysfunction rather than the diagnostic medical model (Smith 1995). This model identifies which activities the injured worker is unable to perform, indicates the therapeutic intervention that will enhance that work function, and establishes goals specific to the return of that function. A report example might read, "The client is unable to lift 20 lb (9 kg) from the floor" as the dysfunction. The functional plan might read, "The client will be able to lift 50 lb (23 kg) from the floor within eight treatments of physical therapy." This assures that employees are being treated with work goals in mind.

JOB ANALYSIS

For injury prevention purposes, it is important to have a detailed, function-focused job description. Testing all applicants to the results of an objective job analysis assures that the employee has the necessary physical capabilities to perform the work, while ensuring nondiscrimination, as all employees will have to meet the same job requirements. No biases that may have been inherent in the industry, no gender, disability, race, or age discrimination, and no biases on how tall or big an individual must be are accommodated. If the person meets all the criteria of the objective job analysis, he or she gets the job. It is clear, objective, and defensible in a court of law.

Receiving adequate training for a job is important for the success of the employee and the employer. The job analysis is used as a guide to design the training and track the progress of the trainees. Injury prevention programs that are brought into the company also should use the job analyses as their basis. Back injury prevention programs that can speak to the actual requirements of the jobs also will be much more accepted by the employees.

There are three primary categories to include in a functional job analysis.

1. Job description.

    a. Shift and responsibilities information.

    b. Skill level and qualifications.

    c. Environment information.

2. Physical demands.

    a. List of the physical requirements of the job—lift, carry, crawl, push, pull, sit, etc.

    b. Specific requirements within each physical demand—30 lb (14 kg), 60 in. (152 cm), etc.

3. Job functions or components.
   a. List of the essential functions of the job.
   b. Functions of the job which are important, but may not be classified as essential.

JOB PLACEMENT ASSESSMENTS

A job placement assessment (JPA) is an assessment of an individual's physical capabilities prior to his or her hire. It provides dramatic cost savings for the company while providing a standardized, valid, and defensible means to assess potential job candidates. The job placement assessment is administered after the job candidate has been offered the position, conditional upon his or her passing this final testing (see Figure 15-1). Although less common, some companies use the JPA as a pre-offer screen to measure all applicants' ability.

The job placement assessment is a form of FCA, covered earlier in this chapter. It generally includes lifting at multiple heights, carrying, and pushing and pulling. Assessing these functional abilities in an objective format, consistent with all other applicants, and specific to the job, provides data to compare with the job requirements. It also provides the legal defense should an applicant contest the results if denied employment. On-site services may be available allowing clear and defensible decisions to be made while the job candidate is still on the premises.

A job placement assessment should follow these principles (Frey 1995):

- Contain specific, objective, and standardized protocols and procedures;
- Components are replications of activities performed on the job;
- The job requirements are based on specific measures of the job;
- Physical performance of the activities is required;
- Standardized equipment is utilized by users;
- A database source for comparison is available;
- The assessment administrator must be neutral and unbiased;
- Outcomes must demonstrate predictability for safe return-to-work; and
- The JPA must be proven legally defensible in its history and references.

JPA OUTCOMES

Financial savings, decreased time lost, decreased worker injury rates, and predictive capability of future injury are the primary issues in the following outcome reports.

Outcome—Case A

In 1988, a paper manufacturer in Minnesota instituted job placement assessments to help stem the costs related to workers' compensation claims and lost work days. In an analysis 2 years later, the experiences of 70 employees hired before the use of JPAs were compared to 70 hired after the initiation of administering JPAs. Use of the JPAs lowered both lost work days and workers' compensation costs (Paper Manufacturer's Savings with Job Placement Assessment Implementation 1990) (see Figure 15-10).

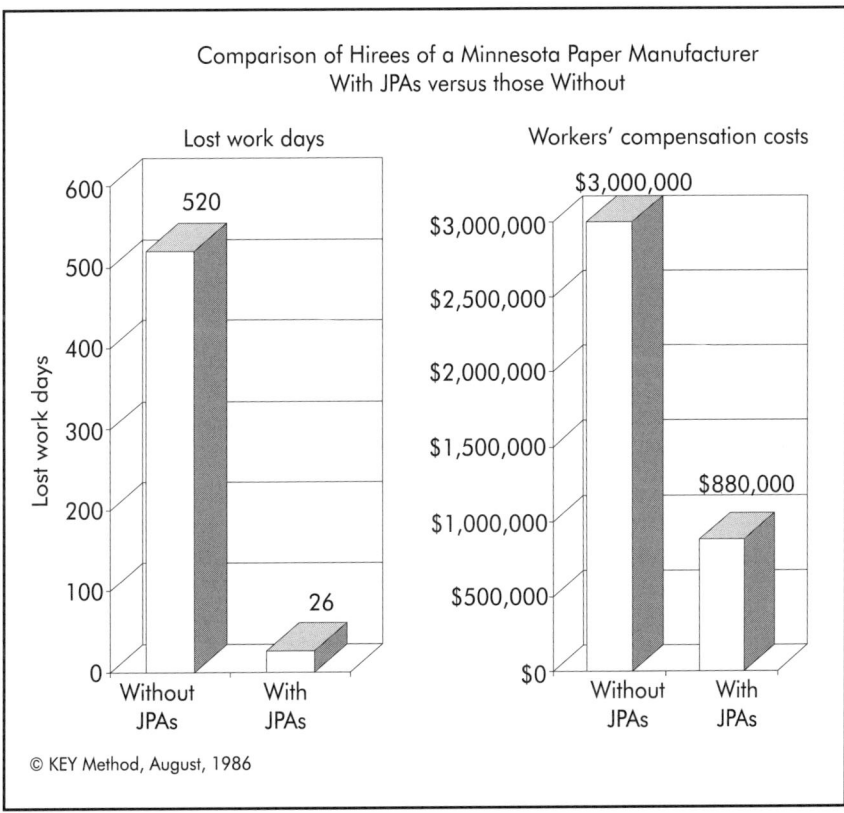

Figure 15-10 Comparison of workers at Minnesota paper manufacturer with JPAs versus those without.

Outcome—Case B

A state transportation authority discovered that job placement assessments enabled them to predict if an employee was at risk of injury. Of the 36 employees injured from 1985 to 1992, 75% had been categorized as at risk by JPAs performed when they were hired. Fourteen providers across the state administered the same system of JPA. While the analysis was based on relatively few cases, the results were statistically significant. The chi square for this cross-tabulation was 15.4, which was significant at $p = 0.00045$ (Personnel Decisions, Inc. 1994).

Outcome—Case C

A major trucking firm used the JPA for two separate hiring locations over a continuous 18-month period. Since administering the JPA on each candidate and hiring based on the results, the locations experienced no new injuries (Worker Data Bank 1994, 1995, 1996).

"TOOL BOX TALKS"

"Tool box talks" may be known as shift meetings, production meetings, safety talks, or point meetings. These are brief sessions, usually from 5 to 15 minutes long. Initially, the consultant conducts the meetings, gradually involving supervisors and employees in running the sessions. These talks must be scheduled and must occur frequently. A message can be initiated in a workshop or seminar, but it is critical that interest and momentum be maintained through interaction of the employees.

The employer needs increasingly more objective information in managing an employee group for the prevention of back injuries. The complex hiring process and the need for defensibility have established the plant manager in a new position of decision making. The tools described in this chapter will assist him or her in assuring that those decisions will serve well for productivity, job satisfaction, and in the courts, should that be necessary.

REFERENCES

Aders, J. E. "Teams Reduce Job Risks." *Workplace Ergonomics* May/June 1996: 24-26.

Aitken, M. J. "Pulse Rate During Submaximal Work." *Journal of Applied Physiology* June 1996, 7: 218-221.

Astrand, P. O., and Ryhming, I. "A Nomogram for Calculation of Aerobic Capacity from Pulse Rate During Submaximal Work." *Journal of Applied Physiolgy* 1954: 7: 218-221.

Benefits of Employee Health Programs. Philadelphia, PA: Cigna, 1991.

Bigos, S., Spengler, D., Martin, N., et al. "Back Injuries in Industry: A Retrospective StudyIII—Employee-related Factors." *Spine* 1986, 11: 252-256.

Block, A.R., Vanharanta, H., et al. "Discogenic Pain Report, Influence of Psychological Factors." *Spine* February 1996, 21(3): 334-338.

Frey, D.H. "Job Placement Assessments and Pre-employment Screening." G.L. Key, ed. *Industrial Therapy.* St. Louis, MO: Mosby, 1995: 110-122.

Gilliand, G. *A Study of Statistical Relationships Among Physical Ability Measures on Injured Workers Undergoing KEY Functional Assessments.* Minneapolis, MN: Personnel Decisions, Inc., 1986.

Grossman, P. "Respiration, Stress, and Cardiovascular Function." *Psychophysiology* 1983: 284-300.

Key, G.L. "Functional Capacity Assessment." G.L. Key, ed. *Industrial Therapy.* St. Louis, MO: Mosby, 1995: 220-254.

Key, G.L. *Key Functional Assessment Training and Resource Manual.* Minneapolis, MN: KEY Functional Assessments, Inc., 1984.

Key, G.L. "The Impact and Outcomes of Industrial Therapy." G.L. Key, ed. *Industrial Therapy.* St. Louis, MO: Mosby, 1995: 31-41.

Key, G.L. "Paper Manufacturer's Savings with Job Placement Assessment Implementation." Minneapolis, MN: KEY Functional Assessments, Inc., September 1990.

Melnik, M.S. "Upper-extremity Injury Prevention." G.L. Key, ed. *Industrial Therapy.* St. Louis, MO: Mosby, 1995: 148-180.

Personnel Decisions, Inc. *Key Functional Assessment Pre-employment Screening Battery as a Predictor of Job-related Injuries.* Minneapolis, MN: Personnel Decisions, Inc., 1994.

Portney, L.G., and Watkins, M.P. *Foundations of Clinical Research.* Norwalk, CT: Appleton & Lange, 1993.

Saunders, H.D., and Saunders, R. *Evaluation, Treatment, and Prevention of Musculoskeletal Disorders,* Vol. 1 and 2. Minneapolis, MN: Educational Opportunities, 1995.

Saunders, H.D., Stultz, M.R., et al. "Back Injury Prevention." G.L. Key, ed. *Industrial Therapy.* St. Louis, MO: Mosby, 1995: 123-147.

Schmidt, A.J.M., Gierlings, E.H., and Madelon, L.P. "Environmental and Interoceptive Influences on Chronic Low-back Pain Behavior." *Pain* 1989, 38: 137-43.

Smith, G.J. "Acute Care and Functional Treatment." G.L. Key, ed. *Industrial Therapy.* St. Louis, MO: Mosby, 1995: 197-219.

Volski, R.V. "Employee Fitness Programs." G.L. Key, ed. *Industrial Therapy.* St. Louis, MO: Mosby, 1995: 181-193.

Waite, H.D. "Use of a New Physical Capacities Assessment Method to Assist in Vocational Rehabilitation of Injured Workers." Thesis. CO: University of Colorado, 1987.

Wheeler, A.H., and Hanley, E.N. "Spine Update: Nonoperative Treatment for Low-back Pain—Rest to Restoration." *Spine* February 1, 1995, 20(3): 375-378.

Worker Data Bank. *KEY Method.* Minneapolis, MN: KEY Functional Assessments, Inc. 1994, 1995, 1996.

BIBLIOGRAPHY

Flynn, G. "Partnerships Contain Costs—Hannaford Bros. Makes Health Care a *Maine* Event." *Personnel Journal* May 1996: 48-55.

"How to Use Effective Litigation Fraud Management." RIMS Toronto '96 Conference Overview. *Risk Management* June 1996.

CHAPTER 16

AN OVERVIEW OF UPPER EXTREMITY DISORDERS

J. Steven Moore
Professor
Department of Nuclear Engineering
Texas A&M University
Co-director of NSF I/UCRC in Ergonomics
College Station, TX

FIVE DISORDERS

This chapter presents medical information about specific medical conditions that affect elbows, forearms, wrists, or hands (called the distal upper extremity). Five specific distal upper extremity disorders are discussed. The first is trigger finger and trigger thumb, which share many similarities with the second condition—de Quervain's tenosynovitis. The third disorder, peritendinitis, is a common diagnosis outside the United States. Lateral epicondylitis (tennis elbow) is the fourth condition. Last, but not least, is the most complicated and controversial of these distal upper extremity disorders—carpal tunnel syndrome (CTS).

The strategy for covering each disorder was based on supplying answers to potential questions that supervisors would most likely ask. Each section was organized according to the following outline topics:

- Normal anatomy and function;
- What is wrong;
- How the condition develops;
- Presentation, symptoms, and signs;
- Who gets it;
- What causes it;
- What is the usual treatment;
- What about an affected worker; and
- What about the work.

TRIGGER FINGER AND TRIGGER THUMB

The formal medical term for trigger finger or trigger thumb is *stenosing tenosynovitis of the digits*. "Stenosing" means that a structure is abnormally narrowed. "Tenosynovitis" means inflammation of a tendon sheath. Thus, the term means that a tendon sheath has narrowed. Many tendon sheaths located throughout the body are susceptible to stenosing tenosynovitis. When stenosing tenosynovitis affects the tendon sheath of one of the fingers, the condition is called *trigger finger*. When it affects the thumb, it is called *trigger thumb*.

Anatomy and Function

The muscles that bend or flex the fingers and one of the muscles that flex the thumb are located on the palm (volar) side of the forearm. The tendons that connect the ends of these muscles to the bones in the fingers are shaped like cords. For each finger, there are two flexor tendons that run together along the palm side of the finger. The anatomy for the thumb is similar.

The tendon sheath is a tubular balloon-like structure filled with a low viscosity fluid (synovial fluid) that reduces friction. Imagine the cords (tendons) pressed into the balloon (the tendon sheath) so that the balloon surrounds the cords. When the cords move back and forth, the fluid inside the balloon reduces friction.

The tendon sheath has a series of ligamentous coverings, called pulleys, looping around the tendons, holding them close to the bones and joints. If you were on the tendon looking toward the pulleys, you would see tunnels with the underlying bone as their floor. Pulleys make up the walls and the roof of the tunnels. Stenosing tenosynovitis is a disorder affecting one of these tunnels, specifically, the tunnel formed by the A1 pulley.

What is Wrong?

With trigger finger or trigger thumb present, the A1 pulley appears thick and fibrous. As the pulley thickens, it reduces the cross-sectional area of the tunnel (stenosis). When the tunnel becomes too narrow, the tendons no longer move freely through the tunnel. Since the flexor muscles are stronger than the extensor muscles, people are usually able to flex the digit, but have difficulty extending it. The result is snapping or locking (called triggering) that occurs when the flexed finger or thumb is straightened.

How Does it Develop?

Current theory suggests that the A1 pulley thickens because it is adapting to repeated or prolonged tension by the flexor tendons.

This tension is primarily related to the degree of bending of the joint and the degree of tension (also called loading) in the tendon, such as "loaded tendons turning corners." Maximum tensions in the A1 pulley appear to occur when the joint at the base of the finger is extremely bent. At this time, it is unknown how high pulley tension must be, how many times it must be experienced, or how long it must last before the A1 pulley begins to thicken.

Presentation, Symptoms, and Signs

For most people, trigger finger or trigger thumb develops gradually, but some cases may follow acute trauma. Snapping, locking, or pain when extending a flexed finger or thumb are the most prominent symptoms. The snapping sensation may be barely perceptible without any actual triggering, or it may be painful, especially when a triggered digit is forcefully extended. The triggering and pain are limited to an area in the palm where the digit joins the hand, the level of the outermost skin crease in the palm for the long, ring, and little fingers. Some people report more difficulties in the morning compared to other times. It may be possible for an examiner to feel a nodule on the tendon in the region of the A1 pulley as well as a clicking or snapping sensation with movement of the digit.

Who Gets It?

Trigger thumb often occurs in children below 6 years of age. Both trigger thumb and trigger finger usually occur in adults above 40. Trigger finger is more common among women than men.

Some people have multiple affected digits. The thumb is the most commonly affected digit, followed by the long and ring fingers. The index and little fingers are rarely affected. Some people with trigger finger or trigger thumb also have other disorders about the hand or wrist. Several studies report that trigger finger or trigger thumb are more common among people with diabetes.

What Causes It?

The exact cause of trigger finger or trigger thumb is unknown. Most knowledge on the subject comes from observations made by health care professionals who have treated the condition. However, these observations may or may not be correct.

The occurrence of trigger thumb in children suggests a congenital factor (present from birth) for those cases. Among adults, a single episode of trauma may account for some cases, but the most common observation points to constant use of the thumb, often described as prolonged overstraining or chronic trauma.

In addition to observations made by health care professionals, there are scientific, epidemiological studies examining the relationship between work and disorders of the elbow, forearm, wrist, and hand. In general, these studies suggest that exertional work demands (intensity of exertion, duration of exertion, frequency of exertion, etc.) are the most important factors.

What Is the Usual Treatment?

Corticosteroid injection, with or without splinting, is usually the treatment of choice. Up to three injections are recommended for the initial management of nonlocking digits for patients at least 10 years old. Surgery is recommended for children under 10 and for any patients with locked digits. Surgery involves making an incision in the palm over the area of the A1 pulley, then cutting the pulley. The edges of the pulley are not sewn back together, with free edges growing back together so that the tunnel is enlarged and free finger movement returns.

What About an Affected Worker?

If a worker with trigger finger or trigger thumb is treated conservatively with corticosteroid injections, no functional limitations are expected. If the affected digit is splinted, however, splints can interfere with grasping, pinching, or pressing with the digit. If surgery is involved, the incision should be kept clean, dry, covered, and free from localized compression or trauma. Since the incision is in the palm, firm gripping or pressure on the palm would not be recommended for a few weeks. Most people return to work on their regular jobs without permanent restrictions or disability.

What About the Work?

Determining when a case is work related is neither easy nor precise. There are, however, several indicators that may be helpful.

1. If several individuals performing the same job develop trigger digit or other conditions involving the elbow, forearm, wrist, and hand, then the job may be part of the problem. This is especially true if workers develop trigger finger or trigger thumb of the same digit(s) repeatedly.

2. Epidemiological studies suggest that a job's exertional demands determine its potential to afflict workers with these types of disorders. Exertional demands can be estimated using a job analysis method called the Strain Index. It involves calculating a score based on six factors: intensity of exertions; duration of exertions; frequency of exertions; posture of the hand and wrist; speed of the work; and duration of the task

(or job) per day. If the Strain Index score is above 7, the job probably contributed to the condition. Jobs with scores below 5 are generally not related to the development of such disorders. By modifying one or more of the six factors in the Strain Index, it may be possible to reduce the exertional demands of the job to a point that the Strain Index score is less than 5.

3. Prolonged flexion of the joint connecting the affected finger or thumb to the hand (the metacarpal-phalangeal or MP joint) might contribute to the development of trigger finger or trigger thumb, such as when a worker constantly holds a tool or object with a tightly closed fist. Reducing the degree of joint flexion or the duration of holding time might be helpful.

4. Localized compression in the area of the A1 pulleys also might be a factor and should be eliminated. Such a circumstance can occur when a square-handled screwdriver is grasped firmly. The edges of the handle could dig into the palm near the A1 pulleys of some fingers.

DE QUERVAIN'S TENOSYNOVITIS

De Quervain's tenosynovitis is a disorder very similar to trigger finger and trigger thumb. It is also a form of stenosing tenosynovitis, but occurs along the thumb side of the wrist instead of the palm side of the finger or thumb.

Anatomy and Function

Most muscles that control the wrist and fingers anchor at the elbow or on the forearm bones. The tendons that connect these muscles to the bones out in the wrist and fingers are, like the finger flexor tendons, shaped like cords. As these tendons cross the wrist joint, they enter tunnels whose floors are formed by the bones, and whose walls and roofs are formed by ligaments (the wrist ligaments are called retinacula). On the back side of the wrist (the same side as the back of your hand), the tendons crossing the wrist pass through six tunnels called the six dorsal compartments. The first dorsal compartment is on the thumb side of the wrist. Using medical terms, de Quervain's tenosynovitis is stenosing tenosynovitis that affects the first dorsal compartment.

The tendon sheath for the first dorsal compartment is located at the end of the radius—the forearm bone on the thumb side. Two muscles that control the thumb, the abductor pollicis longus (APL) and extensor pollicis brevis (EPB), originate on the shaft of the radius in the forearm. The APL inserts on the back side of the first metacarpal bone (the bone that runs between the wrist and the thumb) just beyond the wrist. The EPB inserts on the back side of the proximal phalanx of the thumb (the first bone forming the shaft of the thumb) just beyond the MP joint (at the base of the thumb). These two muscles control the position and orientation of the thumb so the thumb can be used to grip, pinch, or press. These tendons normally glide freely through the tunnel of the first dorsal compartment.

What Is Wrong?

Like trigger finger and trigger thumb, the primary change in de Quervain's tenosynovitis is that the tunnel (the retinaculum) is narrowed because it is thickened. Functional impairment is believed to be caused by impaired gliding of the tendons within the tunnel.

Why Did It Develop?

It is generally believed that, like trigger finger and trigger thumb, the changes related to de Quervain's tenosynovitis result from loaded tendons turning corners. The APL and EPB tendons are loaded whenever the thumb is used. These tendons turn a corner when the wrist or thumb are bent, often resulting from the use of a manual screwdriver.

Presentation, Symptoms, and Signs

The onset of de Quervain's tenosynovitis is usually gradual, with the most common symptom being pain localized to the thumb side of the wrist. The intensity of the pain varies, but it may be severe enough to keep the victim awake at night. It also increases with pinching, grasping, sticking the thumb out to the side (the hitchhiking signal), and bending the wrist toward the little finger. The pain may become severe enough to render the hand useless. There may be slight swelling at the thumb side of the wrist, and sometimes painful ranges of motion of the wrist and thumb. Firm touching may result in tenderness at the thumb side of the wrist. There should be no sensation of creaking (called crepitus), because it suggests a different disorder (called peritendinitis). Stretching or contracting the APL or EPB muscles increases the pain. A maneuver called Finkelstein's test is the most characteristic physical sign. This test involves grasping the thumb with the fingers, then bending the wrist toward the little finger. Increased pain along the thumb side of the wrist indicates a positive test.

Who Gets It?

De Quervain's tenosynovitis primarily affects women (gender ratio approximately 10:1) between the ages of 35 and 55 years. There is no predilection for right versus left side or difference by race.

What Causes It?

The exact cause of de Quervain's tenosynovitis is unknown. Most knowledge comes from observations made by physicians who treated people with the condition. These observations help provide perspective on factors likely to cause de Quervain's tenosynovitis. They do not predict at which point these factors might cause the condition.

There is no apparent cause for some cases. As many as 25% may be related to blunt trauma, such as being struck on the forearm; falling down stairs; sudden wrenching of the hand; and falling on the tip of the thumb. Overexertion of the thumb and using the thumbs a great deal are commonly mentioned circumstances. Examples of such activities include operating a buffing machine, fitting rubber rings on a pipe, typewriting, piano-playing, sewing, knitting, weaving, and cutting. Firm grasp combined with movement of the hand in a radial direction, as in wringing clothes, may be important. The tendons may turn almost 90° at the wrist when being used for strong, unremitting, or repetitive pinching, grasping, pulling, or pushing. Such activities come into play when the worker operates machine keyboards, sewing machines, lathes, drills, presses, grinders, and switchboards. Those at high risk include assemblers, inspectors, spot-welders, and order clerks.

In addition to heavy work, prolonged, monotonous, and tiring activity or persistent repetition of an accustomed task beyond the point of fatigue also may be factors. Examples of this type of activity include prolonged piano-playing; prolonged typing; excessive writing, washing, or wringing out of clothes; carrying heavy objects; and cutting cloth with heavy scissors. Heavy work is not necessarily an etiologic factor.

Repeated unaccustomed activities may be important. Many people start feeling symptoms by the second day of a new job or on resuming an old job after a holiday or illness.

There are no scientific studies in the current literature that focus specifically on the cause(s) of de Quervain's tenosynovitis. In general, disorders of the hand and wrist are primarily related to the exertional demands of the work.

What Is the Usual Treatment?

Initial conservative treatment may begin with immobilization plus an anti-inflammatory medication. Alternatively, the physician may begin with injection of a local anesthetic and long-acting corticosteroid with or without immobilization. When conservative treatment fails, the next step is surgery, involving an incision on the thumb side of the wrist, then cutting the thickened band that forms the tunnel roof (the retinaculum). The edges are left free. When the retinaculum heals, the tunnel is less narrow and the tendons can glide freely.

What About Affected Workers?

A worker being treated conservatively for de Quervain's tenosynovitis may (or may not) have the thumb immobilized in a splint, called a spica splint, that locks the wrist and thumb in one position. If so immobilized, the worker cannot bend the thumb or the wrist without encountering resistance from the splint. If a worker has surgery, the wound should be kept clean, dry, covered, and protected from localized compression and trauma. Splinting may last a short period of time. Forceful use or significant bending of the thumb is generally avoided for several weeks.

What About the Work?

Even though it is generally recognized that de Quervain's tenosynovitis may be related to work, making that determination with confidence may be difficult. Some indicators are identical to those reported for trigger finger and trigger thumb.

1. If other individuals performing the same job have developed de Quervain's tenosynovitis or other conditions involving the elbow, forearm, wrist, and hand, then the job may be part of the problem.

2. The exertional demands of the job may be the most important and can be estimated using the Strain Index.

3. Forceful grasping, pinching, or pressing with the thumb, especially combined with an awkward position of the wrist or thumb, might be a contributing factor. Substitution of grasping for pinching; reducing the degree of wrist or thumb deviation; or reducing the intensity, duration, or frequency of thumb exertions might be helpful.

4. An unaccustomed or unexpected situation, such as a new job assignment, increased work hours, a bad shipment of parts to be assembled, etc., also may be a factor.

PERITENDINITIS

The myotendinous junction is where the muscle joins its tendons. It is a specialized anatomical structure whose purpose is to transmit the muscle's tension generated to its tendons. At each end of muscle cells, which have finger-like projections that match up to similar projections from the tendon, there is a myotendinous junction. To use an analogy from carpentry, the myotendinous junction is like a scarf joint. This structure reduces stresses to the cell membrane of the muscle cells while maximizing the transmission of tension from the muscle to the tendon.

The myotendinous junction appears to be the structure involved in two types of conditions: muscle strains and peritendinitis. Muscle strains will not be discussed further here. Peritendinitis is widely recognized in other parts of the world, but has not been discussed in the United States much since the 1940s. It appears that most cases of peritendinitis in the United States are mislabeled as tendinitis or tenosynovitis.

More than half of the cases of peritendinitis affect the same muscles involved in de Quervain's tenosynovitis, the APL and EPB, but the location of the problem is more in the forearm (an inch or two toward the elbow) rather than at the wrist. Other commonly affected muscles are located on the back side of the mid-forearm.

What is Wrong?

The problem, localized to the myotendinous junction, usually involves swelling and inflammation. The surfaces of both the muscle and tendon may be covered with a sticky substance called fibrin. The tendons and tendon sheaths beyond the myotendinous junction appear normal.

Presentation, Symptoms, and Signs

Pain, aching, soreness, and tenderness localized in the mid-forearm or a few inches above the wrist are the dominant symptoms. Some patients may experience crepitation, a "creaking gate" noise or sensation associated with movement of the affected structures. The affected area may be swollen, red, warm, or tender to the touch. The affected muscle is usually painful when stretched or contracted.

Who Gets it?

Peritendinitis appears to affect men more than women. There is no information about the effect of age. It has been suggested that peritendinitis often occurs in sedentary people who suddenly begin to perform more physically demanding activities.

What Causes it?

It is generally accepted that peritendinitis develops from fatigue and exhaustion of selected muscle groups or direct trauma. Both factors lead to swelling, inflammation, and the deposition of fibrin around the myotendinous junction.

Approximately half the cases may be associated with blunt trauma (contusion). After the acute injury, the person performs his or her usual and accustomed work. Symptoms then appear within 1-14 days. Objective evidence of trauma is usually present, for example, ecchymosis, laceration, or abrasion. The blunt trauma may occur in the course of employment, followed by customary exertions while pursuing a sport or hobby. The opposite sequence may also occur.

Approximately half the cases occur in the context of unaccustomed activity, especially when the activity involves repeated, stereotypical movements. It has been suggested that inexperienced individuals utilize inefficient and overly forceful muscular effort. Examples include renewed employment after long lay-offs or assignment to unfamiliar tasks. Small epidemics of peritendinitis may be precipitated in workplaces by changes from the normal and accustomed routines. There is one story of a flashlight battery manufacturing plant that had a history of few cases of peritendinitis in the years prior to 1931. During 1931, demand for the product increased, so the employees worked longer and faster. In addition, the workers had to modify their usual work methods to accommodate a nonstandard battery size. Within a week, 15 employees developed peritendinitis.

What is the Usual Treatment?

Rest and immobilization are the most effective treatments. If the disorder involves either the APL or EPB, the thumb should be placed into a spica splint. Duration of the disability is usually 1-2 weeks. Disability may be prolonged to one or more months with inadequate rest and immobilization. With proper rest and immobilization, most people recover and return to work without problems. There is no surgical treatment for peritendinitis.

What About an Affected Worker?

In general, a worker being treated for peritendinitis will have a wrist splint and probably take an anti-inflammatory medication. If the condition affects the thumb, it also would be immobilized. Forceful exertions, as well as prolonged exertions, even if not forceful, should be avoided. Returning to the original job appears likely, but easing into the job may be appropriate.

What About the Work?

As noted for other conditions, jobs likely to cause peritendinitis will probably have a history of affecting several workers with a variety of conditions involving the elbow, forearm, wrist, or hand. The exertional demands may be estimated with the Strain Index. Keep an eye out for factors that could contribute to an unaccustomed work situation.

LATERAL EPICONDYLITIS

The medical term for "tennis elbow" is lateral epicondylitis. The lateral epicondyle is the bony prominence located on the outer (lateral) side of the elbow when the arm is held along the side of the body with the palm facing forward. There is also a bony prominence on the inner (medial) side of the elbow, called the medial epicondyle. When someone has pain localized to the medial epicondyle, the condition is called medial epicondylitis (also called "golfer's elbow"). Far more common than medial epicondylitis, lateral epicondylitis has much more published information about it.

Anatomy and Function

There are two muscles that primarily stabilize, extend, and deviate the wrist from side to side. The extensor carpi radialis longus (ECRL) originates just above the elbow and inserts on the back side of the base of the second metacarpal bone (just beyond the wrist). The extensor carpi radialis brevis (ECRB) originates primarily from the bony prominence on the outside of the elbow, called the lateral epicondyle, and inserts on the back side of the base of the third metacarpal bone (also just beyond the wrist). Whenever fingers are used to grasp or pinch, there is simultaneous contraction of the wrist extensor muscles. This stabilizes the wrist joint so that the wrist does not flex when the fingers forcefully grip or press something.

What is Wrong?

Pathology of lateral epicondylitis is not precisely known, but the ECRB seems the most commonly involved structure. The tendon near the origin of the ECRB may appear normal from the outside, but usually has some abnormal tissue on the underside. A tear of the tendon is sometimes observed. The nature of these changes, as well as those observed under a microscope, suggests that something is rubbing the tendon's underside, fraying some of its fibers, and the body is trying to repair this damage.

Presentation, Symptoms, and Signs

Lateral epicondylitis usually presents as pain to the lateral side of the elbow. The onset may be sudden or gradual. The intensity of the pain varies. Relatively minor levels may be described as "discomfort" while more intense levels may be described as "sharp," "severe," or "lightning-like." The pain often limits activities of daily living (such as lifting a coffee cup or jar), leisure pursuits (gardening or sports), and work (both heavy and sedentary). There is usually tenderness localized at or near the lateral epicondyle. Gripping forcefully, pulling the wrist or long finger back against resistance (extension); having the elbow straight (extension) with the forearm turned inward (pronation) and wrist bent forward (flexion); or resisted rotation of the forearm inward (pronation) and outward (supination), increases the pain. Elbow extension or forearm pronation may be limited. Grip strength and wrist extension strength may be reduced.

Who Gets it?

Despite its common name, tennis elbow, only a small percentage (about 5%) of patients with lateral epicondylitis are tennis players. The dominant arm is more often affected than the nondominant arm. Bilateral lateral epicondylitis occurs, but is relatively uncommon.

In general, men and women are equally affected. Most cases occur between the ages of 35-55 years, with an average of approximately 45 years. Lateral epicondylitis appears to be relatively uncommon among African-Americans.

What Causes It?

Why lateral epicondylitis develops is generally unknown. For cases that occur following blunt trauma to the elbow, it is believed that the trauma injured some of the fibers in the ECRB tendon. For nontraumatic cases, it has been suggested that one of the forearm bones (the radial head) may rub the underside of the ECRB tendon. This would most likely occur when the hand grasps an object, the elbow is extended, and the forearm rotates (pronates and supinates), as when using a screwdriver. Lateral epicondylitis does not appear to be a degenerative condition related to aging.

Direct blunt trauma may account for approximately one third of the cases. Unaccustomed repetitive use, related to employment or recreation, is commonly noted as a precipitating factor. Repeated inward and outward forearm rotation movements (pronation/supination) with the elbow almost fully extended, as well as unaccustomed activities involving repeated pronation and supination of the forearm against resistance or while the hand maintains a grip, have been mentioned as factors. Overexertion, forced extension of the wrist, and repeated pronation-supination movements of the forearm have also been mentioned. Some occupations reported to be subject to lateral epicondylitis include piece-workers, masons, gardeners, builders, stevedores, waiters, printers, surgeons, housewives, and painters.

Scientific studies have not established a clear relationship between work and lateral epicondylitis. When lateral epicondylitis appears to be related to work, the exertional demands of the job are probably most influential.

What is the Usual Treatment?

Conservative treatment usually includes limiting use of the hand, avoiding painful activities, and wearing a forearm band. Anti-inflammatory medications are often prescribed, but their effectiveness has not been demonstrated. Some health care professionals recommend resting the hand in a wrist splint. Ultrasound and laser treatments have been demonstrated to be ineffective in the treatment of lateral epicondylitis. Other physical therapy modalities have not been critically evaluated.

Injections of a local anesthetic and corticosteroid are effective in many cases. Most people have increased elbow pain the evening after the injection, but it usually subsides by morning. Limitation of activity following injection is not required. In general, the number of injections rarely exceeds three. Improvement is usually evident in a few days.

Surgery is relatively uncommon. When performed, it usually involves cutting through the tendon at the lateral epicondyle. Postoperatively, the arm is supported by a sling for a few days, then active elbow motion is encouraged. Most people return to work several weeks after the operation.

What About an Affected Worker?

A worker with lateral epicondylitis may do quite well on his or her regular job with oral medications or a forearm band. If there is pain with forceful gripping, check to see if the elbow is extended when gripping. If so, and if possible, redesign the job so the work can be done with the elbow flexed closer to 90°. Another option is to reduce the exertional demands of the job. In particular, look at grip force and forearm twisting (pronation/supination).

What About the Work?

If other individuals performing the same job have developed lateral epicondylitis or other conditions involving the elbow, forearm, wrist, and hand, then the job may be part of the problem. The exertional demands of the job can be estimated with the Strain Index. For lateral epicondylitis specifically, look for forceful gripping or forearm twisting combined with elbow extension.

CARPAL TUNNEL SYNDROME

Carpal tunnel syndrome is the most complex and controversial of the disorders discussed in this chapter. There are controversies about diagnostic criteria, work-relatedness, treatment, job placement during and after treatment, and levels of impairment and disability.

Even though the term *carpal tunnel syndrome* was not used, the first reported case was in 1865. In the late 1800s and early 1900s, the terms *tardy median palsy* and *partial thenar atrophy* were widely recognized and reported. Authorities did not describe these conditions as "occupational" in the contemporary sense, but they were sufficiently thoughtful to report the occupations of their cases. Housewives and tailors were commonly affected. *Occupational neurosis* and *occupational neuritis* were mentioned as conditions associated with jobs that involved repeated motions or exertions in the 1922 edition of the Department of Labor's publication *Occupational Hazards and Diagnostic Signs*. In 1947, physicians from Britain were the first to explicitly mention the condition's relationship to occupation, even though the occupation of five of their six cases was "housewife." The term *carpal tunnel syndrome* appears to have been first used in the 1950s.

Anatomy and Function

The carpal tunnel is a tunnel located on the palm of the hand. The floor and walls are formed by the carpal bones. The roof is formed by a thick ligament, the transverse carpal ligament, that begins near the level of the distal wrist crease (the skin crease nearest the palm) and extends approximately 1.0-1.5 in. (25-38 mm) into the palm. Normal contents of the carpal tunnel include two flexor tendons for each of the four fingers (for a total of eight finger flexor tendons), one flexor tendon for the thumb, and the median nerve. As in other tunnels, these nine tendons within the carpal tunnel are covered by tendon sheaths.

When tendons are relaxed and the wrist is relatively straight (neutral), the pressure in the carpal tunnel, called intracarpal pressure, is at its minimum value. At extremes of flexion and extension, intracarpal pressure increases.

What is Wrong?

The tendon sheaths covering the nine tendons are often reported to be thickened, which appears to be related to swelling (edema) or scarring (fibrosis) within the tendon sheaths. Inflammation does not appear to be involved. The cause(s) of these changes are generally unknown and do not appear to differ according to whether the condition is believed to be work-related or not.

The median nerve often looks normal, but individual nerve fibers inside the nerve may be affected. Most of the individual nerve fibers are covered by an insulation-like material called a myelin sheath. At the site of compression, this insulation appears pushed off the nerve fiber under the area of compression. Since this insulation is necessary for fast conduction of nerve impulses, its loss contributes to slow or delayed nerve conduction, as measured during an electrodiagnostic test.

The symptoms of carpal tunnel syndrome are usually explained on the basis of impaired circulation to the median nerve inside the carpal tunnel when intracarpal pressure is elevated. When intracarpal pressure is elevated to a relatively high level for a sufficient period of time, circulation inside the nerve is stopped. Individual nerve fibers begin to spontaneously discharge and produce unusual sensations of numbness and tingling (called paresthesias). When intracarpal pressure is lowered, blood flow returns, spontaneous nerve discharges end, and the paresthesias end.

Presentation, Symptoms, and Signs

Excluding acute trauma, symptom onset is usually gradual and often related to unaccustomed activity. The dominant symptoms are numbness or tingling, called paresthesias. Actual pain or weakness is uncommon, although intense paresthesias may be reported as painful. Typically, paresthesias affects the thumb, index, long, and part of the ring finger, but should spare the little finger. Symptoms may radiate into the forearm, elbow, arm, or shoulder. Usually, the paresthesias occurs at night or with static grasp, and is most often relieved by changing position or shaking the affected hands. There are no reliable physical findings. Electrodiagnostic studies, often called nerve conduction studies or EMGs, are the best way to confirm the presence of carpal tunnel syndrome.

Who Gets it?

Carpal tunnel syndrome typically affects more women than men and usually people in the 40-60-year old age range. A significant percentage of cases occur in both hands simultaneously, or nearly so. There are numerous personal conditions that may contribute to a person developing carpal tunnel syndrome. Some of the most common include rheumatoid arthritis, diabetes, thyroid disease, obesity, and a history of a fractured wrist.

What Causes it?

At this time, it is not possible to reliably comment on why carpal tunnel syndrome develops in a given person. There are several possible mechanisms that might be related to hand usage and numerous others that would include factors unrelated to hand usage. Some of the possible hand usage models include:

- Thickening of the tendon sheaths inside the carpal tunnel;
- Hypertrophy (enlargement) of the tendons that pass through the carpal tunnel;
- Direct pressure on the median nerve by the flexor tendons when the fingers are used with a flexed wrist;
- Retraction of some small hand muscles (lumbricals) into the carpal tunnel when a tight fist is formed;
- Thickening of the transverse carpal ligament in response to tension from "loaded flexor tendons turning a corner" at the wrist (wrist flexion);
- Alterations within the nerve secondary to repeated or prolonged episodes of elevated intracarpal pressure; and
- Bruising the median nerve within the carpal tunnel secondary to direct trauma or using the palm of the hand as a hammer.

Which of these models is correct, if any, is currently unknown.

The exact cause of a particular person's carpal tunnel syndrome is very difficult to identify. Approximately 40% of cases are idiopathic, meaning no identifiable cause. The work-relatedness of carpal tunnel syndrome is even more controversial. Some state that it either is not, or has not been proven to be, related to work. Others state that the work-relatedness of carpal tunnel syndrome is beyond debate. The truth is probably somewhere in the middle.

Discussion and analysis of the relevant epidemiological literature on the subject of the work-relatedness of carpal tunnel syndrome is beyond the scope of this chapter. The following comments summarize this information.

1. There is evidence that some cases of carpal tunnel syndrome are *associated* with some types of work. However, current studies neither prove nor disprove whether these types of work cause the cases of carpal tunnel syndrome.

2. When carpal tunnel syndrome appears related to work, it occurs in the context of comorbidity, that is, carpal tunnel syndrome is not the only observed disorder and typically accounts for only a small percentage of the elbow, forearm, wrist, or hand disorders associated with such jobs. By contrast, when there is only an isolated case of carpal tunnel syndrome, the job may not be an important factor.

3. The exertional requirements of a job (temporal pattern and intensity of exertions) appear to mediate the increased risk of

having or developing a disorder about the elbow, forearm, wrist, and hand, including carpal tunnel syndrome. The role of wrist posture and type of grasp are questionable. Vibration per se is not a factor, but the exertional demands related to using vibrating tools is important. The temporal characteristics of jobs alone do not adequately characterize the risk of developing an upper extremity disorder.

What is the Usual Treatment?

With rare exception, people with carpal tunnel syndrome should be given a trial of conservative treatment for several weeks or months before surgery is considered. Wrist braces are often recommended for use during sleep. At work, however, wrist braces may hurt more than help. Anti-inflammatory medications and physical therapy are often prescribed, but there is no scientific basis to confirm or refute their effectiveness. At this time, vitamin supplements do not appear to be indicated. Corticosteroid injections appear to help people with carpal tunnel syndrome approximately half of the time.

If conservative treatment fails, surgery is generally effective. Since there are many medical conditions that can cause paresthesias in the hands, confirmation of the diagnosis with an electrodiagnostic study is recommended prior to surgery. Surgery can be expected to relieve numbness and tingling, but not necessarily other symptoms, such as pain. The surgical technique may be the traditional open release or the more recently developed endoscopic technique. Several factors influence the choice of technique. Regardless of technique, the operation involves cutting the ligament that covers the carpal tunnel.

What About an Affected Worker?

A worker with carpal tunnel syndrome suspected to be related to work should be assigned jobs or tasks that place less stress on the hands. One goal is to reduce the occurrence and severity of symptoms. In general, this means performing work that involves less forceful, less frequent, or less prolonged use of the fingers. Bad hand or wrist postures should be avoided. Use of the hand as a hammer should be avoided. For workers who develop carpal tunnel syndrome while working on a problem job, it is possible, but unlikely, that they will do well if returned to that job. In that circumstance, it is recommended that the job be analyzed and, if possible, modified.

What About the Work?

Even though it is difficult to determine the work-relatedness of an individual case of carpal tunnel syndrome, some of the indicators mentioned earlier should be considered.

1. If other individuals performing the same job have developed carpal tunnel syndrome or other conditions involving the elbow, forearm, wrist, and hand, then the job may be part of the problem.
2. The exertional demands of the job can be estimated using the Strain Index.
3. Beware of circumstances where workers may press or hit objects with the palms of their hands. Either eliminate the task, provide a tool, or consider a glove that provides padding to the palm.
4. Look for a circumstance of unaccustomed work, such as a new job assignment, increased work hours, increased productivity, or bad parts.

BIBLIOGRAPHY

Moore, J.S. "Carpal Tunnel Syndrome." *Ergonomics: Low-back Pain, Carpal Tunnel Syndrome, and Upper Extremity Disorders in the Workplace*. J.S. Moore and A. Garg, eds. *Occupational Medicine: State of the Art Reviews*. Philadelphia, PA: Hanley and Belfus, 1992: 741-763.

Moore, J.S. "De Quervain's Tenosynovitis. Stenosing Tenosynovitis of the First Dorsal Compartment." *Journal of Occupational Environmental Medicine* 1997, 39(10): 990-1,002.

Moore, J.S. "Function, Structure, and Responses of Components of the Muscle-tendon Unit." *Ergonomics: Low-back Pain, Carpal Tunnel Syndrome, and Upper Extremity Disorders in the Workplace*. J.S. Moore and A. Garg, eds. *Occupational Medicine: State of the Art Reviews*. Philadelphia, PA: Hanley and Belfus, 1992: 713-739.

Moore, J.S. "Proposed Pathogenetic Models for Specific Distal Upper Extremity Disorders." *Proceedings of the International Conference on Occupational Disorders of the Upper Extremities*. Ann Arbor, MI: 1996.

Moore, J.S. "Tenosynovitis and Peritendinitis." E.A. Emmett, A.L. Frank, M. Gochfeld, and S.M. Hessl, eds. *1996 Year Book in Occupational Medicine*. St. Louis, MO: Mosby Year Book, 1996: xvii-xlvii.

Moore, J.S., and Garg, A. "The Strain Index: A Proposed Method to Evaluate Jobs for Risk of Distal Upper Extremity Morbidity." *American Industrial Hygiene Association Journal* 1995, 56(5): 443-458.

Moore, J.S., and Prezzia, C. "Considerations for Determining the Work-relatedness of Carpal Tunnel Syndrome." O.B. Dickerson and M. Edril, eds. *Cumulative Trauma Disorders: Prevention, Evaluation, and Treatment*. New York: Van Nostrand Reinhold, 1996: 59-97.

CHAPTER 17

CUMULATIVE TRAUMA DISORDERS IN INDUSTRY

Fadi A. Fathallah, Researcher
Patrick G. Dempsey, Researcher
Barbara S. Webster, Researcher
Loss Prevention Department
Liberty Mutual Research Center for Safety and Health
Hopkinton, MA

WHAT ARE CTDS?

Cumulative trauma disorders (CTDs) are soft tissue disorders that affect many workers who perform repetitive industrial jobs. Other terms used for these disorders include "repetitive trauma disorders" (RTDs) and "repetitive strain injuries" (RSIs). The U.S. Department of Labor (USDL), Bureau of Labor Statistics (BLS) classifies CTDs as "Disorders Associated with Repeated Trauma," further described as "conditions due to repeated motion, vibration, or pressure, such as carpal tunnel syndrome, noise-induced hearing loss, synovitis, tenosynovitis, bursitis, and Raynaud's phenomena" (USDL 1986). The soft tissues involved most often include tendons and tendon sheaths, muscles, and nerves of the upper extremities (limbs). CTDs can result in significant suffering, various costs, and decreased productivity and product quality. Therefore, it is important that management take a proactive role in trying to eliminate, or at least minimize these disorders.

Although the exact causes of CTDs are sometimes difficult to ascertain, risk factors associated with them are fairly well identified in the literature. In addition, several approaches are available to assist the safety/ergonomics specialist in implementing an effective strategy for the identification and control of CTDs.

CTD INCIDENCE AND COSTS

There has been a marked increase in the reporting of work-related CTDs since the early 1980s. According to the BLS, the number of disorders associated with repeated trauma rose from 23,000 in 1981 to 332,000 in 1994; a greater than 14-fold increase (USDL 1983; 1995). CTDs represented only 18% of all occupational illnesses in 1981, whereas in 1994 they represented 65% of all illnesses. However, it should be noted that the BLS differentiates occupational illnesses from occupational injuries. For example, CTDs are classified as an occupational illness, whereas low-back pain is classified as an occupational injury. However, up to 7% of low-back pain related conditions were included in the CTD illness category (Brogmus et al. 1996). It should be noted that noise-induced hearing loss cases are also included in the BLS CTD category.

In 1994, occupational illnesses represented 7.6% of the total illnesses and injuries; occupational injuries, 92.4%. Therefore, CTDs represented only 4.9% of the combined total of occupational injuries and illnesses reported by the BLS.

The incidence rate of CTDs varies across different industries. According to 1993 BLS statistics, the overall industry rate of CTDs was 0.383 cases/100 workers; while the highest rates were in the meat-packing industry (12.99 cases/100 workers), knit underwear mills (8.80 cases/100 workers), auto body manufacturing (8.26 cases/100 workers), and poultry processing (7.68 cases/100 workers) (Bureau of National Affairs [BNA] 1995). It should be noted that when reporting the incidence of CTDs, BLS calculates the rate based on 10,000 person years rather than basing it on 100 person years as reported for occupational injuries, thus creating a potential for misunderstanding the scope of the disorder (the rates reported here have been corrected for this problem).

A study of 1993 upper extremity CTD claims from a private workers' compensation insurance carrier found that 78% of the claims occurred within goods-producing jobs and less than 11% in industries associated with significant use of video display units (VDUs) (Brogmus et al. 1996).

Women represented the majority (65%) of workers suffering from CTDs in 1992 according to the BLS (USDL 1994). The majority (71%) had worked more than one year with their employer and 87% of workers with CTDs were older than 24 years. Carpal tunnel syndrome was the condition most often reported (36%) and it required the longest recuperation period of all CTDs—a median of 32 days lost compared to a median of 6 days lost per case for all occupational illnesses and injuries.

While there have been many reports of the increased incidence of CTDs, it has been difficult to determine their costs because often the diagnosis does not associate them with a specific repetitive cause. Another study from a large private compensation insurance carrier analyzed upper extremity CTD claims from 1989, selected through the use of cause, body part, and injury descriptions. The study found that upper extremity CTD cases accounted for 0.83% of all claims and 1.64% of all claims costs (Webster and Snook 1994). The mean cost per case for upper extremity cumulative trauma disorders was $8,070, nearly twice the amount for the average workers' compensation claim ($4,075). The median cost per case for upper extremity CTDs was $824, close to five times the amount for the median of all compensable claims ($168). The large difference between the mean and median costs indicates that upper extremity CTD costs are not evenly distributed; a few cases account for most of the costs. Twenty-five percent of the upper extremity CTDs accounted for 89% of the costs in this study. Almost half of the upper extremity CTD compensable claims were for medical expenses only, indicating that many workers were not disabled long enough to receive wages for lost time. Medical costs represented 32.9% of the total costs; indemnity costs, 65.1%. Based on this study, it was estimated that the total annual compensable cost for upper extremity cumulative trauma disorders in the United States was $563 million.

COMMON CTDs IN INDUSTRY

A wide variety of upper extremity disorders have been recognized and classified as CTDs. Putz-Anderson (1988) classified upper extremity CTDs into three major disorder categories: tendon disorder, nerve disorder, and neurovascular disorder. Table 17-1 gives a brief description of common CTDs within each disorder category. These disorders have been detailed by Cailliet (1982), Armstrong (1983), Kroemer (1989), and Putz-Anderson (1988). It should be noted that the list of disorders in the table is not exhaustive, only the commonly reported disorders are listed.

As mentioned earlier, carpal tunnel syndrome (CTS) is the most reported CTD in industry with the highest average lost work days. CTS is a wrist CTD that is manifested by neurological deficits to the median nerve in the hand. It is commonly caused by pinching or compressing the median nerve and its blood supply as they pass through the carpal region of the wrist.

The following compression mechanisms have been described by Robbins (1963): increase in volume of tendons and sheaths, decrease in tunnel volume due to deformation, or thickening of the transverse carpal ligament. The main signs associated with CTS include various nervous impairments of the first three and a half digits of the hand—thumb, index finger, middle finger, and thumb side of the ring finger (Cailliet 1982). Motor nerve impairment includes

Table 17-1. Common CTDs under tendon, nerve, and neurovascular disorders

Tendon disorders	Description	Symptoms
Tendonitis	Inflammation of the tendon	Localized pain and discomfort
Tenosynovitis	Tendon sheath (covering) inflammation caused by excessive secretion of synovial fluid (tendon lubricant)	Localized pain and discomfort, motion-induced pain
Stenosing tenosynovitis	Persistent pressing of inflamed sheath on the tendon	Localized pain and discomfort, motion-induced pain
De Quervain's disease	Stenosing tenosynovitis affecting tendons of the thumb on the back side of the wrist	Localized pain and discomfort, motion-induced pain
Ganglion cyst	A cyst created by the excessive secretion of synovial fluid	Localized pain and discomfort, motion-induced pain
Lateral epicondilitis (tennis elbow)	Irritation of tendons of the finger extensor muscles which attach on the outside of the elbow	Localized pain and discomfort over the outside of the elbow
Medial epicondilitis (golfer's elbow)	Irritation of tendons of the finger flexor muscles which attach on the inside of the elbow	Localized pain and discomfort over the inside of the elbow
Rotator cuff tendonitis	Thickening of shoulder tendons	Localized pain and discomfort, motion-induced pain, functional impairment
Nerve disorders	*Description*	*Symptoms*
Carpal tunnel syndrome	Entrapment/pinching of the median nerve	Pain, numbness, and tingling of areas of the hand supplied by the median nerve
Guyon tunnel syndrome	Entrapment/pinching of the ulnar nerve	Pain, numbness, and tingling of areas of the hand supplied by the ulnar nerve
Neurovascular disorders	*Description*	*Symptoms*
Thoracic outlet syndrome	Compression of neurovascular bundle as it passes between the neck and shoulder	Pain, numbness, and tingling in the fingers of the hand; arm numbness; weakened wrist pulse
Vibration syndrome (Raynaud's phenomenon, vibration white finger)	Vasospasm due to cold and/or vibration	Finger blanching, pain and numbness, and eventual loss of sensation and control of the hand

reduced motor control and atrophy of the thenar muscles, which results in lack of strength in the hand. Sensory nerve impairment includes diminished sensitivity to stimulation (hypoesthesia), and burning, prickling, and tingling (paresthesia) in the affected area of the hand. Autonomic nerve impairment includes loss of sweat function, and dry and shiny skin (Armstrong 1983; Cailliet 1982).

CTD RISK FACTORS

CTDs can affect workers performing jobs within a wide variety of industries, ranging from assembling small electric motors in a manufacturing plant to processing customer payments in an electric service company. To understand the relationship between work activity and CTDs, it is important to first highlight the known risk factors related to these disorders.

Several CTD risk factors have been identified in the literature. These factors can be divided into two main classes: physical factors and nonphysical factors. The distinction between the two classes lies mainly in the source of the factor. Physical factors commonly involve external factors imposed on the individual by virtue of the nature of the job (job requirements, job layout, equipment/tools used, etc.). On the other hand, nonphysical factors tend to be more focused on the personal attributes, behaviors, and capacities of the individual.

Table 17-2 lists the major physical and nonphysical risk factors believed to be associated with CTDs in industrial settings. The most commonly-cited physical risk factors are excessive force and repetitions, and awkward or non-neutral postures (Putz-Anderson 1988; Armstrong et al. 1987; Silverstein et al. 1987). Workers can be exposed to forceful exertions while performing a variety of tasks, such as lifting heavy objects or stabilizing a hand tool. These exertions require disproportionately high muscle force generation, which could lead to reduced circulation and localized fatigue, even at 15 to 20% of the muscle's maximum capacity (Kroemer 1989). It is common to see jobs in industry where workers repeat the same basic task for an entire work shift. Tasks that require excessive repetitions may impose high demands on the muscles, creating tendon friction that could lead to various tissue disorders (Moore et al. 1991). Sustained awkward or non-neutral postures may result in pinched nerves and blood vessels and can cause impairment in blood circulation. In addition, extreme postures during flexion and extension of the wrist have been shown to reduce grip and pinch strength capabilities, as well as increase pressure in the carpal tunnel region, possibly contributing to CTS (Armstrong et al. 1991).

Table 17-2. Major physical and nonphysical risk factors identified with CTDs

Physical factors	Effects	Examples
Excessive force	Fatigue, decreased circulation	Hand tool operation; lifting heavy objects
Excessive repetition	Reduced muscle capacity, increased tendon friction	Meat de-boning operation, typing
Awkward (non-neutral) postures	Reduced muscle capacity; reduced circulation (static postures); increased pressure in carpal tunnel; nerve pinching	Typing; carpentry; assembly work
Vibration	Neurovascular disturbance	Chain saw/jack hammer operation; use of vibrating tools
Extreme temperatures (cold)	Reduce tactile sensitivity and manipulation ability	Meat packing; food processing; outdoor work during the winter
Velocity/acceleration of the wrist	Increased tendon friction	Meat packing; light assembly; electric wiring

Nonphysical factors	Including	Comments
Individual factors	Endocrinological disorder; gender; pregnancy; oral contraceptives; gynecological surgery; wrist size and shape	Special attention to these factors should be paid especially if physical risk factors are present
Psychosocial/organizational factors	Motivation; personality; peer support; job satisfaction; production pace; work/break schedules; management/worker relationships, etc.	Special attention to these factors should be paid especially if physical risk factors are present

Furthermore, it has been speculated that extreme ulnar deviation of the wrist may trigger de Quervain's disease (Tichauer 1966).

Working under extreme temperatures, especially cold, can affect the tactile sensitivity of the hand and reduce its manipulative ability. For example, lower skin temperature could result in increased forcefulness of exertions to compensate for loss of tactile sensitivity. This could further magnify CTD symptoms. Therefore, it is important to pay special attention to operations performed under cold temperatures, particularly when other CTD risk factors are present. Exposure to vibration can lead to nerve disorders, usually affecting the ability of workers to perform their tasks (Armstrong et al. 1987). Recent studies have also shown that dynamic motion (velocity and acceleration) of the wrist may be related to a history of hand-wrist CTDs in highly repetitive hand-intensive manufacturing jobs (Marras and Shoenmarklin 1993).

In some instances, nonphysical and physical factors receive almost equal attention in relation to CTDs in industry. Traditionally, the focus of nonphysical factors has been on individual traits such as gender, endocrine disorders, and gynecological issues. Individual factors may play an important role, especially for workers who perform jobs with known physical risk factors, since this may increase their risk of developing a CTD. Recently, there has been increased interest in the role of psychosocial and organizational factors in work-related CTDs. These factors include issues such as job satisfaction, employee-employer relationship, social support, and production schedules. Although these factors have not been linked to CTDs in industry, they may act as magnifiers for existing physical factors and should not be overlooked. This is especially true in the case of work-rest schedules where proper break design (frequent, short-duration breaks) could reduce the risk of sustained exertions (Swanson and Sauter 1989). In addition, hormonal stress can lead to increased muscle tension beyond the muscle forces required to perform the job. Job-induced stresses have been postulated to produce a variety of responses from the exposed individuals, including a wide range of emotional (adverse mood states), physiological (increased heart rate), and behavioral (absenteeism) responses (Smith and Carayon 1996).

The strength of the association between a given risk factor and CTDs varies widely, and depends on the factor considered and the specific situation at hand. In addition, these factors do not usually operate independently of each other. Many of them act synergystically and their combined effect is expected to be more detrimental than the sum of their individual effects. Hence, one factor's importance may be greatly influenced by the presence or absence of others. For example, the effect of performing a forceful task with the hand can be altered depending on the adopted hand posture or task frequency.

SURVEILLANCE METHODS

To identify the existence of or potential for CTD risk factors, different "surveillance" methods are available. Surveillance consists of the periodic collection and analysis of data to determine if health and safety problems exist (and the extent of any existing problem), or if risk factors are present. Surveillance can be performed for an entire company (multiple plants), a given plant, certain jobs, or classes of jobs within a plant. While these activities are often done in response to increased awareness or reporting of certain problems, such as CTDs, they should be performed on a regular basis in a proactive manner. When used proactively, surveillance activities have the potential to provide early recognition of problem jobs or tasks so that ergonomic solutions can be implemented.

Surveillance activities are usually classified into passive and active surveillance. Passive surveillance includes activities that use existing data; active surveillance, on the other hand, involves collecting data at the workplace. Relevant data that may be collected include exposure data for various risk factors, checklists, worker interview data, and medical examinations.

PASSIVE SURVEILLANCE

Passive surveillance is typically less expensive and less time consuming than active surveillance because the data already exists. The sources of information include the OSHA 200 Log and Summary of Occupational Injuries and Illnesses, OSHA 100 Supplementary Record of Occupational Injuries and Illnesses, dispensary logs, plant medical records, workers' compensation records, safety and accident reports, transfer requests, grievances, and payroll records (Hagberg et al. 1995; Putz-Anderson 1988). Availability of the different types of data will vary from plant to plant.

Once the relevant forms are collected, the data needs to be extracted. Putz-Anderson (1988) recommends that the minimum information include:

- The total number of CTD cases reported;
- The date each case was reported;
- The department or job of each injured worker; and
- The number of workers on the same job or in the same department.

A factor that can be useful for calculating incidence rates is the number of hours worked by all employees in the previous year, or several years if workers are doing the same job for that long. If this information is not available, it is generally assumed that a full-time worker works 2,000 hours per year. The incidence rate is calculated as follows:

$$\text{Incidence rate} = \frac{\text{Number of new cases} \times 200{,}000 \text{ hours}}{\text{Number of hours worked}}$$

The incidence rate calculation yields a value based on units of cases per 100 workers per year. The denominator is the total number of hours worked by all employees during the observation period (usually a year). As mentioned, the group of workers could be all workers in a plant or workers in specific jobs or departments. Usually, the calculation would be done for each job or department so that the surveillance effort yields specific information concerning where interventions should be directed. According to Putz-Anderson (1988), a plant-wide rate of six cases per 200,000 hours worked is a reasonable acceptable baseline rate with which to evaluate specific jobs or departments. In contrast, Hagberg et al. (1995) recommended that the criteria for no problem existing should be: a plant-wide incidence rate less than one per 200,000 hours, less than a twofold difference in incidence rates between departments, and no-risk factors identified during a walk-through of the plant.

If all employees worked a full year during the observation period, then 2,000 hours can be used to approximate hours worked if the actual number cannot be ascertained. However, if the observation period is less than one year, ensure that the number of hours worked is correct or represents a close approximation. Finally, care must also be taken when interpreting incidence rates if the number of hours in the denominator is low. For example, if only one worker performs a task and is diagnosed with CTS, the incidence rate for the task, using one year of exposure, would be 100 cases per 200,000 hours, which is most likely a gross overestimation. In such cases, active surveillance may be necessary to determine the extent of the potential problem. Active surveillance activities are not expected to be burdensome due to the small number of workers doing the specific task or job.

ACTIVE SURVEILLANCE

Active surveillance options are more extensive than those for passive surveillance. Active surveillance can consist of both worker health and workplace surveillance. The simplest form of worker health surveillance is symptom questionnaires which solicit information concerning pain, discomfort, swelling, etc., for each body part. Putz-Anderson (1988) and Hagberg et al. (1995) provide questionnaires that can be used for such purposes.

The next level of worker health surveillance involves activities like medical screenings, physical exams, and worker interviews. Putz-Anderson (1988) provides illustrations of various tests (for example, Phalen's or Finklestein's) that can be used. Additionally, there has been a recent proliferation of various screening devices that measure parameters, such as median nerve latency or vibratory tactile sensitivity. However, these devices have not been completely validated for screening purposes. Franzblau et al. (1993) found that a self-administered questionnaire soliciting demographic information, prior medical history, occupational history, current medical status, and symptoms of upper-extremity CTDs was the simplest method. Methods such as strength testing, vibration threshold testing, and nerve latency tests, contributed little additional information. With regard to CTS in particular, motor nerve conduction and vibration threshold testing have only limited screening value (Grant et al. 1992).

Finally, another level of active surveillance is workplace risk factor assessment. Hagberg et al. (1995) separate active risk factor surveillance into level 1 (using checklists) and level 2 (job analysis).

Level 1 active surveillance involves using checklists to survey the workplace for apparent CTD risk factors. Such checklists should be filled out by a person knowledgeable about the risk factors and ergonomics. An example of such a checklist is the one provided by Lifshitz and Armstrong (1986).

Job analysis, or level 2 active surveillance, is the most involved aspect of surveillance since it requires measurement of specific levels of various risk factors. As described earlier in this chapter, the primary risk factors for upper-extremity CTDs are force, posture, and repetition as well as cold and vibration for some disorders. Hagberg et al. (1995) present other possible risk factors, such as task invariability, cognitive demands, organizational and psychosocial factors, and static muscle loading, along with a summary of various job analysis techniques. Also, Putz-Anderson (1988) provides a technique complete with an example analysis.

PREVENTION AND CONTROL OF CTDS IN INDUSTRY

Control and prevention of CTDs in industry can be accomplished through two major categories of controls: administrative and engineering. Administrative controls are worker-focused changes where management or medical staff makes an effort to reduce the effects of both physical and nonphysical risk factors.

Administrative controls usually focus on modifying the functions of workers through training, job rotation, and job assignment. On the other hand, engineering controls are job-focused changes where an attempt is made at redesigning the job, equipment used, or workplace layout to control CTD physical risk factors (Putz-Anderson 1988).

The successful implementation of any measure or control greatly depends on the commitment and support of three key groups of personnel: upper management, design staff, and workers. Upper management support is essential since decisions to approve the change are usually initiated by this group. Second, the sincere commitment of the personnel that propose the change can assure its successful implementation. Lastly, it is almost impossible to implement an effective change without including the workers who are directly affected by it. Early involvement of workers in a proposed design is essential since the workers are the most familiar with the details of the job and may provide unique perspectives on the design's potential effectiveness and validity (Drury 1987). Chapter 19, "Managing Work-related Musculoskeletal Injuries," will expand on specific roles of personnel within each of these groups in the process of implementing successful controls.

REFERENCES

Armstrong, T.J. *An Ergonomics Guide to Carpal Tunnel Syndrome.* Akron, OH: American Industrial Hygiene Association (AIHA), 1983.

Armstrong, T.J. "Ergonomics and Cumulative Trauma Disorders." Hand Clinics 1986, 2(3): 481-486.

Armstrong, T.J., Fine, L.J., Goldstein, S.A., Lifshitz, Y.R., and Silverstein, B.A. "Ergonomics Considerations in Hand and Wrist Tendonitis." *Journal of Hand Surgery* 1987, 12A(5, Part2): 830-837.

Armstrong, T.J., Werner, R.A., Waring, W.P., and Foulke, J.A. "Intra-carpal Canal Pressure in Selected Hand Tasks." *Proceedings of the 11th Congress of the International Ergonomics Association.* Paris: International Ergonomics Association, 1991.

Brogmus, G.E., Sorock, G., and Webster, B.S. "Recent Trends in Cumulative Trauma Disorders of the Upper Extremities in the United States: An Evaluation of Possible Reasons." *Journal of Occupational and Environmental Medicine* 1996, 38(4): 401-411.

Bureau of National Affairs (BNA). *Occupational Safety and Health Reporter.* Washington: BNA, 1-4-95: 1,605-1,606.

Cailliet, R. *Hand Pain and Impairment.* Philadelphia, PA: F.A. Davis, 1982.

Drury, C.G. "A Biomechanical Evaluation of the Repetitive Motion Injury Potential of Industrial Jobs." *Seminars in Occupational Medicine* 1987, 2(1): 41-49.

Franzblau, A., Werner, R., Valle, J., and Johnston, E. "Workplace Surveillance for Carpal Tunnel Syndrome: A Comparison of Methods." *Journal of Occupational Rehabilitation* 1993, 3(1): 1-14.

Grant, K.A., Congleton, J.J., Koppa, R.J., Lessard, C.S., and Huchingson, R.D. "Use of Motor Nerve Conduction Testing and Vibration Sensitivity Testing as Screening Tools for Carpal Tunnel Syndrome in Industry." *Journal of Hand Surgery* 1992, 17A: 71-76.

Hagberg, M., Silverstein, B., Wells, R., Smith, M.J., Hendrick, H.W., Carayon, P., and Pérusse, M. *Work-related Musculoskeletal Disorders (WMSDs): A Reference Book for Prevention*. London: Taylor and Francis, 1995.

Kroemer, K.H.E. "Cumulative Trauma Disorders: Their Recognition and Ergonomics Measures to Avoid Them." *Applied Ergonomics* 1989, 20(4): 274-280.

Lifshitz, Y., and Armstrong, T. "A Design Checklist for Control and Prediction of Cumulative Trauma Disorders in Hand Intensive Manual Jobs." *Proceedings of the Human Factors Society 30th Annual Meeting*. Santa Monica, CA: Human Factors Society, 1986.

Marras, W.S., and Shoenmarklin, R.W. "Wrist Motions in Industry." *Ergonomics* 1993, 36(4): 341-351.

Moore, A., Wells, R., and Ranney, D. "Quantifying Exposure in Occupational Manual Tasks with Cumulative Disorder Potential." *Ergonomics* 1991, 34(12): 1,433-1,453.

Putz-Anderson, V., ed. *Cumulative Trauma Disorders: A Manual for Musculoskeletal Diseases of the Upper Limbs*. London: Taylor and Francis, 1988.

Robbins, H. "Anatomical Study of the Median Nerve in the Carpal Tunnel and Etiologies of the Carpal Tunnel Syndrome." *Journal of Bone and Joint Surgery* 1963, 45A(5): 953-966.

Silverstein, B.A., Fine, L.J., and Armstrong, T.J. "Occupational Factors and Carpal Tunnel Syndrome." *American Journal of Industrial Medicine* 1987, 11: 343-358.

Smith, M.J., and Carayon, P. "Work Organization, Stress, and Cumulative Trauma Disorders." S.D. Moon and S.L. Sauter, eds. *Beyond Biomechanics: Psychosocial Aspects of Musculoskeletal Disorders in Office Work*. London: Taylor and Francis, 1996.

Swanson, N.G., and Sauter, S.L. "The Design of Rest Breaks for Video Display Terminal Work: A Review of the Relevant Literature." *Advances in Industrial Ergonomics and Safety I, Proceedings of the Industrial and Ergonomics Conference*. Cincinnati, OH: Taylor and Francis, 1989.

Tichauer, E.R. "Some Aspects of Stress on Forearm and Hand in Industry." *Journal of Occupational Medicine* 1966, 8: 63-71.

U.S. Department of Labor (USDL), Bureau of Labor Statistics (BLS). *Occupational Injuries and Illnesses in the United States by Injury, 1994*. Washington: U.S. Government Printing Office, 1995.

USDL, BLS. *Occupational Injuries and Illnesses in the United States by Injury, 1981*. Washington: U.S. Government Printing Office, 1983.

USDL, BLS. *Record Keeping Guidelines for Occupational Injuries and Illnesses*. Washington: U.S. Government Printing Office, 1986.

USDL, BLS. *Work Injuries and Illnesses by Selected Characteristics, 1992*. Washington: U.S. Government Printing Office, 1994.

Webster, B.S., and Snook, S.H. "The Cost of Compensable Upper-extremity Cumulative Trauma Disorders." *Journal of Occupational Medicine* 1994, 36(7): 713-717.

CHAPTER 18

ANSI-Z365 STANDARD: CONTROL AND PREVENTION OF CUMULATIVE TRAUMA DISORDERS

Marvin J. Dainoff
Professor, Psychology, and
Director, Center for Ergonomic Research
Miami University
Oxford, OH

Published standards represent, at the most basic level, attempts by professionals to achieve consensus on a variety of technical and procedural issues, and to communicate these positions to potentially interested parties. Standards make an important contribution to enhancing commerce by allowing manufacturers to standardize dimensions and configurations so that components can be interchangeable. In addition, standards in the areas of work processes and procedures—such as those related to safety and health—allow organizations to take advantage of already-developed expertise without "reinventing the wheel."

DEVELOPING ANSI-Z365

Working within procedures and guidelines provided by the American National Standards Institute (ANSI), the National Safety Council (NSC) is in the process of developing a consensus-based voluntary standard dealing with the control and prevention of work-related cumulative trauma disorders. The purpose of this standard is to provide procedures (principles and practices) by which upper-extremity CTDs may be managed. Such procedures include ergonomic considerations along with surveillance (identifying several cases of CTDs as well the presence of problematic working conditions), medical management, and training. The proposed standard, ANSI-Z365, is still under development as of this book's publication.

THE ANSI ROLE

ANSI is a private organization that develops both voluntary and consensus-based standards. To quote from its own publication: "Many standards developers and participants support the American National Standards Institute as the central body responsible for the identification of a single, consistent set of voluntary standards called American National Standards" (ANSI 1995). Typically, professional and technical organizations act as standards developers. ANSI is responsible for providing standardized procedures for standard development that ensure principles of openness and due process have been followed. It also verifies in each case, that such principles have, in fact, been adhered to. When this has been accomplished, the published document is accepted by ANSI as an American National Standard.

At the international level, ANSI serves as the United States member of the International Organization for Standardization (ISO). ANSI helps facilitate liaison with organizations developing international standards and is concerned with harmonizing national and international standards efforts. As international trade becomes an increasing part of economic life, these relationships take on greater and greater importance.

ANSI standards are sometimes incorporated into official regulations by governmental organizations, such as the Occupational Safety and Health Administration (OSHA). In the case of OSHA, this is not an automatic process, but requires public hearings, as with any change in regulations.

THE PROCESS

In 1990, the NSC, working through ANSI, organized a committee to develop and produce the voluntary-consensus standard that would come to be known as ASNI-Z365. The participants were organized as an accredited standards committee, with Professor Thomas Armstrong of the University of Michigan, as chair. Participants were divided into task-oriented subcommittees assigned, respectively, to the topics of work analysis and design, surveillance, and medical management. By 1995, several drafts had been prepared and received public comment.

For a document to be accepted by ANSI as an American National Standard, the process by which the standard was developed must have allowed for due process and openness. In particular, standards developers are required to ensure that persons (individuals, organizations) who have a "direct and material" interest in the outcome of the standards process are allowed to have their points of view considered in a fair and open manner. To achieve these goals, ANSI specifies several alternative procedural routes. The route chosen by the NSC was that of the Accredited Standards Committee.

THE ACCREDITED STANDARDS COMMITTEE

The accredited standards committee method is typically used when the perspectives of many different interest groups are involved. It requires that the composition of the voting members of the standards committee reflect an appropriate balance of the interest groups concerned. In the case of Z365, the interest groups were initially defined as:

- Academics/research institutes (academic researchers);
- Employers (companies or trade associations);
- Employees (unions);
- Governmental/regulators (OSHA, National Institute for Occupational Safety and Health [NIOSH]);
- Professional societies (representing occupational medicine, hand surgery, industrial hygiene, ergonomics, occupational therapy, and safety engineering, among others);
- Service providers (ergonomics consulting organizations);
- Insurance companies (in prevention and control roles); and
- Manufacturers/suppliers of ergonomic equipment.

Later on, it was decided to combine the last three categories into a single service category.

Achieving a Balance

The initial committee had 38 voting members. A year later, membership was capped at 54. The goal in accepting new members was to achieve balance among the various categories. The current distribution of members is 13 academics, 13 employers, 6 employees, 3 governmental, 9 societies, and 10 service.

In addition to voting members (and their alternates), all interested individuals may attend meetings as observers. They play important roles in drafting materials and providing feedback, but they do not have a vote.

Drafting and Reviewing

Within the aforementioned structure, consensus is achieved by assigning some committee members (and sometimes observers) to draft sections of the standards, while other members and observers comment and review these sections. On occasion, feedback is requested from a broader audience as well. Responders must have a direct and material interest in the standard.

After each draft has been reviewed, the drafters are obligated to attempt to resolve, or at least address, the concerns expressed. When a final draft has been approved by ballot, and attempts to resolve objections have been made, the draft standard is forwarded to ANSI for final approval.

COMPLIANCE

An ANSI standard consists of a set of requirements. In the case of Z365, these requirements are elements or components of a prevention or control program. The requirements that form the core of ANSI standards are defined at two levels. Statements that use the word *shall* are mandatory requirements, considered fundamental and essential in establishing and maintaining a program. Statements using the word *should* reflect nonmandatory recommendations. These are program elements found to be effective, and would be included in a more comprehensive program.

How compliance is to be carried out is not addressed by ANSI. For many technical standards, private sector testing and evaluation services have been established. Organizations wishing to have their product or service certified as compliant may use such services. There is, at present, no movement toward such procedures for Z365. Presumably, any assessment of compliance will be carried out by the organization itself.

STRUCTURE AND CONTENT

Fundamentally, Z365 consists of methods and procedures for organizing a prevention and control program for managing CTDs. Presently, the core components of the standard include a statement of purpose and scope, a set of definitions, a discussion of compliance, and an overview of the essential program components. In addition, there are several supplementary sections that amplify and extend material in the core components. The format and organization are still under discussion.

COMPONENTS

The first three sections of the core components are relatively brief. The scope and purpose, as previously stated, provides principles and practices for controlling CTDs that arise from a variety of work tasks, including lifting, assembly, and manipulation. This version of Z365 is explicitly limited to CTDs of the upper extremities. Concerns about other portions of the body—particularly lower back disorders—may be dealt with in a subsequent document. Finally, in the process of committee discussions, concerns about the lack of specificity of the CTD label itself were expressed. At one point, it was proposed that the standard be renamed "Control and Prevention of Work-related Musculoskeletal Disorders." However, committee consensus was that, although CTD as a medical diagnosis was problematic, it would be confusing to the public to change a title already approved by ANSI. Therefore, the term CTD was retained.

Definitions and Compliance

The definitions and compliance sections of Z365 are short and relatively straightforward. The definitions section reflects that, although Z365 is aimed at individuals with technical training, a variety of professional and technical backgrounds are involved in CTD control. Thus, terms that are obvious to one professional may be unknown to another. The definitions attempted to remedy this problem. The compliance section essentially elaborates the distinction between *shall* and *should* statements.

Operational Content

The operational content of Z365 consists of the essential program elements described in the fourth section. There are two major components to this section. The first, entitled "Background," consists of a number of statements that, in effect, justify the standard. These statements, resulting from the committee's literature review, assert that, although the causes of CTDs may arise from multiple sources, risk factors associated with working conditions have been associated with CTDs. Such risk factors may include equipment design, work processes, work environment, and work organization. Furthermore, CTDs may often go unreported.

It is further asserted that general design principles can be described that reduce the risk of exposure to CTDs, however, it is not (yet) possible to quantify precise relationships between specific risk factors (specific working conditions) and specific likelihood of CTD occurrence. Nevertheless, it is possible to develop and implement a control program for work-related CTDs.

Process Component

The process component contains the program elements themselves. Operationally, three functions must be accomplished. These are surveillance, job analysis, and medical management. Each function is embedded within an overall organizational structure, which includes management responsibility for defining and supervising the program, establishing appropriate training, and ensuring active participation by employees. The last two elements have some degree of overlap in that employee education regarding CTDs is one aspect of employee participation.

Surveillance

Surveillance has two aspects. The first relates to identifying those employees who have symptoms of CTDs. This typically involves tracking current levels of CTDs, as well as examining records for patterns of disorders across the organization. This component requires that a system of record keeping be in place. A second aspect of surveillance relates to identifying work sites that might have risk factors for CTD. Such identification might be accomplished by job survey procedures. These surveys are proactive in nature in the sense that the goal is to find locations where employees are potentially at risk.

Job Analysis

Job analysis involves an attempt to determine the specific components of risk associated with a given job task, and suggests methods for reducing that risk. Job analyses are more detailed than job surveys. Z365 suggests that the decision to conduct job analyses in specific cases be prioritized on the basis of the surveillance findings. First, priority should be given to jobs with identified cases of CTD. Lower on the list of priorities should be work locations where job surveys had determined high levels of CTD risk factors.

The classes of physical stressors to be identified as risk factors by job analysis include: force, posture and motion, vibration, and temperature. Each of these stressors can, in turn, be characterized by four properties. These properties are: magnitude (the extent of the stressor), repetition (frequency with which the stressor occurs), duration (length of time the stressor is sustained), and recovery (length of time during which the stressor is not present). In addition, work organizational factors (machine-paced work, close performance monitoring, overload, time pressure) have been found to influence the expression of physical symptoms, and should be included as part of job analyses. All of the aforementioned analysis information can be used as a basis for job redesign efforts to lower risk factor levels. Kuorinka and Forcier (1995) summarize the literature documenting the evidence of work-related CTD symptoms.

Medical Management

The final component, medical management, is primarily addressed to health care providers and those in the organization who interact with such providers. This section deals with methods of medical evaluation, medical treatment and follow-up, and record keeping.

Additional sections of the Z365 standard amplify and expand on the basic components. The present draft consists of chapters dealing with surveillance, job analysis and design, medical management, and training.

ISSUES AND CONCERNS

The process of achieving consensus regarding a working draft of a standard can be complicated, particularly in a group as diverse as the members of the Z365 committee. The description of the document reflects current progress toward achieving consensus. However, there are still a number of issues remaining.

There are those who question the basic background assumption, discussed previously, of a link between CTD symptoms and working conditions (Hadler 1990). Moon (1996) and Sauter and Swanson (1996) provide excellent overview discussions of some of these issues. See also Kuorinka and Forcier (1995).

ADDITIONAL FACTORS

There is fairly strong consensus that it is not, at present, possible to define strong quantitative relationships between specific symptom patterns and specific working conditions. This is not surprising. Most occupational disorders involve a single organ system and can be traced to exposure to a rather specific risk factor. For example, asbestosis is linked to the presence of asbestos fibers in the work environment; hearing impairment is linked to prolonged exposure to sounds of a certain frequency and intensity. However, the "exposure" in the case of CTDs is to work itself, taking into account both its physical and organizational component. But, additional factors outside of work may also play an important role, making the potential causal chain much more complex.

Inadequate Analyses

The practical consequence of this complexity, with respect to creating a standard, is that paradoxically, it is not now possible to standardize specific methodologies for functions such as job analysis and job surveys. Accordingly, there is considerable discussion and disagreement among qualified professionals as to the most effective manner of carrying out such analyses.

For example, in the case of assessment of postural risk, it may be recalled that each posture can be characterized with respect to four properties: magnitude, repetition, duration, and recovery. Those working postures, which are typically implicated in upper extremity CTDs include: head/neck angle, trunk angle, right and left shoulder angle, right and left forearm angle, and right and left wrist angle. A thorough analytic study of working posture would require characterizing each of these angles with respect to some critical value or range of values for each of the four properties. However, the resulting analysis is still inadequate since these variables are not independent, but interact with each other. For example, the stress on a bent wrist may be worse if the arm is extended. Thus, the interactions among postures also need to be taken into account.

Moreover, the remaining three physical risk factors (force, vibration, temperature) must be described in terms of their associated characteristics. There is also evidence for interactions among all four groups of physical risk factors, as well as between physical and organizational risk factors. A completely systematic assessment of all relevant risk factor attributes and their interactions would require a procedure so complex as to be completely impractical on the shop floor. This would be true even if the data existed to carry out such an assessment—which it does not.

More practical solutions to the problems of assessment and analysis have been proposed. These are necessarily simplified, reflecting professionals' interpretations of existing scientific evidence, but tempered by the realities of what is realistically feasible to implement. Thus, in the current version of Z365, alternative survey and assessment procedures have been suggested, but not mandated.

In the final analysis, the justification for an imperfect solution is that it is better than no solution at all. There is evidence that some ergonomic solutions effectively reduce CTDs, despite the lack of a precise relationship between exposure and outcome. The current document represents some degree of consensus as to how that relationship might be achieved.

OTHER STANDARDS

ANSI-Z365 is an important standard relating to ergonomics. However, several other ergonomic standards exist or are in progress. The Human Factors and Ergonomics Society (HFES), the professional society for ergonomists in the United States, is helping develop three ANSI standards. ANSI/HFES 100-1988 is an approved ANSI standard that provides technical requirements for computer terminals and workstations. This standard has been in place since 1988, and is now being revised. ANSI/HFES 200, currently under development, deals with ergonomic considerations in software development. ANSI/HFES 300, also under development, focuses on issues of anthropometry and biomechanics. In addition, ISO has an extensive series of ergonomic standards in various stages of development. The 9241 series relates to work with computers and computer workstations. Finally, the American Society for Testing and Materials (ASTM) is developing an ergonomics standard.

REFERENCES

American National Standards Institute (ANSI). "Procedures for the Development and Coordination of American National Standards." New York: ANSI, 1995.

Hadler, N.M. "Cumulative Trauma Disorders." *Journal of Occupational Medicine* 1990, 32(1): 38-41.

Kuorinka, I., and Forcier, L. *Work-related Musculoskeletal Disorders (WMSDS): A Reference Book for Prevention*. London: Taylor and Francis, 1995.

Moon, S.D. S.D. Moon and S.L. Sauter, eds. "A Psychosocial View of Cumulative Trauma Disorders: Implications for Occupational Health and Prevention." *Beyond Biomechanics: Psychosocial Aspects of Musculoskeletal Disorders in Office Work*. London: Taylor and Francis, 1996.

Sauter, S.L., and Swanson, N.G. S.D. Moon and S.L. Sauter, eds. "An Ecological Model of Musculoskeletal Disorders in Office Work." *Beyond Biomechanics: Psychosocial Aspects of Musculoskeletal Disorders in Office Work*. London: Taylor and Francis, 1996.

CHAPTER 19

MANAGING WORK-RELATED MUSCULOSKELETAL INJURIES

Gary A. Mirka
Assistant Professor

Carolyn M. Sommerich
Assistant Professor

Department of Industrial Engineering
North Carolina State University
Raleigh, NC

THREE-TIER SYSTEM

Work-related musculoskeletal injuries can be managed through a three-tier control system, based on the classic epidemiological scheme of primary, secondary, and tertiary prevention. Primary prevention is proactive in nature, with the goals of preventing the occurrence of musculoskeletal injuries and keeping employees healthy and on the job.

By contrast, secondary and tertiary prevention are reactive in nature. Secondary prevention measures intervene early, when injury or illness symptoms first appear, in order to cure, or at least slow development. Tertiary prevention is designed to minimize an employee's problems once injury or illness has occurred. Limiting sequelae and disability, and promoting patient rehabilitation and early return-to-work are the goals of tertiary prevention. To be effective in the long run, an ergonomics program created to address the problems of work-related musculoskeletal disorders needs to function at all three levels.

WIDESPREAD COMMITMENT

The long-term success of an ergonomics program depends upon the participation and commitment of many different departments and individuals within an organization. These include managers, human resource personnel, engineers, supervisors, medical specialists, safety specialists, and hourly employees. Some companies will not have representation from all of these groups on staff. In such cases, individuals are often asked to develop skills or perform functions outside their areas of training, or the company may need to hire an external consultant.

An important factor is that, in a fully-functioning ergonomics program, musculoskeletal injuries will be managed through a network of individuals from all areas within a company, and this system will only be as strong as its weakest component. A failure in any area can result in disinterest, frustration, or program failure.

Each employee must recognize the importance of his or her role and follow through on tasks associated with the program. An effective management approach to the control of musculoskeletal disorders takes advantage of each individual's duties and skills, and integrates them to form the facility's ergonomics program. By aligning an employee's training and new responsibilities in musculoskeletal management with his or her experience and talents, the increased work load should be marginal, fitting comfortably with that individual's present duties.

MANAGEMENT'S ROLE

Managers provide vision for a company. They make decisions affecting the organization's overall performance and long-term profitability. Managers should not treat the costs associated with musculoskeletal injuries and illnesses as just another "cost of doing business," but should provide an appropriate level of commitment and resources to reduce these costs. They need to be aware that costs are not limited to medical costs and other related workers' compensation costs (often referred to as direct costs), but can also include costs associated with replacing workers. Also, the inexperience of new hires can lead to costs associated with reduced quality and productivity. Even when those suffering from musculoskeletal pain remain on the job, there may be costs associated with reduced quality and productivity directly related to their pain. These indirect costs are often difficult to quantify and require some data collection, but when they are considered in conjunction with the more direct costs, they impact management's decision-making process. When managers consider all these costs, they often decide that supporting a company ergonomics program to reduce the incidence of these injuries and illnesses makes sound business sense. Management commitment is one of two make-or-break factors in such an effort, the other being operator involvement.

One form of management commitment that must be in place at all three levels of prevention is simply a highly visible show of support for the ergonomics process and the ergonomics team (a small, diverse group of employees charged with carrying out many activities within the ergonomics program). Providing everyone in the facility with the time and encouragement to participate in the work of the ergonomics team and being visible at important ergonomics team meetings or other team functions also demonstrates management's recognition of the importance of the work being done. Some team members may feel uncomfortable or inadequate when faced with the task of performing job analysis and job redesign. An expression of support and encouragement from management can help these individuals overcome these feelings and free them to contribute to the success of the program.

PRIMARY PREVENTION

Training

A large part of the commitment that management must make in the primary prevention mode comes in the form of training and education, which should be company-wide. Hourly workers need to learn about the risk factors and early warning symptoms of musculoskeletal problems. Engineers and supervisors must be trained in ergonomic workplace design. Although the time that these employees spend in training is time away from their usual jobs, in the long run many companies have found this to be a good use of, and investment in, their human resources.

Funding

Another type of management commitment comes in the form of funds for purchasing tools and equipment designed upon sound ergonomic principles. A good example seen in industries everywhere is the issue of industrial seating. Well-designed chairs are not inexpensive, but the advantages and long-term savings from this up-front investment can be compelling. Another example would be appropriate hand tools. Ergonomically-designed hand tools can often cost 50-100% more than those purchased in the local hardware store. However, the cost of one case of carpal tunnel syndrome can quickly justify the higher expense of a well-designed hand tool.

Staff Requirements

A manager needs to assess the current staff and decide whether any members can adequately perform their required tasks or if contract assistance is required, either part-time or full-time. This may include ergonomics consultants or medical management specialists (occupational health nurses, occupational physicians, physical therapists, etc.). Recognizing the limitations of current staff and contracting with outside help is another form of commitment from management.

Another management function that can affect the ergonomics process is negotiating with employees or labor unions. Typically, task modifications and the methods used to determine employee compensation require some negotiation between management and labor. Modifications may include changes in the pacing of lines (human-paced versus machine-paced), pay schedules (piece rate versus hourly wage), placing limits on maximum piece rates, or development of appropriate work and break schedules.

If a manager understands the musculoskeletal consequences of various pacing and compensation strategies, he or she will be better equipped to discuss with labor the problems associated with piece rate or machine-paced work, and to arrive at a pacing and compensation strategy that will, in the long run, save the company money while reducing the risk of employee musculoskeletal problems.

SECONDARY PREVENTION

Management commitment in the secondary prevention mode is similar to that in the primary prevention mode, except it calls for a more reactive response to existing problems. The allocations of funds and time, as well as negotiations with employees, are often required at this level. Rather than purchasing new equipment, funds for secondary prevention may be needed to modify existing equipment. Instead of being allocated for training, time must be allocated for engineering redesign. Negotiations with employees or unions would focus on how issues could be addressed by altering previous agreements or current practices.

External personnel may be needed to address specific issues raised by the current situation. For example, an ergonomics consultant may be called upon to evaluate a job and make specific recommendations based on concerns raised by the operators. An existing musculoskeletal problem would provide strong incentive to develop such a relationship if there was no formal relationship established with a local health care provider.

TERTIARY PREVENTION

At the tertiary level, management commitment may require some philosophical changes. Unfortunately, some companies will simply not allow the existence of light- or alternate-duty jobs. Studies have shown that the duration of time spent away from work due to injury or illness is inversely related to the likelihood that an injured employee will ever return to work. Bringing an employee back on alternate duty (as established in consultation with a medical practitioner) can improve morale and return the employee to work in a timely fashion.

Appropriate compensation is an important issue in establishing a light-duty program. Should the employee be compensated at the same rate as before the injury? If the employee was on piece rate, how should the pay rate be set for the alternate-duty job? The answers to these questions will often require negotiation and consultation between management, Human Resources, and labor. If handled improperly, this issue can generate friction between employees on alternate duty and their peers. However, developing a time schedule for leaving the alternate duty can often help alleviate these problems.

A change in the methods used to account for work-related musculoskeletal problems can also aid in the identification of problem areas within a facility. Often the costs associated with musculoskeletal injury and illness are buried in the budget of the Human Resources department. Why not place costs associated with each work-related musculoskeletal disorder on the line or in the department where the injury occurs? This accounting method would show that a line that has a high volume of production, but at the cost of a high incidence of musculoskeletal injuries, may in fact be less productive in the larger sense, and therefore rational modifications in production levels may be in order. This is another example of a possible philosophical change that may need to be considered when addressing the problems of musculoskeletal stress at the tertiary level of prevention.

HUMAN RESOURCES' ROLE

The traditional roles of the Human Resources department are hiring and firing, dealing with workers' compensation claims, and interacting with outside care providers in cases of accident, injury, or illness. This group's day-to-day appreciation for the human side of business makes it an invaluable component in a program to control musculoskeletal problems. Human Resources personnel typically have the ability to quickly access much of the raw data (employment, injury, and compensation records), identifying "hot spots" in a facility, and tracking long-term trends in the effectiveness of efforts to reduce musculoskeletal injuries.

Based on relative familiarity with the majority of the work force and access to personnel and injury records, a representative from the Human Resources department makes a good addition to the plant ergonomics team. That individual can often coordinate the composition of the team by helping choose employees who can be of the greatest service and influence on the team. He or she can also provide timely updates of the number of employees away from work due to work-related musculoskeletal problems, as well as the number of individuals on alternate duty.

PRIMARY PREVENTION

One of the key roles Human Resources can play in the primary prevention mode, is to help administer discomfort surveys to the employees on the shop floor. Discomfort surveys measure the musculoskeletal stress of the employees. In contrast to an analysis of the Occupational Safety and Health Administration (OSHA) Form 200 Logs, which document existing injury and illness, discomfort surveys allow a company to be more proactive by addressing early symptoms, thereby addressing problems before they become OSHA Form 200 recordable. Discomfort surveys administered on an annual basis can be used to examine trends within departments, as well as identify those departments in greatest need of help. This can be an invaluable method for tracking the effectiveness of an ergonomics program.

The Survey

An example of a discomfort survey is shown in Figure 19-1 (parts a, b, and c). Figure 19-1a outlines the type of background questions typically asked, including some historical and nonwork-related information from the individual, in addition to information about the job he or she is currently performing. The issue of the discomfort survey's anonymity is a sensitive one. Fear of supervisor or management reprisal can limit the respondent's honesty and forthrightness. Therefore, unless there is some overriding reason for including respondents' names, surveys are usually anonymous. There does, however, need to be enough information on these surveys so that the reviewer can identify the specific job or task the respondent performs.

Since many musculoskeletal disorders develop over time, it is important to gather data on a respondent's previous jobs within the company, as well as jobs performed for previous employers. Additionally, since the importance of adequate rest periods has been revealed as part of the recovery mechanism of musculoskeletal tissues, it also helps to understand the types of activities in which individuals engage away from their jobs, such as hobbies or second jobs.

Ergonomics Discomfort Survey

Name _____ Date _____ / _____ / _____

Building and room # _____ Job title _____

Average hours worked/week _____ Hours/day spent at monitor/scope _____

Hr/day standing _____ Do you wear protective/cleanroom clothing? Y __ N __

How many years have you been employed by this company? _____

How many years have you been doing your current job? _____

What other jobs have you done in the past 10 years? Include jobs within this company as well as other places of employment.

_____
_____
_____
_____
_____

When you are away from work what are your hobbies/activities (for example, sports, crafts, hobbies, odd jobs, etc.)?

_____
_____
_____
_____

Break down your daily work activities and give an approximate percentage of time each is performed.

Activity	%
_____	_____ %
_____	_____ %
_____	_____ %
_____	_____ %
_____	_____ %

On the following page, please place an "X" on those areas that have caused you discomfort in the past six months. For each of these Xs, rate the severity of the discomfort on a scale of 1 (slight discomfort) to 10 (unbearable pain).

Figure 19-1a First page of discomfort survey.

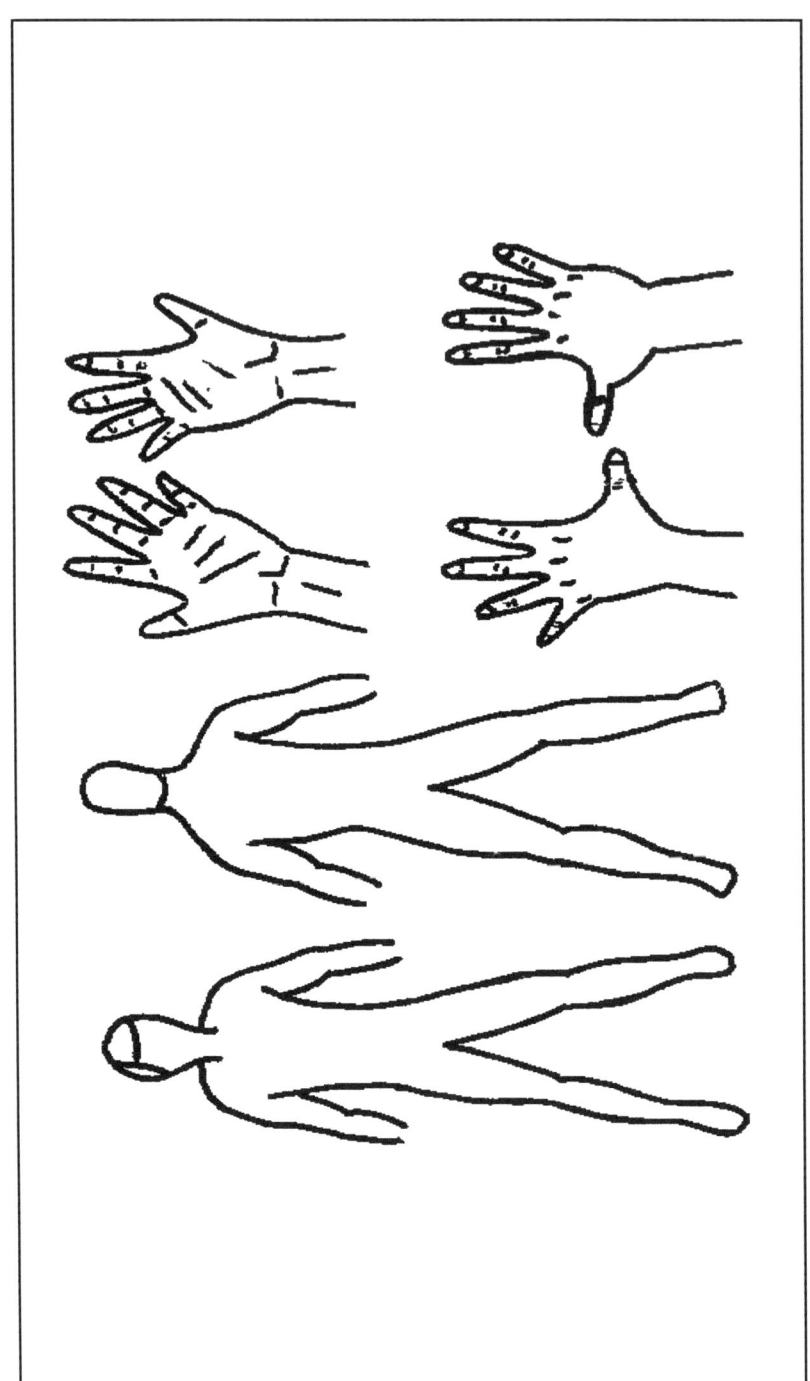

Figure 19-1b Second page of discomfort survey.

Fill out one of these forms for each of the Xs.

Part of body _____

1. Check the box(es) which best describe your problem.

    ☐ Tingling   ☐ Burning

    ☐ Numbness   ☐ Stiffness

    ☐ Loss of strength   ☐ Pain

    ☐ Cramping

2. Do you currently have this discomfort? Y __ N __

3. Usually, how long does the discomfort last? _____

4. How long have you had this discomfort? _____

5. At what time of the day do these symptoms occur?

    ☐ Morning

    ☐ Afternoon

    ☐ Evening

    ☐ At night

6. Is there any activity which makes these symptoms reappear? _____
   _____
   _____

7. Have you seen a doctor or other medical person about this problem?
   Y __ N __

   If yes, what was the diagnosis? _____
   _____
   _____

8. Have you lost any time from the job because of this problem? Y __ N __
   If yes, how many days? _____

Figure 19-1c Third page of discomfort survey

The second page of the survey (Figure 19-1b) shows the form of the human body. Employees are asked to mark with an 'X' those parts of the body causing them discomfort. They are then asked to rate this discomfort on a scale of 1-10, with one being a mild irritation and ten being near unbearable pain. For each of these Xs the individual is asked to elaborate on the description of the discomfort. Done on multiple copies of page three (Figure 19-1c), this detailed information can help the ergonomics team understand the severity of an employee's symptoms.

Taken individually, discomfort surveys do not always lead to the most accurate description of work-related musculoskeletal stress. This may be due to an individual's feelings towards a supervisor, how the employee feels the day the survey is conducted, or a variety of other factors. However, if several people performing the same job have similar complaints, the analyst should note the trend and pursue a more in-depth investigation. Even if the severity of the complaints is low, consistent reports from several individuals may indicate the early stages of a significant problem.

Further, it is a good idea to visit any job that has elevated (greater than five) values of discomfort. Speak with the operators to gain insight into the origins of their complaints. The investigator can try to correlate the information on the discomfort survey with the types of tasks the individuals perform to identify the source of the discomfort. When considering reports of employee discomfort, illness, or injury, it must be understood that there is a relationship between an employee's overall happiness and the likelihood that he or she will file a workers' compensation claim. Therefore, identifying other problems that may interact with physical workplace stress or the employee's recovery also can be helpful in reducing the overall cost of musculoskeletal injuries.

There are a couple of other indices that Human Resources personnel can track to help identify jobs or departments requiring the attention of the ergonomics team. These indices include turnover rates and movement of employees within a company. Employees tend to move away from jobs that they find unpleasant. There are many possible sources of an individual's dislike for a particular job or department, including the likelihood that the job may be causing some musculoskeletal discomfort.

Another domain of Human Resources that can be considered primary prevention is the development of sound hiring practices. Individuals who have a history of low-back injuries or upper-extremity cumulative trauma disorders are at greater risk of having recurrent problems, and therefore need to be identified and placed in jobs that will not expose them to stresses likely to cause re-injury. This, however, can be a very sensitive topic because of issues related to discrimination. The Americans with Disabilities Act (ADA) has some fairly strict guidelines that must be followed to reduce a company's exposure to possible lawsuits from job applicants.

Two key provisions of the ADA are that functional testing must be job specific, and that medical tests can only be required after a job has been offered. Strict protocols must be developed in conjunction with engineering to identify those aspects of the job that are critical and that cannot be eliminated without unreasonable accommodation. For further information, refer to the ADA documentation (Equal Employment Opportunity Commission 1991).

SECONDARY PREVENTION

Once an occupational injury or illness has occurred, the Human Resources department plays a vital role in managing and controlling the subsequent events that play a major role in the ultimate outcome of the recovery process. Unfortunately, the Human Resources department is often not the first point of contact for an injured worker. Typically the injured worker will bring the problem to the attention of a supervisor, who often will make a decision as to whether or not to send the person to the Human Resources department. Therefore, it is important that supervisors know the formal procedures to be followed in the event of a complaint by one of their workers.

In the case of an emergency, the employee should receive immediate medical attention at a first aid station or emergency room. If there is no urgency, the Human Resources representative should speak with the individual and gather information about what the employee was doing at the time of the incident. Memories of the specifics of an incident tend to fade rather quickly. Gathering and documenting this information at the time of the incident can aid in understanding the problem in the future.

Once it has been determined that an injury or illness has occurred and that it can be attributed to the workplace, it is recorded on OSHA Form 200 Logs. In addition to being a requirement for a majority of businesses in the United States, these logs also can be a valuable tool for the plant ergonomics team in identifying jobs or departments that require evaluation. This is especially true if the person who fills out the forms understands musculoskeletal disorders and includes enough information to provide insight into how the injury occurred. This is another good reason for including a Human Resources representative on the ergonomics team.

Often, individual states require the submission of supplemental forms that include more detailed documentation of all circumstances regarding the incident. This additional information can be a great help in identifying the root cause of a problem.

OSHA Logs can be helpful from two different perspectives. First, they provide data for quantifying the incidence of injury. Second, data is also available for calculating a severity of injury index. The incidence of a disorder is the number of new cases of the disorder that develop in a given time period. An incidence rate is a normalized representation of that value, which can be used to compare rates among departments, companies, or industries, regardless of work force size.

The standard way to represent incidence rate is to express it as the number of new cases of a specified disorder that would have occurred in a group of 100 people working for one year (40 hours per week, 50 weeks per year). For the work group of interest (work cell, line, department, plant, division, etc.), calculate the incidence rate (IR) for the time period of interest as follows:

$$IR = \frac{\text{no. of new cases in work group for time period} \times 200{,}000}{\text{total hours worked in the time period by the work group}}$$

For example, if 14 people worked in a department for 60 weeks, they worked traditional 40-hour work weeks, and there was a total of two new OSHA recordable back injuries during that time period, the incidence rate for back injuries would be:

$$IR = \frac{2 \text{ cases} \times 200{,}000}{14 \text{ people} \times 40 \text{ hours per week} \times 60 \text{ weeks}} = 11.9$$

This rate is interpreted as follows: if there were 100 workers in a department and they incurred back injuries at a rate similar to the hypothetical department of 14 people, approximately 12 of the 100 people might be expected to incur a back injury over a one-year time period. The normalized representation facilitates comparisons between jobs with many employees and those with only a few. Often states can provide a company with reference information about incidence rates for various occupational injuries and illnesses by Standard Industrial Classification (SIC) codes. Typically, these data are not categorized by type of injury or illness, but this can still help companies benchmark with others in their industry.

Another, often more practical, way to represent a problem is to describe it in terms of severity. Unlike incidence rate, severity has no standard form of representation. Some examples of plausible severity indices include number of days lost or restricted per recordable injury, workers' compensation cost per recordable injury, or total cost of all recordable injuries. Each of these metrics can be used to define the severity of a condition in a facility. While it is true that incidence rates significantly above or below industry averages can lead to a visit from an OSHA inspector, an OSHA-recordable incident does not cost a company anything. One of the potential outcomes of training and education of an hourly work force is a short-term increase in the OSHA-recordables. This reflects heightened awareness of symptoms and situations that were there all along. Earlier detection provides the opportunity to correct problems before they can impact any severity index, which is where the significant monetary costs are incurred.

TERTIARY PREVENTION

The role of Human Resources in the tertiary prevention of musculoskeletal disorders involves enforcement and communication. Often Human Resources personnel are the main communication channel between the medical provider and the company. Along with this role comes the responsibility for keeping track of an individual's recovery, both at home and at work. Those individuals away from their jobs under doctor's orders, should be contacted periodically to check on the progress of their recovery. It is all too easy for a person to fall through the cracks of a bureaucracy as complicated and frustrating as the workers' compensation system. The Human Resources department plays a vital role in seeing that this does not happen.

Regular interaction with an employee and his or her doctor communicates both concern and interest in returning the individual to productive employment. Human Resources personnel should coordinate with engineers and supervisors to make sure that any light-duty restrictions placed on an individual are fully enforced, and the injured employee is following doctor's orders. It is critical that supervisors are made aware of any work restrictions and given an indication as to how long the restrictions are to be in place. Human Resources personnel need to be this communication channel.

ENGINEERING'S ROLE

When supplemented by appropriate ergonomics training, engineers have the skills and opportunity to evaluate the basic design of a workstation and make appropriate modifications to reduce or eliminate physical stressors. Unfortunately, in their college curricula very few engineering graduates were exposed to issues related to the human operator. Many talented and capable working engineers may not have the training necessary to recognize potential musculoskeletal hazards in their designs, or to recognize existing hazards in the workplace. As a result, some engineers may feel threatened by an ergonomics program, perceiving that they will be blamed for designing lines or equipment that may be deficient in terms of ergonomics design principles.

Every effort should be made to ensure that all plant personnel recognize that managing musculoskeletal problems occurs through a positive process based on both proactive and reactive prevention strategies. Engineers should receive ergonomics training that will augment their existing knowledge of sound design principles, helping them to make appropriate engineering design decisions in the future.

PRIMARY PREVENTION

Because of their daily interaction with and control over the workplace, engineers can be the best source of primary prevention of musculoskeletal disorders. Further, as a result of their training, engineers have the analytical tools to evaluate a work environment, identify problems, and devise solutions. They can easily learn to apply quantitative assessment techniques, such as the National Institute for Occupational Safety and Health (NIOSH) lifting equation (Waters et al. 1994), to evaluate manual material handling tasks. They are also capable of applying any of the qualitative or semi-quantitative checklists that have been developed to evaluate risk of upper-extremity disorders.

To make the most of their quantitative and analytical abilities, engineers should receive ergonomics training. This training should emphasize basic body mechanics, anthropometry, and the identification of workplace risk factors for musculoskeletal disorders. After they have received this awareness training, engineers are able to apply these concepts when they lay out a new work area or order a new piece of equipment.

Engineers need to be aware of three major workplace risk factors: requirements for exertion of sustained or high levels of force, awkward or sustained body postures, and high levels of repetition.

With proper physical workstation design, forces and body postures often can be controlled. Keeping work near elbow height, developing appropriate support fixtures, providing appropriate hand tools, and eliminating lifting from floor level or to overhead positions are examples of some simple guidelines that engineers should consider when designing or evaluating a workstation. Often, an ergonomically-designed workstation also becomes more efficient. This happens because some of the important issues in ergonomics, such as minimizing material handling and awkward postures, are important issues in the work measurement and design efficiency context. Engineers also often determine how raw materials are delivered to an operator and, therefore, need to consider the ergonomic implications of their material handling practices when designing a production system.

Engineers are also able to place some of these safety concerns within the framework of the production process to control the third risk factor, repetition. Engineers understand the implications of slowing down a line or adding another operator. While either of these options will decrease the number of repetitions any one individual performs, they have quite different effects on line output.

Many engineers have the training to cost-justify the purchase of a new piece of machinery or tool, after considering the risks associated with an existing design and the probability that those risks will result in a musculoskeletal problem. While managers are able to look at the big picture and cost-justify the program as a whole, engineers can look at individual projects and cost-justify changes by looking at such variables as decreased injury costs, increased quality, increased production, and decreased scrap.

SECONDARY PREVENTION

An engineer's role in secondary prevention is to evaluate existing workplace designs in light of new information he or she obtains on workplace ergonomics. Typically, a concern will be raised by an operator or a supervisor. This then may be brought before the plant's ergonomics team, and an engineer will be called in to help understand and quantify the problem. The engineer will then be asked to assist in the redesign process to eliminate the identified physical stressor, be it excessive force, awkward postures, high repetition rates, or something else. During this process, the engineer should work closely with the operators who have first-hand knowledge of the workplace and process, and who must work with any changes that are made. Design changes and acceptance of those changes are usually more successful when they result from a cooperative effort between engineers and operators.

TERTIARY PREVENTION

One of the tasks an engineer can perform in the tertiary mode of injury prevention comes in the form of identification or generation of light-duty jobs. Given a set of limitations (on force exertion, repetition rates, or body postures) recommended by a medical practitioner, the engineer can evaluate the jobs in a facility and identify those with requirements that do not exceed the physical limitations of a returning employee. Much of the information needed by the engineer to make this evaluation already exists. It is the product of work measurement and time study processes industrial engineers traditionally perform.

One more way the engineer can play a role in the ergonomics program is by designing accommodations for individuals with physical limitations. This may be one of the more interesting, stimulating, and rewarding tasks the engineer performs. If an employee returns from sick leave with permanent or long-term medical limitations that prohibit a return to the job, either an alternative job should be identified or the old job modified to comply with the worker's limitations. Working together, creative engineers and skilled trades employees may provide an employee with the ability to rejoin the work force and enjoy all the benefits of gainful employment.

THE SUPERVISOR'S ROLE

Line supervisors are often the gateway of communication between operators and the rest of the company. They have intimate knowledge of production processes and jobs in their areas, the skills and aptitudes of their operators, and management's production and quality goals. The line supervisor can be an important source of information for operators, engineers, Human Resources personnel, upper management, and the ergonomics team. As such, line supervisors should participate at each level of prevention of work-related musculoskeletal injuries.

PRIMARY PREVENTION

One of the most important ways in which the supervisor can assist with primary prevention of work-related musculoskeletal injuries is to foster good communications with the operators in the area. Operators should feel comfortable approaching the supervisor with questions related to ergonomics or musculoskeletal injuries, and supervisors should be able to answer such questions. As the operators' link to the ergonomics team, the supervisor can bring operators' questions, concerns, and ideas about work methods, tools, or workstation design to the ergonomics team. The supervisor also should be able to direct the operator to appropriate sources of additional information, such as medical management, Human Resources personnel, or the ergonomics team.

Another way supervisors contribute is by their knowledge of jobs and employees in their work areas. To reduce the risk of future injuries, the supervisor can bring historically high-risk jobs to the attention of the ergonomics committee or safety personnel. With some ergonomics training, the supervisor may be able to make decisions regarding placement of operators to match skills and physical limitations with job requirements. For example, it may be possible to temporarily move a pregnant employee from a job requiring forceful hand activity and awkward wrist postures in an effort to reduce the employee's likelihood of developing carpal tunnel syndrome during her pregnancy.

The line supervisor has knowledge of the day-to-day operations in his or her area, but can also take a big-picture view and observe temporal trends. The supervisor, therefore, is well positioned to identify jobs with high turnover rates. High turnover may be the result of a cycle of injury and operator replacement or it may indicate that operators are attempting to cycle out of a particular job before they become injured. In the latter case, the problem would not be detected through review of OSHA logs or medical records.

Production Trends

The supervisor can also identify trends in production, alerting fellow supervisors, engineers, and purchasing personnel when problems arise, as well as when processes and parts run particularly well. Plants with existing quality programs are familiar with the idea of providing a high-quality product at every stage of production. Such a program may have ergonomic benefits as well, such as a reduction in the force required to assemble parts when they are consistently well made.

SECONDARY PREVENTION

If an employee does develop symptoms of a musculoskeletal injury, he or she may inform the supervisor before informing anyone else. The supervisor should be able to direct the employee into the medical management process and alert the ergonomics team that the employee's workstation should be surveyed for ergonomic hazards.

Controlling Exposure

There are essentially two ways to control exposure to ergonomics hazards. The preferred method is through engineering controls that result in physical changes to the workstation, tools, product, or task. The other, administrative controls, are essentially rules developed to reduce the amount of time a worker is exposed to a hazard, although the hazard is still present. This is accomplished through job rotation, extending the proportion of rest time in work/rest cycles, or procedural rules. By combining knowledge of the jobs in the area with some ergonomics training, the supervisor can help design and implement many types of administrative controls. For example, job rotation is beneficial only if the jobs through which workers rotate are truly different in terms of the physical stressors encountered by the workers. Moving from one hand-intensive task to another will not reduce a worker's exposure to risk factors associated with upper extremity musculoskeletal disorders. Supervisors can help identify combinations of jobs that will reduce worker exposure to musculoskeletal injury risk factors. Supervisors also may be required to enforce procedural rules, such as those having to do with approved work methods or maintaining appropriate line speeds in piece rate facilities.

TERTIARY PREVENTION

If an operator does incur a lost time injury, the supervisor can maintain contact with that individual while he or she recuperates. Upon the operator's return to work, the supervisor can play an important role in reintegrating the employee into the work team. If there are restrictions placed on the operator, the supervisor can work with Human Resources and medical management personnel to ensure that work restrictions are not violated. This may be accomplished through some temporary job or task reassignments, identification of light-duty jobs in the area, or creation of a light-duty job. If the returning operator is to receive some form of regular treatment (occupational or physical therapy, for example), the supervisor can arrange for a stand-in worker to fill in during the employee's treatment sessions.

THE OPERATOR'S ROLE

Operators in a plant typically have two primary responsibilities: performing their own tasks and training new operators. When a facility initiates a plan to manage work-related musculoskeletal injuries, operators may be asked to expand their existing responsibilities and take on new ones.

PRIMARY PREVENTION

There are a number of ways in which operators can participate at the level of primary prevention. One important way is to get the most out of their ergonomics training by being attentive, open to new ideas, and asking questions. When ergonomics-related concerns arise about workstations, tools, or work methods, operators should bring their questions to supervisors or the ergonomics team.

Operators, themselves, may be able to suggest improvements to elements of their work when they combine basic ergonomics training with existing knowledge of their jobs. If changes in work are considered or implemented, operators should give thoughtful feedback to supervisors, engineers, or the ergonomics team. In their role as job trainers, operators can utilize their basic ergonomics training when instructing new operators about work methods and tool usage. Another way operators can participate in primary prevention is to volunteer to be a member of the plant's ergonomics team.

Exposure to physical stressors can occur off the job, as well as on. Employees need to recognize these exposures and control them if possible. Personal factors can predispose individuals to developing certain musculoskeletal disorders. Once an employee is influenced by these factors, he or she may wish to discuss them with a personal physician or medical management personnel at work, so that efforts can be made to offset or compensate for their existence.

SECONDARY PREVENTION

If an operator experiences early warning signs or symptoms of musculoskeletal injury, he or she should follow the company's reporting procedures. Early reporting of symptoms should be encouraged. As with primary prevention, operators experiencing symptoms should assess their exposure to ergonomics hazards outside of work, and take steps to control those exposures. They may receive help with this assessment from medical management, the ergonomics team, or their personal physician.

TERTIARY PREVENTION

Probably the most important thing an operator can do once an injury occurs is to seek and comply with the advice of knowledgeable health care providers. Unfortunately, once tissue damage occurs, a full recovery (full strength and endurance, full range of motion, lack of discomfort) may not be possible. However, in conjunction with appropriate medical treatment, assessment and control of exposure to hazards both on and off the job will help to ensure that the injured worker's recovery progresses as far as possible. Workplace or job changes tailored to the employee's residual limitations can then be designed and implemented to allow the employee to continue to be a productive member of the company's work force.

MEDICAL MANAGEMENT'S ROLE

Medical management is one of the four major program elements in an ergonomics program. The other three are work site analysis, hazard prevention and control, and training and education (OSHA 1991). The tasks that come under the auspices of medical management include training and education, establishment of treatment protocols, and assisting with return-to-work issues. Any or all of these tasks may be handled in-house by a full- or part-time occupational health physician or nurse, or on a contract basis through a local hospital, rehabilitation facility, or private physician who is knowledgeable about work-related musculoskeletal injuries.

PRIMARY PREVENTION

The number one prevention task of medical management personnel is to establish a medical management plan. The plan is a written document that specifies the procedures to follow when an employee reports symptoms of a musculoskeletal injury or when he or she loses time due to a musculoskeletal injury. The plan also contains specifics regarding the second main primary prevention task of medical management—employee training and education.

Medical management personnel inform plant personnel about symptoms and early warning signs of musculoskeletal injuries, the physical hazards (on and off the job) that may put individuals at risk for developing musculoskeletal injuries, and procedures the company has established for employees to follow if they have questions or concerns about potential symptoms. When a facility initiates an ergonomics program, all employees receive some amount of training, based upon their responsibilities. In addition to large scale training and education, if the facility has a health care professional on site, that individual should be available to address employee questions and provide education to new hires.

SECONDARY PREVENTION

Should an employee develop symptoms, the medical management plan specifies the steps that should be taken by the employee, the employee's supervisor, and other specific individuals to report and treat those symptoms. If a health care provider is on site, the plan may specify that pre-approved diagnostic testing, conservative treatment protocols, or both, be initiated by the health care provider. If arrangements for health care services have been made with an outside provider, the plan would specify how and when the employee would be seen by that provider.

TERTIARY PREVENTION

If an employee incurs a lost-time musculoskeletal injury, medical management personnel can improve the odds for the employee's successful return to work. They can work closely with the employee's health care provider to track the employee's recovery and ensure that any work restrictions are adhered to when the employee returns to work. Medical management personnel can work with supervisors to put work restrictions in place. They also provide information to Human Resources personnel regarding appropriate treatments and recovery periods for various types of musculoskeletal disorders and help ensure that each injured employee receives appropriate treatment and returns to work after an appropriate recovery period (neither premature, nor protracted).

THE SAFETY SPECIALIST'S ROLE

To a great extent, the role of the safety specialist in a program designed to manage musculoskeletal problems depends on the skills, training, and current scope of the job. In a smaller company, one employee may have responsibility for all safety-related issues, while a larger company may employ several individuals specializing in one or more occupational health areas, such as safety, industrial hygiene, or ergonomics. Regardless of their training, however, safety specialists typically will have the tools to quantify the severity of a problem and identify areas of greatest urgency. Much of the work outlined for the Human Resources department could be carried out by a safety specialist with appropriate training.

PRIMARY PREVENTION

Periodic inspections or "walkthroughs" of each production line can often highlight areas of concern. A safety specialist with the appropriate training will be able to quickly identify risk factors for a job and communicate those concerns to the ergonomics team and line engineers. Operator discomfort surveys also may be administered by the safety specialist. If train-the-trainer methods are used in ergonomics training, the safety specialist may become the ergonomics trainer and educator for the plant's operators. The safety professional may be well positioned to keep other personnel informed of new ergonomics-related information and legislative efforts, because of the literature which he or she receives (product literature, safety journals, and trade publications).

SECONDARY PREVENTION

The safety specialist may take primary responsibility for data entry and analysis of OSHA Form 200 Logs. The safety specialist may already have responsibility for tracking safety and industrial hygiene concerns using the logs. Tracking musculoskeletal problems would then be a minor addition to that existing duty. The safety specialist also should be involved in job evaluation and redesign to ensure that other safety-related standards are being followed, such as identifying pinch points in machinery or scissors lifts, or providing adequate machine guarding.

TERTIARY PREVENTION

At this level, and at the secondary level of prevention, the safety specialist may act as an advocate for expenditures associated with modifications to workstations or tools. Safety specialists can also provide insight for establishing or identifying light- or alternate-duty jobs, and may interact with the medical care provider to document and maintain physical capacity restrictions for injured workers.

ROLES FOR OTHER PRODUCTION SUPPORT GROUPS

Purchasing and Maintenance are just two of the support groups that can contribute to the prevention effort. Purchasing personnel can work closely with engineers and the ergonomics team to identify equipment and tools designed to reduce exposure to ergonomic hazards, while still meeting the company's economic objectives. Additionally, Purchasing personnel can work directly with suppliers to encourage them to provide equipment and tools with ergonomic design features. A large company has sufficient buying power to do this alone. Smaller companies may work through trade associations to bring pressure to bear on suppliers.

Maintenance personnel can contribute to the prevention effort in at least three ways. One would be through performance of normal tasks. It is generally easier for operators to work with equipment that is running well. Maintenance personnel also may be able to provide insight into differences between seemingly similar pieces of equipment in situations where one operator is having problems and others are not. Second, Maintenance personnel often have the skills, experience, and inventive nature to develop new tools or modify existing equipment. By working with engineers, the ergonomics team, and operators, Maintenance personnel may be able to implement workstation modifications that are identified through the ergonomics program. Third, Maintenance personnel may participate on the ergonomics team, either as regular members or on an as-needed basis.

INTEGRATION

While individual roles are important, it may be the integration of individuals and groups into an ergonomics team that makes the critical difference as to whether a facility's ergonomics program succeeds or fails. The ergonomics team is a core group of people who meet on a weekly or biweekly basis to discuss current areas of concern, develop solutions to existing problems, and evaluate control measures they have put in place.

The team is typically composed of 4 to 8 people, and should be a mix of salaried and hourly employees. From the hourly side, ideally the team would have members who have firsthand knowledge of most of the major lines or departments in the facility. This arrangement facilitates communication between the ergonomics team and shop floor employees, one of the more important aspects of this type of program.

On the salaried side, a representative from the Human Resources department, one from engineering, and one from the medical or safety areas would be ideal. In identifying potential team members, the goal is to identify individuals who either have information to provide or are in a position to make things happen. Follow-through is important to the success of any ergonomics program. Identify problem areas (through surveys, OSHA log reviews, or operator inquiries), then follow-up with the corrective actions necessary to eliminate the problems.

In addition to this core group, it is helpful, from time to time, to bring other individuals into the ergonomics team meetings, including some of the operators performing jobs that are being analyzed, as well as the line engineers and line supervisors for these jobs. Their input can be invaluable when considering current problems, as well as potential solutions.

CONCLUSIONS

Every individual in a company has some role to play in the management of work-related musculoskeletal disorders. The overall effectiveness of an ergonomics program will depend, in large part, upon the quality of work from each member, how well representatives from the different parts of the company work together, and how well the team's ideas are communicated and disseminated throughout the organization. With time, the effectiveness of a system for managing musculoskeletal injuries can be measured in terms of decreased costs for workers' compensation, recruitment, and training, and increases in product quality and employee morale and productivity.

REFERENCES

Equal Employment Opportunity Commission (EEOC). *Americans with Disabilities Act Handbook,* Publication EEOC-BK-19. Washington, DC: EEOC, 1991.

Occupational Safety and Health Administration (OSHA). *Ergonomics Program Management Guidelines for Meatpacking Plants,* OSHA 3123. Washington, DC: OSHA, 1990.

Waters, T.L., Putz-Anderson, V., and Garg, A. *Applications Manual for the Revised NIOSH Lifting Equation.* DHHS (NIOSH) Publication No. 94-110. Washington, DC: National Institute for Occupational Safety and Health, 1994.

CHAPTER 20

ERGONOMICS: PART OF CONTINUOUS IMPROVEMENT

Steven L. Johnson
Professor of Industrial Engineering
University of Arkansas
Fayetteville, AR

American industry has experienced a widely publicized increase in the reported incidence of disorders associated with the upper extremities (hands, wrists, arms, and shoulders). This increase affects both worker and company. There are direct financial costs to the company, such as increased workers' compensation costs and health insurance premiums. The worker feels the cost in lost wages. However, there are also indirect financial costs that are potentially much greater, although they are not as easily documented. They include high turnover and absenteeism rates, increased training requirements, as well as the reduced efficiency and decreased product quality that go along with continually changing personnel.

Effective implementation of ergonomics can lead to substantial benefits in terms of increased productivity, improved product quality, reduced absenteeism and lower turnover rates, as well as lower occupational safety and health costs. This can be done with an overall total quality management (continuous improvement) philosophy.

Use of automation, rather than improvement of manual operations, has taken precedence traditionally in the United States. However, that trend is rapidly reversing with the increased requirement of more flexible manufacturing methods.

The trend toward more effective manual or mechanized operations could possibly increase attention to work-related injuries. The highly publicized increase in the incidence of work-related disorders associated with repetitive movements has recently increased the visibility of ergonomics to management, labor, and government. Whether the increased attention in the popular press is due to the accelerated incidence of work-related disorders, or whether the increase in reported disorders is due to the popular press, is a valid area of discussion.

TERMINOLOGY

The Occupational Safety and Health Administration's (OSHA) approach and terminology has significantly restricted the definition of ergonomics to *work-related musculoskeletal disorders* (MSD). In fact, this phrase is really much more representative of OSHA's concentration than is the term *ergonomics*. For example, OSHA uses the term *ergonomic hazards* to represent workplace and work method characteristics considered to be related to musculoskeletal disorders. This is a poor choice of words in that ergonomics is the solution to, not the cause of, the hazard. The term *biomechanical hazard* would seem to be much more descriptive and appropriate. A more important aspect of the current characterization of ergonomics is that it does not address the primary benefits companies can experience by utilizing good ergonomic design—better labor productivity and improved product quality.

RIGHT AND WRONG TERMS

The popular press often uses terms such as *repetitive strain injury, repetitive motion disorder*, and *overuse syndrome*, to label the types of disorders addressed in this chapter. These terms have very limited utility and, in fact, focus unjustifiable attention on the repetition rate or cycle time of the task. The ANSI committee addressing this area (ANSI-365) more appropriately labels its efforts as the *control of work-related cumulative trauma disorders*.

The term *cumulative trauma disorder* is a more appropriate term for a number of reasons. First, one of the most important characteristics of this disorder is that it does not result from instantaneous events like a slip and fall, laceration, or an amputation. Rather, it develops over a period of time so the effects are cumulative. Note that this results in the disorder being categorized as an illness rather than an injury in the OSHA reporting system (OSHA 200 Logs).

The term *trauma* is also descriptive since it indicates that the task requires the human body to be used in a manner for which it is ill suited. For example, the human hand is very effective in exerting a relatively high force or precise dexterity when the wrist is straight. However, when the wrist is bent, the biomechanics are seriously affected and both the effectiveness of the effort and the risk of injury increase dramatically.

Disorder, the last term in the phrase, is descriptive in that some people have a propensity to experience problems while others do not. A high-risk task for one operator may not be high-risk for a different operator. A disadvantage of the term *cumulative trauma disorder* is that, within the current classification system, noise-induced hearing loss is included. This has increased the confusion occurring in much of the popular press. Throughout this chapter, the terms *biomechanical hazards* and *musculoskeletal disorders* will be used to represent what is frequently referred to as *ergonomic hazards* and *repetitive motion disorders*, respectively.

EFFECTIVE IMPLEMENTATION OF ERGONOMICS

It is worth noting that the material distributed by OSHA increasingly refers to the concepts of total quality management (TQM). For example, the *Draft Ergonomic Protection Standard Summary of Key Provisions* (August, 1994) states that the process is, "consistent with international quality assurance activities (International Organization for Standardization [ISO] 9000)." Similarly, terminology relating to *continuous improvement* (another frequently used TQM concept) is increasingly found in OSHA documents. This is important since one primary factor impacting effectiveness of an ergonomic effort is that progress is occurring. It is also important, both externally and internally, to completely document the progress. That is, recording "where you came from" can be as important as documenting "where you are presently" or "where you are going." Beyond documenting the activities that illustrate progress, there are specific components of an ergonomics effort necessary to experience positive results. As with any effective program implementation, the format and scope of the program are dependent upon characteristics of the organization such as size, products, and tasks.

POLICY AND PROCEDURES DOCUMENT

It is essential that the goals and objectives of the effort be fully understood and supported throughout the entire organization from the shop floor worker to top management. Although some companies find it beneficial to separate the ergonomics effort from the safety effort, ergonomics is often integrated as an element of safety policy. One objective of the policy statement calls for ergonomics to be considered as important as production and product quality.

There are two goals of an ergonomics policy. First is to prevent work-related musculoskeletal injuries by using engineering controls to reduce or eliminate the task characteristics associated with the disorders. This is accomplished by applying the science and technology of ergonomics to the design of work spaces, work methods, tools, and equipment. The objective is to reduce the risk of injury while simultaneously increasing the effectiveness of manual operations.

The second goal is to prevent the progression of a disorder, if one does occur. This involves early detection and treatment of disorders so they do not develop into more severe cases. The total effort obviously involves the multidisciplinary efforts of Engineering, Operations management, Human Resources management, and health care providers (internal and external). The policy and procedures document is necessary to delineate the authority and responsibility of each of these contributors in an integrated effort. In addition to these staff and line functions, it is important that input be obtained from workers directly affected by both the occupational risks and the potential modifications that are designed to reduce those risks.

EMPLOYEE INVOLVEMENT

Inclusion of affected employees is an important component of an effective ergonomics effort. A suggestion system that allows the operator to make input as to potential or perceived hazards can be accomplished through a card/box system on the plant floor or in the cafeteria or break room. As with any suggestion system, the mechanism to collect and record the concerns or recommendations is much easier to implement than are the methods of evaluating them. For an employee involvement system to be effective, it must provide feedback about the suggestion in a reliable and timely manner. This process is even more complicated by the increasingly multilingual nature of many manufacturing and processing facilities.

One aspect of employee involvement, as defined by OSHA, is that the employees are to be provided information about musculoskeletal disorders. However, one method of addressing the degree of discomfort that may be experienced, but not reported to the medical facility, is through the use of an active surveillance instrument. This is often referred to by OSHA as a *symptom survey*. This term, itself, may have the unfortunate consequence of predisposing the worker to having "symptoms."

A better approach is the *job improvement survey* (Figure 20-1). This form asks the responder to evaluate the potential for improvements, as well as indicating discomfort. It is important to note that the utility of the information received is very different if the survey asks about possible improvements prior to questions about discomfort, as opposed to posing the discomfort question first.

Ergonomics: Part of Continuous Improvement

```
                                                              # 0867
                        Job Improvement Form

   Primary Job_____           Department _____
   Shift_____

   How long have you been doing this job?
   ☐ Less than 1 month    ☐ 1-3 months      ☐ 3 months-1 year
   ☐ 1-5 years            ☐ more than 5 years

   Do you regularly do any other jobs?    ☐ No        ☐ Yes
      If you do, what jobs do you do?   (1)_____
      (2)_____        (3)_____
   How can any of these jobs be improved?_____
   _____

   Check the areas below that could be improved.
   ☐ long forward reach   ☐ high work level   ☐ awkward postures
   ☐ quality of materials ☐ hand tools        ☐ machine design
   ☐ machine maintenance  ☐ job training

   What other things make your job harder?_____
   _____

   Have you had any soreness within the last month?  ☐ No   ☐ Yes

   If yes, put a check for each part of your body that has felt sore (L-Left
   and R-Right).
```

Body Part	Barely Noticeable	Moderate	Very Noticeable
Neck	☐	☐	☐
Shoulder (L)	☐	☐	☐
(R)	☐	☐	☐
Arm (L)	☐	☐	☐
(R)	☐	☐	☐
Wrist (L)	☐	☐	☐
(R)	☐	☐	☐
Hands (L)	☐	☐	☐
(R)	☐	☐	☐
Fingers	☐	☐	☐
Upper back	☐	☐	☐
Lower back	☐	☐	☐

Figure 20-1 Job improvement form.

Please answer all questions.

When was the last occurrence?
☐ today
☐ yesterday
☐ last week
☐ more than a week ago
☐ more than a month ago

When does it occur?
☐ during the shift
☐ after work
☐ during the night
☐ all the time
☐ it varies

How often does it occur?
☐ every day
☐ once a week
☐ once a month
☐ less than once a month
☐ it varies

How long does it last?
☐ 1 hour
☐ all day
☐ 1-7 days
☐ 8-30 days
☐ more than 30 days

If you do more than one job, which one causes the most problems? _____

Have you reported this soreness to your...?   ☐ Supervisor   ☐ Plant nurse
                                               ☐ Not reported

Have you seen a doctor about this?   ☐ No   ☐ Yes
  If yes, what treatment did you receive? _____

Were you put on another job because of this?   ☐ No   ☐ Yes
  If yes, what was that job? _____

Have you lost any time from work because of this?   ☐ No   ☐ Yes
  If yes, how long were you away from work? _____

What have you found that reduces the soreness? _____

What have you found that makes the soreness worse? _____

How many times have you worked 6 days a week in the last month?
☐ 0   ☐ 1   ☐ 2   ☐ 3   ☐ 4

Figure 20-1 Job improvement form. (continued)

```
┌─────────────────────────────────────────────────────────────────┐
│  How many times have you worked 7 days a week in the last month?│
│    ☐ 0       ☐ 1       ☐ 2       ☐ 3       ☐ 4                 │
│                                                                 │
│  How many hours per day do you work?                            │
│    ☐ Always 8   ☐ Sometimes less than 8   ☐ Sometimes more than 8│
│                                                                 │
│  If you worked more than 8 hours a day in the last month, how   │
│  long did you work?                                             │
│    ☐ 9 hours  ☐ 10 hours  ☐ 11 hours  ☐ 12 hours  ☐ more than  │
│                                                          12 hours│
│                                                                 │
│  How many days in the last month did you work more than 8 hours?│
│    ☐ 0       ☐ 1       ☐ 2       ☐ 3-5     ☐ 6-10              │
│    ☐ 11-20   ☐ regularly                                        │
│                                                                 │
│        Reviewed by plant nurse          Date____/____/____      │
│        Reviewed by ergonomics committee Date____/____/____      │
│                                                                 │
│  ─ ─ ─ ─ ─ ─ ─ ─ ─ ─ ─ ─ ─ ─ ─ ─ ─ ─ ─ ─ ─ ─ ─ ─ ─ ─ ─ ─ ─ ─    │
│                   (separate along perforated line)              │
│                                                                 │
│      This information is confidential and your name will be     │
│  separated from the form. The medical staff will be the only    │
│  individuals that will have access to your name.                │
│                                                                 │
│  _____       _____ │
│  Please print your name                Date                     │
│                                                                 │
│  _____                                │
│  Signature                                                      │
│                                                      # 0867     │
└─────────────────────────────────────────────────────────────────┘
```

Figure 20-1 Job improvement form. (continued)

These forms are anonymous with a code number on the form and a tear-off signature sheet. Due to medical information being included, the only person with access to the names is the plant health care provider. The forms, without the names, are reviewed by the ergonomics committee to evaluate the technical and economic feasibility of the recommendations.

The natural concern of some managers that there will be a significant number of intentional or unintentional false indications of problems from the survey appears to be unwarranted. Although there is sometimes an initial transient increase in visits to the medical facility, this dissipates after the administration of the survey. In addition, some of the visits result in catching conditions very early, rather than later when the cost to the company and the individual would be much higher. The job improvement survey provides valuable information about potential problem areas and possible solutions; however, it is the trend information developed from subsequent applications of the survey that is useful in evaluating the effectiveness of the ergonomics effort.

ERGONOMICS COMMITTEE

The ergonomics committee should be separate from the safety committee, although they do interact and naturally share some members. The committee is made up of representatives from Production (hourly and supervisory), Safety, Medical, Human Resources, Maintenance, and Engineering. As with other continuous improvement teams, the ability to communicate and interact effectively to develop consensus is important.

The committee assists in the prioritization of recommendations from the ergonomic job analysis (discussed later). As previously mentioned, they also analyze and make recommendations based on the active surveillance information (job improvement survey). By performing periodic walkthrough surveys of the facilities, they can track the progress of the recommended modifications and identify potential improvements. An important responsibility of the ergonomics committee is to maintain documentation of the ergonomics effort and ensure that there is periodic review by them and upper management.

SURVEILLANCE METHODS

Surveillance can be divided into active methods that utilize surveys of current employees and passive surveillance methods that utilize archival records. The job improvement survey is an example of an active surveillance procedure.

A passive surveillance method of detecting jobs associated with musculoskeletal discomfort involves the review of archival occupational safety and health records (OSHA 200 Logs). Although this data is readily available and often discussed in the context of an ergonomics analysis, its usefulness is limited when the incidence rate is as low as it is for musculoskeletal injuries.

The fact that the injury (illness) is cumulative makes the temporal nature of the worker's job history important. However, this information is not recorded on the log. That is, the job that created or contributed to the problem may not be the job that the operator is performing when he or she reported it to the medical facility. Other factors, such as job rotation, also make analysis of safety data tenuous.

Lastly, almost all manufacturing and processing facilities go through continuous modifications to process layouts, equipment, tools, methods, etc. This can lead to erroneous conclusions as to the causal relationships that are based on historical data. Given these drawbacks, archival data provides a starting point for establishing which tasks have been associated with problems in the past. However, the analysis of archival documents is no substitute for interactive discussions with health care providers, Production supervision, and the workers themselves.

A source of information that can be useful in interpreting the relative severity of disorders is the workers' compensation file. In addition, absenteeism records and the amount of turnover (due to termination or bidding-out) from particular jobs can assist in determining if they involve biomechanical stress. However, it is important to note that this information, which relates to individuals, is confidential and appropriate procedures should be documented and followed for its use.

JOB-SITE ANALYSIS METHODS

A disadvantage of passive surveillance methods is that they evaluate the operation from a historical perspective, rather than analyzing present conditions. The rapid changes that occur in many production operations can result in an analysis that misrepresents current conditions. An ergonomic job-site analysis addresses the operations as they currently exist. Again, the issue is not simply to find the problem jobs from a safety perspective, but to make recommendations in the workplace design, work methods, tools, and equipment that can improve the effectiveness and efficiency of the whole production process.

Checklists and Narratives

Two forms of ergonomic job-site analysis traditionally used to address musculoskeletal disorders are checklists and narratives. The advantage of checklists is that they require less time and can be performed by individuals with less technical understanding of ergonomics. However, this simplicity inherently causes the checklist approach to be less complete, sometimes resulting in serious misrepresentation of the task requirements. In particular, the temporal nature of the task is very difficult to capture with a checklist (for example, tracking exposure across task elements, tasks, or even jobs if rotation occurs). In addition, checklists are only descriptive, revealing a potential problem, rather than prescriptively indicating alternative methods of alleviating it. A small portion of a checklist distributed by OSHA for the purpose of calculating risk factor scores is shown in Figure 20-2. The scores (demerits) are summed across all risk factor categories. As stated by the instructions for the use of this checklist, "a score above 5 is a problem job."

Risk Factor Category	Signal Risk Factors	Time			Total Score
		2-4 hours	4-8 hours	81 hours add 0.5/hours	
Repetition	Identical or similar motion(s) performed every few seconds	1	3		
Hand force	Grip more than 10 lb (4.5 kg) per load	1	3		

Figure 20-2 Section of OSHA risk factor checklist.

Although this type of checklist appears to be very attractive from the standpoint of data collection, it still requires a significant amount of judgment with respect to both the documentation and interpretation. For example, if a person's entire job, for an eight-hour period, is to lift an 11-lb (5-kg) item and put it back down every 14 seconds (resting between movements), the job would inappropriately be considered a "problem job" using this checklist (repetitive [3] plus hand force [3], resulting in a score of 6).

The second general method of conducting an ergonomic job-site evaluation by documenting conditions and recommendations in a narrative form, requires more time and is generally performed by an ergonomics professional. However, it is generally far more complete, valid, and useful in the process of improving tasks that involve undesirable characteristics. It is often as important to document the positive characteristics of a job without problems as it is to note the problem areas. The narrative approach provides this information, whereas the checklist does not.

The narrative analysis method documents the job (for example, case packing), that is divided into tasks, such as palletizing, that are subsequently divided into task elements (lifting cases from the conveyor). The task element level is where effective ergonomic analysis, documentation, and improvement occurs. A beneficial characteristic of the narrative form of analysis is that it includes a discussion of the technical and economic feasibility of alternative modifications. Conducting any ergonomic job analysis, particularly using a narrative, prescriptive approach, requires the analyst to have a relatively high level of technical expertise in ergonomics.

A Computer-based System

A computer-based job analysis system has been developed to address the disadvantages of both the checklist and narrative approaches. The objective of this system is to provide the completeness and validity of the narrative method, along with the speed and simplicity of the checklist approach. It is especially useful for the individual with little or no training in ergonomics.

An ergonomic job-site analysis has two functions in this context. First, it is an evaluation tool to document the characteristics of the tasks associated with musculoskeletal disorders, that is, to signal risk factors. Second, it is used by operational personnel to prescriptively evaluate their tasks and suggest effective modifications. The data entry portion of the computer-based job analysis includes three major sections, each with subsections. Each section and subsection involve information critical to performing a complete evaluation of the physical, physiological, and psychological characteristics of the job. Sections include:

- Job and task identification and descriptions;
- Identification of the job, operator, and analyst;
- Work organization;
- Workplace layout;
- Equipment, tools, and product characteristics;
- Task element descriptions;
- Body posture;
- Motion patterns and force requirements;
- Temporal characteristics of the activities;
- Manual material handling characteristics (if applicable);
- NIOSH lifting guidelines (1981 and 1991); and
- Liberty Mutual Insurance data for lifting, lowering, pushing, pulling, and carrying.

After the job characteristic information is entered, the job is analyzed with respect to the postural- and motion-related risks of musculoskeletal disorders. The risk factor categories are listed, along with the total number of occurrences and duration of exposure for each category. These results are then used to recommend potential modifications to the workplace layout, work methods, tools, and equipment that could reduce exposure to the particular risk factors. The recommendations are the beginning of the problem-solving process, not the end. That is, there is no substitute for the knowledge and experience of operational personnel. However, documented problem areas (task elements) and recommended improvements give operational personnel the information necessary to reduce job characteristics that result in unnecessary effort, fatigue, and injury.

ERGONOMICS TRAINING

Employees

The amount and type of training in ergonomics varies for different groups within the organization. General awareness training is presented to all employees frequently exposed to biomechanical hazards and includes information on how to detect problems early, before they become permanent or disabling. This involves understanding the early signs and symptoms of disorders and being able to differentiate these from normally expected sensations experienced when becoming accustomed to a new job. If a worker can detect problems before they become serious, it has a positive impact on both the operator and the company.

Although there is a general impression that training leads to a flood of workers "developing" symptoms, this generally does not occur if the presentation is adequate. In addition, to the extent that it does occur, some of the increase represents previously unreported, although real, disorders. The others seem to dissipate quite rapidly. Many organizations have found it beneficial to videotape this training for repeated use with new employees.

Part of this training is to ensure that the workers understand the advantages of good biomechanics, such as postures and motions, and the consequences of poor biomechanics. Training addresses job-specific instruction on correct methods of performing a particular task.

Although training appears beneficial to the company, it is amazing how inadequate it is for unskilled jobs. In general, the procedure is "go sit beside Sally or John who have been doing the job for 27 years, and they will show you how to do it." This, in combination with the incorrect assumption that operators strive to do the job better and easier, is the cause of many musculoskeletal disorders. The instruction should stress that reporting early symptoms will not result in negative repercussions for the operator, such as lost time or termination.

Supervisors

The next group that must receive training on ergonomic principles is Production supervision. First-line supervisors can be the most effective ergonomists in the facility. It is very important that supervisors receive training before the general training of workers. An effective ergonomics effort can only occur when supervisors understand the benefits of good biomechanics for both company and worker (occupational health, productivity, product quality, absenteeism, and turnover rates). In general, supervisors can easily recognize biomechanical hazards like awkward posture, and with the assistance of Maintenance and Engineering, they can frequently suggest effective methods of reducing those hazards. It is a matter of viewing the production process with an eye for ergonomics, along with the traditional concern for production numbers. For example, when supervisors discover that the absenteeism and turnover rates in their area are reduced, they will likely become positive toward ergonomics, independent of its contribution to reducing injuries.

The supervisors' training also includes information on how they can contribute to the early recognition of problems. For example, addressing muscle soreness and providing an opportunity for conditioning an employee for a physically demanding task can result in retraining the employee and reducing the potential onset of a disorder.

Another portion of the training deals with OSHA reporting rules, as they pertain to supervisors. For example, if the supervisor moves an operator from one job to another due to discomfort, this is an OSHA-reportable event, even if the individual did not visit the medical facility. An important part of supervisors' training addresses the benefits of the job improvement survey, stressing that the information is used for improvements rather than in a punitive manner.

Engineering and Maintenance

Training Engineering and Maintenance personnel encompasses the characteristics of workplaces, work methods, tools, and equipment that can affect the risk of musculoskeletal disorders. One individual who should be included in this training is the Maintenance manager since ergonomic modifications are one part of the numerous responsibilities balanced by the Maintenance department. In addition to understanding the factors that have been associated with musculoskeletal disorders (risk factors), engineering controls shown to reduce worker exposure to these factors are also presented during this training.

It is important to understand that the vast majority of modifications require very little time and virtually no capital expense. In particular, it is much easier and less costly to design the system initially, rather than being required to retrofit the configuration after installation. This means that many of the ergonomic considerations should occur at the equipment design and specification stage.

The most effective training for engineers, as well as production supervisors, involves the use of examples from their own facilities. Often, examples can be selected from the ergonomic job analysis and used as the basis for problem-solving exercises during the training sessions. Training fosters communication among the various functions being trained, a very profitable side effect. Again, it is important to understand that the purpose of the training is to facilitate continuous improvement in the system, rather than waiting until a situation degenerates to the point of being a problem. As with the Production supervisors, the Engineering staff must understand the benefits of the job improvement survey. They should not be intimidated by or resistant to recommendations that originate from the line workers.

Health Care Providers

The health care providers, both internally and externally, require training as part of an ergonomics effort. It is important that there is consistency among all parts of the organization with respect to the protocol for detecting and treating musculoskeletal disorders. Specifically, the signs and symptoms that indicate disorders and the actions to be taken based on those signs and symptoms must be standard. To make work-relatedness and return-to-work decisions, it is important that the risk factors associated with each operation be consistently documented. For example, *alternative assignments* (inappropriately referred to as light duty) should indicate the types of restrictions that can be accommodated, including standing, bending, and hand work.

Training also involves imparting a knowledge of the jargon used in the facility. A job can have one title in the Human Resources office, another according to production, and yet a third used by the worker when visiting the medical facility. The different vocabularies among Engineering, health care, Human Resources, and ergonomists can be a serious barrier to effective communication.

An important aspect of health care provider training, in the context of an ergonomics effort, is that they understand their responsibilities related to the active surveillance instrument (job improvement survey). Collecting data, but not taking action based on the information, can have more serious consequences than not taking the data in the first place.

Another important component of this training is a full understanding of OSHA's reporting rules as they pertain to musculoskeletal disorders.

Ergonomics Committee

It is necessary that individual members of the ergonomics committee have in-depth training in all aspects of the ergonomics effort. As with Production supervision and Engineering, material generated during the ergonomic job-site analysis can be a good basis for training the ergonomics committee. The committee must be familiar with the engineering controls that can reduce disorders, as well as technical and financial constraints that must be considered. Communication skills are also an important part of the training for any continuous improvement group.

Training is probably the most important component of an effective ergonomics effort. Without adequate and appropriate training, the ergonomic job-site analysis is often rendered useless. A frequent scenario is that a company hires an external ergonomics consultant to perform an analysis and make recommendations that would reduce the incidence of disorders. Without the training discussed in this section, the analysis and recommendations are likely to be placed (or misplaced) on a shelf. The management, however, feels that they have "done ergonomics." Again, performing an analysis without making the feasible modifications has more severe consequences than not having conducted the analysis.

MEDICAL MANAGEMENT PROGRAM

The ergonomic job analysis and the majority of training are intended to prevent musculoskeletal disorders by ergonomically designing the workplace layout, work methods, tools, and equipment. However, even in the most advanced technically and economically feasible system, some individuals may experience musculoskeletal problems as a result of their job. Therefore, the second objective of an ergonomics program is to provide early detection and treatment to reduce the severity of a disorder.

PREVENTION

The medical management system facilitates the prevention effort by using information from the passive surveillance (analysis of the OSHA Logs) and active surveillance (medical aspects of the job improvement survey). By informing the ergonomics committee that a particular operation has a high incidence of fatigue and discomfort, even if the operators do not visit the medical facility, modifications can be made before injuries occur. Addressing trends from the job improvement survey is potentially much more sensitive than trying to detect trends in the OSHA Logs. Another interrelationship between the prevention efforts and medical management system is in the designation of alternative, light-duty assignments that are appropriate for various physical restrictions.

Documenting and consistently following a strict protocol for diagnosing and treating musculoskeletal disorders is important to control liability. For example, the specific physical signs, such as Tinel's sign and Phalen's test, and symptoms that are considered positive indicators, must be documented. It is also important to fully document treatment protocols, so they are fully understood and consistently followed.

Documentation

In the context of total quality management systems, complete and accurate documentation is very important. Part of the program procedure should include its periodic review. To document the engineering controls resulting from either the job site or job improvement survey analysis, an ergonomics notebook has been used effectively by many organizations. Each operation for which modifications were suggested should be included, along with an evaluation of the technical and economic feasibility of the modifications, and an estimate of their potential effectiveness. Based on this information, the proposed modifications are prioritized.

There are three categories into which the modifications can be placed. The first includes the changes that are easy, quick, and inexpensive. For example, a change may aim to improve the operator's posture by lowering the work height, which simply requires cutting the legs on a workbench.

Many improvements in the facility that occur as part of normal operations are taken for granted and may not be recognized as being associated with ergonomics. Part of the training for supervisors and engineers is to ensure that these changes get recorded in the ergonomics documentation.

The second category of modification involves changes that are prioritized, scheduled, and have an expected completion date. Indicating that modifications are in progress is not sufficient. This category is probably the most important for the purpose of indicating to OSHA that ergonomics is receiving serious attention and that continuous improvement is occurring.

The third category involves modifications not technically or economically feasible at the current time. These should be included with an indication that they are recognized and that, if they become feasible by changes in the facility, they will be implemented.

Part of the documentation often ignored or minimized is the follow-up and evaluation to determine the effectiveness of the modification. It is not sufficient to make the modifications and assume that the problem has been corrected. In all likelihood, some of the changes will not have the desired results. In fact, as with any design effort, unexpected (and sometimes undesirable) consequences can arise. It is important to document that the modification was not successful and, to the extent possible, the reasons.

A full description of the diagnosis and treatment protocols also should be included in the ergonomics documentation. This has both the positive effect of ensuring consistency and the negative effect of reducing the flexibility of the health care providers. Documentation should also include a summary of the results of the passive and active surveillance efforts.

CHAPTER 21

VIBRATION-INDUCED CUMULATIVE TRAUMA DISORDERS

Donald E. Wasserman
Human Vibration Consultant
Cincinnati, OH

THE NATURE OF VIBRATION

Vibration appears as an integral part of our everyday lives. As we drive our cars, trucks, and motorcycles we feel vibration; clothes washers and dryers vibrate; lawn mowers vibrate; powered hedge clippers and brush cutters vibrate; power boats vibrate; powered shop tools vibrate. At work, machinery vibrates, lift trucks vibrate, pneumatic- and electrically-powered tools all vibrate, and the list goes on and on. For the most part, little is thought about vibration since it is so common; until something happens to make us take notice.

You may be surprised to know that vibration, especially occupational vibration, is certainly not as harmless as one would initially think. Some of the effects of occupational vibration exposure were discovered in the very early 1900s and are still being studied today.

Vibration refers to the directional motion of an object. There are actually up to six directions at any one point (Wasserman 1987) front-to-back, side-to-side, up or down, and three corresponding rotations: pitch, yaw, and roll. What we see with our eyes as an object moves is called *displacement*; what we don't see is the object's speed, or velocity, the time rate of change of a moving object. Nor do we see acceleration, which is the time rate of change of speed of the moving object.

Rotation is not measured in human vibration work. Usually, the aforementioned three linear accelerations are simultaneously measured. Acceleration represents the vibration intensity impinging on people as they work and is given in terms of gravitational (g) units or in feet sec/sec (ft/s^2) or meters/sec/sec (m/s^2), where 1g = 9.8 m/s^2. The repeatability or cyclic nature of the vibration is called *vibration frequency* and is expressed in Hertz (Hz).

Did you ever wonder why soldiers walk across a bridge rather than march across it? If they marched across a bridge, it could collapse due to resonance. The intense vibration set up by the mechanical marching excites the bridge structure so that it actually absorbs the vibration and internally amplifies it to the point where the structure could collapse. This phenomenon is called *vibration resonance*.

Similarly and unfortunately, when we come into contact with certain types of vibration commonly found in the workplace, our bodies unwillingly act in concert with this vibration, amplifying it and exacerbating the effects of the exposure. This is known as *human resonance*. If, for example, vertical vibration came into contact with the body at about 5 Hz at the buttocks, one could expect to measure as much as 2.5 times that incoming vibration at the person's head in the same direction. Thus, resonant vibration is to be avoided as much as possible.

Finally, be aware that there are some 8 million workers in the United States alone, that are occupationally exposed to vibration (Wasserman et al. 1974); of which 7 million are exposed to whole-body vibration (WBV) and 1 million are exposed to hand-arm vibration (HAV). The medical effects of WBV and HAV are not the same.

WHOLE-BODY VIBRATION

Whole-body vibration (WBV) or head-to-toe vibration is usually experienced by operators of trucks, buses, locomotives, lift trucks, heavy equipment operation, farm vehicle operation, overhead cranes, and found near vibrating machinery such as punch presses or mold shakeout areas in foundries.

Studies of diseases in large worker populations (called epidemiology studies) (Milby and Spear 1974; Gruber and Zipperman 1974; Gruber 1976; Hulshof and van Zantan 1987) and laboratory studies (Dupuis and Zerlett 1986; Wilder et al. 1994; Wasserman 1995) have indicated that WBV exposure is associated with various musculoskeletal diseases including, but not limited to, low-back pain, degenerative intervertebral disc diseases, and herniated and slipped discs. In addition, some studies show that females exposed to WBV have additional gynecological risks, especially during pregnancy (Seidel and Heide 1986; Riazanov 1985).

Some medical consequences of WBV exposure appear as cumulative trauma disorders (CTDs), where WBV exposure is experienced by the worker with no apparent difficulties for an extended period of time. Then, problems such as a slipped disc might occur for no apparent reason or from an innocuous event like leaning over to pick up a light object. Research work, in the United States and elsewhere, is seeking to elucidate these actual injury mechanisms (Wilder et al. 1985).

WBV exposure can cause both safety and health problems. These problems are more likely at human resonance frequencies where humans are especially vulnerable. At that point, a small amount of impinging vibration can produce a large effect because of the internal involuntary amplification of this vibration by the human body. For example, a truck driver experiencing 5 Hz vertical WBV could possibly lose control of his truck and cause an accident because the vibration interferes with his grasp on the steering wheel (Wasserman 1987; Griffin 1990).

Finally, it should be noted that measured WBV acceleration levels are usually low (usually much less than 1 g), but not necessarily harmless.

HAND-ARM VIBRATION

Hand-arm vibration (HAV) exposure usually arises when workers use vibrating pneumatic-, electrical-, or gasoline-powered hand tools such as chain saws, grinders, chippers, drills, nut tighteners, jack hammers, demolition tools, etc.

Some of the medical effects of HAV exposure were discovered in the early 1900s when a famous occupational physician, Dr. Alice Hamilton, was called in the dead of winter of 1918 to the stone quarries of Bedford, Indiana, because the stone cutters who used pneumatic cutting hammers were experiencing tingling and numbness in their fingers. Due to increased vibrating tool exposure, and triggered by cold temperatures, these workers next experienced a far worse stage of the disease, episodic attacks of finger blanching or whitening, resulting from a loss of finger blood supply.

The disease is known as Raynaud's Disease, for a French physician who, in 1862, first reported it in female housewives not exposed to vibration. To differentiate between the few people who have this condition naturally and those who acquire it by vibration, the former is called Raynaud's Disease, the latter Raynaud's Phenomenon or, more popularly, *vibration white finger* (VWF). It has been thoroughly documented through the years (Wasserman 1987; Griffin 1990; Pelmear et al. 1992). For the most part, this disease is irreversible if the vibration and frequent finger blanching continues. If this condition is left untreated and the worker is not removed from the HAV exposure, in the extreme case it can eventually result in finger gangrene and possible digit amputation (Hamilton 1918).

Today, VWF is known as hand-arm vibration syndrome, or HAVS, since we now know that hand-arm vibration can not only irreversibly damage finger circulation, but also affect the bones and cartilage in the fingers and hands. Table 21-1 is the original Taylor-Pelmear scale first introduced in the 1960s for medically classifying the extent of HAVS in a worker's fingers and hands.

Tables 21-2 and 21-3 show the latest medical scale for HAVS adopted in 1986. Called the modified Taylor-Pelmear scale or "Stockholm Classification," it is used by physicians worldwide. HAVS is a very painful condition and, although not curable, is treatable using so-called calcium channel blocker medications (Pelmear et al. 1992). Common prevalence of HAVS in vibrating, pneumatic-tool user populations is approximately 40-50%, with about 1-2.4 years latent interval time range to the appearance of the initial finger tip blanching after beginning vibrating tool use (Pelmear et al. 1992; Wasserman et al. 1982).

Table 21-1. Stage assessment for HAVS
(Taylor-Pelmear classification system)

Stage*	Condition of Fingers	Work and Social Interference
OO	No tingling, numbness, or blanching of fingers	No complaints
OT	Intermittent tingling	No interference with activities
ON	Intermittent numbness	No interference with activities
OTN	Intermittent tingling and numbness	No interference with activities
1	Blanching of a fingertip with or without tingling and/or numbness	No interference with activities
2	Blanching of one or more fingers beyond tips, usually during winter	Possible interference with activities outside work; no interference at work
3	Extensive blanching of fingers; frequent episodes in both summer and winter	Definite interference at work, at home, and with social activities; restriction of hobbies
4	Extensive blanching of most fingers; frequent episodes in both summer and winter	Occupation usually changed because of severity of signs and symptoms

*Complications are not used in this grading.

Table 21-2. Stockholm Workshop Scale for the classification of cold-induced Raynaud's phenomenon in the hand-arm vibration syndrome

Stage*	Grade	Description
0		No attacks
1	Mild	Occasional attacks affecting the tips of one or more fingers
2	Moderate	Occasional attacks affecting distal and middle (rarely also proximal) phalanges of one or more fingers
3	Severe	Frequent attacks affecting all phalanges of most fingers
4	Very severe	As in stage 3, with trophic skin changes in finger tips

*The staging is made separately for each hand. In the evaluation of the subject, the grade of the disorder is indicated by the stages of both hands and the number of affected fingers on each hand; example: "2L(2)/1R(1),"–/3R(4)," etc.

Table 21-3. Stockholm Workshop Scale for the classification of sensorineural affects of hand-arm vibration syndrome

Stage*	Symptoms
0SN	Exposed to vibration, but no symptoms
1SN	Intermittent numbness, with or without tingling
2SN	Intermittent or persistent numbness, reduced sensory perception
3SN	Intermittent or persistent numbness, reduced tactile discrimination and/or manipulative dexterity

*The sensorineural stage is to be established for each hand.

Finally, it should be noted that for the most part, HAV-measured tool acceleration levels are usually very high, many times in excess of 1 g, which in part accounts for the short latent intervals and thus the quick onset of HAVS attacks in workers.

VIBRATION MEASUREMENTS

To ensure that vibration measurements are uniformly made so that results are comparable, the international scientific community has established measurement standards. Figure 21-1 shows the

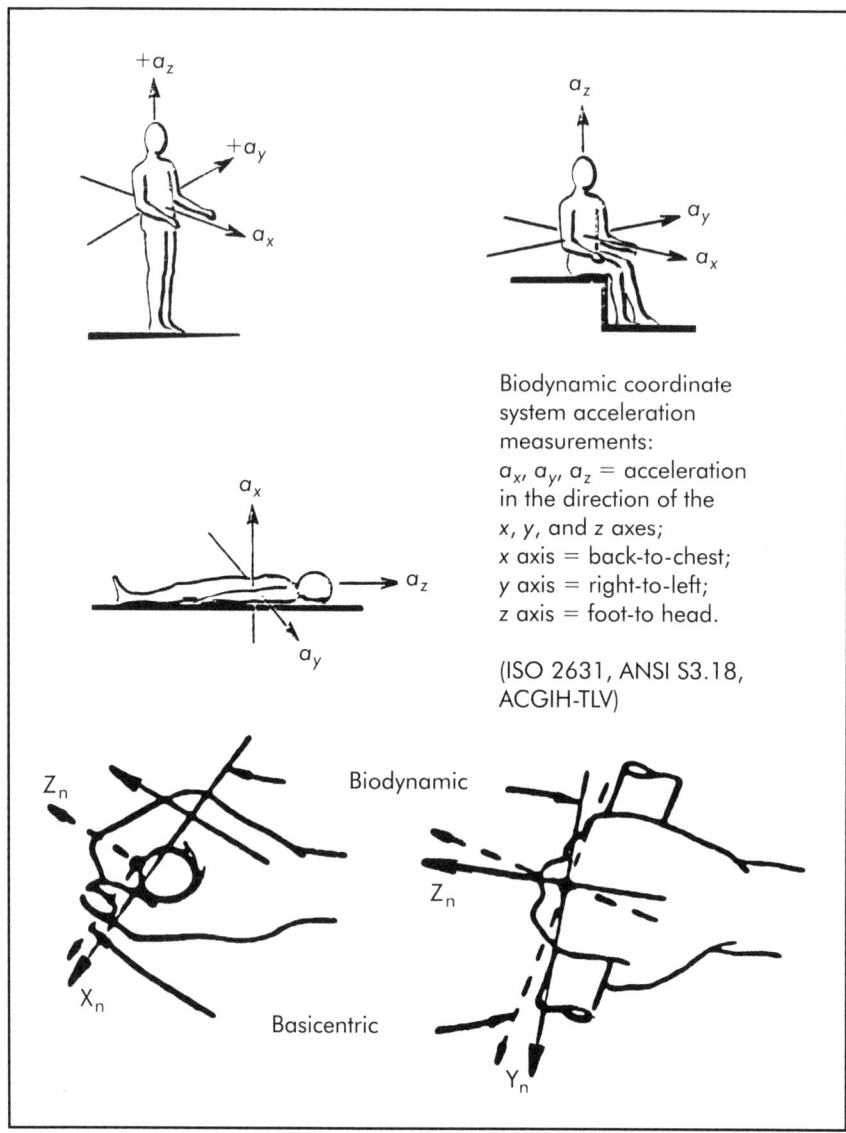

Figure 21-1 Coordinate system used for human whole-body vibration measurements.

triple vector (triaxial), mutually perpendicular coordinate directions respectively used for WBV and HAV measurements. For WBV, the z axis is the long axis (head-to-toe) of the body; the y axis is from side-to-side (across the shoulders); and the x axis is front-to-back (through the sternum). Similarly, for HAV measurements, the z axis is the long bone axis on the hand; the y axis is across the knuckles; and the x axis is through the palm of the hand.

Actual WBV measurements are simultaneously obtained using three accelerometers attached to a small metal cube and placed in the center of a hard rubber pie-plate-type disk. This disk is then placed on top of a seat cushion, for example, and measures the road vibration coming up through the seat to the driver's buttocks. Also, the measurement disk can be placed on the floor near a vibrating machine where a worker stands.

Hand-arm vibration measurements are usually obtained by mounting a small cube containing three perpendicular accelerometers to an automotive hose clamp. The hose clamp and accelerometers are then clamped to the tool handle.

STANDARDS

All triaxial measurements are simultaneously tape recorded and the vibration data is computer analyzed and compared to national and international health and safety standards for either WBV or HAV, to determine if these acceleration measurements have exceeded the respective standards. Currently in use in the United States are three WBV standards: ISO 2631, ANSI S3.18 (1983), and TLV 1995 (International Organization for Standardization [ISO] 1978; American National Standards Institute [ANSI] 1979; ANSI 1983; American Conference of Government Industrial Hygienists 1995); and three HAV standards: ANSI S3.34 (1986), TLV 1984-1998, and NIOSH #89-106 (ANSI 1986; American Conference of Government Industrial Hygienists 1984-1995; National Institute for Occupational Safety and Health [NIOSH] 1989).

It is not the intent of this chapter to detail the aforementioned WBV and HAV standards, which are detailed elsewhere (Wasserman 1987; Griffin 1990; Pelmear et al. 1992). Rather, the reader is advised to obtain the actual standard(s)* needed and read them thoroughly before attempting to use them. Suffice to say, all the aforementioned WBV and HAV standards are not identical; there are some significant differences.

---

*ISO and ANSI standards available from: Acoustical Society of America, 120 Wall Street, 32nd Floor, New York, NY 10005. ACGIH-TLV is available from: ACGIH, 1330 Kemper Meadow Drive, Cincinnati, OH 45240. NIOSH standard available from: Publications Department, NIOSH, 4676 Columbia Parkway, Cincinnati, OH 45226.

CONTROLLING VIBRATION IN THE WORKPLACE

Controlling vibration is usually multifaceted and includes vibration reduction and ergonomics principles (the latter is defined as adapting the workplace situation to the worker, not the reverse). General guidelines for WBV work situations (Wasserman 1987; Griffin 1990) include:

1. Use air-ride seats as a first choice, or mechanically suspended seats as a second choice, on vehicles to reduce WBV exposure to drivers.
2. Check and maintain vehicle suspension systems, tires, and tire pressure.
3. After long periods of driving/riding in vehicles, do not lift or bend immediately. Rather, first walk around and stretch for a few minutes. Use minimum twisting when exiting a vehicle (American Conference of Government Industrial Hygienists 1995; Wilder 1993).
4. In fixed plant situations, mechanically isolate vibrating equipment, machinery, etc., from floors and workers' bodies.
5. Where possible, keep workers away from vibrating equipment by using remote controls, switches, closed circuit TV, etc.
6. As appropriate, use WBV standards and guides.
7. If signs and symptoms of back pain and back disorders occur, consult a physician.

Guidelines for HAV work situations (Wasserman 1987; Pelmear et al. 1992; ISO 1978; ANSI 1979; ANSI 1983; American Conference of Government Industrial Hygienists 1995) include:

1. If possible, use antivibration (A/V) tools.
2. Try not to use vibration-damping materials externally wrapped around conventional vibrating tool handles. Many of these materials provide minimum (low frequency) vibration damping, and the resulting increased tool handle diameter, due to the wrap, could elicit additional cumulative trauma disorders (CTDs) of the hands and upper limbs (Pelmear et al. 1992; ANSI 1995).
3. A properly designed ergonomic tool handle places minimum strain on the operator's hand and arm, while keeping the wrist straight (in the neutral position).
4. Use A/V tools with a high power-to-weight ratio.

5. Various types of tool-hand configurations are available for specific tasks (Lindquist 1986). These include bow handles used on large rivet and chipping hammers where it is necessary to transmit high feed forces with minimum wrist torque (Figure 21-2a). The hammer is used at an approximate natural angle of 70°, the centerline of the arm and wrist, as the tool is gripped. The pistol handle is used for precision tasks by keeping the tool length short and minimizing the bending forces on the wrists by retaining the tool's center of mass on top of the handle (Figure 21-2b). When using a 70° natural angle, it has been found that low-grip forces transmit twist forces to the hand comfortably, whereas high-grip forces can keep the arm and wrist straight and, simultaneously, transmit high tool-feed forces to the workpiece. Screwdrivers, drills, grinders, and nut runners use straight handles (Figure 21-2c). A tool diameter of 1.50 in. (38 mm) for males and 1.34 in. (34 mm) for females appears optimum to maintain the strongest grip strength at minimum hand strain. To minimize wrist loading while performing precision work, the tool should be held at the shortest distance between the tool and workpiece.
6. Use tools suspended from the ceiling and/or mechanical arm balancers to minimize tool weight and fatigue on the worker.
7. Reroute air exhaust tool lines away from the worker's fingers and hands to minimize cold air exposure that can induce finger blanching attacks.
8. Maintain vibrating tools and implements according to the manufacturer's instructions. Replace worn out tools and implements.
9. There are several suggestions for workers to follow. Let the tool do the work, gripping it only when necessary and as lightly as possible, safely minimizing the vibration coupling into the hand. Also, keep the hand and body warm and dry. Avoid smoking, since nicotine constricts the blood vessels. Use only full-finger antivibration (A/V) gloves that fit well and keep the hands and fingers warm and dry. Where possible, take vibration-free breaks.
10. As appropriate, use the HAV standards.
11. See a physician if signs and symptoms of hand-arm vibration syndrome appear.

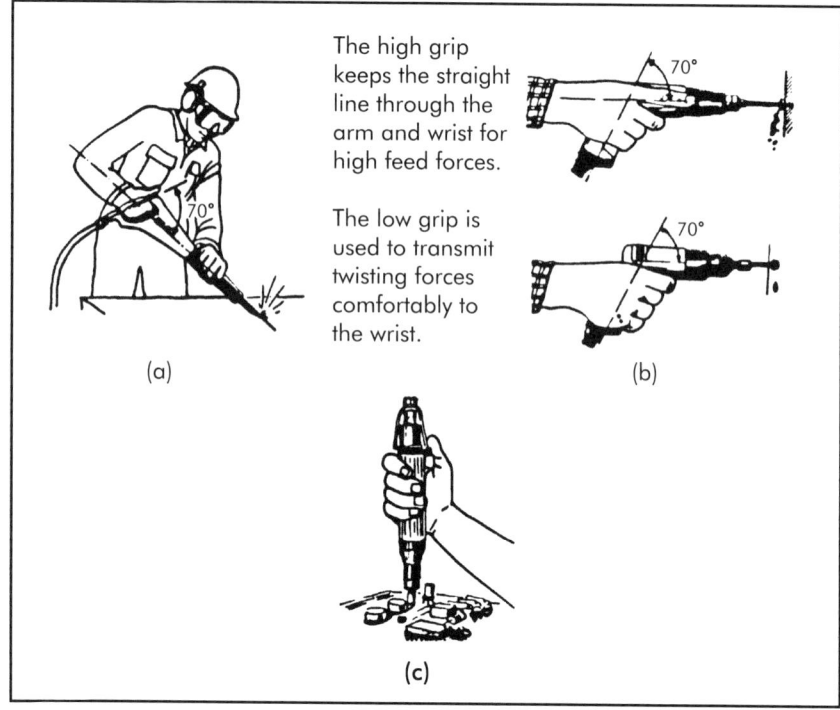

Figure 21-2 Ergonomically designed handles used with antivibration tools (Courtesy Atlas-Copco Co.).

REFERENCES

American Conference of Government Industrial Hygienists (ACGIH). *Threshold Limit Values for Hand-arm Vibration.* Cincinnati, OH: ACGIH, 1984-1995.

American National Standards Institute (ANSI) S3.18. *Guide for the Evaluation of Human Exposure to Whole-body Vibration.* New York: ANSI, 1979.

ANSI S3.29. *Guide for the Evaluation of Human Exposure to Whole-body Vibration in Buildings.* New York: ANSI, 1983.

ANSI S3.34. *Guide for the Measurement and Evaluation of Human Exposure to Vibration Transmitted to the Hand.* New York: ANSI, 1986.

ANSI Z365. *The Control of Work-related Cumulative Trauma Disorder, Part 1: Upper Extremities.* New York: ANSI, 1995.

Dupuis, H., and Zerlett, G. *The Effects of Whole-body Vibration.* Heidelberg, Germany: Springer-Verlag Publishers, 1986.

Griffin, M. *Handbook of Human Vibration.* London: Academic Press, 1990.

Gruber, G. "Relationship between Whole-body Vibration and Morbidity Patterns among Interstate Truck Drivers." *U.S. Department of Health, Education, and Welfare (DHEW)/National Institute for Occupational Safety and Health (NIOSH) Report* #77-167. Cincinnati, OH: DHEW/NIOSH, 1976.

Gruber, G., and Zipperman, H. "Relationship between Whole-body Vibration and Morbidity Patterns among Motor Coach Operators." *DHEW/NIOSH Report #74-104*. Cincinnati, OH: DHEW/NIOSH, 1974.

Hamilton, A. "The Effects of the Air Hammer on the Hands of Stone Cutters." *Industrial Accidents and Hygiene Series*, Report 236, #19. Washington, DC: U.S. Bureau of Labor Statistics, 1918.

Hulshof, C., and van Zantan, B.V. "Whole-body Vibration and Low-back Pain—A Review of Epidemiologic Studies." *International Archives of Occupational and Environmental Health* 1987, 59: 205-220.

International Organization for Standardization (ISO) 2631. *Guide for the Evaluation of Human Exposure to Whole-body Vibration*. Geneva, Switzerland: ISO, 1978.

Lindquist, B. "How to Design Vibration-controlled Power Tools." *Ergonomic Tools in Our Time*. Atlas Copco and B. Lindquist, eds. Stockholm, Sweden: TR Tryck, 1986.

Milby, T., and Spear, R. "Relationship between Whole-body Vibration and Morbidity Patterns among Heavy Equipment Operators." *DHEW/NIOSH Report #74-131*. Cincinnati, OH: DHEW/NIOSH, 1974.

National Institute for Occupational Safety and Health (NIOSH). *Criteria for a Recommended Standard for Hand-arm Vibration*, Report #89-106. Cincinnati, OH: NIOSH, 1989.

Pelmear, P., Taylor, W., and Wasserman, I. *Hand-arm Vibration: A Comprehensive Guide for Occupational Health Professionals*. New York: Van Nostrand-Reinhold Publishers, 1992.

Riazanov, V. *The Effects of Whole-body Vibration on the Female Organism*. Moscow, USSR: Academy of Medical Sciences/Institute of Industrial Hygiene and Occupational Diseases, 1985.

Seidel, H., and Heide, R. "Long-term Effects of Whole-body Vibration." *International Archives of Occupational-Enviromental Health* 1986, 58: 1-8.

Wasserman, D.E. *Human Aspects of Occupational Vibration*. Amsterdam, The Netherlands: Elsevier Publishers, 1987.

Wasserman, D. "Whole-body Vibration Exposure and the Human Spine." *Sound and Vibration* 1995, 29: 5.

Wasserman, D., Badger, D., et al. "Industrial Vibration—An Overview." *Journal of the American Society of Safety Engineers* 1974, 10: 38-40.

Wasserman, D., Taylor, W., Behrens, V., et al. "VWF Disease in U.S. Workers Using Chipping and Grinding Hand Tools." Volume 1, *Epidemiology*. DHHS/NIOSH Report #82-118. Volume 2, *Engineering*. DHHS/NIOSH Report #82-101. Cincinnati, OH: DHHS/NIOSH, 1982.

Wilder, D. "The Biomechanics of Low-back Pain." *American Journal of Industrial Medicine* 1993, 23: 577-588.

Wilder, D., Frymoyer, J., and Pope, M. "The Effects of Vibration on the Spine of the Seated Individual." *Automedica* 1985, 6: 5-8.

Wilder, D., Wasserman, D., Pope, M., Pelmear, P., and Taylor, W. "Occupational Vibration Exposure, Measurement, Standards, and Control." *Physical and Biological Hazards of the Workplace*. P. Wald and G. Stave, eds. New York: Van Nostrand-Reinhold Publishers, 1994.

CHAPTER 22

EVALUATING ERGONOMICS PROGRAMS

Gary B. Orr
Industrial Engineer, OSHA
Office of Ergonomic Support
Department of Labor-Occupational Safety and Health Administration
Washington, DC

David C. Alexander
President
Auburn Engineers
Auburn, AL

There is not a common standard to evaluate every ergonomics program. Whether a program is a success or failure depends partly on the measures an organization uses. To the employee working at a hazardous job, ergonomics should eliminate or reduce risk. To a stockholder, money spent on ergonomics should improve the stock's value. A successful ergonomics program will do both.

Each organization should decide the objectives of its ergonomics program. Unfortunately, some organizations try to copy programs from other organizations, with the predictable outcome of too few results obtained at too high a cost. The past is prologue, and people initiating a new program can learn from the efforts of those who have gone before them. Since the start of the Industrial Revolution, fitting work to people has been a goal of management. Consequently, there is a rich history of what works (and doesn't work) regarding ergonomics.

In the last half of the 20th century, a growing number of companies have started ergonomic programs. The lessons learned from past as well as future directions for evaluating ergonomic programs will be explored in this chapter. A common experience of many ergonomics coordinators is that hard work and good intentions are important, but the perception that a program is working is as important as the effort that is expended. The difference between successful and unsuccessful ergonomics efforts can be traced to the expectations of management, ergonomics coordinator(s), and workers.

A common misconception is that ergonomics is an applied science for repetitive assembly work. Organizations that employ *people* or have a product used by *people* have ergonomic issues. Ergonomic applications outside of manufacturing can be found in construction, hospitals, agriculture, military, nuclear plants, and schools. A successful ergonomics program is like building a rock wall; you have to start with what is available and shape it into what you want.

THE ERGONOMICS PROCESS AND ITS EVALUATION

Ergonomics is a process that just keeps expanding deeper and broader into the organization. At some point, it will become so imbedded that it is not visible as a separate process. The term "ergonomics program" can be useful in describing the ongoing activities that support the overall ergonomics process. As the ergonomics process grows, it encompasses larger spheres of influence or domains, as shown in Figure 22-1.

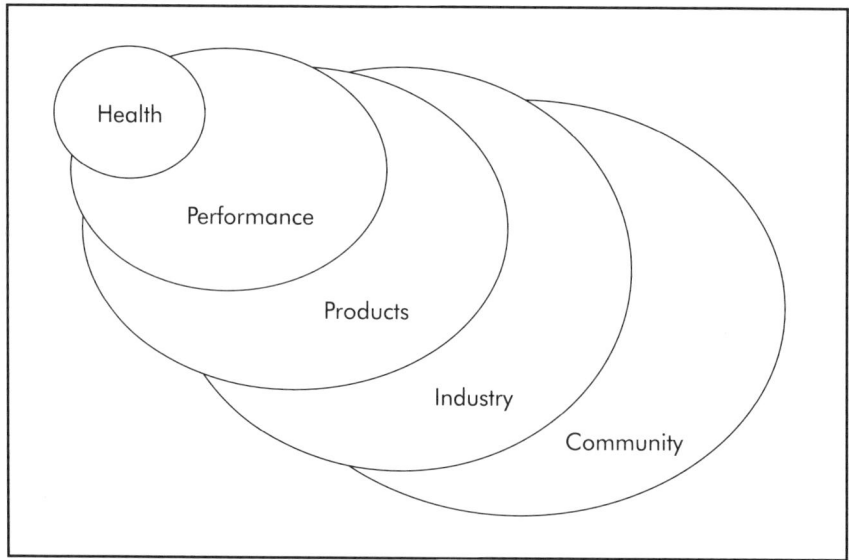

Figure 22-1 Domain influenced by ergonomics.

Often, moving a stagnating program from one domain to the next will breathe life into it. After several years of working in one area of ergonomics, such as health hazards, people in the organization become bored; the ergonomics effort stagnates as the organization tries to push measures into levels of diminishing return. For example, some organizations concentrate only on workers' compensation claims (where the measure is to drive these costs to zero). However, as the costs decline, so do ergonomics program efforts, resulting in a failed program and ultimately increased compensation costs. This cycle may be repeated several times before the organization learns to enhance the ergonomics program rather than let it deteriorate.

Success within one domain usually provides a foundation for success in the next. However, it is not always necessary to start at the health and safety level. Some organizations have started at the products level and used a successful human factors program to resolve employee health concerns.

The movement from controlling health hazards to performance reflects a shift in perceived importance from an inward look at the organization to an outward look at the impact of the organization on the world it serves.

While the evaluation criteria vary for each phase of the ergonomics process, there are two factors that must be consistent: management commitment and employee involvement.

MANAGEMENT COMMITMENT

All initiatives in an organization that require resources, either capital or labor, must have management approval. Approval, however, should not be interpreted as commitment. Support from managers is required to ensure ergonomics is properly integrated into the organization. Management can subvert the ergonomics effort with words or actions. When high-level managers comment that ergonomics is expensive overhead or back injury complaints are usually made by malingerers, the program is doomed, and the person responsible for implementing ergonomics should look for another project. Many corporate managers continue to believe that careless workers are really to blame for injuries and illnesses. A 1967 survey of industrial injuries in Pennsylvania concluded that only 26% of accidents resulted from employee carelessness, which leaves 74% with a direct link to the job or work environment.

Management involvement can be spurred by personal circumstance, a need to reduce costs, a desire to improve performance, or a commitment to a quality work environment. Managers who have suffered an overuse injury often become advocates of ergonomics. Ergonomic efforts have often started after an executive secretary developed a musculoskeletal disorder. Managers have also become strong advocates of ergonomics when a successful implementation has resulted in improved business performance at their plant or a competitor's. Even the most recalcitrant manager will go along when there has been a series of successes.

Management commitment includes a vision of how ergonomics will shape the organization. Effective managers understand the culture of the organization, often looking to reform areas that do not conform to business needs. Introducing ergonomics will support some existing goals of the organization and conflict with others. For the investment in ergonomics to yield positive results, managers must develop a vision for the organization to move from the status quo to an organization with ergonomics integrated into its culture (see Table 22-1).

Table 22-1. An example of models used for problem solving

Business as Usual	Ergonomics Culture
Ergonomics is an expense	Ergonomics is an investment
Ergonomics is required (OSHA worker compensation)	Ergonomics is used because it is good business
Limited number of people understand ergonomics	Infrastructure supporting ergonomics is broad and deep
Ergonomics is not a consideration in changes	Ergonomics is a consideration for all new equipment, tools, facilities, information, and jobs/tasks
People wait for an ergonomist or committee to resolve problems	People are knowledgeable and encouraged to resolve problems themselves or in their natural work groups
Workers do not know about ergonomics	Ergonomics is commonplace
Ergonomics is not commonly used at work	Ergonomics is taken home and used to protect self and family members

Source: Auburn Engineers: *Why Ergonomics Programs Fail*, Auburn, AL. With permission.

Management commitment can be evaluated as follows:
- Good—there is management accountability and a strategy for implementing ergonomics;
- Better—actively involved management assists in developing a strategy and short term objectives;
- Best—an advocate in upper management assists in developing a vision for ergonomics and understands that implementation requires an infusion of various resources as the process develops. Management reviews the adequacy of the following:
- Budget;
- Personnel;
- Assigned responsibility;
- Expertise and authority;
- Means to hold responsible persons accountable; and
- Program review procedures.

On the road to failure:
- Programs do not have visible support from upper management;
- Ergonomics is perceived by the employees as another flavor of the month program;
- No policy statement on ergonomics and no written program; and
- No requirement for measurable short-term results.

EMPLOYEE INVOLVEMENT

The ergonomics coordinator must be a technical resource with skills in ergonomics. This person is a company employee or contractor with a thorough knowledge of the structure, information flow, and key players in the organization. The technical skills may vary with the size of the site and type of activities of the organization.

Supervisors and workers also must be involved in ergonomics. Ergonomic interventions will affect the way workers perform their jobs. Consequently, they should be involved in reviews and changes. There are several problems with relying on an ergonomics expert to do all the work:

1. No one is a better expert on the job than the person who does the job.
2. Convincing the worker that an expert can solve the problem is a tough sell.

3. Solutions that create the perception that the job will take more time, is dangerous, or will significantly change the routine are hard to sell.

Employee involvement can be evaluated as follows:

- Good—employee with strong technical skills in ergonomics leads a team composed of management and labor;
- Better—matrix team led by a person with strong management skills is able to address concerns across the site. A person with technical skills in ergonomics participates with other team members;
- Best—ergonomic knowledge and problem solving is developed in each functional area of the site (office, shipping/receiving, maintenance, etc.). A person with technical skills in ergonomics acts as a coach. Ergonomic issues are considered in safety, production, and quality meetings at all levels. Workers and their representatives participate fully in developing the ergonomics program and conducting training, education, and audits. Employees enthusiastically support the ergonomics program.

On the road to failure:

- Programs depend totally on people outside the organization;
- Packaged programs where little time is spent adapting to the needs of the organization;
- Employees are not aware of the policy on ergonomics and reporting; and
- The policies and attitudes of management discourage reporting.

CONTROL AND PREVENTION OF OCCUPATIONAL HEALTH HAZARDS

Increasing injuries, rising workers' compensation costs, compliance with the Americans with Disabilities Act, and citations from the Occupational Safety and Health Administration (OSHA) have prompted many organizations to address musculoskeletal disorders in the workplace. Organizations have used a variety of models to address musculoskeletal disorders (see Table 22-2).

A review of safety and health programs resulted in the development of a four-component program that was presented in OSHA's guidelines for voluntary safety programs. The same model was used in the ergonomics program management guidelines for meatpack-

Table 22-2. Models used for problem solving

Hygiene Model	Engineering Model	Quick and Dirty Model
1. Anticipate 2. Recognize 3. Evaluate 4. Control	1. Identify the problem 2. Analyze the problem 3. Identify alternatives 4. Select the best solution 5. Implement the solution 6. Follow up	1. Find 'em 2. Fix 'em

ing plants. The program has been used extensively within the red meat industry and in other organizations. The four components are:

- Work site analysis;
- Hazard control;
- Medical management; and
- Training.

WORK SITE ANALYSIS

Necessary elements in finding existing problems:

- Surveillance techniques include reported cases as well as observation of the workplace;
- Injury and illness information is reviewed periodically, with trends signaling the need for controls;
- Accident and near miss reports are reviewed to find links to musculoskeletal disorders;
- Investigations involving musculoskeletal disorders include work methods review, work history, health history, and off-the-job information;
- Surveys conducted periodically to drive appropriate corrective action; and
- Current hazard analyses are documented for all work areas and communicated and available to all the work force as well as designers, suppliers, and buyers.

Warning signs include:

- Problem jobs not prioritized;
- No timetable for corrective actions;
- Acute and chronic disorders investigated with the same form; and
- Employees reluctant to report hazards.

HAZARD PREVENTION

Necessary elements of hazard prevention include:

Organization.

- Focal point for ergonomics within the organization; and
- Temporary employees and contractors are included in the ergonomics program.

A problem-solving process.

- A problem-solving method has been implemented, allowing various people to participate, with the process describing the risk factors and encouraging employee input;
- Engineering controls implemented for most problem jobs;
- Initial efforts concentrate on jobs where success can be easily measured; and
- The effectiveness of corrections are reviewed in follow-up audits.

Engineering.

- Plans operations and procedures to prevent exposure to hazards;
- Notifies the ergonomics coordinator of planned changes to jobs or the environment; and
- Involves the ergonomics coordinator in new installations at the design phase of the project.

Purchasing.

- Ensures that only equipment, tools, and materials that use ergonomics principles are purchased.

Warning signs include:

- The ergonomic coordinator or outside expert is relied upon to control all problem jobs;
- Ergonomic coordinator is not familiar with job assessment tools (such as the NIOSH lift equation);
- There is work on problem jobs, but no results;
- No follow-up to determine the effectiveness of controls; and
- Mostly high-cost solutions, with implementation two or more years in the future.

MEDICAL MANAGEMENT

Health care providers, when treating injured workers and detecting musculoskeletal disorders, must:

- Use a protocol for diagnosing musculoskeletal disorders;
- Be trained to treat musculoskeletal disorders;
- Recommend placement of employees whose physical and mental capabilities are within the requirements of the job;
- Periodically examine employees working at hazardous jobs, and periodically observe their jobs;
- Be available for all shifts and fully involved in hazard identification and training;
- Provide employees with multiple physician review at organization's expense;
- Formulate a musculoskeletal disorder management plan for each employee with a work-related musculoskeletal disorder;
- Have regular contact with employees who are injured and are away from work or on restricted duty;
- Make contact with the persons who develop return-to-work plans;
- Recognize symptoms that indicate a problem job and communicate the needs for a job review.

Warning signs include:

- Employees having surgery without attempting conservative treatment; and
- Splints and supports commonly used without review by a health care provider.

TRAINING

Necessary elements for imparting awareness, education, and job skills:

- Employees involved in training development and implementation;
- Employees (including temporary employees and contractors) can recognize hazards associated with their job and recognize symptoms of musculoskeletal disorders;
- Supervisors and workers know the policy on musculoskeletal disorders;
- Observational and problem-solving skills are developed within a work group, such as accounts receivable, warehouse, quality lab, assembly line, and maintenance;
- Targeted education for management, designers, engineers, and purchasing employees;

- An ergonomic coordinator with knowledge appropriate for the responsibilities and organization; and
- Time provided to allow employees (new hires and relocated) to acclimate to jobs.

Warning signs include:

- Training employees to recognize musculoskeletal hazards, without a problem-solving strategy;
- Ergonomics is considered only as common sense and no training is provided for management or technical employees; and
- Training not audited for effectiveness.

THE BOTTOM LINE

In addition to the four components, a test of a program's success can be found by examining two specific items. First, has there been a significant change in the statistics of importance to the organization (for example, have injuries or compensation costs been significantly reduced)? Second, is there a specific system or set of systems in place that leads to these changes, and helps to ensure the continuation of resulting gains? If there have been significant changes without accompanying changes in the system, then the results are likely to cease in the future.

Success in controlling occupational health hazards and additional items within the other domains is based on both activity measures and impact measures of the process. By looking at these measures, a balanced review of the ergonomics process and its associated program can be achieved.

Necessary elements include:

- A plan for allocating resources and coordinating work site analysis, hazard prevention, medical management, and training;
- Reports and measures implemented to monitor the ergonomics process;
- Random audits of the workplace, medical management, and training;
- Observation analysis to evaluate progress;
- Tracking data such as incidence rates, severity rates, workers' compensation costs, number of problem jobs that are controlled, and the quality of the problem-solving effort; and
- Auditing the ergonomics program at least annually.

Warning signs include:

- No short-term measures; and
- No mix of activity and impact measures.

CONTROL OF CONDITIONS AFFECTING PERFORMANCE

Ergonomic principles are used not only to control health issues, but also business performance issues. The ergonomic process impacts policy decisions. For example, workstations may be changed to control problems among computer users in the health phase; in the performance phase, breaks and job enlargement become policy issues. There is not a distinct line separating many health and performance issues, but there are work groups within the organization that are able to identify performance issues and implement successful controls.

Without an ergonomics process in place, new jobs are not designed with the worker in mind. Job designers and engineers have a variety of reasons for not using ergonomic principles, but the bottom line is that the majority of engineers are not taught ergonomics principles in school. Engineers and job designers need to become aware of these principles and their practical application if the organization is to eliminate hazards from new workstations. Table 22-2 lists several reasons why designers and engineers do not use ergonomics.

Table 22-2. Reasons why designers and engineers don't use ergonomics.

1. There is no real science.
2. People don't want to change.
3. We don't have ergonomic problems with new designs.
4. People should go home tired after work.
5. Indoor work without heavy lifting can't cause an injury.
6. You can find workers who don't complain about the job.
7. That job will be automated soon.
8. If we mention ergonomics concerns, we'll get more complaints.
9. Workers don't know what is best for them.
10. Ergonomics costs too much.

Necessary elements to control conditions affecting performance:

- Measures of the ergonomics process are expanded to include turnover, absenteeism, rework, and customer complaints;
- Reports of accidents and near misses are distributed to engineering and purchasing, as well as to safety and medical personnel;
- Surveys and interviews are used to detect problems before a medical visit;
- Ergonomic reviews are integrated into existing audits of performance and quality;
- A comprehensive safety and preventative maintenance program that maximizes equipment reliability and performance; and
- Information on successes is shared with other sites.

Warning signs include:

- Measures of the effort to control problem jobs is eliminated; and
- Incidents and health trends not reviewed.

EVALUATING PRODUCTS OF THE ORGANIZATION

Manufacturers who develop consumer or military products may have started their ergonomics effort with a human factors program. The aim of many of these programs is to improve the reliability or usability of products (make them user-friendly). While the ergonomic principles are the same for health, performance, and products, the ergonomic skills and location of expertise in the organization may be different for the products phase. Today, many organizations integrate marketing, design, production, sales, and service, so that ergonomics principles can be applied throughout the organization. In the products phase, ergonomics must be considered in design.

In addition to fit and usability issues, people in marketing and design may want to know how a user perceives the product. For this type of analysis, the ergonomist must have psychological and sociological skills.

Products for external and internal customers should be considered. An organization may spend time and effort to develop a customer survey form to obtain information about a product. The same techniques can be used to obtain information on the effectiveness of forms and procedures used within the organization.

Necessary elements:

- An ergonomic review as part of the audit of instructions and procedures for external (for example, usability, instructions) and internal products (such as forms, computer systems);
- Ergonomic expertise is expanded to include psychological and sociological analysis;
- Audits measure the health and performance impacts of the ergonomics process; and
- Ergonomics considered in job content and organizational changes.

Warning sign:

- Ergonomics is added, rather than integrated, into the job demands of designers.

OUTREACH WITHIN THE TRADE

In the quality process, a certain level of maturity is reached when an organization begins to involve its customers and suppliers in its quality program. The same is true of an ergonomics program. Within this domain, the organization will begin an outreach beyond its own boundaries.

Initially, this will involve suppliers, customers, and changes in the manufacturing process to reduce or eliminate ergonomics hazards and risk factors that cannot be resolved any other way. For example, a manufacturer of automobile parts may approach a customer about changes to the part, which will enhance the ergonomics of manufacturing (not unlike current suggestions made to reduce costs or improve quality and reliability).

In other cases, the organization may approach its suppliers with changes that make manufacturing easier. For example, an automobile assembly plant may request changes that make the installation of certain parts safer—just as they might request changes to lower costs or improve quality in today's market.

Finally, as organizations become more confident in their own ergonomics process and program, they will become willing to share it with others within the trade to enhance overall safety. Many trade groups sponsor industry-wide safety and health meetings, and the sharing of ergonomics information will be handled in a similar way.

Necessary elements:

- Meetings held within the site, within the company, within the industry, and within the supplier community to discuss ergonomic issues;
- Cost of supplies considered a measure of the ergonomics process; and
- Audits of health, performance, and products conducted and the results assessed by management.

Warning sign:

- Requiring suppliers to accept measures or structure for the ergonomics process developed at your organization.

CONTRIBUTIONS TO THE TECHNICAL/LEGISLATIVE COMMUNITY

Many organizations hesitate to allocate resources to an ergonomics process because they don't know of anyone in their industry who has been successful at applying ergonomic principles. The technical and personnel barriers seem insurmountable to many organizations. The knowledge base on ergonomics, however, continues to grow. Many of the rumors about ergonomics can be dispelled by someone in the industry who can say, "we tried it and it worked."

Where there is a need for legislation or policy decisions on ergonomics, the organization should be prepared to share its experience with the legislative committees. This process allows the legislative body to make informed decisions.

Ergonomics fits the job to the workers and results in the elimination or reduction of discomfort. Therefore, the ergonomics process plays an important role in resolving quality of work life issues. Ergonomic principles can be used away from work to recognize symptoms and influence changes in lifestyle.

Necessary elements:

- People who have been involved in the ergonomic process are given time to present talks and papers on their experience;
- Forms and procedures used for ergonomics are shared with other industries;
- Employees discuss the use of ergonomic principles on applications away from work; and
- Industry issues affected by ergonomics become management and trade association action items.

CONCLUSION

There are many intangibles in the evolution and evaluation of a successful ergonomics process. Many elements of the ergonomics process are difficult to measure and require judgment, which may lead to a variety of opinions on the degree of success. The process can be likened to a voyage into a new frontier.

The elements of each organization, such as management philosophy, goals, available resources, and labor relations, create unique environments that preclude one right answer for all. Each organization must find its own comfort level with ergonomics, which will ultimately enable it to design jobs with people in mind.

The opinions expressed in this chapter are those of the author and do not reflect the intent or position of OSHA on the evaluation of ergonomics programs.

BIBLIOGRAPHY

Alexander, D.C, and Pulat, B.M. *Industrial Ergonomics—A Practitioner's Guide*. Norcross, GA: Industrial Engineering and Management Press, 1985.

Auburn Engineers. *Why Ergonomic Programs Fail, Baseline Compliance Audit of an Occupational Ergonomics Program*. Auburn, AL: Auburn Engineers Press, Auburn Engineers Inc.

Auburn Engineers. *Baseline Compliance Audit of an Occupational Ergonomic Program*. Auburn, AL: Auburn Engineers Press, Auburn Engineers Inc.

Cordtz, Dan. "Safety on the Job Becomes a Major Job for Management." *Fortune*, November 1972: 112.

Occupational Safety and Health Administration (OSHA) 3123. *Ergonomics Program Management Guidelines for Meatpacking Plants*. Washington: U.S. Department of Labor, 1993.

Occupational Safety and Health Voluntary Safety Program Management Guidelines. *Federal Register* 54: 16, January 26, 1989: 3,904-3,916.

Olishifski, J.B., and McElroy, F.E. *Fundamentals of Industrial Hygiene*. Chicago, IL: National Safety Council, 1971.

Rabinbach, Anson. *The Human Motor*. Boulder, CO: Basic Books, A division of Harper Collins Publishers, 1990.

CHAPTER 23

AUDITING ERGONOMICS

Colin G. Drury
Professor
Department of Industrial Engineering
State University of New York at Buffalo
Buffalo, NY

MEASUREMENT, CHANGE, AND AUDITING

Audits are used for checking that certain acceptable policies and practices have been consistently followed (Carson and Carlson 1977). In this chapter, auditing methods are applied to ergonomics.

Modern management is encouraged to be "measurement-driven." Deming's cycle of "plan, do, check, act" has measurement at its core and, indeed, he repeatedly stressed "management by fact" (Deming 1986). After many years in which companies insisted that the only numbers were financial, we are returning to the concept that broadly defined measurement is essential to the change process. Companies embarking on one of the newer quality methodologies are encouraged to use more detailed measurements of performance and more global comparison standards. These new methodologies have in common the implementation of change, and ergonomics is no exception. Thus, it is only appropriate that the same standards of measurement be applied to ergonomics and human factors.

Fortunately, ergonomics has developed many useful measurement techniques (Wilson and Corlett 1995). The requirement is to turn the wealth of ergonomics evaluation techniques into practical, usable ergonomics audit methodologies. This chapter shows how audit methodologies can be designed, evaluated, and used in industry. It draws on a more detailed treatment, which is the primary reference source for further reading (Drury 1997).

WHY AUDIT ERGONOMICS?

Although measurement is *de rigeur* in industry today, we need to have a clear idea of why an ergonomics audit in particular may be valuable. Assuming we are auditing the ergonomics of a production system rather than a product, then our main reason for an ergonomics audit is to measure how well ergonomics has been applied to that production system. Whether we are measuring a whole company, an assembly line, a control room, a shipping department, or a hospital radiology laboratory, the true test of ergonomics application is that measurably more features will be ergonomically correct. An obvious analogy is to chair design: no matter what the sales brochure says about ergonomics, if the chair is the wrong height or wrong shape it will not be an ergonomic chair. Ergonomics is as ergonomics does!

A Program Rather than a Project

An ergonomics audit takes a snapshot of the production system at a point in time, rather than, for example, analyzing injury or quality records. Because the audit evaluates a whole production system, it can be used as an assessment tool for the organization that performed the ergonomics. In this way, the impact of an ergonomics effort can be measured beyond a single project.

Most ergonomics evaluations use matched measurements before and after an ergonomics project to measure the project's success. For example, to demonstrate successful ergonomics efforts in reducing the cumulative trauma disorder potential of leather sewing operators, Drury and Wick (1984) measured daily damaging wrist motions for each wrist before and after ergonomics changes were made at several workplaces. Other examples of measuring the success of individual projects, from manufacturing cells to production lines, can be found in Drury (1991). In each, the project was the focus, so that an audit really was not needed. But as we move from ergonomics projects to ergonomics programs, audits become valuable to assess the results of a range of ergonomics activities.

As a human factors group becomes more involved in strategic decisions about identifying and choosing the projects it performs, evaluation of individual projects is less revealing. All projects could have a positive impact, but the group could still achieve more with an astute choice of projects. The group could conceivably have a more beneficial impact on the company's strategic objectives by stopping all projects for a period to concentrate on training the management, work force, and engineering staff to make more use of ergonomics. An audit answers the question, "how well is the overall ergonomics program performing," rather than "what was the benefit of that (single) ergonomics project."

Moving Picture

An ergonomics audit proceeds by measuring key indicators of an ergonomics application at a single point in time. It can be repeated at intervals to measure progress and trends, giving a moving picture rather than a snapshot. Indeed, an ergonomics audit can be applied before and after an ergonomics program to measure its impact in the same way that individual ergonomics projects are evaluated.

Beyond the evaluation of ergonomics programs, we can add features to an ergonomics audit that allow it to evaluate the overall ergonomics *process* within a company. The issues here are how well an ongoing ergonomics function is utilized, whether its existence and expertise are known to the appropriate people, and so on. An example of such an audit of a multiplant company is given in Drury (1990). A more recent example will be presented later in this chapter. For evaluating the ergonomics process, the techniques are somewhat expanded beyond those used in program evaluation. In addition to examining the ergonomic quality of the production system, we need to probe the networks of contacts and information surrounding the ergonomics function, usually using interview techniques.

CHOICES FOR AN AUDIT SYSTEM

Now that we have decided why to audit, the obvious next question is how to do it. However, before giving examples of audit techniques and methodologies, some basic choices must be made. What is the audited unit? How is it sampled? How much depth and breadth is needed?

To determine how well ergonomics has been applied, we must decide, "applied to what." For jobs where an operator remains relatively fixed at one workplace (or a few workplaces), then the workplace represents the obvious unit of audit. Thus, our audit samples from the list of all workplaces. This is the most common audit assumption (Mir 1982; Drury 1990). However, for some jobs, such as maintenance, it may not be possible to enumerate all the workplaces because they only come into being when a particular item needs maintaining. For such cases, the natural unit is the daily activity of the (maintenance) operator, and there must be a sampling of activities rather than workplaces. Audit programs for such jobs have been developed for aviation maintenance and will be given as examples later (Chervak and Drury 1995; Koli 1994).

Ideally, an audit system would be broad enough to cover any task in any industry. It would provide highly detailed analysis and recommendations, and be applied rapidly. Unfortunately, the three variables of breadth, depth, and application time are likely to trade off in a practical system. Thus, a thermal audit sacrifices breadth to provide considerable depth based on the heat balance equation, but requires measurement of seven variables (Parsons 1992). Some can be obtained rapidly (air temperature, relative humidity), but some take longer (clothing insulation value, metabolic rate). Conversely, structured interviews with participants in an ergonomics program can be broad and rapid, but quite deficient in depth (Drury 1990).

TYPES OF CHECKLISTS

At the level of audit instruments, such as questionnaires or checklists, there are comprehensive surveys such as the Position Analysis Questionnaire (McCormick 1979), the Arbeitswissenschaftliche Erhebungsverfahren zur Tätikgkeitsanalyse (AET), which takes two to three hours to complete, or the simpler Work Analysis Checklist (McCormick 1979; Rohmert and Landau 1983; Pulat 1992). Alternatively, there are simple single-page checklists, such as the Ergonomics Working Position—Sitting Checklist that can be completed in a few minutes (SHARE 1990).

We need to choose the most appropriate trade-offs between depth, breadth, and time. Often, it is better to hedge our bets and conduct a broad rapid audit that points to specific areas of need for more detailed (and longer) audits.

We must also deal with the quality of the measurements we obtain from the audit. We must consider those attributes of a measurement that allow it to be issued with confidence by practitioners. These attributes are conventionally defined by:

1. Validity—how well the measure relates to the underlying phenomenon. For example, how does a standard surface roughness measurement relate to the customer's perception of finish quality? We evaluate validity by correlating our measure with an independent measure of the phenomena.

2. Reliability—if we take the same measurement many times, does it give the same value? Unless we have reliable measures, we are introducing unnecessary noise into the measurement process. Thus, we should have frequently calibrated gages to measure physical phenomena or well-constructed psychophysical scales to evaluate human-derived phenomena.

3. Sensitivity—if the measurement does not vary much despite known changes in the process, its sensitivity is in doubt. We must be able to observe the variation in the measurement as the process is changed. Thus, a stop watch is inappropriate for assessing the performance of an electronic timer if the variability between timer examples is of the order of 0.1 seconds.

Before using an existing audit system, see that all criteria have been met. For example, the Upper Extremity Checklist was evaluated on these criteria and found to be valid, reliable, and sensitive (Keyserling et al. 1993).

AUDIT DESIGN

Having asked the basic questions about audit systems, we can move on to good practice in audit design. The two basic data collection methods are the survey form or checklist used at each workplace, and the interview format used to collect process data from other stakeholders. Both should follow established questionnaire/checklist/interview design principles, for example, those in Sinclair (1995) or Patel, Drury, and Prabhu (1993). These principles cover layout, working, and sampling, and are valuable in ensuring reliable data collection.

Some points specific to survey forms need to be raised. First, as far as possible, we should use ergonomics standards and verified best practices as the basis for workplace evaluations. A good review of available standards is in the special issue of *Applied Ergonomics* (Parsons 1995). There are other best practice sources that do not have the weight of standards, but are nonetheless valuable; for example, the NIOSH lifting equation (Waters et al. 1993) and the military human factors standard MIL-STD-1472D (Department of Defense 1989).

If standards and other good practices are used in a human factors audit, they provide a quantitative basis for decision making. Measurement reliability can be high and validity self-evident for legal standards. However, it is good auditing practice to record only the measurement used, and not its relationship to the standard, which can be established later. This removes any temptation by the analyst to bend the measurement to reach a predetermined conclusion. Illumination measurements, for example, can vary considerably over a work space, so that an audit question:

Work surface illumination >750 Lux   ☐ yes   ☐ no

could be legitimately answered either way for some work spaces by choice of sampling point. Such temptation can be removed, for example, by the following question.

Illumination at four points on workstation:
☐   ☐   ☐   ☐   Lux

Later analysis can establish whether, for example, the mean exceeds 750 Lux, or whether any of the four points fall below this level.

HOW TO AUDIT ERGONOMICS

The auditor has a choice of existing audit programs, from simple checklists to fully developed computer-based sampling systems. Only two general-purpose systems will be presented here to illustrate the range available, augmented by a small case study to show typical results. More examples of audit systems and their use can be found in Drury (1997).

The first example is a general-purpose, workplace-based audit, the Ergonomics Audit Program (Mir 1982; Drury 1990). This system was developed for a multinational company, and comprises a cluster sampling plan to select departments and workplaces, and a workplace survey.

The workplace survey was designed based on ergonomic aspects derived from a task/operator/machine/environment model of the person at work. Each aspect formed a section of the audit, and sections could be omitted if they were clearly not relevant, such as manual material handling aspects for data entry clerks. Questions within each section were based on standards, guidelines, and models. Table 23-1 shows the major sections and typical questions.

Data was entered into the computer program and a rule-based logic evaluated each section to provide messages to the user in the form of a "Section shows no ergonomic problems" message, for example, "Results from analysis of auditory aspects—everything OK in this section"; or a "Discrepancies from a single input" message, for example, "Seats should be padded, covered with nonslip materials, and have front edge rounded"; or a "Discrepancies based on the integration of several inputs" message, for example, "The total metabolic workload is 174 watts, intrinsic clothing insulation is 0.56 clo, initial rectal temperature is predicted to be 97° F (36° C), final rectal temperature is predicted to be 99° F (37° C)."

Counts of discrepancies were used to evaluate departments by ergonomics aspect, while the messages were used to alert company

Table 23-1. Workplace survey: structure and typical questions

Section	Major Classification	Examples of Questions
1. Visual aspects		• Nature of task • Measure illuminance at task (midfield, outer field)
2. Auditory aspects		• Noise level, dBA • Main source of noise
3. Thermal aspects		• Strong radiant sources present? • Wet bulb temperature • Clothing inventory
4. Instruments, controls, displays	• Standing vs. seated • Displays • Labeling • Coding • Scales, dials, counters • Control/display relationships • Controls	• Are controls mounted between 30-70 in. (76-178 cm)? • Signals for crucial visual checks • Are trade names deleted? • Color codes same for control and display? • All numbers upright on fixed scales? • Grouping by sequence or subsystem? • Emergency button diameter > 0.75 in. (19 mm)?
5. Design of workplaces	• Desks • Chairs • Posture	• Seat to underside of desk > 6.7 in. (17 cm)? • Height easily adjustable 15-21 in. (38-53 cm)? • Upper arms vertical?
6. Manual material handling	• NIOSH Lifting Guide, 1981	• Task, H, V, D, F
7. Energy expenditure		• Cycle time • Object weight • Type of work
8. Assembly/repetitive aspects		• Seated, standing, or both? • If heavy work, is bench 6-16 in. (15-41 cm) below elbow height?
9. Inspection aspects		• Number of fault types? • Training time until unsupervised?

personnel to potential design changes. This latter use of the output as a training device for nonergonomic personnel was seen as desirable in a multinational company rapidly expanding its ergonomics program.

Reliability and validity have not been assessed, although the checklist has been used in a number of industries (Drury 1990). The workplace survey has been included here because, despite its lack of measured reliability and validity, it shows the relationship between the audit as methodology and the checklist as technique.

Aircraft Checklists

The ERGO, EEAM, and ERNAP checklists are part of complete audit systems for different aspects of civil aircraft hangar activities (Koli et al. 1993; Chervak and Drury 1995). They were developed for the Federal Aviation Administration (FAA) to provide tools for assessing human factors in aircraft inspection (ERGO) and maintenance (EEAM) activities, respectively. Inspection and maintenance activities are nonrepetitive in nature, controlled by task cards issued to technicians at the start of each shift. Thus, the sampling unit is the task card, not the workplace, which is highly variable between task cards. Their structure was based on extensive task analyses of inspection and maintenance tasks, which led to generic function descriptions of both types of work (Drury et al. 1990). Both systems have sampling schemes and checklists, and are computer-based with initial data collection on either hard copy or input directly into a portable computer. Recently, both have been combined into a single program (ERNAP) distributed by the FAA's Office of Aviation Medicine. The structure of ERNAP and typical questions are given in Table 23-2.

As in Mir's Ergonomics Audit Program, the ERNAP checklist is modular, and the software allows formation of data files, selection of required modules, analysis after data entry is completed, and printing of audit reports. Similarly, the ERGO, EEAM, and ERNAP instruments use quantitative or yes/no questions to compare the entered value with standards and good practice guides. Each takes about 30 minutes per task. Output is in the form of an audit report for each workplace, similar to the messages given by Mir's workplace survey, but in narrative form. Output in this form was chosen for compatibility with existing performance and compliance audits used by the aviation maintenance community.

Reliability of a first version of ERGO was measured by comparing the output of two auditors on three tasks. Significant differences were found at $p < 0.05$ on all three tasks, showing a lack of

Auditing Ergonomics

Table 23-2. ERNAP structure and typical questions

Audit Phase	Major Classification	Examples of Questions
I. Pre-maintenance	Documentation	Is feed forward information on faults given?
	Communication	Is shift change documented?
	Visual characteristics	If fluorescent bulbs are used, does flicker exist?
	Electric/pneumatic equipment	Do push buttons prevent slipping of fingers?
	Access equipment	Do ladders have nonskid surfaces on landings?
II. Maintenance	Documentation (M)	Does inspector sign off work card after each task?
	Communication (M)	Explicit verbal instructions from supervisor?
	Task lighting	Light levels in four zones during task (fc).
	Thermal issues	Wet bulb temperature in hanger bay (°F [°C])
	Operator perception	Satisfied with summer thermal environment?
	Auditory issues	Noise levels at five times during task (dBA)
	Electrical and pneumatic	Are controls easily differentiated by touch?
	Access equipment (M)	Is correct access equipment available?
	Hand tools	Does the tool handle end in the palm?
	Force measurements	What force is being applied (lb [kg])?
	Manual material handling	Does task require pushing or pulling forces?
	Vibration	What is total duration of exposure on this shift?
	Repetitive motion	Does the task require flexion of the wrist?

Table 23-2. ERNAP structure and typical questions *(continued)*

Audit Phase	Major Classification	Examples of Questions
II. Maintenance *(continued)*	Access	How often was access equipment repositioned?
	Posture	How often were following postures adopted?
	Safety	Is inspection area adequately cleaned for inspection?
	Hazardous material	Were hazardous materials signed out and in?
III. Post-maintenance	Buy back	Are discrepancy work sheets readable?

inter-auditor reliability. Analysis of these differences showed them to be largely due to errors on questions requiring auditor judgment. When such questions were replaced with more quantitative questions, the two auditors had no significant disagreements on a later test. Validity was measured using concurrent validation against six Ph.D. human factors engineers who were asked to list all ergonomic issues on a powerplant inspection task. The checklist found more ergonomic issues than did the human factors engineers. Only a small number of issues were raised by the engineers that were missed by the checklist.

For the EEAM checklist, again an initial version was tested for reliability with two auditors, who answered the same on only 85% of the questions. A modified version was tested and the reliability was considered satisfactory with 93% agreement. Validity was again tested against four human factors engineers, and this time the checklist found significantly more ergonomics issues than the engineers, without missing any issues they raised.

The ERNAP audits have been included here to provide examples of checklists embedded in an audit system where the workplace is *not* the sampling unit. They show that nonrepetitive tasks can be audited in a valid and reliable manner. In addition, they demonstrate how domain-specific audits can be designed to take advantage of human factors analyses already made in the domain.

At this point, it should be mentioned that the term *ergonomics audit* also has been used to describe the classification of recorded human errors in complex systems. A novel interview-based audit system was proposed by Fox (1992) based on methods developed at British Coal (Simpson 1994). Here, an error-based approach was taken, using interviews and archival records to obtain a sampling of actual and possible errors. These were then orthogonally classified using Reason's (1990) active/latent failure scheme by Rasmussen's (1987) skill-, rule-, knowledge-based framework. Each active error is thus a conjunction of slip/mistake/violation with skill/rule/knowledge. Within each conjunction, performance-shaping factors can be deduced and sources of management intervention listed. This methodology has been used in a number of mining-related studies and has been highly successful in reducing human errors (Simpson 1994).

AN ERGONOMICS AUDIT EXAMPLE

An ergonomics audit program was used to evaluate an ongoing ergonomics program in an electronic equipment manufacturing plant. The following is adapted from the audit report to the company (Fish 1996).

The audit took place at a three-year-old plant where ergonomics had been used in the original design. In the year prior to the audit, an ergonomics program had been introduced to alleviate musculoskeletal injury problems, with an on-site nurse-ergonomist and a visiting corporate ergonomist as ergonomics providers. An ergonomics process was implemented whereby operators having musculoskeletal difficulties on their job were encouraged to seek help via their supervisors. Ergonomic awareness was provided by a two-hour course on personal ergonomics offered to operators, an eight-hour required course for supervisors, and a 40-hour required course for engineers. An ergonomics committee was formed under the safety committee to foster awareness and address specific ergonomic problems.

To evaluate the ergonomics program, the workplace survey of the ergonomics audit program was used on 15 workplaces, covering a variety of jobs. Interviews were held with six decision makers (managers, supervisors), five users (one was a union representative), and three providers (nurse-ergonomist, corporate ergonomist, safety specialist). All interviews probed the definition and use of ergonomics, asked for specific examples of its use, and encouraged open-ended comments using questions similar to those reported in Drury (1997).

The workplace survey was presented in the same order as appears in Table 23-1. The visual environment was generally inadequate, with a mean illuminance level of 379 Lux at the task. This was somewhat below typical minimum recommendations of 500-1,000 Lux for industrial assembly tasks. Task lights were rarely used, so that for most workplaces the work point was the darkest part of the visual field, rather than the brightest.

In contrast, the auditory and thermal environments were good, with a mean sound pressure level of 68 dBA, and a generally warm, dry atmosphere. Only three of the workplaces were outside the American Society of Heating, Refrigerating, and Air-conditioning Engineers (ASHRAE) comfort zone, all in the direction of low humidity.

Few workplaces needed instrument, control, or display evaluation. The only problem was found in the packaging manipulators, where the toggle controls were of the wrong type, too high for regular use, and inadequately labeled.

The workplace design audit showed good and poor features. Chairs were well-designed and adjustable, but could be used only for breaks, as the workplaces required all-day standing. The electrostatic discharge straps on the operators' feet were found to be quite restrictive in prolonged standing. The main discovery in the assembly and repetitive aspects of the audit was the high frequency of damaging wrist motions at almost all workplaces. This was due, in part, to product design and workstation layout, and also to operators not adopting the most advantageous positions. Product and workplace redesign should be accompanied by operator training in good posture. In addition, the working points at workstations were usually too high because product size was not considered when fixing conveyor heights. Excessive shoulder work was required at the current heights. Finally, many components were located far from the operator's standing position, leading to excessive reaching. Again, workplace redesign was in order.

Workplaces with a manual material handling component were analyzed using the NIOSH (1981) lifting formula. For two tasks, the actual weight exceeded the acceptable limit, mainly due to excessive reaching. Energy expenditure limits were not exceeded. Inspection tasks were not sampled.

For the more process-oriented aspects of the audit, results from the decision makers, users, and providers are considered together to form a picture of the ergonomic culture of the plant. Perhaps the most surprising result was that five of the six decision makers and four of the five operators defined ergonomics in terms of "fitting the worker to the job," rather than the expected "fitting the job to the worker." Their comments bore out this bias, coming mainly from an occupational medicine perspective rather than from an engineering perspective. Ergonomics involvement was entirely reactive, with providers responding to problems perceived by the work force. The main person called upon as provider was the nurse-ergonomist, who was perceived as much more receptive than supervisors or engineers.

Ergonomics providers were in short supply, with only the nurse-ergonomist admitting much ergonomic activity. She had been trained on the 40-hour course two years previously, and now spent about 50% of her time on ergonomics. Supervisors, who were officially the key first respondents for the ergonomics process, were not well informed or particularly receptive. Typical supervisor comments were that ergonomics was "good in theory," "good if used legitimately," and "a tool for getting out of work." Users perceived that there was little response from supervisors beyond immediate productivity and safety concerns. Interestingly, all comments and examples showed ergonomics as a safety concern, with no mention made of its positive effects on quality or productivity.

Overall, the audit showed a continuing need for ergonomics (workplace survey), but a culture in which ergonomics was quite distorted (interviews). The ergonomics process was seen as safety-related, reactive, and resulting in medical accommodations of the operator to the job. Indeed, the audit process itself prompted a broader definition of ergonomics in the plant and managerial steps toward more effective use of ergonomics resources.

LESSONS FROM AUDITING

As may be expected from the small aforementioned audit, the results of audits can be quite surprising to management. This stems, in part, from our using a definition of ergonomics going beyond musculoskeletal injury reduction. But even within the well-known areas of wrist and back ergonomics, the results can startle as the audit goes beyond the counting of currently-reported injury and illness to examine the workplaces that are key causal factors in wrist and back injuries. Where there is no current ergonomics program, the audit can provide both a baseline for ongoing measurements and detailed direction to management. Is manual material handling really the problem? Can conveyor design be a contributing factor? Is the lighting adequate, or do operators have to compromise their posture by bending close to the work point to achieve adequate vision? Whether evaluating a facility to determine the need for an ergonomics program or measuring the performance of an ongoing program or process, the ergonomics audit is becoming a recognized first ergonomic step for plant management.

REFERENCES

Carson, A.B., and Carlson, A.E. *Secretarial Accounting*, 10th Edition. Cincinnati, OH: South Western Publishing Company, 1977.

Chervak, S., and Drury, C.G. "Simplified English Validation." *Human Factors in Aviation Maintenance—Phase 6 Progress Report. DOT/FAA/AM-95/xx*. Springfield, VA: Federal Aviation Administration, Office of Aviation Medicine, National Technical Information Service, 1995.

Deming, W.E. *Out of the Crisis*. Cambridge, MA: Massachusetts Institute of Technology, 1986.

Department of Defense. *Human Engineering Design Criteria for Military Systems, Equipment, and Facilities* (MIL-STD-1472D). Washington, DC: Department of Defense, 1989.

Drury, C.G. "Ergonomics Practice in Manufacturing." *Ergonomics* 1991, 34: 825-839.

Drury, C.G. "The Ergonomics Audit." E.J. Lovesey, ed. *Contemporary Ergonomics* 1990: 400-405.

Drury, C.G. "Human Factors Audit." G. Salvendy, ed. *Handbook of Human Factors*. New York: John Wiley and Sons, 1997.

Drury, C.G., Prabhu, P., and Gramopadhye, A. "Task Analysis of Aircraft Inspection Activities: Methods and Findings." *Proceedings of the Human Factors Society 34th Annual Conference*. Santa Monica, CA: Human Factors Society, 1990: 1,181-1,185.

Drury, C.G., and Wick, J. "Ergonomic Applications in the Shoe Industry." *Proceedings of the International Conference on Occupational Ergonomics* 1984, 1: 489-493.

Fish, L. Private communication. 1996.

Fox, J.G. "The Ergonomics Audit as an Everyday Factor in Safe and Efficient Working." *Progress in Coal, Steel, and Related Social Research* 1992: 10-14.

Keyserling, W.M., Stetson, D.S., Silverstein, B.A., and Brouwer, M.L. "A Checklist for Evaluating Ergonomic Risk Factors Associated with Upper-extremity Cumulative Trauma Disorders." *Ergonomics* 1993, 36(7): 807-831.

Koli, S.T. *Ergonomic Audit for Non-repetitive Task*. Unpublished M.S. thesis. Buffalo, NY: State University of New York at Buffalo, 1994.

Koli, S., Drury, C.G., Cuneo, J., and Lofgren, J. "Ergonomics Audit for Visual Inspection of Aircraft." *Human Factors in Aviation Maintenance—Phase Four, Progress Report, DOT/FAA/AM-93/xx*. Springfield, VA: National Technical Information Service, 1993.

McCormick, E.J. *Job Analysis: Methods and Applications*. New York: AMACOM, 1979.

Mir, A.H. *Development of Ergonomic Audit System and Training Scheme*. Unpublished M.S. thesis. Buffalo, NY: State University of New York at Buffalo, 1982.

National Institute for Occupational Safety and Health (NIOSH). *Work Practices Guide for Manual Lifting*. Cincinnati, OH: U.S. Department of Health, Education, and Welfare (DHEW)/NIOSH, 1981.

Parsons, K.C. "Ergonomics Standards." Special issue of *Applied Ergonomics* 1995, 26(4): 235-305.

Parsons, K.C. "The Thermal Audit. A Fundamental Stage in the Ergonomics Assessment of Thermal Environment." E.J. Lovesey, ed. *Contemporary Ergonomics* 1992, 85-90.

Patel, S., Drury, C.G., and Prabhu, P. "Design and Usability Evaluation of Work Control Documentation." *Proceedings of the Human Factors and Ergonomics Society 37th Annual Meeting.* Seattle, WA: Human Factors and Ergonomics Society, 1993: 1,156-1,160.

Pulat, B.M. *Fundamentals of Industrial Ergonomics.* Englewood Cliffs, NJ: Prentice Hall, 1992.

Rasmussen, J. "Reasons, Causes, and Human Error." J. Rasmussen, K. Duncan, and J. Leplat, eds. *New Technology and Human Error* 1987, 293-301.

Reason, J. *Human Error.* New York: Cambridge University Press, 1990.

Rohmert, W., and Landau, K. *A New Technique for Job Analysis.* London: Taylor & Francis, Ltd., 1983.

Safety and Health Applications and Resources Exchange (SHARE). *Inspecting the Workplace.* SHARE Information Booklet. Australia: Occupational Health and Safety Authority, 1990.

Simpson, G.C. "Ergonomic Aspects in Improvement of Safe and Efficient Work in Shafts." *Ergonomics Action in Mining,* Eur 14831. Luxembourg: Commission of the European Communities, 1994.

Sinclair, M.A. "Subjective Assessment." *Evaluation of Human Work*, 2nd Edition. J.R. Wilson and E.N. Corlett, eds. London: Taylor & Francis, 1995.

Waters, T.R., Putz-Anderson, V., Garg, A., and Fine, L.J. "Revised NIOSH Equation for the Design and Evaluation of Manual Lifting Tasks." *Rapid Communications, Ergonomics* 1993, 36(7): 749-776.

Wilson, J.R., and Corlett, E.N., eds. *Evaluation of Human Work.* London: Taylor & Francis, 1995.

CHAPTER 24

ECONOMIC EVALUATIONS OF ERGONOMIC INTERVENTIONS

James R. Buck
Professor
Department of Industrial Engineering
University of Iowa
Iowa City, IA

Proposed ergonomic projects are no different than others in terms of managerial acceptance. If the projects appear to be economically viable and money is available, then they will likely be undertaken. Otherwise not. Perhaps a more distinct statement is: "To succeed in introducing a new, ergonomically better, working method, that method also must have economic advantages. Those advantages can be reductions in cumulative trauma syndrome problems, lower turnover of the work force, fewer expected accidents, greater production efficiency, and many other features that have economic implications" (Rose et al. 1992). This chapter describes some features of making that economic analysis based on cost or revenue estimates that are found in industry and elsewhere. A wide variety of economic situations are shown to demonstrate the breadth of the need to perform economic analyses associated with ergonomic designs.

Another statement about the role of economic analysis in ergonomic design is, "The economics of the problems addressed cannot be ignored" (Simpson 1993). For many years ergonomists behaved (implicitly if not explicitly) as though their interests in the promotion of health and safety gave them a "moral right" that took them beyond any concern with the economic implications of their work. This attitude is totally untenable. The nonsense of such a position was encapsulated nicely in a conversation with a manager from the manufacturing sector who said: "What is the point in spending vast sums of money to create the healthiest, and safest factory in the country, if the only people who work in it are liquidators?"

While this is clearly an overstatement, it indicates that changes, however desirable and for even the most laudable of motives, require investment. The investment potential of even the most profitable organization is finite and therefore needs to be justified. Also the investment must generate a return, not just in altruistic terms, but in financial terms.

BENEFITS AND COSTS

To present a proposal and defend its economic quality, one should first identify the benefits and costs and, secondly, assure that the benefits exceed the costs. While that is not the end, it is a good beginning. As a simple example, suppose that the Ergonomics department recommends the installation of an electric screwdriver to speed up manual assembly operations at a workstation in which numerous stud screws are required. The Ergonomics department estimates that the addition of this screwdriver will reduce the 20 different assembly times by three minutes and reduce operator fatigue. An industrial-grade electric screwdriver costs about $150 and is expected to last about two years in the working environment.

Consequently, the benefits derived from this power tool must repay the investment and costs of operation with a reasonable return on the investment over that two-year time duration. Part of those costs are the electrical power usage, generously estimated at $0.20 per hour of use, and minor maintenance of $2.00 per week. The Ergonomics department could not think of how much would be saved with regard to fatigue reduction, so they agreed to ignore that feature. With the investigative work completed on identifying benefits and costs, the remaining job was to describe this proposal in economic terms, clearly understandable to management.

Most economic information comes tagged with different time or unit bases as the foregoing example illustrates. No one can mentally put such numbers together meaningfully. Therefore, the ergonomics personnel must do some paperwork to justify this information so that it can be digested. One way is to select a useful time in the future for presenting all benefits and costs accrued by that time. In the example, a two-year time frame is an excellent choice because the power screwdriver is expected to last that duration.

Suppose one starts with the benefits on a per shift basis. Without the electric screwdriver, each assembly requires 20 minutes. A shift of eight hours consists of 480 minutes, but typically 15% of that time is lost to various allowances (personal time, resting from fatigue, and delays). That leaves 408 minutes per shift. The number of those special assemblies completed per shift with and without the power screwdriver are respectively

$$\frac{408 \text{ minutes}}{\text{shift}} \times \frac{1 \text{ assembly}}{20 \text{ minutes}} = \frac{20.4 \text{ assemblies per shift without}}{\text{power screwdriver}}$$

$$\frac{408 \text{ minutes}}{\text{shift}} \times \frac{1 \text{ assembly}}{17 \text{ minutes}} = \frac{24.0 \text{ assemblies per shift with}}{\text{power screwdriver}}$$

A power screwdriver allows 3.6 more assemblies per shift. With a $10.00 per hour labor rate for this job (including employee benefits and other indirect costs), the cost savings per shift is:

$$\frac{3.6 \text{ assemblies}}{\text{shift}} \times \frac{17 \text{ minutes}}{\text{assembly}} \times \frac{\$10.00}{\text{Hour}} \times \frac{1 \text{ hour}}{60 \text{ minutes}} =$$

$$\frac{3.6 \times 17 \times 10}{60} = \$10.20 \text{ savings per shift}$$

Assuming a single shift operation over seven shifts each week and 50 weeks of work per year, the benefits over the two years accumulate to:

$$\frac{\$10.20 \text{ savings}}{\text{shift}} \times \frac{7 \text{ shifts}}{\text{week}} \times \frac{50 \text{ weeks}}{\text{year}} \times 2 \text{ years} = \$7,140.00$$

This calculation of benefits is conservative because it ignores any interest earned over the two years as the benefits stream in. But it is assumed here that the persons performing the assembly operations have ample work so that these savings continue over the full two years.

Now turning to costs, one should first look at the investment and an ample return on that investment. If 10% per year is considered an adequate return on the investment, then the accumulated costs over the two years are:

$$\$150.00 \, (1 + 0.1)^2 = \$181.50$$

This calculation is exactly as a banker would view a loan to the Ergonomics department for two years, where the loan repayment is $181.50 at the end of the two years. Another cost is power, which is calculated as:

$$\frac{\$0.20}{\text{hour}} \times \frac{8 \text{ hours}}{\text{shift}} \times \frac{7 \text{ shifts}}{\text{week}} \times \frac{50 \text{ weeks}}{\text{year}} \times 2 \text{ years} = \$1,120.00$$

The final cost component is maintenance, which accumulates over two years:

$$\frac{\$2.00}{\text{week}} \times \frac{50 \text{ weeks}}{\text{year}} \times 2 \text{ years} = \$200.00$$

The costs total to:

$$\$181.50 + \$1,120.00 + \$200.00 = \$1,501.50$$

The net benefits over costs are estimated at:

$$\$7,140.00 - \$1,501.50 = \$5,638.50$$

saved over the two years or $2,819.25 per year. Any manager would consider this project a very lucrative economic opportunity, especially since the interest earned on savings in addition to the fatigue reduction would increase the benefits even more.

This economic illustration describes an elementary future worth development. That is, all benefits and costs are brought forward to the end of the life of an asset and combined. While some niceties were ignored in the aforementioned calculations, that was the general intent. The principal reason to bring all costs to a specific point in time is because money has time value. Accordingly, dollars at one point in time have a different value at another point. Adding dollar values at different points in time is meaningless. The following interest calculations and discounted cash flow discussion show how to convert dollars at one point in time to those at another, based on an interest rate that is considered a minimum attractive rate of return (MARR).

INTEREST CALCULATIONS AND DISCOUNTED CASH FLOWS

The material that follows is based on concepts and calculations shown in typical engineering economics textbooks (DeGarmo et al. 1984; Fleischer 1994; Newnan 1991; Theusen and Fabrycky 1989; White et al. 1989; Bussey 1978; Park and Sharpe-Bete 1990).

When capital is invested and left for a specified period of time, say one year, then interest is normally earned on that investment. Traditionally, upper management specifies that interest rate to reflect the MARR. An investment principle of P dollars at the start of the investment earns interest i at the end of the year and so the value of the account one year hence is $F_1 > P$. The relationship between P and F_1 in this case is:

$$F_1 = P(1 + i) \tag{1}$$

When interest is compounded over k time periods, then the relationship between F_k and P is:

$$P(1+i)^k = F_k \qquad (2)$$

It follows directly that:

$$P = F_k [1+i]^{-k} \qquad (3)$$

If P in the preceding equation is the investment, i is the minimum attractive rate of interest, and F_k is a cost savings k time periods in the future, then P is the largest investment that cost savings can economically justify. P is also called the *present worth* value of the cost saving F_k. It also follows that F_k is the *future worth* of the investment P.

Present Worth Analysis

The present worth of a project is a monetary value that occurs at the beginning of a project and is economically equivalent to the algebraic sum of all expenditures (negative values) and income (positive values) from the project. Equation 3 describes the single expenditure case. If there was only a single cost saving at one particular future time, then the total present worth equals P less the actual investment needed to launch the project. When that difference is negative, the project costs more than the savings it generates. When the difference is positive, the positive amount is the surplus amount over the minimally required amount. More surplus is better.

In most cases, there are cost savings over a number of different years. If equation 3 is applied to each different cost (signed negatively to represent a cash outflow) and also to each different cost savings (signed positive to represent a cash inflow) at various future years (for example, n, n', n'', etc.), then an equivalent P for each cash flow is found and the algebraic sum of each resulting P is the present worth less the investment needed to launch the project. That sum of the Ps less the investment is the *net present worth*. It follows that each future cost or cost saving is brought back to the present by the operator $(1+i)^{-n}$ to reflect the time value of money, and the present time equivalents are summed algebraically. Since the investment is also a cash outflow at the present, it too has a negative present worth. The algebraic sum of all positive and negative present worth denotes the project's present worth. When the project's present worth is positive, then a project is economically viable and those with more present worth are more viable. Being economically viable does not mean it will be selected by rational people because other potential projects may be better. But a negative total present worth means that the project is not economically viable and should not be selected.*

Uniform Cash Flows Over Time and Annual Worths

While it is possible to compute present worths in the manner previously indicated, it is computationally much easier to recognize that cost savings and costs usually come in some functional pattern over the future. A common concept is to consider saving a uniform series of $A each year over the next k years. The present worth equivalent of that entire pattern of cost savings is:

$$P(i) = A\left[\frac{1 - (1 + 0.1)^{-k}}{i}\right] \qquad (4)$$

Now, instead of making k separate calculations, one can make a single calculation with the same result. For example, suppose that the minimum attractive rate of return is 10% and the estimated cost savings were $100 each year for three years. Then, the equivalent present worth of these three future cost savings is:

$$P(10\%) = \$100\left[\frac{1 - (1 + 0.1)^{-3}}{0.10}\right] = \$248.68$$

Computing these three present worths individually, one obtains:

$$P = 100\,(1.1)^{-1} + 100\,(1.1)^{-2} + 100\,(1.1)^{-3} =$$
$$90.91 + 82.64 + 75.13 = \$248.68$$

With a larger number of future time periods, this computational savings is substantial.

A third basis for examining alternative prospects over time is to convert all cash flows into equivalent annual worths. An illustration is when there is an initial investment P and after k years there is a substantial future cost savings F_k. The investment P has an equivalent annual worth described by a minor variation of equation 4:

---

*This statement is not without exception if it is related to other projects. It is correct when all projects are independent of each other. A project with negative present worth can have such strategic values that when it is coupled with other projects, it is economically superior.

$$A(i) = \left[\frac{i}{1-(1+i)^{-k}}\right] P \qquad (5)$$

In this case P and A are negative to reflect an outflow of cash. The other component of annual worth is the cost savings at the end of the asset life of k years. That cost savings has an equivalent P based on equations 3 and 5 and shows that there is an equivalent A. Putting those two equations together we have

$$F_k (1+i)^{-k} \left[\frac{i}{1-(1+i)^{-k}}\right] = F_k \left[\frac{i(1+i)^{-k}}{1-(1+i)^{-k}}\right] = A' \qquad (6)$$

The annual worth for this example would be the $-A$ resulting from the initial investment as shown by equation 5, plus the A' value of equation 6. Naturally, a negative result would show that the costs exceeded the benefits and the project should not be undertaken. A positive result denotes that the return recovers the investment P and provides a reasonable return on that investment in the form of a minimum attractive return on it. A more typical case of annual worth consists of two or more potential projects where the expected lives of the projects under consideration are different. One cannot compare potential projects with different lives using present or future worth because longer lives increase the present or future worth. However, annual worth is the same period of cash flow for each project—one year. For example, suppose that a $1,000.00 investment was made and cost savings (C) were expected at the end of a five-year time period. It follows that a 10% return on that investment over five years has an equivalent annual worth of

$$C = P \frac{i}{1-(1+i)^{-k}} = -\$1,000.00 \frac{0.1}{1-(1.1)^{-5}} =$$

$$-\$1,000.00 \, (0.2638) = -\$263.80$$

and a future cost savings that generates a positive annual worth of $263.80 is

$$C' = +\$263.80 = F_5 \frac{0.1 \, (1.1)^{-5}}{1-(1.1)^{-5}} = F_5 \, (0.1638)$$

and $F_5 = +\$1,610.51$.

In most textbooks on the subject of engineering economics, the symbols P, F, and A mean present, future, and annual value, respectively. However, those books typically use the symbol n to specify the duration of the project (rather than k). Many books also give conversion factors in tables with factor symbols $(P/F, i, n)$ to mean find P, given F, with an interest rate, and a project duration n. Usually, a separate page of tables is given for each different interest rate and rows in those tables correspond to different n values. Each column is a different factor: $(P/F, i, n)$, $(F/P, i, n)$, $(A/F, i, n)$, $(A/P, i, n)$, $(P/A, i, n)$, and $(F/A, i, n)$. For example, part of a table may look similar to

$i = 10\%$ Discrete Compounding Table

N	$(F/P, i, n)$	$(P/F, i, n)$	$(F/A, i, n)$	$(P/A, i, n)$	$(A/F, i, n)$	$(A/P, i, n)$
4	1.4641	0.6830	4.6410	3.1699	0.2155	0.3155
5	1.6105	0.6209	6.1051	3.7908	0.1638	0.2638
6	1.7716	0.5645	7.7156	4.3553	0.1297	0.2296

Note that the last two columns of the row for five periods are the two factors used in the preceding calculations. Such tables are convenient for performing economic calculations. Formulae corresponding to these factors are:

$(F/P, i, n)$	$(P/F, i, n)$	$(F/A, i, n)$
$(1+i)^k$	$\dfrac{1}{(1+i)^k}$	$\dfrac{(1+i)^k - 1}{i}$
$\dfrac{1}{(1+i)^{-k}}$	$(1+i)^{-k}$	$\dfrac{1-(1+i)^{-k}}{i(1+i)^{-k}}$

$(P/A, i, n)$	$(A/F, i, n)$	$(A/P, i, n)$
$\dfrac{(1+i)^k - 1}{i(1+i)^k}$	$\dfrac{i}{(1+i)^k - 1}$	$\dfrac{i(1+i)^k}{i(1+i)^k}$
$\dfrac{1-(1+i)^{-k}}{i}$	$\dfrac{i(1+i)^{-k}}{1-(1+i)^{-k}}$	$\dfrac{i}{1-(1+i)^{-k}}$

The upper row in each set of formulae shows these factors with positive exponents and the lower row shows equivalent factors with negative exponents. These six formulae are strictly for relating single cash flows at time 0 (corresponding to P) and at time k (corresponding to F), and the uniform series of $\$A$ per period over the k (or n) time periods.

Using Functional Series for Estimating and Analyzing

Up to this point, the only functional series discussed was the uniform or step series. A step series is, just as the step notion suggests, a constant quantity of $A cash flow occurring at the end of each time period and flowing for k time periods. It is clear that the step series can greatly simplify computations in the analysis. Since there is $A occurring each time period, the total cash flow over k years is simply Ak, which is purely the sum of the cash flow without discounting for time. It is, therefore, easy to estimate the magnitude of A from available records of similar costs. Simply find the total expenditure and divide by the duration of time the expenditure occurs.

Two additional functional forms useful in economic analysis are the ramp functions. As the name implies, these cash flows increase or decrease in a uniform manner over time. A case in point is the maintenance required of hand tools or machines acquired for production operations. Tools and other machines need maintenance and with increasing age, they need more maintenance. Accordingly, costs increase over time and the up ramp can be used to approximate those cost increases. The *up-ramp cash flow formula* for computing present worth is:

$$P(i) = \frac{C}{i^2}[1 - (1+i)^{-k}] + \frac{C}{i}[1 - (k+1)(1+i)^{-k}] \quad (7)$$

A traditionally recognized functional series, known as a *gradient*, is similar except the first cash flow of a gradient occurs at time 2 rather than time period 1. Most records of a company showing expenditures describe those expenditures without discounting. Those records should describe up-ramp expenditures individually as Ck after k time periods and they should accumulate over k periods to $Ck(k+1)/2$. Accordingly, a simple but effective way to estimate the cash flow increase $C is to find the accumulations over several time periods and divide through each by $k(k+1)/2$. An average resulting C will occur and can be used in equation 7. Consider the following cash flow series, where the average increase in costs is about $50 each year, but not exactly. Also shown is the present worth of each individual cash flow based on MARR equal to 12%, along with the cumulative present worth. Over the five years, the present worth accumulates to slightly over $500 (Table

24-1). Equation 7 is shown with the average increase of $50 more each year and $i = 12\%$. While the computation of the equation is considerably less, the present worth turns out to be about $500.

$$P(.12) = \frac{50}{0.12^2}[1 - (1.12)^{-5}] + \frac{50}{0.12}[1 - (5+1)(1.12)^{-5}] =$$

$$1{,}501.99 - 1{,}001.90 = 500.09$$

Table 24-1. Cash flow series

k=	1	2	3	4	5	Average
Cash flow (F_K)	47	102	158	195	248	
Cumulative Cash flow	47	149	307	502	750	
× 2/k (k+1)=	47	49.67	51.16	50.20	50.00	49.55
F_K (1.12)$^{-k}$	41.96	81.31	112.46	123.93	140.72	
PW cumulative	41.96	123.28	235.74	359.67	500.39	

A highly-related form is the linearly decreasing cost savings over time or the *down-ramp series*. This case frequently occurs when a large cost savings starts out, but erodes over time. Decreases could be caused by reduced public demand for a product during certain times of year* or added maintenance or deterioration. Suppose that the cost savings started out at $R per time period and then decreased each time period by $C. So long as $Ck < R$, then the present worth is:

$$P(i) = \frac{R}{i}[1 - (1+i)^{-k}] - \frac{C}{i}[1 - (k+1)(1+i)^k] - \frac{C}{i^2}[1 - (1+i)^{-k}]$$

or   (8)

$$P(i) = \frac{1}{i}\{R[1 - (1+i)^k] - C[1 - (k+1)(1+i)^k] - \frac{C}{i}[1 - (1+i)^k]\}$$

Equation 8 is merely the step series present worth less the up-ramp present worth. For example, consider a step series of $5,000.00 per year, which is decreased by $100.00 each year. Accounting records can be checked over the years to test if the estimated $100.00

---

*Lawn mowers in the winter or snow shovels in the summer for those locales where these products are off-season.

decrease per year is correct. The cumulative values in Table 24-1 aid in this purpose. If the minimum attractive rate of return is 10%, then the present worth over a five year time period is:

$$P(i) = \frac{1}{0.1}\{5{,}000\,[1-(1.1)^{-5}] - 100\,[1-(5+1)(1.1)^{-5}] - \frac{100}{0.1}[1-(1.1)^{-5}]\}$$

$$P(i) = \{18{,}953.93 - 2{,}725.53 - 3{,}790.79\} = \$12{,}437.61$$

Accordingly, the downward ramp is easy to compute as well, verifying the same result except for a little round-off difference. It also should be pointed out that an equivalent annual worth can be found by multiplying the present worth by (A/P, 10%, 5) or 0.26380 and the result is $3,281.01. In the case of a future worth at the end of five years, simply multiply by $(1 + 0.10)^5$ or 1.61051, where the result is $20,030.90. Consequently, one can perform either of these alternative forms of analysis.

If a future worth is desired instead, one simply needs to multiply the present worth by $(1 + i)^k$ to find the future worth at the future time k periods later. Also, for those wanting to compute the equivalent annual worth, the future and present worth can be multiplied by the respective factors

$$\left[\frac{i}{1-(1+i)^{-k}}\right] \quad \text{or} \quad \left[\frac{i(1+i)^k}{(1+i)^k - 1}\right] \tag{9}$$

to create the uniform C value. In the case of the up-ramp numerical example, the equivalent future worth and annual worth is respectively:

$$F(.12) = 500.09\,(1.12)^5 = 500.09\,(1.76234) = \$881.33$$

$$A(.12) = 500.09\left[\frac{.12}{1-1.12^{-5}}\right] = 500.09\,[0.27741] = \$138.73 \text{ per year}$$

It similarly follows for the numerical case of the down ramp that the future and annual worth are:

$$F(.10) = 12{,}437.61\,(1.10)^5 = 12{,}437.61\,(1.61051) = \$20{,}030.90$$

$$A(.10) = 12{,}437.61\left[\frac{0.1\,(1.10)^5}{(1.10)^5 - 1}\right] = 12{,}437.61\,(0.26380) = \$3{,}281.01$$

per year, verifying the foregoing formulae.

Net present worth analysis should be made at the minimum attractive rate of return (MARR) as specified by management to determine economic viability. Management usually sets MARR at the average cost of borrowed capital plus an allowance for risk in overall estimation. It was also stated earlier that when cash inflows are signed positively and outflows negatively, then more net present worth is better than less present worth. These same remarks apply to future and annual worth. Clearly, a negative net present worth is an economically-unacceptable project and one that should not be accepted unless there are only worse choices. Typically, there is the choice of doing nothing different at zero cost or gain. So it would seem that when facing two or more exclusive alternative projects, one should select the project with the greatest net present worth. If the alternatives have equal lives, then the project with the greater net present worth is the correct choice. However, with different lives, it is economically inappropriate to select the project with the greatest net present worth. The reason is that more present worth occurs with longer lives. Accordingly, the way to make comparisons between alternative projects with different lives is to assume a time frame equal to the least common multiple (LCM) of the contending projects and each project is to be repeated within the LCM years. For example, project A may be repeated twice and project B may be repeated three times and the total net present worth of each case can be compared because the total time period is the same.

Consider one option where the project is expected to require a $1,000.00 investment, yield $600.00 a year for a life of five years, and where a $300.00 salvage value is anticipated. An alternative requires a $1,100.00 investment, but it has a ten-year life where it is expected to return $500 each year in cost savings and, at the end of its life, yield a salvage value of $400. The LCM of these two alternative projects is 10 years. The first alternative with a repetition after five years has a present worth of:

$$-1{,}000\,[1 + 1.1^{-5}] + 600\left[\frac{1 - 1.1^{-10}}{0.1}\right] + 300\,(1.1^{-5})\,[1 + 1.1^{-5}] =$$
$$-1{,}620.92 + 3{,}683.74 + 307.94 = \$2{,}368.00$$

The alternative project over ten years has a net present worth of:

$$\{-1{,}100 + 500\left[\frac{1 - (1.1)^{-10}}{0.1}\right] + 300\,(1.1)^{-10}\} = \$2{,}088.00$$

It follows that the shorter project is more economical and should be selected over the project with the longer life. An alternative method of calculation that saves the use of the least common multiple is to use the annual worth and compare the annual worths directly as:

$$\{-1{,}000 + 300\,(1.1)^{-5}\}\left[\frac{0.1}{1-(1.1)^{-5}}\right] + 600 = \$385.34/\text{year}$$

$$\{-1{,}100 + 300\,(1.1)^{-10}\}\left[\frac{0.1}{1-(1.1)^{-10}}\right] + 500 = \$339.80/\text{year}$$

These two annual worths can be converted back to a present worth over ten years by using the reciprocal of equation 9 for annual worth or 6.14457 and the two results are respectively 385.34 × 6.1446 = $2,368.00 and 339.80 × 6.1446 = $2,088.00, which is exactly the earlier result, except for a little round-off difference. This numerical illustration shows that when the lives of alternative projects are different, it is easier to compare them using annual worth analysis, and that both annual worth and present worth provide the same answers each time.

INSPECTION ECONOMICS

Since inspection is principally an information-gathering process, the primary concern of design for inspection is the elimination of production errors. To do this well, one needs to reduce errors in inspection, as they mislead quality management. A second, but important, consideration is to minimize the cost aspects due to inspection. Economic considerations, particularly, need to be considered when there is a chance of product liability problems.

A simple cost model of inspection may be constructed using the cost of the inspection station per unit of time, costs of the two types of errors, and the cost associated with downsteam production slowdowns due to an insufficient number of items passing the inspection station per unit of time. Figure 24-1 shows a decision tree of the expected number of items per unit of time. In this figure, q is the fraction of defectives or fraction-lot defectives, as the term is typically applied. The reader will likely recall that type-1 inspection errors occur when the inspector rejects items that meet specification and type-2 errors occur when the inspector accepts defective items. Accordingly, a type-1 error only occurs when the item is good, so the probability of a type-1 error or P_1 is a conditional probability given that the item being inspected is good. Thus, the joint probability that a good item is presently being inspected and that a mistake is made is $(1-q)\,P_1$. It similarly follows that P_2 is the conditional probability of a type-2 error and the joint probability

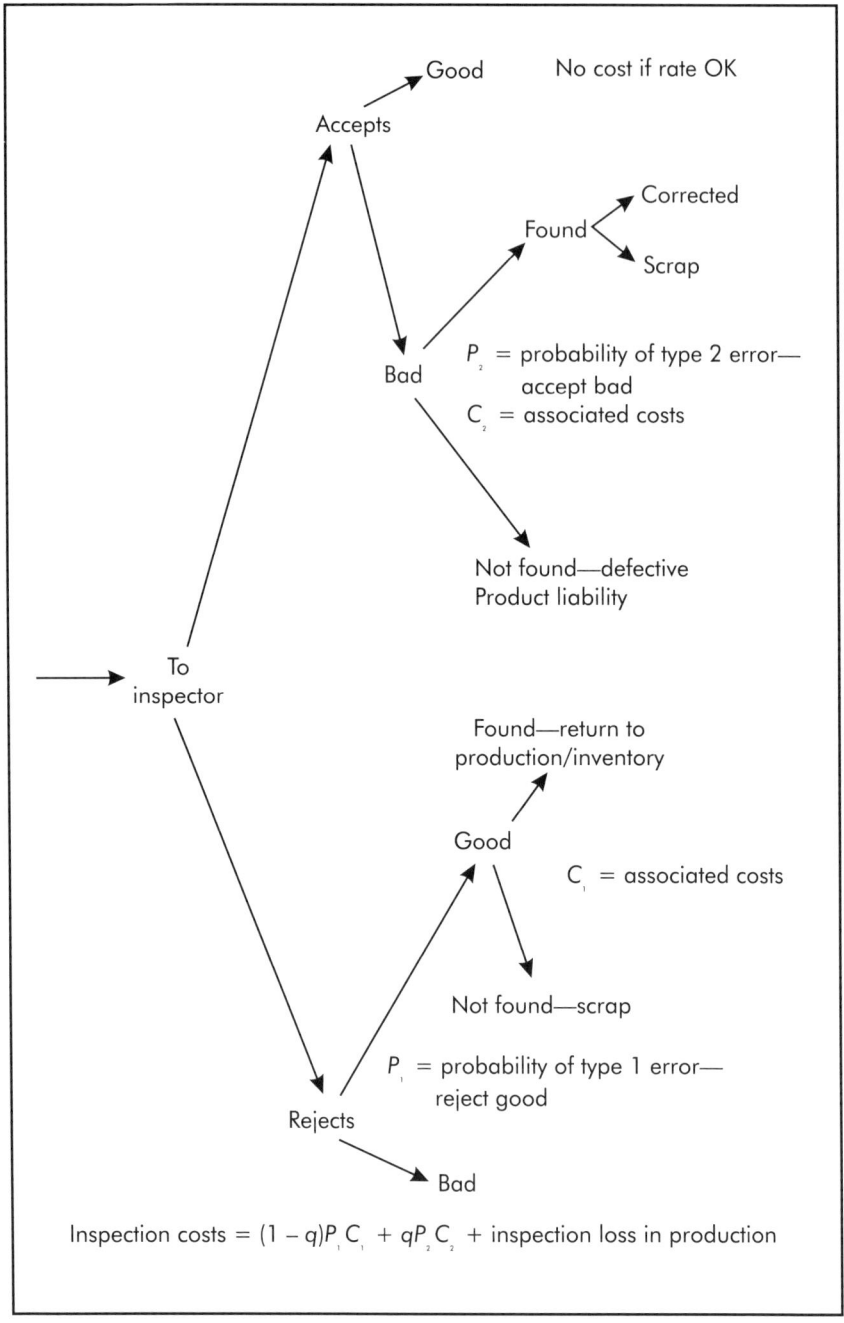

Figure 24-1 Inspector decisions, inspection errors, and associated costs.

that a type-2 error will be made is qP_2. The N units to be inspected come to the inspector, who either accepts or rejects each item, and those N items fall into four piles: N_1 rejected good items (type-1 errors), N_2 accepted bad items (type-2 errors), N_3 accepted good items, and N_4 rejected bad items. That is,

$$N = N_1 + N_2 + N_3 + N_4 =$$
$$N = N(1-q)P_1 + N(1-q)P_2 + Nq(1-P_1) + Nq(1-P_2)$$

It follows that N_1 and N_2 are the expected number of errors of types 1 and 2 respectively.

INSPECTION COSTS

Figure 24-1 also shows some of the features of inspection error costs. In the case of type-1 errors, there is a cost of finding out an inspected item is good and bringing it back into production or inventory. There is also the cost of not finding good items, which is the cost of materials and all work that went into the item before that inspection, less any scrap values realized. Even if those two separate costs can be separately estimated, there is also the need to estimate the fraction f_1 of good items found in the rejects. Accordingly, the average cost of a type-1 error is:

$$C_1 = f_1 \text{ (cost of finding and returning)} +$$
$$(1 - f_1) \text{ (materials + labor costs - salvage)}$$

There are two elements to the cost of a type-2 error as shown in Figure 24-1. One of those components is the cost of finding a bad item in the accepted group and either correcting or scrapping it. The other component is the defective item that goes out to the public. Clearly, some items that get into the hands of the public are simply returned to dealers who replace the items and return them to the company. Other defective products may get out and hurt someone and a fraction of these can wind up as product liability cases. While those products that cause product liability cases become extremely costly, the fraction that get this far is excessively low. Assuming that f_2 is the fraction of type-2 errors that are discovered prior to leaving the factory and that a single expected cost can be estimated for all product units leaving the factory, then the cost of a typical type-2 error can be found:

$$C_2 = f_2 \text{ (cost finding and repair/scrap)} +$$
$$(1 - f_2) \text{ (average product liability item)}$$

It follows that:

$$\text{Inspection error costs} = N \{(1 - q) P_1 C_1 + q P_2 C_2 \} \qquad (10)$$

which does not cover the cost of the inspector, any effects of slow inspection on reducing production rates, and inspection station costs, which are additional. It is obvious from equation 10 that reductions in P_1 and P_2 will reduce these inspection error costs. It is known from the literature (Smith and Barany 1971) that P_1 and P_2 change with q. However, neither P_1 nor P_2 are known to change with the variable N, which is the number of items inspected as long as time is not an associated factor.

RATE CONSIDERATIONS IN INSPECTION

When delivery to the inspector is by belt, V is the velocity in distance units per unit of time and S is the average interspacing between successive items for inspection. Both the velocity and distance units should be the same. In the case of dynamic visual inspection, where items are to be inspected as they are delivered by a moving belt, both P_1 and P_2 increase with greater belt velocities V and/or shorter distances S between successive items. A number of studies verify that type-1 and type-2 error probabilities change with the amount of time spent per item, regardless of dynamic or static visual inspection.

The number of accepted good (nondefective) items in this stream of product or component items is N_3 per unit of time. If those items go immediately back into production operations, then production can move along provided that N_3 is sufficiently large. Otherwise, costs will increase with the shortfall of production items. If type-1 errors are all correctly acted upon during a second inspection, then an expected N_1 more items will resume production, but after a delay for re-inspection and repair. In any case, it would be interesting to embed into a computer simulation, estimates* of the effects of the belt velocity V and item interspacing distance S on type-1 and type-2 inspection errors, along with any effects of these variables on the time allowed per item and any added effect of that time variable on inspection-type errors. Once estimates of these costs are further added, then the resulting simulation would provide a vehicle to determine how to operate the inspection system.

---

*If there are rework activities in parallel with the inspection station, then that rework and repair station will get the rejects from the inspection station and repaired items could be reinserted into the production stream in parallel with the inspection station. Since rework and repair is expected to be highly varied with regard to time, the added reflow could be quite varied, but dependent upon the rate of inspection.

LOCATION OF THE INSPECTION STATION

Another question in inspection operations is where to locate inspection stations if those stations are different from ordinary production operations. One clear approach to this analysis is to use digital computer simulation. Five different production operations are indicated in Figure 24-2 by squares numbered 1 through 6. Operation 5, denoted by a diamond shape, is an operation where components are joined in assembly. The two different inspection operations are shown by circles. Four different schemes are pre-

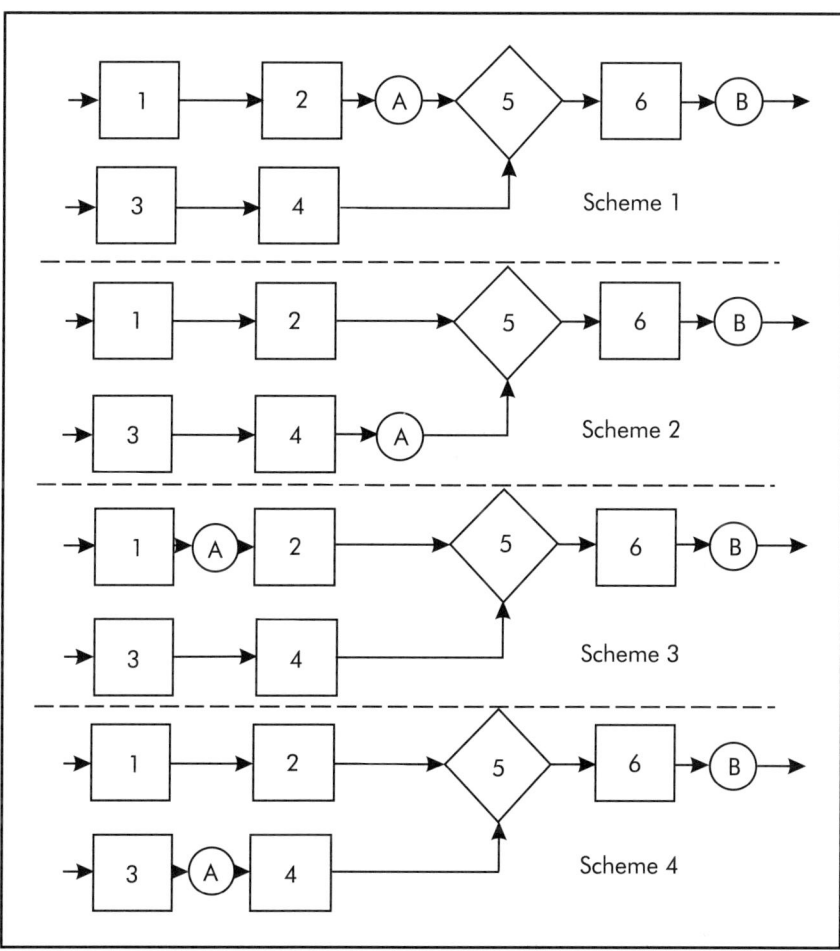

Figure 24-2 Alternative locations in a production layout for placement of an intermediate inspection station.

sented with different locations of inspection station A, and all other operations are the same. Numerous simulations of these four schemes were made using different probabilities of type-1 errors and associated type-2 errors based on the results of several studies.

Figure 24-3 presents the economic results of these simulations in terms of an equivalent annual cost.[†] Since type-1 inspection errors reject good items, many such errors will starve downsteam operations and create added costs. Note that in the cases shown here, no matter what the probability of type-1 errors is, scheme 3 has a lower cost than any other scheme. Also, scheme 4 is similar to scheme 3 but quite different from schemes 1 and 2 where inspection station A is directly in front of the assembly operation (no. 5). This result suggests that starving the assembly operation is more costly. This problem is only intended to show how simulation can be used to evaluate the economics of ergonomics problems.

ECONOMICS OF LEARNING

One learning curve model[††], proposed earlier, was the exponential model. In this model, the time required on the nth trial (t_n) is related to the time on the $n^{th} + 1$ trial (t_{n+1}) as:

$$t_{n+1} = \alpha t_n = \beta \tag{11}$$

where α and β are constant parameters of this model. Notice that the time values form a series where the current values are multiplied by the constant α ($0 < \alpha < 1$) and then the constant β is added. Accordingly, the series is sometimes known as the α/β series. The first forward difference is:

$$\Delta t_n = t_{n+1} - t_n = [\alpha - 1] t_n + \beta \tag{12}$$

An alternative expression to equation 11 is:

$$t_n = \alpha^{n-1} [t_1 + t^*] + t^* \tag{13}$$

[†] *If the present worth of equation 4 is equated to the actual present worth and over the actual duration, then the A parameter value of the uniform series gives the annual cost. Alternatively, substitute the present worth in equation 5 for the P parameter and compute A.*

[††] *This learning curve model is known as a discrete exponential model. The more popular power form model is $t_n = t_1 n^{-s}$, where s is the constant learning rate constant, t_1 is the time on the first trial, and n is the number of trials up to this point.*

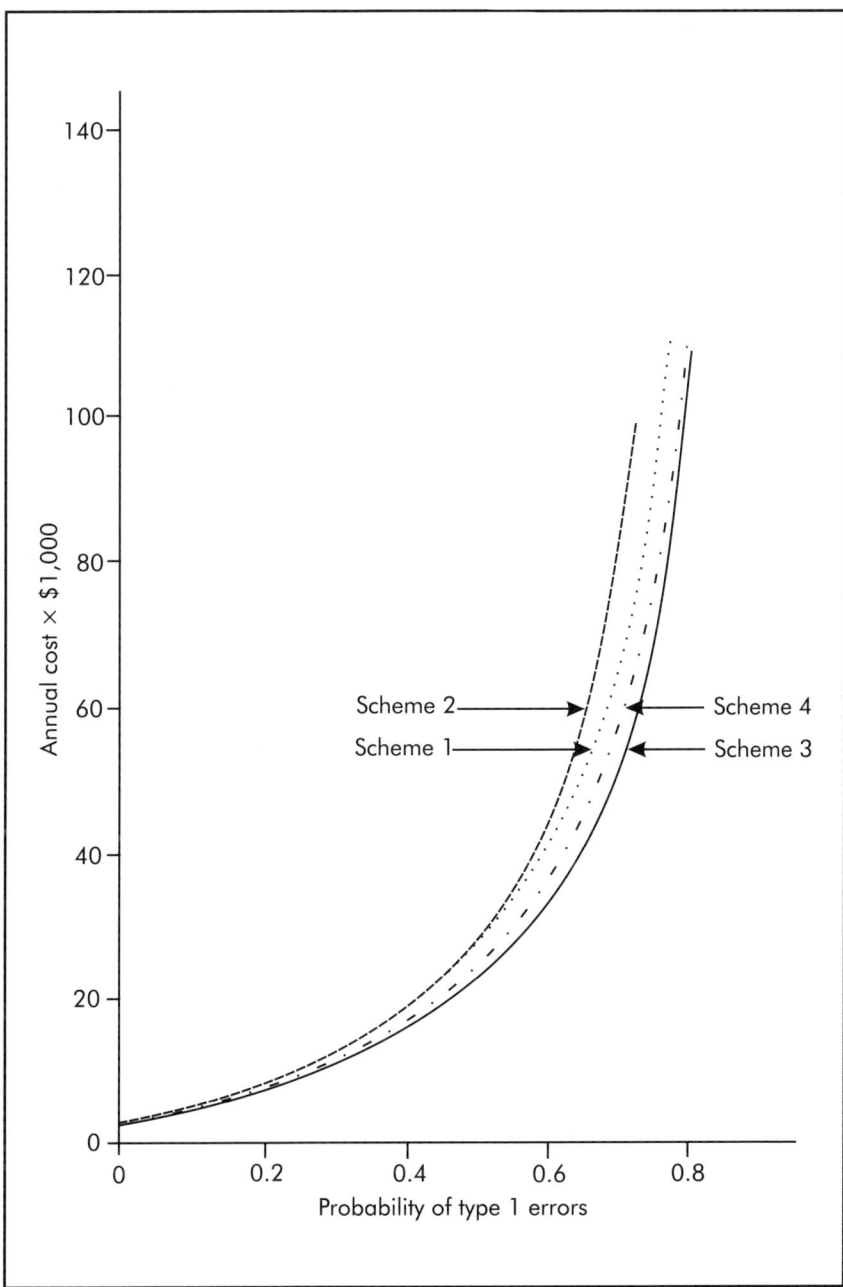

Figure 24-3 Annual costs of inspection operations using the four alternative locations of the inspection station as a function of the probability of type-1 inspection errors.

where t^* is the asymptote of performance times with this exponential model. As α is a fraction ($0 < \alpha < 1$), a larger number of trials n makes the first term smaller and smaller so that only the t^* value remains. This α/β series model‡ has a sum from the first through the kth element, which is:

$$\sum_{n=1}^{k} t_n = \frac{\alpha^k - 1}{\alpha - 1}[t_1 - t^*] + kt^* \tag{14}$$

It follows that the *cumulative average* over k elements of the series is equation 14 divided by k. Note that there are two terms on the right-hand side of equation 14. The first term describes the extra time required by the learning process (sometimes called "transitory" time) and the second term describes the time required after all learning effects are removed (called "steady state" time). Consider the example where normal time measurements were made on a human operator and converted by speed rating to a standard operator making successive product units. Those time values were:

$n =$	1	2	3	4	5
$t_n =$	4.000	3.960	3.921	3.882	3.845
$n =$	6	7	8	9	10
$t_n =$	3.808	3.772	3.736	3.702	3.668

where the time units of t_n are hours. Fitting the exponential learning curve to these data can be done by first computing Δt_n based on equation 12:

$n =$	1	2	3	4	5
$\Delta t_n =$	−0.040	−0.039	−0.039	−0.037	−0.037
$n =$	6	7	8	9	10
$\Delta t_n =$	−0.036	−0.036	−0.034	−0.034	—

When Δt_n is made a straight-line function of t_n, then the resulting linear equation‡‡ is:

---

‡The reader may also recognize this model as a *first-order, forward-difference with constant coefficients*.

‡‡A good way to perform the line fitting is by using linear regression analysis. Many statistical computer packages find linear regression equations.

$$\Delta t_n = 0.0423 - 0.0206 t_n$$

As equation 12 shows, the vertical intercept of this equation is equal to the parameter β, which is 0.0423, and the slope of this equation equals α − 1. Therefore[†], α is 0.9794. In this fitted model, the asymptote t^* is β divided by 1 − α or 2.0534. Since t^* is also the horizontal intercept, that value can be computed as:

$$0.0423 - 0.0206\, t^* = 0$$

and the result is the same. Parameter β may be simply computed as (1 − α) times t^*, or the product 0.206 (2.053), which is 0.0423. Normal practice in the production of this product is to have two operators work on it directly until the order size has been completed and then the whole order is shipped. With 50 product units produced by each operator, each individual standard operator would need to spend:

$$\frac{.9794^{50} - 1}{.9794 - 1} [4.0 - 2.053] + 50\,(2.053) = 163.7 \text{ hours}$$

Since these time units are in actual working hours, distinct from standard hours, and there are 6.8 actual hours in a shift, the time required per operator is 24.07 shifts of 8 standard hours each, or a total of 192.56 standard hours. Based on a five day work week and a single eight standard-hour shift each day, it will take four weeks, four days, and about 0.6 hours to complete the production run of 50 units. Note that at full production, there are 52 weeks per year and 40 standard hours per week, for a total of about 2,080 standard hours per year. If each operator earns $20.00 per hour, including fringe benefits, then the two assigned operators are costing the company:

2 operators × $20/operator/hr × 40 hrs/wk = $1,600.00 per wk
$1,600.00/wk × 52 wk/yr = $83,200.00/yr[††]

[†]*Note: this annual amount is the cost for two operators over the entire year neglecting vacations, time off, etc. From the company's viewpoint, the $83,200.00 is the annual cost of manual labor for this job.*

[††]*As Buck and Cheng (1993) show (see Bibliography), sometimes there is enough noise in the learning data that it does not turn out to be a fraction. Other learning curve models can be used in that case or the performance time data can be converted to cumulative average times by adding up t_n values and dividing by n.*

If the minimum attractive rate of return is stated to be 15% before taxes, then the present worth of this operation is:

$$\frac{\$83{,}200.00}{0.15}[1-(1.15)^{-.09258}] = \$7{,}131.00$$

This formulation of the present worth is based on the 192.56 standard hours as being 0.09259 years (using 2,080 standard hours per year). It is assumed in this formulation that the two operators are new employees who start producing the items in question and then ship them as soon as completed.

An alternative approach would be to use one operator and just deliver the production later. In that case, the required number of hours under this learning curve can be determined using equation 14:

$$\frac{.9794^{100}-1}{.9794-1}[1.947] + 100\,(2.053) = 288.02 \text{ working hours}$$

and those hours require slightly over 42 shifts or 0.16292 standard years at the same rate per operator as above or $41,600. The equivalent present worth is:

$$\frac{\$41{,}600}{0.15}[1-(1.15)^{-.13849}] = \$6{,}243.00$$

Use of a single operator saves about $888.00 (that is, $7,131.00 less $6,243.00) in present worth. If the single operator could start production early enough so that there is no late delivery costs, it is economically better to use a single operator because of the continued learning advantage. However, if production cannot begin earlier and there is a cost of late delivery because of other production requirements, then there is a trade-off between continued learning and the delayed production costs.

TRAINING AND TRANSFER

Suppose that an on-the-job training program was being considered for the job noted previously. The learning curve for the standard operator is an exponential learning curve where $\alpha = 0.9794$, $t^* = 2.053$, $\beta = 0.0423$, and $t_1 = 4.000$ hours. This planned training program was empirically tested on a couple of people who were initially rated to reflect the standard human operator. It was estimated that the training program lowered several parameters of this learning curve. One of those parameters was the learning rate, reducing α in the learning curve from 0.9794 to 0.95 because individuals knew what to look for and what to do to improve their own performance. This training program also provided information about the job so that individuals could start the on-the-job training phase with lower initial cycle times of 3.5 hours. In addition, the training program illustrated a newly proposed method for performing the task, which was expected to improve ultimate performance times to 2.0 hours per unit. Thus, the new learning curve following the training program was estimated for 100 cycles to take

$$\frac{.95^{100} - 1}{.95 - 1} [3.5 - 2.0] + 100 (2.0) = 229.82 \text{ working hours}$$

In terms of standard time after the training program, these working hours require 229.82/0.85 or 270.40 standard hours in 33.80 standard shifts over 0.1296 years. The present worth of the on-the-job learning based on 2,086 hours/year and $20.00/hour is:

$$\frac{\$41,600}{.15}[1 - (1.15)^{-.13}] = \$4,993.38$$

Time requirements without the training program are shown to be 288.05 hours compared with 229.82 working hours. This training program saves $6,243.53 minus $4,993.38 or $1,250.15 less the cost of the training and salary during the training. The training program required four standard hours and $100 to run it. Also, there is a cost of the trainees at

$$4 \text{ hours} \times \$20/\text{hour} + \$100 = \$180.00$$

giving a total present worth of $5,173.38 compared to the $6,243.53 without the training program. Thus, the expected present worth savings are $1,250.15 less $180.00, or $1,070.15. While the transfer effectiveness ratio (TER)* of this training program is:

$$\text{TER} = [338.88 - 274.40]/4 \text{ standard hours training} = 16.12$$

which states that the training was over 16 times as effective as on-the-job experience. Unfortunately, the TER measurement is not as meaningful to managers as cost savings.

Frequently, an operator performs an assembly operation until the quantity of the production run is completed. That operator works on other jobs until a new production run is started. During the time between sequential runs, there is an erosion of operator

---

*See Wickens (1992) p. 239 for TER calculations, which are hours saved divided by training hours. Since training was in standard hours, so were the hours saved.

skills specific to the particular product. Forgetting curves are estimates of that skill erosion for various durations between runs. Some recent studies show that operators who have acquired the skills with a sufficient amount of production will follow the forgetting curve, but once back on the same task, their learning rate on the second production run is much greater and the second learning curve tends to return to the original learning curve where performance time per piece ended on the first production run. While more research is needed to establish the relationships between repeated learning curves, the economic implications are very interesting.

LIFTING BELTS AND ECONOMICS

Should management require all persons who lift things to wear lifting belts when it is still unclear how much good these belts do? At the writing of this chapter, there was about as much positive evidence for these belts in the literature as negative evidence. Many companies were equally unclear about the efficacy of belts. However, the cost of a belt is about $20 to $40.

What should the ergonomist recommend? From an economics viewpoint, this may be a tough call because it is outside of typical engineering economics analysis. Cost magnitudes are very complex and there is risk. In the case of low-back pain due to lifting, a few chronic cases constitute about 90% of most industrial medical disabilities, and that makes a case of low-back pain potentially a very expensive legal situation. If one of those people sues, the costs may seem prohibitive. Accordingly, belts appear to many companies to be inexpensive insurance against an attorney who says that the company was insensitive to the industrial health of the workers. Most companies recognize that it will be several years before the efficacy of lifting belts will really be known. Consequently, it is better to make the small $20 to $40 investment now, than to have a potentially difficult lawsuit erode many thousands of dollars because the company appeared irresponsible by not insisting that employees wear lifting belts. While some people might call this "legal blackmail," it is quite true that past histories have shown the costs of low-back pains to be extreme. The point is that situations with small costs at the front end and huge possible costs at the rear, tend to slip by economics analysis. On-the-other-hand, if future research proves lifting belts to be effective in reducing the incidence or severity of low-back pain, then these executives will act as if they knew it all the time. Otherwise, they will say they insured to protect the company.

FINAL REMARKS

As professionals in industry, ergonomists propose projects that require investments. If those investments earn or save enough money over the project life and the money is available, those projects will be undertaken by management. Exceptions to this rule occur when the companies are required by law to do something or there is a large legal or financial risk associated with a relatively small investment. Some of these legal problems occur with Occupational Safety and Health Administration (OSHA) regulations or the new Americans with Disabilities Act (ADA), which states that industry shall make industrial jobs accessible by handicapped persons unless it is uneconomical to do so. There again, the burden of economic proof is on the company.

REFERENCES

Bussey, L.E. *The Economic Analysis of Industrial Projects*. Englewood Cliffs, NJ: Prentice-Hall, 1978.

DeGarmo, E.P., Sullivan, W.G., and Canada, J.R. *Engineering Economy*, 7th. Edition. New York: Macmillan, 1984.

Fleischer, G. *An Introduction to Engineering Economy*. Boston, MA: PWS Publishing, 1994.

Newnan, D.G. *Engineering Economic Analysis*, 4th Edition. San Jose, CA: Engineering Press, 1991.

Park, Chan S., and Sharpe-Bete, Gunter P. *Advanced Engineering Economics*. New York: John Wiley & Sons, 1990.

Rose, L., Ericson, M., Ghlimskar, B., Nordgren, B., and Ortengren, R. "Ergo-index—Development of a Model to Determine Pause Needs After Fatigue and Pain Reactions During Work." M. Mattila and W. Karwowski, eds. *Computer Applications in Ergonomics, Occupational Safety and Health*. Amsterdam, The Netherlands: Elsevier Science Publishers R.V., 1992: 461-46.

Simpson, G.C. "Applying Ergonomics in Industry: Some Lessons from the Mining Industry." E.J. Lovesey, ed. *Contemporary Ergonomics* 1993. Edinburgh, Scotland: Ergonomics Society, 13-16 April, 1993: 490-503.

Smith, L.A., and Barany, J.W. "An Elementary Model of Human Performance on Paced Visual Inspection." *American Institute of Industrial Engineers Transactions*. Norcross, GA: Institute of Industrial Engineers 1971, 4: 298-308.

Theusen, G.J., and Fabrycky, W.J. *Engineering Economy*, 7th Edition. Englewood Cliffs, NJ: Prentice-Hall, 1989.

White, J.A., Agee, M.H., and Case, K.E. *Principles of Engineering Economic Analysis*, 3rd Edition. New York: John Wiley & Sons, 1989.

BIBLIOGRAPHY

Buck, J.R., and Hill, T.W., Jr. "Alpha/Beta Difference Equations and their Zeta Transforms in Economic Analysis." *American Institute of Industrial Engineers Transactions*. Norcross, GA: American Institute of Industrial Engineers, 1975, 7(3): 330-358.

Buck, J.R. *Economic Risk Decisions in Engineering and Management*. Ames, IA: Iowa State University Press, 1989.

Buck, J.R., and Cheng, S.W.J. "Instructions and Feedback Effects on Speed and Accuracy with Different Learning Curve Models." *I.I.E. Transactions*. Norcross, GA: Institute of Industrial Engineers 1989, 25(6): 34-47.

Buck, James R. *Economic Risk Decisions in Engineering and Management*. Ames, IA: Iowa State University Press, 1993.

Cronback, L.J., and Cleser, G.C. *Psychological Tests and Personnel Decisions*, 2nd Edition. Urbana, IL: University of Illinois Press, 1965.

Drury, C.G. "The Effect of Speed of Working on Industrial Inspection Accuracy." *Applied Ergonomics* 1973, 4: 2-7.

Ezey, M. Dar-el, and Vollichman, R. "Speed vs. Accuracy Under Continuous and Intermittent Learning." A.F. Ozok and G. Salvendy, eds. *Advances in Applied Ergonomics*. 1st International Conference on Applied Ergonomics. West Lafayette, IN: U.S.A. Publishing Corporation, 1996, May 21-24: 676-682.

Goldberg, S. *An Introduction to Difference Equations, Science Editions*. New York: John Wiley & Sons, 1961.

Hill, T.W., Jr., and Buck, J.R. "Zeta Transforms, Present Value, and Economic Analysis." *American Institute of Industrial Engineers Transactions*. Norcross, GA: Institute of Industrial Engineers, 1974, 16(2): 120-124.

Imker, F.W. "The Back Support Myth." *Ergonomics in Design* April, 8-12, 1994.

McGill, S.M, Norman, R.W., and Sharratt, M.T. "The Effect of an Abdominal Belt on Trunk Muscle Activity and Intra-abdominal Pressure During Squat Lifts." *Ergonomics* 1990, 33: 147-160.

Pope, M.H., Andersson, G.B.J., Frymoyer, J.W., and Chaffin, D.B. *Occupational Lowback Pain: Assessment, Treatment, and Prevention*. Saint Louis, MO: Mosby, 1991.

Rizzi, A.M., Buck, J.R., and Anderson, V.L. "Performance Effects of Variables in Dynamic Visual Inspection." *American Institute of Industrial Engineers Transactions*. Norcross, GA: Institute of Industrial Engineers 1979, 11(4): 278-285.

Tanchoco, J.M.A., Buck, J.R., and Egbelu, P.J. "Analysis of Capital Expenditure Projects with Uncertain Discrete Cash Flows." *Omega* 1982, 10(3): 1-9.

Wentworth, R.N., and Buck, J.R. "Presentation Effects and Eye-motion Behaviors in Dynamic Visual Inspection." *Human Factors* 1982, 24(6): 643-658.

Wickens, C.D. *Engineering Psychology and Human Performance*, 2nd Edition. New York: Harper Collins Publishers Inc., 1992.

CHAPTER 25

WORLDWIDE CORPORATE ERGONOMICS EFFORTS—USA

Brian Peacock
Manager, Manufacturing Ergonomics Laboratory
General Motors Corporation
Warren, MI

THE GROWTH OF ERGONOMICS

The first half of this century saw a demand for human science for the purpose of productivity. Scientific management focused on manual skills, as the pursuit of efficiency saw the change from craft-based industry to the production line. The medical profession and physiologists became concerned with human physical capabilities and limitations with particular attention to manual work and fatigue. Industrial psychologists addressed the effects of organizational, social, and environmental factors on employee well-being and motivation. The second world war drew particular attention to anthropometry and cognitive performance, with a focus on display and control design. The engineers and psychologists joined forces to study control and communication theory under the collective name of *cybernetics*.

The evolution of ergonomics over the second half of this century has been influenced by a series of significant events that have changed its academic basis, professional practice, and public image. The consumer movement first precipitated a rise in interest in product safety and later in opportunities for marketing products that have "ergonomic" characteristics. The power generation and transportation industries became interested in the factors that cause human error. One major stimulus for this activity was the Three Mile Island nuclear plant incident, which drew attention to human fallibility when faced with complex information handling under stressful conditions. The development of mental work load measurement resulted from increased complexity in such areas as military systems and air traffic control. The quality movement addressed problems of inspection and equipment maintenance through application of vigilance and attention research. Throughout this period, the ubiquitous issues of anthropometry became the most tangible indication of human variability and the need for the application of probability and statistics to most ergonomics problems. The 70s and 80s saw a surge in interest in social factors as they affected individual and group productivity.

Two major factors stimulated the growth and division of ergonomics during the 80s and 90s. The personal computer and modern telecommunications—the information highway—created an unprecedented ability to collect, analyze, and transfer information. This stimulated a massive surge in interest in interfaces between the user and the plethora of electronic devices. "Usability analysis" became a cornerstone of human factors practice.

The second factor—the "epidemic" of cumulative trauma disorders—arose from two sources. Demands for product quality and productivity saw a resurgence in the scientific management approach to machine-paced, short-cycle, standardized work in both factory and office. Simultaneously, the information revolution gave rise to a phenomenal increase in the number of people tied to computer and telecommunications terminals—examples being the data processing and telemarketing industries. These two issues—informational technology and repetitive work—created polarization in the profession. The psychologists saw their crock of gold in computer icons and the health scientists in lower back and the carpal tunnel. The industrial engineers attempted to bridge the gap. Meanwhile, anthropometry and chair design continued to be the common ground for all concerned as it attempted to deal with the continuum from comfort to postural health.

THE PROFESSION

Technical developments were paralleled by growth in the associated professions. Academic interest continued with the parallel activities of psychologists, health scientists, and industrial engineers. The legal profession found its niche, not so much in the physical aspects of products, but rather in the absence or inadequate design of warnings. The legal profession also joined the debate surrounding (ill) health effects of work, attempting to categorize the multidimensional problem of the dose response relationship for the purpose of assigning blame. They clearly identified the irony that the principal focus of the profession—human variability—is also its Achilles heel. The military and space industries spanned the physical and cognitive domains with interests ranging from human error in complex vehicle or missile control to the physical demands of combat.

Organizations

The principal United States ergonomics organization is the Human Factors and Ergonomics Society, which has a membership of more than 5,000. The Industrial Ergonomics Technical Group is one of the largest subdivisions of this society and consists of professional ergonomists from academia, industry, consulting, and government. The Institute of Industrial Engineers, which has a membership of over 20,000, also has a strong ergonomics component. Recently, the ergonomics division of that society merged with the Work Measurement Division, which underscores the strong relationship between these traditional industrial activities. The American Industrial Hygiene Association (AIHA) membership also has a strong interest in physical ergonomics and has been particularly active in the standards-setting debate. Most industrialized and many developing countries also have ergonomics societies. The oldest society is the British Ergonomics Society, formed in the 1940s. Its developments paralleled those of the U.S. Human Factors and Ergonomics Society with a blend of physical, informational, and macro ergonomics interests. Every three years the International Ergonomics Society holds a conference that serves as a meeting ground for academics and practitioners throughout the world.

There have been increasing efforts over recent years to establish professional certification for ergonomists. In the United States, the Board of Certification of Professional Ergonomists (BCPE) was established in the early 1990s. The first wave of membership involved the submission of credentials, which included a master's degree in ergonomics or a related subject, seven years professional or academic practice, and evidence of contributions to the analysis, test, and design components of ergonomics. Over 500 people qualified under this phase. The second phase included the requirement for a master's degree, five years practice, and an examination in the theory and practice of ergonomics. About 40 professionals per year are certified by these criteria. Professionally-certified ergonomics practitioners carry the designation CPE (Certified Professional Ergonomist) or CHFP (Certified Human Factors Professional) after their name. In Europe the Center for the Registration of European Ergonomists (CREE) was also formed in the early 1990s, with similar purposes and standards to the BCPE.

There are numerous journals in the field of ergonomics. The Human Factors and Ergonomics Society publishes the journals *Human Factors* and *Ergonomics in Design* as well as a monthly newsletter. In Britain, *Ergonomics* and *Applied Ergonomics* are the premier journals. *The Journal of Human Ergology* is published in Japan. Other journals include *The International Journal of Industrial Ergonomics* and *Human Factors in Manufacturing*. In addition, there are many journals with ergonomics content in the areas of biomechanics, psychology, occupational medicine, management, and engineering. Recently, a number of practitioner magazines have been introduced, including *Workplace Ergonomics* and *CTD News*.

POLARIZATION OF PHYSICAL AND COGNITIVE ERGONOMICS

The division of the profession into physical, informational, and macro compartments is further complicated by an orthogonal division into consumer product and occupational applications (Table 25-1). The consumer should buy a product or service with regard to its function, usability, and ergonomic characteristics. Unfortunately, these characteristics may be differentially weighted and frequently overruled by preference or fashion factors. The automotive design industry continually discusses these function and form trade-offs. Most human-made things are ergonomic most of the time, but all things are unergonomic under certain conditions. Consumers in our competitive society can vote with their checkbooks; in the long run they will become more demanding and informed with regard to function.

Table 25-1. Categorizations of ergonomics

Technical Area	Application Area	
	Consumer Products	Industry
Physical ergonomics	Comfort, convenience	Health, productivity
Information ergonomics	Human performance, error	Safety, quality
Macro ergonomics	Marketing, product liability	Motivation

The objects of occupational ergonomics are usually employees who have less control over the choice of their equipment and over their working conditions. The ergonomics decisions affecting their jobs are in the hands of surrogates, such as managers, purchasing officers, unions, and governments. Most professional ergonomists will advise a participatory approach to job and equipment design—turning the objects of interest into subjects. However, this participatory process is best used for fine-tuning purposes once the basic ergonomics rules have been applied and conditions of use and potential abuse have been described.

PHYSICAL ERGONOMICS IN INDUSTRY

The remainder of this discussion will focus on the physical ergonomics domain in industry. Table 25-1 indicates that the principal objective of physical ergonomics in industry is the simultaneous improvement of health and productivity, with an emphasis on the prevention of cumulative trauma disorders. People familiar with the modern manufacturing industry will be aware of the resurgence of enormous pressures to become more efficient and competitive in production. This emphasis has, to some extent, replaced the focus on quality that occurred in the 70s and 80s, although nowadays high quality is the price of market entry.

ORGANIZATIONAL FACTORS

Issues of quality are not usually seen as being the domain of contemporary ergonomics—rather they are dealt with by engineering approaches to error proofing. These involve not only engineering design to reduce error, but also the trend toward short-cycle work so that operators can become highly skilled very rapidly. This trend perpetuates the vicious circle of repetitive motion strain and the more insidious problems of mindless work, leading to motivational and unwanted compensatory problems. Despite these peripheral problems of quality and motivation, the practice of industrial ergonomics currently converges on the physical problems of repetitive work.

In most companies, ergonomics is found in the Health and Safety department, usually part of the personnel activity. This arrangement may be appropriate for reactive programs but, for proactive programs, there is a stronger argument to locate ergonomics within the engineering activity. The principal source of ergonomics education in universities is within the Industrial Engineering departments and this is the logical location in manufacturing industry, if such a department exists. In recent years, industrial engineering in many companies has been amalgamated with manufacturing engineering and the focus on time and method study has been passed on to the production organization, often through the medium of work teams.

In organizations with short cycle production lines and labor costs that are a relatively high portion of overall product cost, the traditional industrial engineering time study and line balance activities still remain. For example, machining and mechanical assembly of small components may involve cycles of 20 seconds or less, press shops and food processing may have 2-5 second cycles, whereas automobile assembly typically employs 1-5 minute cycles. These kinds of organizations require substantial input from traditional industrial engineering, and the opportunity to balance the spatial and force stressors with temporal exposure is best co-located within that department. Such arrangements may appear as a conflict of interest and purpose if there is not sufficient input from the production organization and employees themselves. In unionized organizations, this conflict often escalates to a major confrontation over staffing levels, with ergonomics being at the center of the debate.

Contemporary arrangements of manufacturing cells and work teams may help to resolve the conflicts. Such arrangements are conducive to job enlargement and rotation, and provide substantial autonomy for the work team to carry out ergonomics and task balance activity. In plants with these arrangements, the industrial engineer or ergonomics specialist fulfills an advisory role to the work team. Prime examples of such arrangements are to be found in Japanese transplants.

ERGONOMICS PROGRAMS

A contemporary view of ergonomics is that the benefits can be achieved by the implementation of ergonomics programs. For example, OSHA's "Meatpackers Guidelines," (OSHA 3121, 1990) the agreements that OSHA implemented in 1990 with the automobile manufacturers and their unions, and the more recent efforts by ANSI to develop a musculoskeletal disorders prevention standard, are generally programmatic in nature. They contain descriptions of the types of activities that "shall" or "should" be performed by organizations to reduce the incidence, severity, and cost of cumulative trauma disorders. The general content of these programs includes organizational approaches to obtain management commitment and employee involvement. They involve extensive training efforts, procedures for job analysis, and rules for hazard abatement. They also prescribe detailed medical management protocols, including record linkage between job analysis, intervention, and outcomes. A particularly controversial aspect of these programs is the level of reporting necessary to ensure that the program is being implemented and the desired outcomes are achieved. Observers generally appreciate that there may be many alternative approaches to solving the CTD problem. Consequently, the details of programs should be tailored to particular industries and situations.

Most programs at the time of this writing are reactive in nature. They address individual existing jobs. In reality, reactive ergonomics will continue, for the foreseeable future, to provide the major opportunity for intervention. Most companies do not have long development programs with new products and processes. They tend to use and modify existing facilities, equipment, and work methods. Some larger organizations, notably in the automotive industry, have a regular cycle of new product and process development. It is common for extensive retooling to take place at the time of model change. These changeover periods also provide the occasion for balancing production lines by arranging the content and sequence of jobs for both production and ergonomics purposes. The automotive companies, therefore, are somewhat advanced in their systematic integration of ergonomics concepts into their new product and process development cycle. Perhaps the greatest opportunity for proactive ergonomics lies in the use of the computer as an integrating device for product and manufacturing process design, production simulation, and anthropomorphic modeling.

The generally recognized contemporary purpose of manufacturing ergonomics is to reduce the incidence and severity of cumulative trauma disorders. The public view of the profession and standards is one of protectionism. The logical direction, therefore, is the reduction of physical stress, particularly in terms of posture, movement, force, repetition, and static activities. Clearly, this simplistic view is diametrically opposed to the generally accepted practices of physical fitness that seek to increase all of these stresses. The problem is confounded by the wide variability in human capabilities, limitations, and vulnerabilities. This dilemma is central to the debate regarding ergonomics standards and guidelines.

CUMULATIVE TRAUMA DISORDERS

Cumulative Trauma Disorders (CTDs) have many synonyms, each reflecting a particular characteristic of the clinical condition, or its etiology (Bernard 1997). Cumulative trauma implies repeated micro trauma as also does repetitive strain injury (RSI). A semantic and legal problem lies in the use of the words "disorder" and "injury." To some, an injury results from a single acute incident from a well-defined agent. On the other hand, CTDs are often classified as illnesses that may have a less well-defined cause. The expression *overuse injuries* (OUIs) gets closer to the point that the tissues may be damaged simply by exceeding normal use.

More recently, the term *musculoskeletal disorders* (MSDs) has focused attention on the site rather than the cause of damage. This site approach may be helpful in explaining cause and effect.

A fairly common experience is a blister or callous, associated with the skin, usually directly attributable to a source of external friction. Where the friction is particularly intense (acute), the lesion may be called an abrasion, which involves a well-described circulatory response and repair process. The formation of a callous clearly happens over time and involves a reaction of the skin tissue (thickening) to a repeated insult, such as walking in tight shoes.

Tendons (connections of muscles to bone), ligaments (connections between bones around joints), joint capsules (sheets of fibrous tissue surrounding synovial joints), bursae (fluid-filled sacks between tendons and bones), and tendon sheaths (lubricated tubes around tendons where they rub over each other or over bones) are, like the outer layers of the skin, less endowed with capillaries. Rather, they receive their nutrients through such structures as synovial membranes and general pervasion of fluid. A similar phenomenon to blister or callous formation may occur in the musculoskeletal systems when a particular structure is subject to repeated friction, tension, compression, or micro tearing. The particular site will be directly related to the mechanical stress site and form of damage. Because of the widespread differences in anatomical structure and micromotions between individuals, the precise sites and the form and rapidity of reaction may be different. Common sites include the junctions between tendons, bones, and muscles and where tendons are restrained by ligaments (retinaculae) or bones.

Muscle tissue and the intramuscular connective tissue, on the other hand, are more generously supplied with capillaries and, when subjected to strain or micro trauma, repair more quickly than tendons. A common observation is that muscles usually respond to repeated activity by getting bigger and stronger, if sufficient rest is provided between bouts of exercise. Similar growth responses may be observed at the junction between tendons and bone and in the tendons themselves. However, these sites may also succumb to chronic disorders such as "tennis elbow."

Other forms of cumulative trauma may affect the cartilaginous surfaces of bone ends in the joints, such as osteoarthritis. However, there is some evidence that this disorder may be as much a result of underuse as overuse. Damage to the peripheral nerves may be more indirect. For example, one theory of carpal tunnel syndrome is that the finger flexor tendons thicken either through overuse or in response to micro trauma. These thickened tendons, in turn, compress the median nerve as it passes through the narrow fibro-osseos canal in the wrist and interferes with the circulation to the nerve, resulting in interference with sensory and motor conduction.

A key characteristic of cumulative musculoskeletal disorders is that they vary in severity and that some people become more incapacitated than others. Some people have lower tolerance thresholds than others. It is not uncommon for highly motivated athletes to ignore such injuries until they either repair naturally (with judicious rest) or become intolerable. On the other hand, employees who may be less motivated to withstand discomfort, and who have less opportunity to avoid the external pacing of a production line, may have a lower threshold and be more likely to respond by seeking a legitimate medical relief. These so-called psychosocial contributions to cumulative musculoskeletal disorders have received considerable anecdotal attention, but they are not easily subject to unequivocal empirical study.

The comparison between industrial employees and athletes with regard to cumulative trauma is perhaps the most fruitful path to a solution to this enormous medical, economic, legal, and political problem. Baseball pitchers don't pitch every day and marathon runners don't compete every week. Most athletes adopt strategies of cross training, if only to reduce boredom. It is clear that production line employees, as well as data processing and checkout clerks, must benefit from assignment strategies that create physical job variety. Job enlargement and rotation are two such strategies. However, these strategies require careful planning to address both the physical content of the alternative work and the psychosocial response to change.

COMMERCIAL OPPORTUNITIES

The accelerated growth of physical ergonomics over the past few years has been fueled by many commercial opportunities. These opportunities have three main forms: training, job analysis, and physical devices.

TRAINING

The training surge started with the offering of short courses by university professors, then spread to the specialized consultants, and finally to a much broader range of related consultants with a narrower focus on physical task analysis and mechanical solutions.

It is now possible to obtain courses that last from one hour to over a year, as described in Table 25-2.

Table 25-2. Ergonomics training courses

Duration	Type	Audience	Content
1 hour	Awareness	Executives, employees	Politics, philosophy, and economics Personal involvement in job design
1 day	Introductory	Managers, engineers	The right questions to ask, rules and tools
1 week	Technical	Engineers	Analytical tools and solutions
1 month	Advanced	Practitioners	Theory, in-depth analysis
1 year	Professional	Specialists	Ergonomics leadership

Legislative and standards efforts have stimulated this major expansion in the ergonomics training field. It is generally understood that ergonomics cannot be done solely by ergonomists. Rather, product, manufacturing, and industrial engineers need to be trained to incorporate ergonomics concepts in their routine activities along with traditional concepts of engineering, cost, the environment, mass, and lead times. These engineers are often supported by ergonomics practitioners of varying levels of training, particularly where large amounts of detailed job analysis are deemed appropriate. It is becoming common for large companies to provide one specialist per location to fulfill the role of ergonomics specialist. Finally, supervisors, managers, and executives need training in the purpose, philosophies, politics, processes, and economics of ergonomics.

A general structure of ergonomics training involves course durations of one hour, one day, one week, one month, and one year. The one-hour, or awareness level, is usually appropriate for most company employees including management, engineering, union leadership, production, and support employees. The content and presentation form needs to be tailored to the particular audience with an emphasis on purpose, process, and outcomes. The one-day, or introductory level, is generally aimed at managers and engineers with some direct responsibility for managing or supervising ergonomics activity. The content includes an outline of the rules, tools, and techniques of ergonomics as applied to the particular industry. The one-week, or technical training, is aimed at individuals with routine responsibility for ergonomics, possibly as part of their product or manufacturing engineering jobs. The one-month advanced ergonomics training is aimed at individuals expecting to have full-time responsibility for ergonomics at the plant, divisional, or corporate levels. Most large companies employ ergonomists with professional training and certification.

JOB ANALYSIS

Contemporary ergonomics analysis methods involve checklists and work sheets of various levels of sophistication. They range from simple univariate screening devices to complex models involving the integration of such stressors as posture, force, and time. When these methods fail to solve the perceived or predicted problem, then more complex analyses may be used, including computer models or biomechanical, physiological, psychophysical, and survey devices. On rare occasions, the classical experimental approaches may be attempted. However, the complexity of manufacturing settings is such that these ideal approaches are usually not feasible, particularly when there are time and production constraints.

Checklists

By far the most widespread device is the checklist. There are probably more ergonomics checklists than ergonomists. Most cover a range of spatial, force, and exposure (time) factors, usually in one-dimensional form.

The answers to checklist questions range from qualitative and subjective to quantitative and objective. For example, a subjective checklist may ask the question: "Does the worker appear to be out of breath?" A more objective, but more difficult to answer question would be: "What is the average energy expenditure of the employee in kilocalories per minute over the working day?"

Some checklists attempt to integrate the various forms of stress by addition or multiplication methods that require answers to different questions. It is important to note, however, that complex analyses usually have to be decomposed into their contributing elements for solution prioritization and implementation. For example, most force factors are the province of a product engineer, most spatial factors—how the workplace is arranged—are the responsibility of a manufacturing engineer, and most time (exposure) factors—how much work is contained in the job cycle—are decided by the industrial engineer.

The checklist approach is akin to the stereotypical health maintenance organization (HMO) approach in health care where standard responses are applied to standard sets of conditions. For the most part, these approaches, particularly if they are quantitative, satisfy the expectations of management, unions, and employees. Furthermore, they can be applied by *monitors*—engineers or production employees with minimal background training in ergonomics. However, the complex interactions between the work conditions, individual susceptibility, and the circumstances of a task is such that standardized ergonomics analysis and intervention is not always successful. An alternative approach is to employ more highly trained and experienced ergonomists or interdisciplinary teams who use less structured and more rapidly-convergent analysis methods. Such approaches, in both the reactive and proactive mode, are more likely than the former structured approach to identify complex interactions.

The NIOSH Lift Equation

A widely-used and abused analytical device is the National Institute for Occupational Safety and Health (NIOSH) Lift Equation, which has 1981 and 1991 versions (Waters et al. 1993). This is a remarkably useful device in that it integrates the major physical factors in manual material handling, and indicates which factors have the greatest contribution to the overall stress index. The equation, developed by a professional consensus of leading ergonomists, is based on the best available epidemiological, biomechanical, physiological, and psychophysical evidence.

A major advantage, and ultimate disadvantage, is that the equation is very easy to use. This has led to widespread misuse in conditions beyond the scope of the equation by practitioners who are not cognizant of its theoretical basis and limitations. In the hands of well-trained individuals, the equation can contribute to a very useful understanding of the physical stresses involved in manual material handling—arguably the most important cause of occupationally-induced, low-back pain.

Anthropomorphic Models

The more advanced analytic techniques range from computer-based anthropometric, biomechanical, and physiological models of human beings at work to video, electrogoniometers, and electrophysiological techniques. Anthropomorphic modeling is experiencing a surge in interest and is finding its roots in robotics programming. Unfortunately, the devices available to date are more like robots than people in that they do not address the variability of shape and size, and their joints and joint interactions are much less complex than human joints. However, if the purpose is limited to the simple visualization of a human figure in a three-dimensional spatial work environment, then the current models may be adequate.

A second shortcoming of these current, simple models is that they are quite difficult to develop and require a good knowledge of robot simulation languages, as well as human form and function. The challenge for companies that aspire to lead the application of these techniques is the concurrent development of the technology (hardware and software), people trained in ergonomics and computer graphics, and an organizational structure that integrates this technology into the existing product and manufacturing system development processes.

Biomechanical Models and Methods

The leading contemporary biomechanical models have been developed by the University of Michigan Center for Ergonomics. These two- and three-dimensional devices are essentially static models in that they do not account for the complex effects of inertia on human function and vulnerability. However, in trained hands, they are remarkably powerful analytic tools. The biomechanical device world encompasses video and electrogoniometers, such as the "lumbar motion monitor" developed at Ohio State University, which describes joint movement and electromyography. These sophisticated tools are backed by a wide variety of data reduction and analytic software, which aspire to identify conditions that could approach the limitations of the human musculoskeletal system. They do not, however, address the difficult issue of human variability. They are not calibrated to measure conditions of actual musculoskeletal failure and require advanced training for their users.

PHYSICAL DEVICES

The physical device opportunities range from external mechanisms that place materials in convenient locations, to devices that assist the movement of materials, to orthotic devices that support various joints. The placement devices include lift, tilt, and rotating tables; adjustable work benches; sloping storage racks; and a plethora of computer workstation devices. The movement of materials is assisted by articulating arms and balancers.

Perhaps the greatest ergonomics material handling aid is the attachment of wheels to carry-on luggage. Ironically, the market, rather than the ergonomics profession, should claim credit for expanding this form of assist. Before the reader confuses ergonomics with engineering, it must be pointed out that these are complementary professions—the ergonomists articulate the human interface requirements in detail and the engineers implement the solutions.

Orthotic Devices

Perhaps the most infamous orthotic device that has had widespread commercial success is the back belt. These come in rigid, flexible, and pneumatic varieties. Their promotion is based on two theories. First, it is claimed that they support and restrict the range of motion of the spine, either directly or through increasing intra-abdominal pressure. Second, it is suggested that they act as training or reminder devices. Empirical evidence regarding their effects ranges from indications that they reduce injuries to suggestions that they encourage risky behavior, cause atrophy, and increase the incidence of back injury. The consensus of medical and ergonomics experts on these devices is that they may be useful in protecting injured tissue during rehabilitation, but that they have no place in the protection of healthy tissue (NIOSH 1994).

Similar theories are pursued for limb support devices, including wrist rests, foot orthotics, and tendon bands. Again the empirical support is equivocal. The ski boot provides something of an exception. Its rigidity certainly protects the ankle, although there is the strong possibility that the loss of mobility at the ankle places greater demands on the knees for movement control and balance adjustment, thus giving rise to a greater incidence of knee injuries. Unfortunately, trends such as these are not subject to clear epidemiological investigation. Perhaps the most famous orthotic of all is the chair. It started with three legs and a small flat surface to support the ischii. Now it has sprouted various configurations of legs and lateral, vertical, and horizontal support.

The prevalent theories of support in the context of preventing cumulative trauma disorders deal simply with the postural and force elements. They ignore the fundamental contributions of time and the undoubted fact that the human being is a dynamic organism that thrives on an optimal level of physical stress. The chair is the prime example. Most chairs—even ones improvised from the edge of a desk—are comfortable for short periods of time, especially if one has been standing for a long time. Conversely, all chairs are uncomfortable if one is constrained to them for long periods. It is a relief to stand up and stretch one's legs or get the circulation going. In other words, comfort, discomfort, fatigue, or even cumulative trauma result from the distribution, over time, of static and dynamic physical activity.

THE PROBLEM OF SCIENCE AND STANDARDS

The process of science is quite straightforward: ideas, based on theory, experience, or intuition, are translated into hypotheses, which are often mathematical. These hypotheses are then tested empirically and rejected, confirmed, or modified. Science is iterative—it moves in a slow evolutionary way, sometimes punctuated by revolutionary insights. Technology lags science. The application of technology, for investment reasons, lags even further behind. This gap between science and applications is exacerbated by the educational and motivational gaps between scientists and practitioners. These truisms are clearly evident in ergonomics.

Most ergonomics science is reductionist. Typical laboratory or field experiments involve a handful of subjects and a limited number of conditions. The results of experiments often point in some general direction but, because of the enormous problem of human, situational, and temporal variability, many theories and derived technologies are subject to invalid extrapolation and unfair criticism. There are particular problems with selecting sample subjects for experiments and the differences, due to adaptation, of target populations to whom the experimental findings are applied.

The challenge to practitioners is to take the available literature and, using their professional experience, interpret, interpolate, and extrapolate it to suit the particular situation. A pertinent example is the practitioner's need to answer the question: What should be the weight limit of an object that has to be handled manually? The scientist will say, "It all depends...," but that does not help the practitioner who may not have sufficient information regarding the population of lifters and conditions of lifting.

The practitioner may be pressured by factors other than the capabilities of his population, such as traditional object weight ranges in his industry. The practitioner also may have to contend with an aging cohort of lifters whose capabilities are less than average.

CONSENSUS

Perhaps the only reasonable approach to ergonomics is to base decisions on a consensus of interested parties. The different parties will inevitably have different backgrounds and biases, and will weight sources of evidence differently. For example, employers of manual material handlers would rather pay only one person to lift an object, whereas the more conservative health scientist may opt for two people, or one and an expensive handling device. The challenge to the practitioner, given advice from the scientists and technologists, is to come up with a compromise that will please most of his or her customers, most of the time.

This simple, but very pertinent example, clearly highlights the problems caused by human variability. It also leads to the discussion of the general problem of ergonomics standards.

ERGONOMICS STANDARDS

Operational Definitions

In recent years, there have been many attempts to consolidate ergonomics knowledge into the form of standards. A major initial hurdle has been the semantic discussion regarding the term "standard" and its synonyms. The following operational definitions are offered to help prevent misunderstandings.

Standard—a standard is usually quantitative and based either on science or on a consensus. Failure to adhere to the standard may result in some penalty under the law or an incompatibility between different systems. For example, the spacing of railway tracks or the size of household electrical plugs are standards that allow manufacturers to produce compatible items. Similarly, a traffic speed limit is a standard developed by consensus (some would say negotiation) that facilitates the control of driver behavior.

Specification—a specification is a clear quantitative statement. Commonly, however, a specification may be accompanied by an allowable range of deviation that reflects the capability of a manufacturing process. For example, the quantity of fluid in a container may be one liter plus or minus 1%. An ergonomics specification for a work bench height may be 35 in. (89 cm), with plus or minus 5 in. (13 cm). In general, deviation from the target value of a specification may be indicated by a (nonlinear) loss function associated with increasing costs.

Guideline—a guideline may be quantitative or qualitative and provides information to assist designers or decision makers. In general, failure to follow a guideline may not necessarily result in a penalty or incompatibility. Rather, it will increase the probability of such occurrences. For example, a guideline (NIOSH Lifting Equation) for manual material handling will not guarantee that no one will be hurt. Rather, if followed, it will reduce the probability of injury. A guideline may be conservative or liberal. The NIOSH lift guideline may be an appropriate protective device for healthy adults accustomed to physical work, but it may be inappropriate for older people or physical work in extreme heat conditions.

Requirement—a requirement is usually an internal company statement regarding adherence to government or industry standards or guidelines. For example, a company may require that all its manual material handling jobs have a NIOSH lift index below 3.0, whereas another company, that hires fewer robust people, may set its requirement at 2.0.

Design, Performance, Programs

Another way to classify standards is to differentiate between design, performance, and programmatic forms.

Design—as described earlier, a design specification is an engineering concept. It defines, in a quantitative way, the inputs to, or the design parameters of, a system. A school zone speed limit of 25 mph (40 km/hour) or a parcel weight limit of 50 lb (23 kg) are specification standards. Such standards are appealing to both engineers and enforcement officials as they are clear and objective. They may not, however, be reliably predictive of process outcomes, as accidents may happen below 25 mph (40 km/hour), and 50 lb (23 kg) may be too heavy for some people in some situations.

Performance—a performance measure represents an outcome of a process. A common injury/illness performance measure is the number of diagnosed occurrences per 100 employees per year. Similarly, a quality measure may be the number of defects per 1,000 products, and a productivity index may be the number of items produced per day. It should be noted that these performance measures commonly contain a contextual component or denominator. The performance standard may be a particular level or a commitment for the measure to change at a prescribed rate.

Programs—a programmatic standard describes the characteristics or elements of a process (program). For example, a healthy lifestyle will include attention to diet and exercise, minimization of bad habits, and regular checkups. A musculoskeletal-disorder programmatic standard will include such elements as management commitment, employee participation, training, job analysis, hazard abatement, medical management, record keeping, and reporting. It should be noted that such programmatic standards are generally qualitative. However, it is possible to specify in detail the precise nature of these components. For example, one could specify a particular analysis checklist or medical management protocol. A program performance measure may be the number of open jobs six months after the completion of an analysis phase. The degree of specificity of programmatic standards (or guidelines) has been a major issue over the past decade.

Recent History of Standards

In the late 1980s, OSHA issued a set of "Meatpackers Guidelines." These guidelines were programmatic in nature and were aimed at improving the performance of meat processing plants as measured by injury-illness rate. The logic of this approach was that, if the companies were encouraged to do certain things, including the application of ergonomics concepts, the rates would go down. During the early 1990s, OSHA developed agreements with the Big Three automotive companies and unions. Like the Meatpackers Guidelines, these agreements were programmatic in nature, although considerably more detailed. They also included various program performance measures, including timelines.

Following this model, OSHA made a very ambitious attempt to issue a programmatic standard for general industry, including explicit risk factor specifications in the form of a checklist. The draft form of this standard contained voluminous details of how to implement a musculoskeletal disorder reduction program, with extensive reference to the scientific literature. These activities became the focus of intense political debate, causing the standard's issuance to be suspended. The principal and successful objections were the criticisms of the scientific evidence and the failure of OSHA and NIOSH to prove a clear "dose response" relationship.

The latest efforts to promulgate a standard have been through ANSI and various states, notably California. These efforts have generally restricted themselves to programmatic guidelines or requirements, with the debate related to the degree of specificity and, in some instances, to the justification for a standard of any form. The standards' opponents argue that humanistic, economic, and legal forces will serve to obviate the need for government intervention in this complex area.

Accommodation Standards

The traditional accommodation standard in anthropometry involves the fifth or 95th percentiles for reach and fit respectively. Where the outcome is less critical or the use is less frequent than the 25th and 75th percentile, accommodation levels may be adopted. Where comfort and convenience can be achieved at relatively low cost, the first or 99th percentiles may be used. When risk of severe outcomes are associated with complex human performance, then it is common to discuss accommodation levels of 0.01 or 99.99. Many of these accommodation standards assume that the underlying human characteristic is normally or symmetrically distributed. However, many human performance (psychophysical) and vulnerability (to cumulative trauma disorders) characteristics are skewed, and some may not be continuous.

In most human performance situations, it is only possible to approximate these measures because of sampling and investigation design constraints. Consequently, an attempt to accommodate the 95th percentile, for example, may involve considerable error where the original investigation samples and application populations are inherently different. These issues lead to a fundamental policy in ergonomics standards development. Ergonomists should be responsible for supplying the evidence, but managers, politicians, the courts, and society at large should be responsible for setting accommodation or risk levels. Standards and laws in the United States are often supported by a cost-benefit analysis that amalgamates the performance probabilities with estimates of outcome severity. An example of such standard setting can be seen in the establishment of speed limits on various roads under various conditions. There is a clear advantage of high speed; one arrives at one's destination more quickly. The probability of accidents is very low—most people drive for a lifetime with no or few untoward events. However, when accidents do occur, the outcome is often very severe and costly. The annual death rate on U.S. roads is over 40,000, but speed per se is only one of many factors contributing to this statistic. Society and its representatives—the law makers, law enforcers, and the courts—set, implement, and interpret the standards.

The debate regarding physical ergonomics standards has not yet reached this level of sophistication. However, devices such as the NIOSH lifting guidelines and "Snook" tables (Snook and Ciriello 1991) have gone a long way in attempting to link the practicalities of standards setting to the complex scientific basis. The sister profession of industrial hygiene is more advanced in that it has set values such as permissible exposure limits (PELs) for many individual chemical, particulate, and physical environmental stressors. Many of these stressors are similar to the mechanical stressors that cause cumulative trauma disorders—there are certain advantages associated with the stressor, but at high levels the outcomes may reach intolerable levels. The difference between these (industrial hygiene) stressors and those commonly encountered by physical ergonomists is that they are extrinsic to human work. The stress of physical work is intrinsic—it is the thing the individual is paid to do!

The recent flurry of activity regarding ergonomics standards has done a lot to articulate the many and complex issues. Eventually the political debate, backed by the multiple sources of scientific, logical, and epidemiological evidence, will converge on societally acceptable standards or guidelines.

EXPOSURE—THE TIME FACTOR

In the case of physical ergonomics, the principal system characteristics include object location, orientation, size, shape, weight, and other forces associated with object manipulation and the various temporal elements such as repetition, duration of static exertions, and length of the work shift. The weight and force factors are usually the responsibility of the product engineer. The spatial factors that impose postural and movement stress on the operator are the responsibility of the tooling or manufacturing engineer. Finally, the temporal factors are usually the responsibility of the industrial or operations engineer. As far as the operator is concerned, it is his or her interaction with all factors that affect process outcomes, such as productivity or cumulative trauma. Most recent ergonomics interventions focus on reducing force and spatial stressors. If the object is heavy, provide a hoist; if the job location is too low, provide an adjustable-height work surface.

Too little attention has been paid formally to the temporal issues, although they are at least equal contributors to such outcomes as product quality, productivity, and cumulative trauma. Furthermore, the simplistic reduction of spatial and force stressors in the design of ergonomic workplaces may increase the opportunity for greater productivity, but may result in greater temporal stress, such as high repetition rates or reduced within-cycle physiological recovery time. The age-old adage, "a change is as good as a rest," is clearly pertinent in physical work. The provision of change (cross training) is central to most athletic conditioning. The challenge in industry is to provide appropriate change.

Job enlargement, from the physical viewpoint, involves the provision of a greater variety of physical activities within a defined job cycle. A broader view of enlargement includes increasing the cognitive and organizational content of a job. For example, contemporary industrial practice includes total productive maintenance concepts in which assembly operators have responsibility for the service and maintenance of their equipment, to the extent that their knowledge and training permit. A derivative of job enlargement is job rotation, in which operators perform one job for a short period of time before moving on to another. The athletic analogy is circuit training. It is essential that the alternative jobs are as physically (or cognitively) different as possible. Another derivative of enlargement is the concept of enrichment. One theory is that broader participation will bring with it greater feelings of ownership in the overall organization and hence lead to greater levels of motivation. However individuals are notoriously different in their appreciation of "enrichment."

Job enlargement and rotation concepts have a number of drawbacks. The first issue is training. A number of operators in a particular area or rotation must be trained to the same level, and it can only be assumed that they all have sufficient physical and cognitive capacity for performing the broader variety of jobs to the required quality standards.

A second drawback relates to the problem of change from the traditional practice of narrow job responsibilities to broader ones. These changes may result in a loss of ownership of a particular job or job location. Furthermore, where there are traditional seniority agreements that allow individuals some degree of job choice, those who have graduated to better jobs may be very reluctant to revert to less desirable ones. These seniority agreements are based on a concept of fairness in that length of service should bring with it rewards of greater choice. A major drawback of all of these concepts of rotation, enlargement, and enrichment is the inherent variability of individuals. There is no doubt, from the physical or mechanistic point of view, that these concepts have merit. It is also clear that they also have many operational and motivational advantages. However, misapplication of these concepts may lead to insufficient reduction of physical stress, and unthinking imposition may result in emphatic opposition. Individual enrichment is a perception of the individual.

TEAMS

One step beyond these individual enlargement, rotation, and enrichment programs is the organizational approach to teams, which includes many of the benefits described in previous paragraphs. An additional benefit is that the team can assume much greater responsibility than an unlinked group of individuals for various operational and managerial aspects of the job.

A simplistic view of the team is based on the concept of fairness and equal rewards for equal work. Each team member shares the work equally, for example, by job rotation. The athletic equivalent is found in volleyball where all team members rotate around the different positions. Other team sports present a different view—the team, team leader, or the coach assigns particular players to particular responsibilities, according to their abilities. Of course, individuals may differ in their perceptions of their own capabilities and those of others, but, with time, teams either gel or fall apart.

In the athletic setting, the reward for being a team player is a share in the success of the team. In the industrial setting, team concepts may have less clear extrinsic rewards. The intrinsic social and organizational value of teams has been shown to have substantial productivity, product quality, and motivational benefits. From the physical ergonomics viewpoint, the team concept can employ work assignment strategies that reduce the intensity of physical stress and, thus, the incidence of cumulative trauma. However, work teams are seen by organized labor as usurping some of their traditional responsibilities as representatives of a large collection of individual workers. Consequently, team work, a potentially powerful concept, must face the realities of the overriding political climate.

OBJECTIVES OF ERGONOMICS

FITTING THE TASKS

One traditional view of ergonomics is that it should focus on fitting the task to the man, woman, or child. The complementary activity involves selection and training of individuals to fit the task. It is clear that both approaches are based on the same scientific knowledge. Selection and training are likely to fail unless the characteristics of the task system are well defined. Conversely, task system design demands some knowledge of the capabilities and limitations of the particular population of users. It would be unthinkable to attempt to design an airplane without assuming a certain training level of a pilot. However, the training of an individual in the correct method of lifting may be irrelevant if the weights and conditions involved are beyond his capacity. Competitive sports represents the extreme form of selection and training, whereas the design of consumer products, such as the telephone and most grocery packages, represents a successful attempt to accommodate the capabilities and limitations of a very wide audience. The notorious VCR represents a failed attempt to design for the broad range of potential users. In the long run, the challenge for ergonomics-human factors is to blend hardware and software design with humanware selection and training.

Facilitators

Where there is a gap between system design and user capability, there is the opportunity for the use of facilitators. Indeed, physical and cognitive facilitators represent the ultimate opportunity for interface design. A facilitator is a device that may or may not be used to achieve system function, depending on user capability. In the physical world, manual material handling devices may be used or not, depending on the strength of the user. In the cognitive arena, a label or warning may be used by first-time users, but more experienced users do not need them. The same reasoning is true for telephone lists, bicycle training wheels, golf carts, and daily living aids for elderly and disabled people. The concept that a system should be designed for all users is inefficient. More ergonomics-human factors focus should be addressed to the use of facilitators for subsets of the user population.

PREVENTING UNWANTED OUTCOMES

Physical Failure

One major component of ergonomics deals with analysis of the reasons why people make errors, leading to the design of systems to prevent the occurrence of human error or reduce the severity of unwanted outcomes. Human beings fail for many reasons. They may fail physically or in their ability to process information appropriately. Physical failure may vary from the simple inability to reach an object or fit into a space through biomechanical strain, sprain, or fracture to a physiological or pathological failure—such as a heart attack or fall. Such failures, however, are relatively rare causes of accidents.

Cognitive Failure

Failure in information processing is both the most frequent form of failure and the form that leads to the greatest severity of unwanted outcome. For example, failure to pay attention to appropriate information while driving a car, controlling a power plant, diagnosing an electrical fault, or making a financial decision may result in catastrophic outcomes. Sometimes the failure of attention may be due to the individual being distracted, or the relevant information not being displayed appropriately. Other information processing failures may occur due to inappropriate perception, forgetting, or adopting a wrong problem-solving strategy. The opportunities for human failure are carried full circle back to the physical arena when one considers those accidents that are due to lack of skill, as in manipulating a tool or crane control. These types of informational handling failure are at the root of most industrial accidents and can loosely be called cognitive failures.

Behavioral Failure

Another type of failure can be described as behavioral. In these cases, individuals or groups habitually do things in a way that is conducive to accidents or containment of the outcomes of accidents. Typical examples would include driving too fast, not fastening a seat belt, not wearing eye protection, not switching off the mains when repairing an electrical fault, or disabling a machine guard. Such behaviors all have perceived positive rewards, and generally the perceived risk of failure is remote, although the outcomes may be catastrophic. In recent years, many major companies have successfully instituted programs that create a general safety climate to combat this type of human failure. Such programs may require constant reinforcement to maintain an acceptable level of good behavior.

A common approach to the safety problem is the "engineering approach," which aspires to develop foolproof systems. A shortcoming of this approach occurs when engineers fail to understand that users of their systems do not think as they do. Of course, engineers should aspire to build safe systems, but such designs should be human-centered in that they comprehend the many physical, cognitive, and behavioral causes of human failure.

ANTHROPOLOGY OF WORK

The anthropology of work is a topic of great interest to theoretical ergonomists and those interested in the history of the subject. Managers, practitioners, and legislators may also benefit from a brief discussion of this macro view of the profession. Traditional work, as typified by agriculture, primary industry, and building, requires the human being to exert force to move materials. This work typically involves the selection or conditioning of people to perform tasks that require near maximal efforts. Such tasks are enhanced by the design of levers, wheels, ramps, and external power sources. However, a common characteristic of these traditional tasks is that the location and orientation of the handled objects are varied. In these high-demand tasks, typical injuries are acute in nature, commonly involving the back or shoulder.

Over the past 100 years, the use of external power has become the norm and individuals have filled the gaps between mechanized functions. One result of this mechanization and automation has been a reduction in the variability of shape and size of objects in a particular process. Another result is that many handling tasks have become machine paced. Parcel and luggage handling, warehouse operations, and retail checkout operation are typical of contemporary manual material handling tasks. Occupational hazards are now more likely to be cumulative in nature, with minor injuries exacerbated by continued work.

Traditional craft-type jobs involved the skillful use of tools to form and attach wooden and metal objects. Such jobs required greater use of the arms and hands for the control of relatively light objects. Although many such jobs still exist, the advent of more consistent forming machinery (sawing, casting, stamping, turning, etc.), together with conveyors, has increased the throughput of such operations. The result is that the requirement for precise manipulative skill has diminished and has been replaced by high-intensity manipulation as typified by the Henry Ford production line and contemporary machining work cells. One result is that the amount of walking involved has been greatly reduced. The delicate and versatile control devices—the hands—have been relegated to highly repetitive, but still precise, work. The result is that individual muscle groups and motor units become overstressed due to insufficient recovery.

The most recent development in physical work is due to the meteoric expansion of computer applications. Presently, the typical operator performs rapid, but similar, manipulations of some input device, such as a keyboard or mouse. Such devices typically confine the individual to a fixed sitting position with long-term static postures and reduced metabolic stimulation. The manual operations are also typically highly precise, similar, and repetitive, with the result that individual micro structures become overstressed.

The advent of formal ergonomics in recent years has focused, inappropriately, on the physical (spatial and force) rather than the temporal aspects of these material handling and manipulation tasks. The result is that physical work has become less varied and the injuries that were once acute, and often substantial, are now more likely to be chronic and less extreme. Thus, ergonomic disorders may indeed be the iatrogenic result of the misapplication of ergonomics. The medical profession is now faced with problems complicated by varied reporting thresholds, which may often be affected by psychosocial (for example, boredom), rather that simply physical factors.

This somewhat brief and simplistic account of the trends in physical work is offered to explain, in part, the role that engineering and ergonomics have played over the years to reduce, rather than optimize, physical stress. It is hoped that a more holistic view of employment will reverse these unfortunate (from the operator's physical viewpoint) trends.

REFERENCES

Bernard, Bruce P. (ed.), NIOSH 1997. "Musculoskeletal Disorders and Workplace Factors," U.S. Department of Health and Human Services, 1997.

NIOSH 1994. "Backbelts: Do They Prevent Injury?" Cincinnati, OH: Department of Health and Human Services, 1994: 94-127.

Snook S. H., and Ciriello V. M. "The Design of Manual Handling Tasks: revised tables of maximum acceptable weights and forces," *Ergonomics* Volume 36, No. 9, 1991: 1197-1213.

Waters, T. R., Putz-Anderson, V., Garg, A., and Fine, L. J. "Revised NIOSH Equation for the Design and Evaluation of Manual Lifting Tasks." *Ergonomics* Volume 36, No. 7, 1993: 749-776.

BIBLIOGRAPHY

Chaffin, Don B., and Gunnar, B. J. Andersson. "Occupational Biomechanics," 2nd Edition. New York: John Wiley and Sons, 1991.

Eastman Kodak Company. *Ergonomic Design for People at Work*, Volumes 1 and 2. New York: Van Nostrand Reinhold, 1986.

Gjessing, Christopher C., Schoenborn, Theodore F., and Cohen, Alexander. *Participatory Ergonomic Interventions in Meatpacking Plants*. Washington, DC: U.S. Department of Health and Human Services, 1994: 94-124.

Helander, M., and Nagamachi, M. *Design for Manufacturing*. Bristol, PA: Taylor and Francis, 1992.

Kroemer, Karl, Kroemer, Henrike, and Kroemer-Elbert, Katrin. *Ergonomics*. Englewood Cliffs, NJ: Prentice Hall, 1994.

Konz, Stephan. *Work Design*, 4th Edition. Scottsdale, AZ: Publishing Horizons, 1995.

Lewis, H.W. *Technological Risk*. New York: W.W. Norton and Co., 1990.

Mital, Anil, and Karwowski, Waldemar. *Workspace, Equipment, and Tool Design*. New York: Elsevier Science, Inc., 1991.

Mital, A., Nicholson A.S., and Ayoub, M.M. *Manual Materials Handling*. Bristol, PA: Taylor and Francis, 1993.

Majchrzak, Ann. *The Human Side of Factory Automation*. San Francisco, CA: Jossey-Bass Publishers, 1988.

Peacock, Brian, and Karwowski, Waldemar. *Automotive Ergonomics*. Bristol, PA: Taylor and Francis, 1993.

Putz-Anderson, Vern. *Cumulative Trauma Disorders*. Bristol, PA: Taylor and Francis, 1992.

Salvendy, Gavriel. *Handbook of Human Factors*. New York: John Wiley and Sons, 1987.

CHAPTER 26

CORPORATE ERGONOMIC EFFORTS IN GERMANY

Hans-Jörg Bullinger
Professor and Head
Fraunhofer Institute for Industrial Engineering (IAO),
Stuttgart, and Head
Institute for Human Factors and
Technology Management (IAT)
University of Stuttgart, Germany

Martin Braun
Research Scientist
Institute for Human Factors and
Technology Management (IAT)
University of Stuttgart, Germany

Rainer Schopp
Head Product Design
Fraunhofer Institute for
Industrial Engineering (IAO)
Stuttgart, Germany

EVOLUTION OF ERGONOMICS

Ergonomics is considered a higher-ranking designation for the study of human work and the exploration of its regularities (Rohmert and Luczak 1989). The term *human factors* is virtually a synonym for ergonomics. According to Luczak et al. (1987), *human factors engineering* can be defined as a systematic approach to the analysis, order, and design of the technical, organizational, and social conditions of work processes. The scientific knowledge of ergonomics is translated into practice in the course of ergonomic work design. By adapting engineering and man to each other, ergonomics, in its practical application, contributes to keeping the load imposed on the working man as balanced as possible. This is done by securing the highest possible economic benefit from the work system.

Initial ergonomics activities in Germany resulted from a growing consciousness of human work in the early 19th century. However, there were no major research activities in the field of ergonomics for decades. It was not until the foundation of the Reichsausschuss für Arbeitsstudien (REFA) (Association for Work Studies) in 1924 that the systematic and institutional development of research activities in ergonomics started. There has been a marked increase in the number of research institutes representing the ideas of ergonomics since about 1950. Now, it can be assumed that research and teaching are conducted in the field of ergonomics at nearly all technical universities in Germany.

MOTIVES OF ERGONOMICS

The reasons that lead to the application of measures for ergonomic design are manifold. Ultimately, credence must be given to the unstoppable development of technology, the constantly changing market conditions, and the changes of values in society. Consequently, the measures for ergonomic work design are primarily determined by human, technical, and economic factors.

ECONOMIC EFFICIENCY

For work systems designed to fulfill a specific purpose, the economic efficiency will always be a high-ranking goal. An optimal proportion of expenditure and earning will be aimed at it. Ergonomic design can make a decisive contribution to the increase of the economic efficiency of a work system. This contribution is substantiated by the following criteria:

1. Reduction of social costs. Ergonomic design can contribute to the maintenance of health and well-being of people and, thus, to a reduction in the social costs to be borne by companies and society.

2. Economic utilization. Ergonomic design can contribute to the economic utilization of human working power and, thus, to an increase in the effectiveness of the work system.

3. Competitive advantages. In view of the sensitized consciousness of users and buyers, the ergonomic design of products can contribute to economic advantages.

HUMANIZATION

Due to a change in the values of society, it has been recognized in recent years that work systems must be designed to account for human needs in work performance. In the end, people alone create benefits within a work system, but they also claim these benefits for themselves. With their work contribution, humans can exert considerable influence on how a work system fulfills its economic purpose. Hence, the human factors must be considered as having the same weighting as technical and economic factors in the planning and operation of work systems. Conceptive ergonomic measures that conserve the well-being, health, and safety of the employed and promote their motivation, make a decisive contribution to human potential.

NORMATIVE AND LEGAL ASPECTS

Contents of ergonomic work design are traditionally regulated by laws, rules, and directives on a national level. The Works Council Constitution Act, the Safety at Work Act, the Place of Work Rules, regulations for the prevention of accidents, standards, and directives of associations can be mentioned as examples. These legal directives are supplemented by in-plant regulations from case to case.

New laws and rules came into force with the completion of the European Community internal market in 1993, and made more exacting requirements on the ergonomic design of workplaces and technical products. Article 100a (Internal Market Directive) and Article 188a (Social Welfare Provisions) of the "European Economic Community (EEC) Treaty according to the Single European Act" will create an enforceable standard of health and security in the future. Consequently, those engaged in development, manufacture, and trade will bear the full responsibility for the human-oriented design of their products. The law of the European Community must be transformed into the German law and its observance enforced. The structure of the European work and health protection directive is shown in Figure 26-1.

SOCIODEMOGRAPHIC DEVELOPMENT

Apart from technological innovations and new production and service concepts, corporate planning activities for the future must also include expected changes in the composition of staff members (Bullinger et al. 1993). According to the demographic and socioeconomic change expected in the coming decades, the age and social structures of the gainfully employed will change in line with developments in the population structure of the Federal Republic of Germany. Consequently, there will be more older persons employed in the future (Figure 26-2). The amount of employed persons with higher-value professional training, who perform jobs of accordingly higher qualification, will increase; whereas, the share of positions requiring lower qualification will decrease. Moreover, the share of those engaged in the service sector will increase in the

Internal Market Directives Article 100a EEC Treaty	Work Protection Directives Article 118a EEC Treaty
Objectives: • Harmonization within the EU • Dismantling of trade barriers Examples: Directives on the condition of products, for example, machinery Realization: Binding standards, realization without any deviations below or above	Objectives: • Harmonization within the EU • Improvement of safety and health protection for workers Examples: Directives for safety and health protection at work • Outline directive 89/391/EEC • 12 individual directives Realization: Minimum requirements, deviations over higher-level standards are permissible

Figure 26-1 Structure of European Union's (EU) Work and Health Protection Directive.

future, while the share of those in production will decrease (Klauder 1993).

In view of these future developments, work processes and equipment will have to be designed by conceptive and preventive measures so that employees will have the possibility of qualified and productive working until they reach their pensionable age. This is where ergonomics can make valuable contributions in the field of age-appropriate work and technology design—by creating reasonable job standards.

INSTITUTIONS INVOLVED IN ERGONOMICS

Ergonomics in Germany has evolved through the natural and engineering sciences and has manifested itself in application-related research (Rohmert and Luczak 1989). Current ergonomics practice can be classified into a technical-economic discipline, a medical-physiological discipline, and a psychological discipline. Ergonomic research and application activities can be found in the university sector, at public and private institutions, in associations,

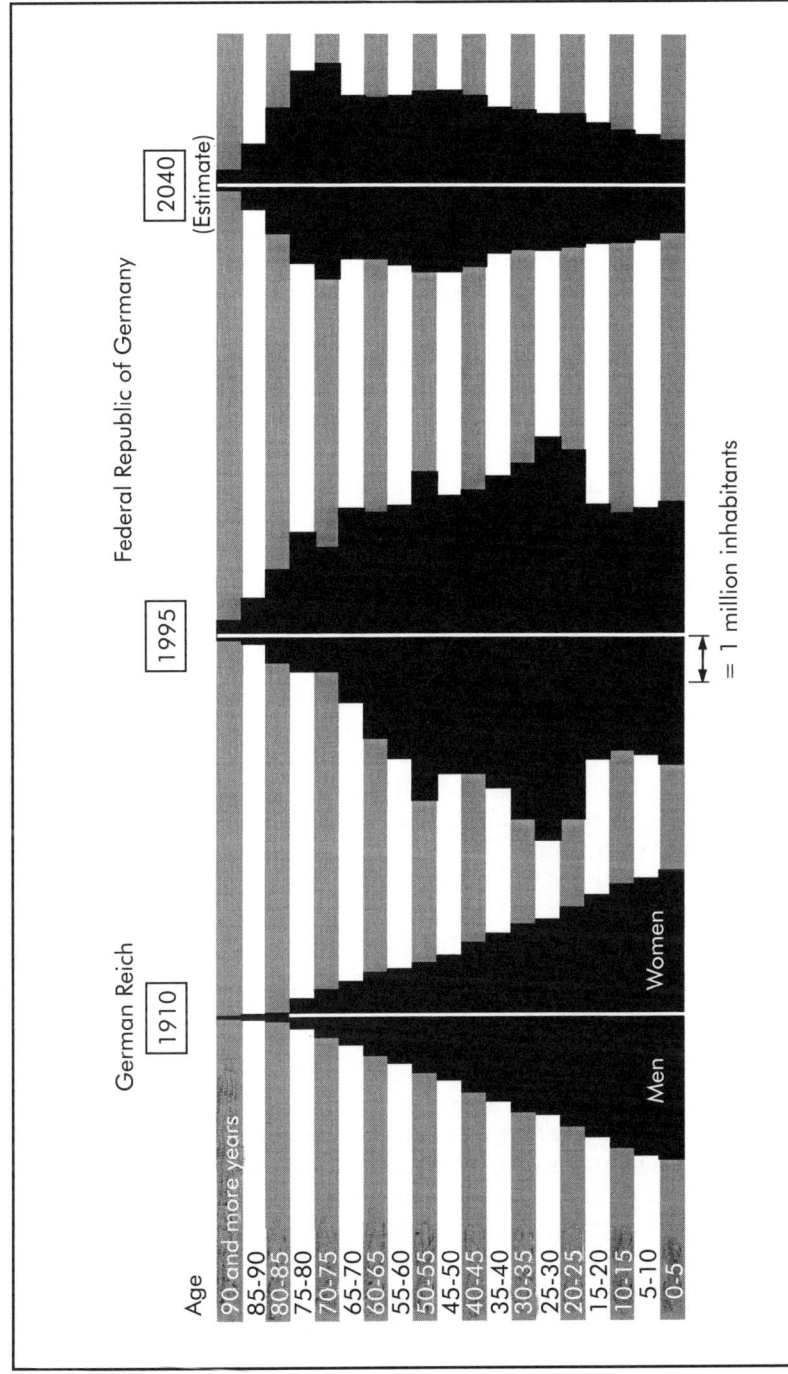

Figure 26-2 Age structure in Germany (Statistisches Bundesamt 1995; Statistisches Landesamt Baden-Würtemberg 1991).

and in industry. There is an intensive exchange of knowledge and experience among the various institutions. Examples of authoritative institutions and bodies of ergonomic research and development follow. Institutions that regulate, promote, and supervise the application of ergonomics knowledge are not discussed.

PUBLIC INSTITUTIONS AND PROGRAMS

Bundesanstalt für Arbeitsschutz und Arbeitsmedizin (BAuA)

As a central public law institution in the field of ergonomics, the Bundesanstalt für Arbeitsschutz und Arbeitsmedizin, Federal Employment Protection Office, with its registered office in Dortmund, is directly subordinate to the Federal Ministry of Labor and Social Affairs. The task of the BAuA is to support the Federal Ministry in the field of work environment and health protection, for which this office maintains numerous cooperation relations with other institutions. Under its public mandate, the BAuA analyzes the job conditions in companies and administrative authorities. It develops solutions to problems and promotes the practical application of scientific knowledge by organizing appropriate information and training events, and by means of various publications. Moreover, it participates in developing standards.

Humanisierung des Arbeitslebens Action Program

The ergonomics efforts made in Germany have been integrated in a comprehensive state action program, Humanisierung des Arbeitslebens (Humanization of Working Life), since the beginning of the 1970s. In its early years, this program ranked high in the catalog of goals of political decisions, which, among other things, brought about a research program, Arbeit und Technik (Work and Technology), by the federal government. It also helped establish respective goals in different collective agreements (EKD 1990). The subject matter is still of great importance, even though the essential goals of the action program have been reached.

SCIENCE AND RESEARCH INSTITUTIONS

Universities

The growing importance of ergonomics within recent decades also has been accounted for at universities. At the present time, ergonomics is represented at numerous German universities that offer programs in research and teaching. Human factors teaching and specialized subjects extend to the engineering and economic sciences, industrial medicine, physiology, psychology, sociology, and vocational education. Due to the wide variety of disciplines involved in these allied subjects, university education proves to be heterogeneous (Luczak et al. 1987).

Fraunhofer-Gesellschaft

Fraunhofer-Gesellschaft (FhG) is a leading organization for applied research in Germany. Founded in 1949, it has 46 research institutes. Its activities include applications such as engineering and process studies, work and product design, as well as information and communications technologies. Some institutes of the FhG are engaged in the research, planning, and development of human-oriented structures for production and office areas under the premise of ergonomic design. The target groups of the FhG are industry and public institutions that conduct research projects of interest to society.

Gesellschaft für Arbeitswissenschaft

The Gesellschaft für Arbeitswissenschaft (GfA) was founded in 1953, and took its place among other national scientific societies in Germany. This human factors society sees itself as a scientific association primarily concerned with issues of education and research in human factors and ergonomics. The goal of the society is the interdisciplinary exchange of views for promoting scientific work. The GfA is a member of the International Ergonomics Association.

REFA-Verband für Arbeitsstudien und Betriebsorganisation

The Verband für Arbeitsstudien und Betriebsorganisation (Association for Work Studies and Plant Organization) is not engaged in research activities, but commits itself intensively to transforming ergonomics knowledge into practice. The base-oriented organizational structure of this association guarantees close contact with practicians in the production and service sectors. The REFA association has several training facilities, including working manufacturing facilities.

Industry and Service Sector

Nearly all large German industrial and business companies have facilities engaged in the research and application of ergonomics. The coexistence of scientific and applied research facilities lies in the different research concentrations and application needs. Practitioners locate deficits of scientific knowledge through practicable and economical methods of analysis, planning, and design in the field of physiology of effort, information technology and mental job design, and personnel deployment (Rohmert and Luczak 1989). This divergence requires the establishment of company-specific research facilities and, in the end, leads to a balanced and productive synthesis of scientific and application-oriented activities within the field of ergonomics. The primary task of the company-internal facilities is to identify ergonomic deficits relevant to the company, develop design solutions, and apply them to in-plant practice.

Apart from academic and industrial research facilities, there are numerous small and medium-size engineering and consultant offices in the service sector engaged in the field of knowledge transfer, planning, design, implementation, and testing of work systems and consumer products based on ergonomics criteria.

ERGONOMIC DESIGN

GOALS

The main purpose of ergonomics is design. Design activities aim to adapt systems, organizations, jobs, machines, products, and environments to human physical and mental abilities. This improves work conditions and promotes productivity. Systems should be designed so that they are easy to use, and contribute to a balanced work load. One way to make a system function is by matching the people to the system by training or education.

In recent years, there has been a worldwide paradigm shift in economics. This is seen in the globalization of production and trade markets, in broader market awareness through modern information and communication technology, and in increased customer focus. To meet growing customer demand for quality and performance, and to preserve Germany's competiteveness, some industries implemented a process of ongoing innovation and rationalism. Nevertheless, people, the actual performers, retain their key positions despite automated production. By designing and implementing innovative ergonomic production structures, companies can meet the future market challenges.

APPLICATIONS

Main applications of ergonomics are safety, work load assessment, industrial engineering, human-computer interaction, as well as anthropometry, biomechanics, physiology, and psychology. In industrial practice, ergonomics activities are seen in job design, workplace and equipment design, and the design of consumer products.

As a result of increased automation in production systems, workers are subjected to changed load situations. Much physical labor in manufacturing has been replaced by material handling aids and mechanical processes. Physical work has been greatly reduced in

Germany. On the other hand, the introduction of computers in the workplace—manufacturing as well as office—has led to raised demands for human information processing within highly complex systems. And since workers are increasingly confronted with psychomental loads, suitable measures of ergonomic job design are intended to contribute to a balanced stress and strain situation. Organizational strategies are applied covering not only work process and job design, but also qualification and human resource management, working time design, and remuneration.

Due to raised consciousness of possible damage of one-sided work, an ergonomically responsible human behavior is gaining importance. Under behavior ergonomics, placed within the context of job design, human behavior patterns are investigated. Findings help avoid ineffective and injurious patterns.

Regarding workplaces, ergonomic measures primarily concern anthropometric and biomechanic design. Also considered are such environmental factors as lighting, climate, sound, and pollutants. Workplace and environmental design include office workplaces; visual display unit workplaces; assembly workplaces; workplaces for control, checking, and supervision; drivers' workplaces in road and rail vehicles, ships, aircraft, and building machines; workplaces in the health care sector; and workplaces in the household. Ergonomics affects the design of equipment and consumer products primarily in the man-machine interface. It thus comprises the activity-oriented design of hand-sided handles, steering elements, controls, and displays. Product design applications can be seen in such items as machine tools, hand tools, medical apparatus, measuring instruments, controls and inspection means, seating furniture, means of transport, leisure time and sports outfitting, and vehicle components. The design of software is also an example.

A major reason for ergonomically designed workplaces is to help minimize high absenteeism—and its accompanying financial losses—stemming from disabilities sustained in poorly designed work areas. Figure 26-3 shows that physically straining working conditions are still widespread in Germany, despite relief introduced by automation measures. In the future, ergonomics will play an even greater role in human-oriented work system design.

DESIGN REQUIREMENTS AND METHODS

Due to the public's high quality awareness, the design of products and workplaces has to meet a great number of functional, ecologic, and economic requirements. Apart from technical criteria,

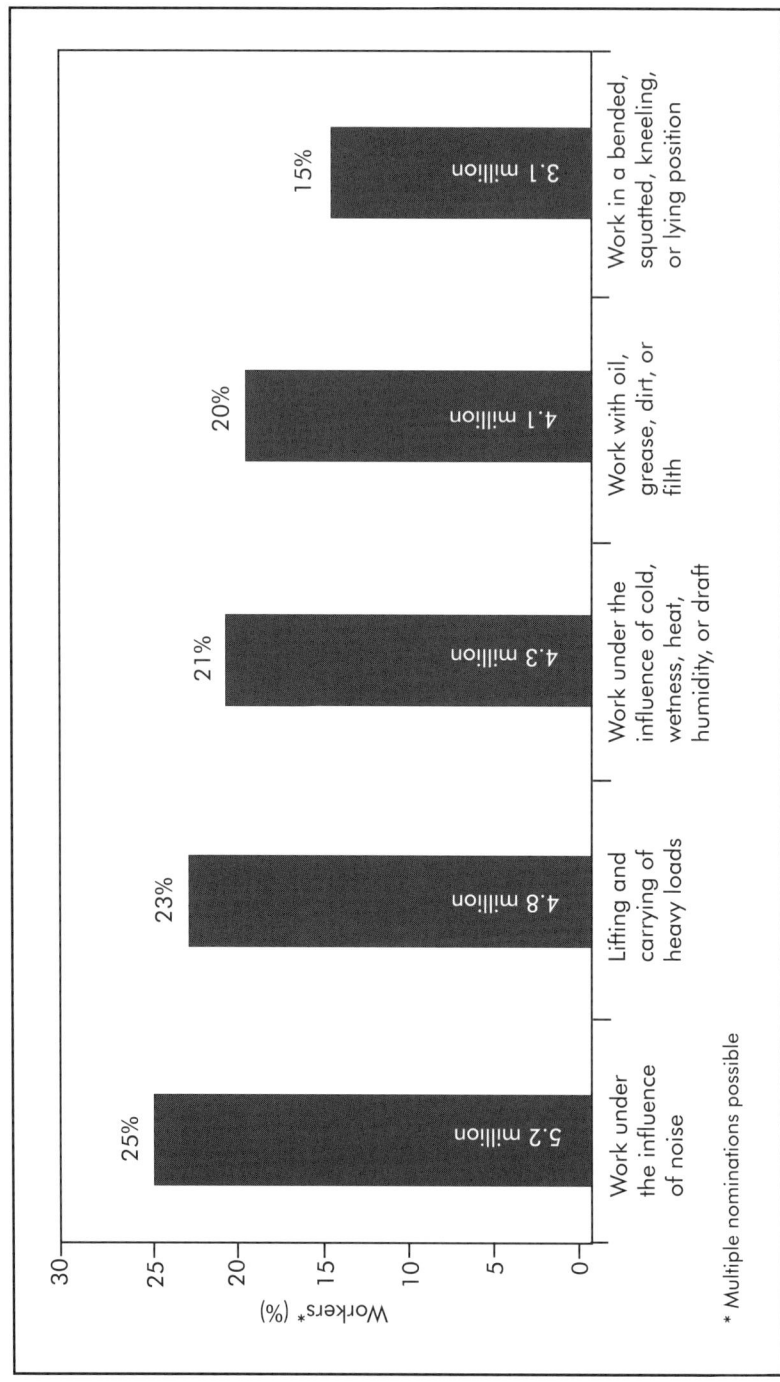

Figure 26-3 Physically-straining work conditions in West Germany (EKD 1990).

an integrative ergonomic design of equipment and workplaces must include such factors as safety and esthetics.

Each set of ergonomic design tasks must incorporate methodical approaches, adequate tools, and experience. Commonly, ergonomic design happens within the scope of a superior sociotechnological system design, in which the following steps have proven successful:

- Target and definition of the design task;
- Analysis of the object to be designed;
- Development of alternative design solutions;
- Evaluation and selection of design alternatives;
- Realization of the selected design solution; and
- Evaluation and implementation.

For the realization of ergonomic design measures, a variety of standardized and non-standardized methods exists. They can be divided into data survey, analysis, design development, and evaluation. Figure 26-4 shows a set of anthropometric design methods.

Traditionally, ergonomic design has an application-oriented engineering background, therefore, often using objective methods like measuring, counting, or calculating. Now, resulting from interdisciplinary approaches, objective methods are supplemented by subjective methods of a sociological background like observing, questionnaires, and scenario techniques.

Current information technology provides a wide range of computer-based tools and methods for ergonomic design within all application fields. These systems often show the development of manual design methods. Data-based and human-model oriented systems are used for analysis, development, and evaluation in the application fields of anthropometry, biomechanics, and interface design. Merits of current computer-based design tools include high performance and user-friendliness. Latest research activities intend to combine virtual reality techniques with methods of ergonomic design, for gaining a most efficient approach to complex design tasks.

EXAMPLES OF ERGONOMIC DESIGN

The following design cases are used as examples for demonstrating the requirements and dimensions in ergonomic design. In so doing, it becomes obvious that a design process—in line with the complex problems inherent in humans and work—is dependent on many varied factors, with technical, physical, formal-esthetical,

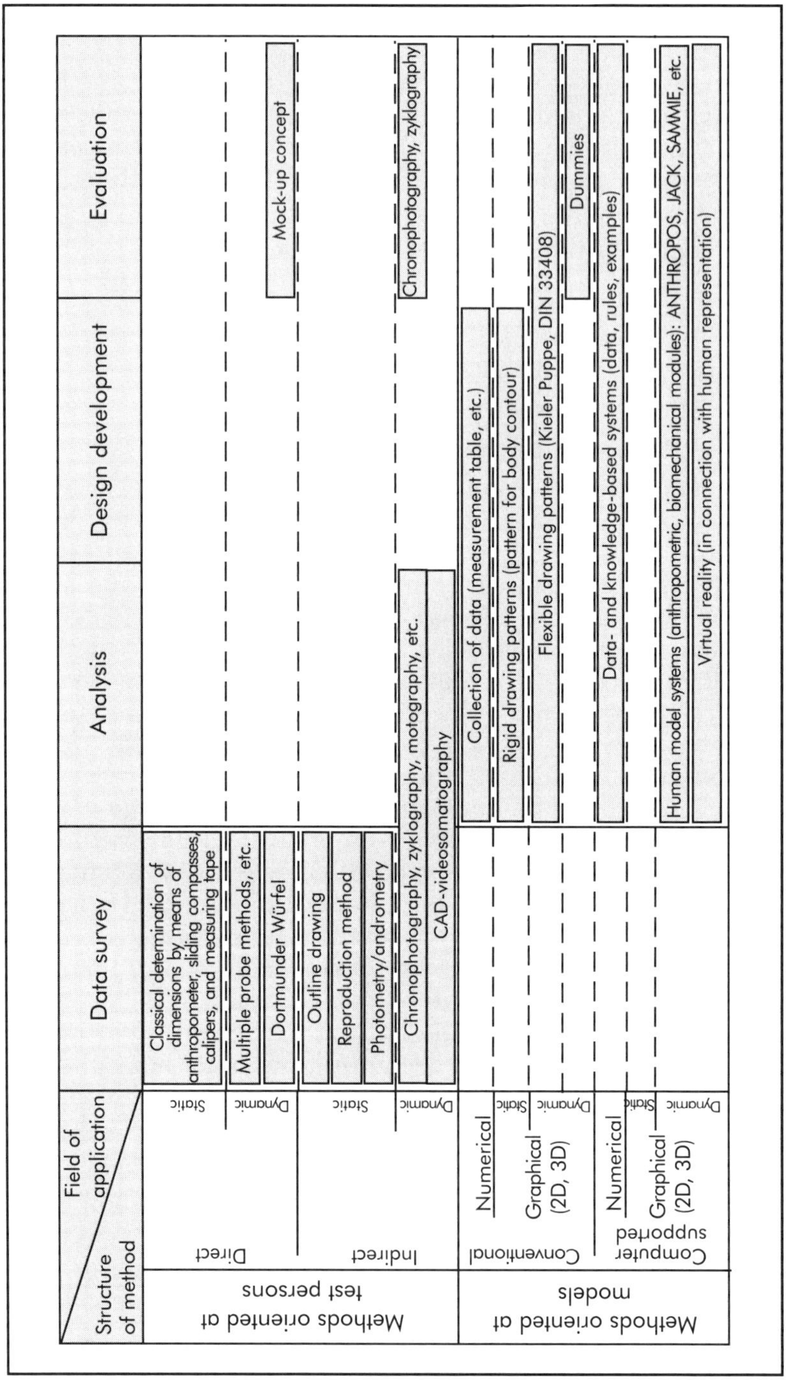

Figure 26-4 An anthropometric design method.

and ecological design criteria considerations in addition to ergonomic design criteria.

Hand Drill

Tools can be designed as either hand-side (facing the user) or work-side (serving the work). The hand-side, as interface to the user, is of greater importance for the ergonomic design of work equipment than the work side (Bullinger 1994).

The design of a hand drill is, above all, intended to improve the general handling of the tool for holding and guiding. A reduction in the total weight and tilting moment, by displacing the tool's center of gravity near the handle or by arranging a supporting handle, make a marked improvement in the handling conditions.

By its arrangement and shape, the handle of the hand drill makes possible the manual exertion of high-feed forces, damping developing reaction forces and torques at the same time. In so doing, the feed force is favorably generated by positive power transmission. The dimensional design of the handle and controls is oriented at the anthropometric and physiological properties of the hand-arm system.

The assignment of motion aims for the best possible agreement between the functional direction of a tool and the anatomically favorable movements of the worker. The kinematics of anatomically correct modes of movement favor the exertion of force and prevent forced motions and postures. Figure 26-5 shows the correct and incorrect assignments of motion for a hand drill. By placing the handle in the rear motor region, the direction of force will be in alignment with the functional direction resulting from the drilling axis.

Circular Grinding Machine

Hand tools and machine tools can only be used for their intended purposes if their respective total and part dimensions are adapted to the body dimensions of users. The dimensioning of work equipment and arranging of manipulation places, controls, and indicators must be oriented at the action and functional spaces of all users and their posture and position. As early as the design phase of machine tools, length dimensions and force-related dimensioning must be aimed at minimizing muscular strains by avoiding physiologically unfavorable forced postures and static holding work.

Circular grinding machines are used for external, internal, and contour grinding of turned parts. During the tool change operation, the operator stands by the front door of the grinding machine. The dressing tool mounting screws to be loosened are not directly visible from where the machine operator stands and are accessible

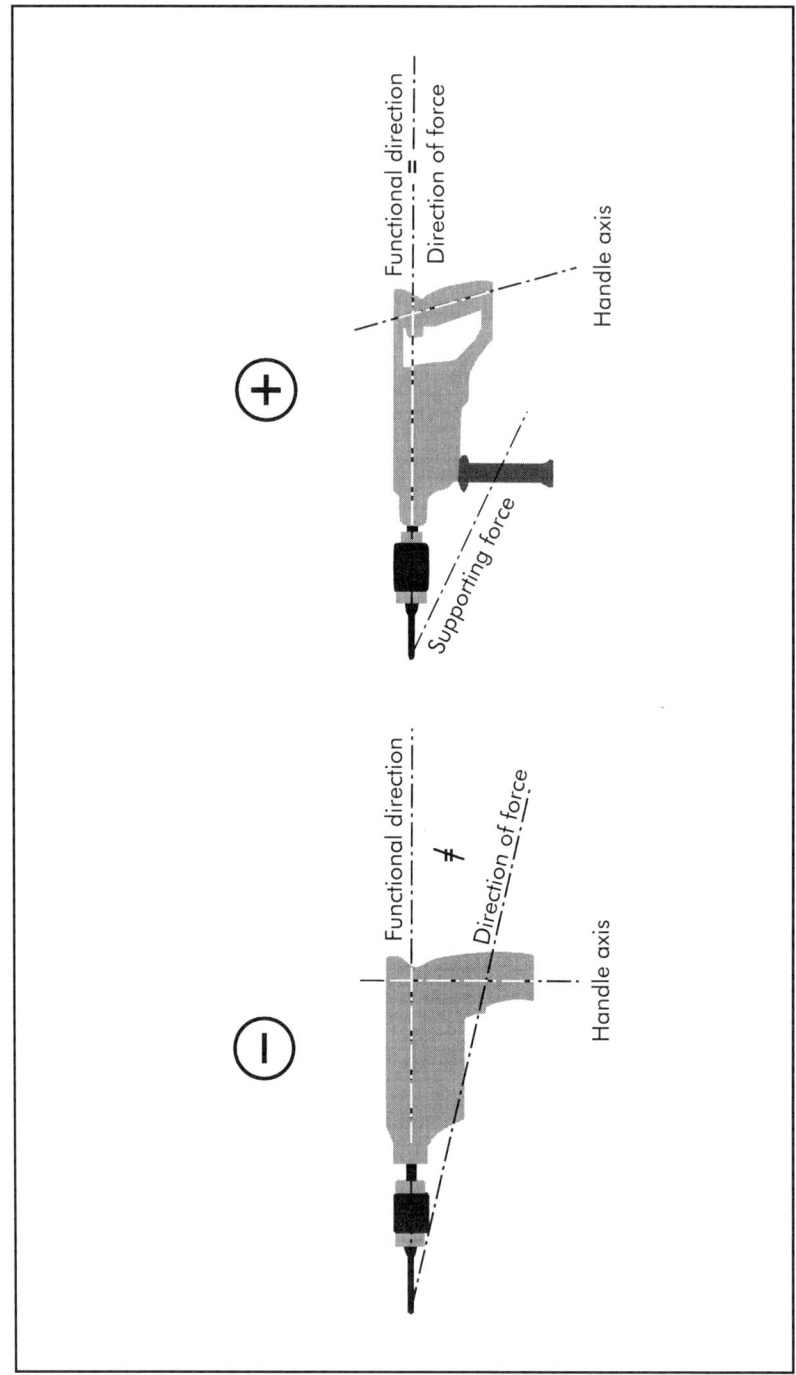

Figure 26-5 Correct and incorrect assignment of motion in the ergonomic design of a hand drill.

only from the grinding wheel side. Figure 26-6 shows that a small operator can look at the screw points only if standing on tiptoe and bending far across the spindlestock. The strains resulting from this physiologically unfavorable posture are intensified in that the operator knocks against the water duct with his knee and is prevented from fully approaching the machine bed.

In a redesign of the machine, the front door became a multipart sliding door, with the spindlestock accessible directly from the front. A new mounting module with overhead screw points was conceived for the dressing tool, permitting direct visibility and tool change in an upright posture. Reduced width of the water duct near the front door and a foot entrance space allowed the operator to fully approach the machine bed.

Grinding Workplace

A large number of workers absent due to sickness from an unbalanced strain situation motivated the corrective ergonomic redesign of a grinding workplace. Women were exclusively working at this workplace in two shifts.

The grinding workplace is used for chamfering cutting tip inserts. In its initial condition, it consists of a work table not adjustable in height, with a carrier stand comprising a functionally self-contained grinding unit, including a grinding spindle, spindle bearing housing, and a drive. The workplace has a table surface that is not sufficiently dimensioned for the various work and transport operations or material storage. In addition, there are various material feeding, buffering, and auxiliary devices, one work chair, and one foot rest.

The constructively determined arrangement of the grinding unit beneath the tabletop and the restriction of the legroom associated with it, leads to distinct forced postures for the operators. In addition, a floor bar obstructs moving the chair under the tabletop, intensifying the effect of the forced posture. Further strains are compressions on the upper sides of the thighs, and hand and arm contact with hard, metallic edges. The foot rest is either used improperly or not at all. This situation leads to great discomfort and an increased risk of musculoskeletal diseases for the workers. Figure 26-7 shows the typical postures encountered in the initial condition of the grinding workplace.

A mandatory constructive requirement made on the redesign of the grinding workplace was not to modify the grinding unit. An analysis of various grinding unit positions revealed that only the arrangement of the spindle bearing housing in the legroom

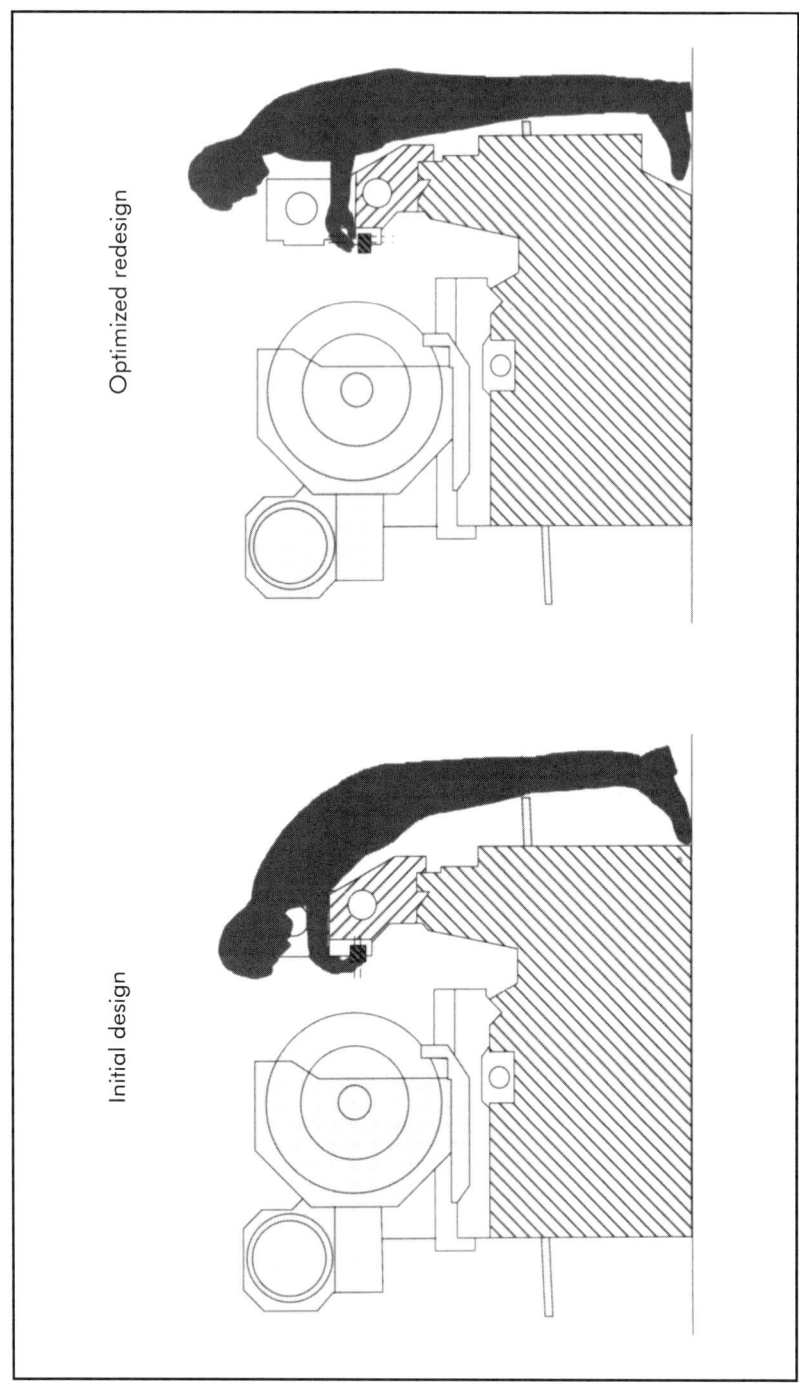

Figure 26-6 The optimized redesign of a circular grinding machine.

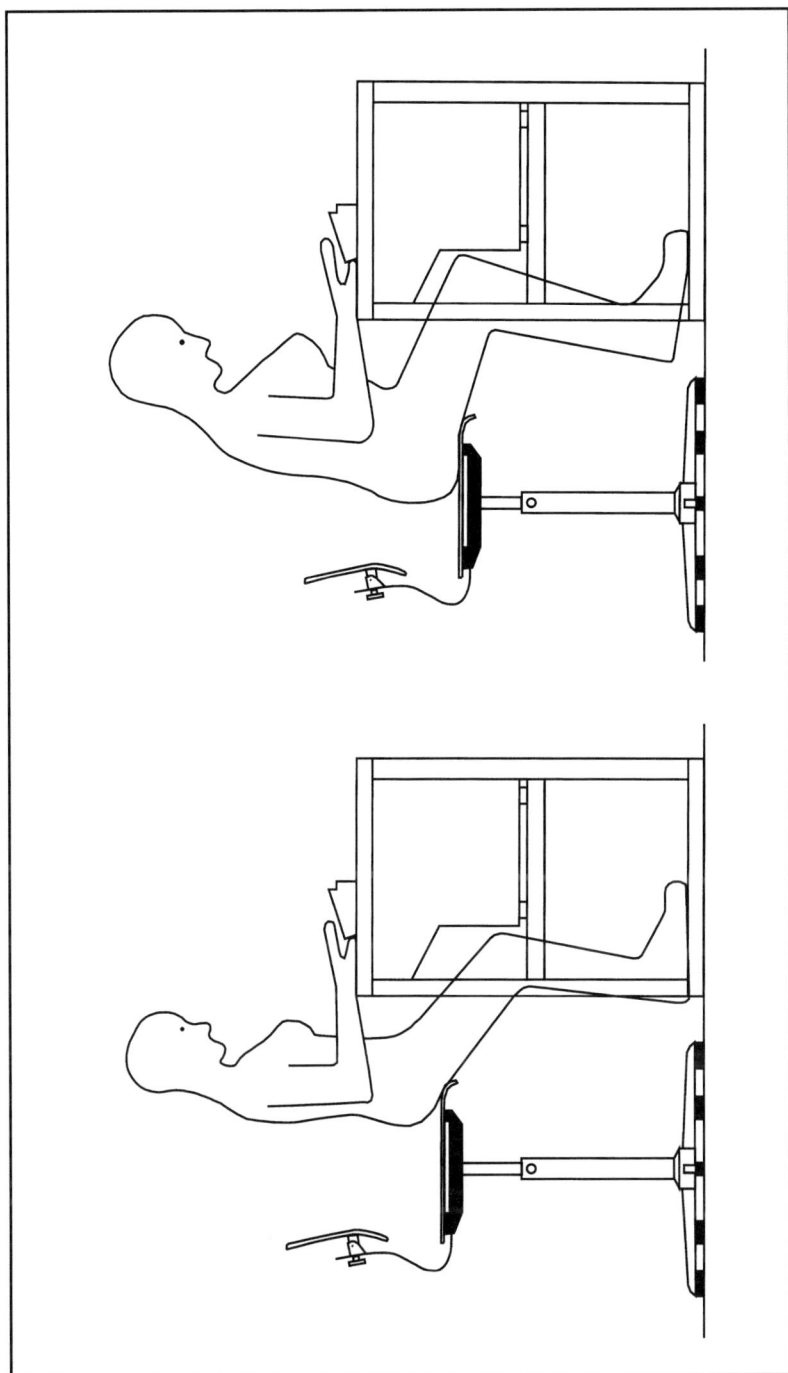

Figure 26-7 Analysis of the grinding workplace in its initial condition.

between the knees could meet this requirement without any forced postures dangerous to health. This made it possible for the constructive modifications in the arrangement of the grinding unit to be restricted to the carrier stand. Nevertheless, such a redesign of the grinding workplace was only a compromise. By ensuring an optimal posture, and accounting for the flexibility of the sitting posture, there is no doubt that the new design reduced the forced postures. However, the women working at these workplaces must continuously sit with both legs wide apart, and this must be rated critically regarding acceptance and reasonableness. For reasons of visual protection, the work table is enclosed at its rear and on its sides, providing protection from draught at the same time. Figure 26-8 shows the basic posture when working at the redesigned grinding workplace.

Apart from the modification of the carrier stand, the corrective design measures comprise a completely new table construction, including a larger tabletop with adequate working and stacking surfaces. The table, manufactured from aluminum sections, is not height-adjustable, therefore, a height-adjustable foot rest and a height-adjustable upholstered work chair are used. For supporting the hand-arm system for fine-motor activities, the workplace has two upholstered armrests arranged along the front edge of the table. In addition, the bottom end of the grinding spindle and drive belt is protected by an enclosure. All contact surfaces of the legs are upholstered to provide protection against striking and touching cold metallic surfaces. The workplace is made complete by appropriate auxiliary devices, lighting equipment, energy supply equipment, and lockers for personal belongings. Figure 26-9 shows the corrective redesign of the working surface.

SUMMARY

Ergonomics activities in Germany are founded on long tradition, although their scientific and practical importance did not grow until about 1950. In the course of ergonomic design, scientific knowledge of ergonomics is translated into practice.

Current motives for ergonomic design activities include economic, humane, legal, and sociodemographic factors. In view of the growing relevance of these factors, increasing importance of ergonomics can be expected.

Research activities in ergonomics can be found at numerous public and private institutions. Such activities are conducted by

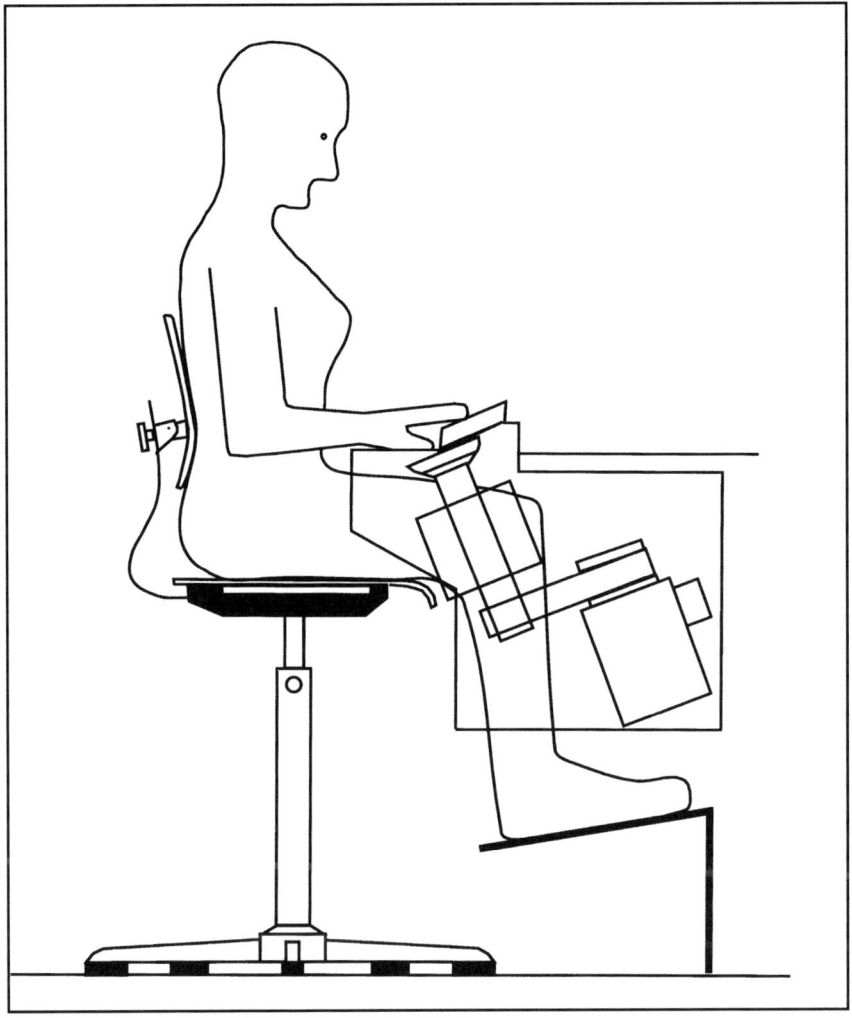

Figure 26-8 Basic posture when working at the redesigned grinding workplace with the grinding unit in its original position.

university facilities, the research facilities of industry and economy, and in the widespread service sector. Applications of ergonomic design invariably concern all kinds of companies and sectors. Job design, as well as workplace, equipment, and consumer product design constitute essential fields of application.

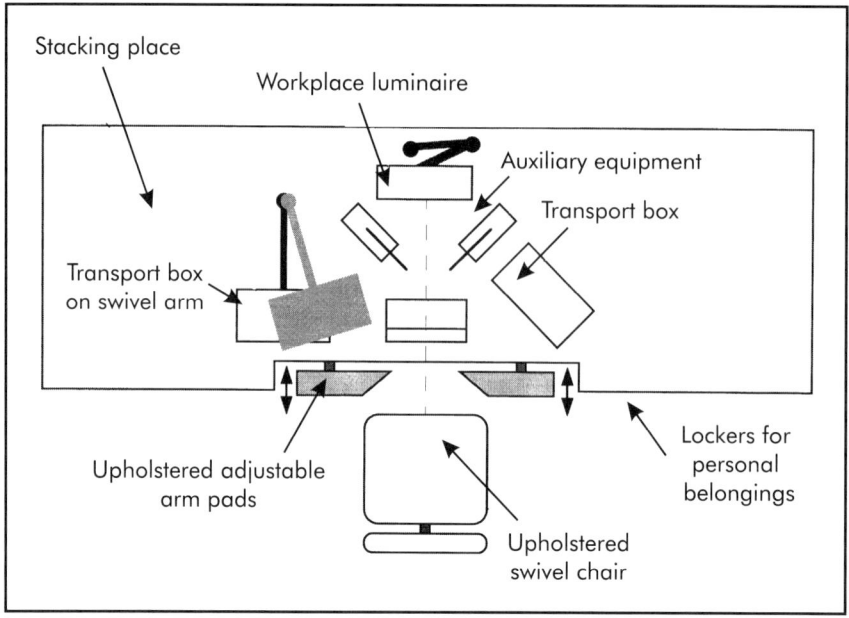

Figure 26-9 Corrective design of the working surface.

REFERENCES

Bullinger, H.J., Volkholz, V., Betzl, K., Köchling, A., and Risch, W., eds. *Alter und Erwerbsarbeit der Zukunft.* Heidelberg, Germany: Springer, 1993.

Bullinger, H.J. *Ergonomie-Produkt-und Arbeitsplatzgestaltung.* Stuttgart, Germany: Teubner, 1994.

EKD, ed. "Arbeit, Leben und Gesundheit; Perspektiven, Forderungen und Empfehlungen zum Gesundheitsschutz am Arbeitsplatz." *Kirchenamt der Evangelischen Kirche in Deutschland (EKD).* Gütersloh, Germany: Mohn, 1990.

Kirchner, J.-H. "Arbeitswissenschaft in Forschung und Lehre." *Rationalisierung* 1973, 24: 31-33.

Klauder, W. "Ausreichend Mitarbeiter für Tätigkeiten von morgen?" *Alter und Erwerbsarbeit der Zukunft.* H.-J. Bullinger, V. Volkholz, K. Betzl, A. Köchling, and W. Risch, eds. Heidelberg, Germany: Springer, 1993.

Luczak, H., Volpert, W., Raeithel, A., and Schwier, W. *Arbeitswissenschaft, Kerndefinition-Gegenstandskatalog-Forschungsgebiete.* Eschborn, Germany: Rationalisierungs-Kuratorium der deutschen Wirtschaft, 1987.

Rohmert, W., and Luczak, H. "Einführung in die Ergonomie." *Handbuch der Ergonomie,* 2nd Edition. Bundesamt für Wehrtechnik und Beschaffung, ed. München, Germany: Hanser, 1989.

Statistisches Bundesamt, ed. *Statistisches Jahrbuch 1995 für die Bundesrepublik Deutschland.* Stuttgart, Germany: Metzler Poeschel, 1996.

Statistisches Landesamt, ed. *Zahlen zur Bevölkerungsentwicklung.* Stuttgart, Germany: Statistisches Landesamt Baden-Würtemberg, 1991.

CHAPTER 27

CORPORATE ERGONOMIC EFFORTS IN SWEDEN

Åsa Gabrielsson
Project Leader, Quality and Process Development
Division of Industrial Ergonomics, and
Center for Studies of Humans,
Technology and Organization

Jörgen Eklund
Associate Professor
Division of Industrial Ergonomics, and
Center for Studies of Humans,
Technology and Organization

Gunnela Westlander
Professor
Division of Industrial Ergonomics

Linköping University
Sweden

There is a long-held tradition in Sweden and the other Scandinavian countries of taking into account the well-being of the work force. During the 17th and 18th centuries, forestry, mining, and metal production were expanding. Because of the small population, there was, and is now, a shortage of workers. So, the consideration of worker well-being not only came from respect for people, but also from economic realism.

The character of work tasks in raw material production (wood and iron) emphasized the focus on physical working conditions, which is also reflected in the Swedish definition of *ergonomics*, which is "a cross-disciplinary research and application field dealing with integrated knowledge about human characteristics and needs in the interaction between humans, technology, and environment when designing technical components and work systems."

Historical and cultural factors have been of great importance for the adoption and development of the sociotechnical tradition.

The relatively good collaborative climate between management and workers has been one important element for what is often referred to as the "Swedish model." The unions and the employers' association play a dominant role. Participative, semiautonomous work groups, decentralized decisions, and collaborative efforts for change have become dominant ways of industrial development. Only a small minority of the labor force has chosen not to become union members. Ergonomics and work environment issues have been very important issues in the discussions between the parties in the labor market, both centrally and locally. Today, organization is seen as an important means to obtain improvements in ergonomics and physical work conditions.

STRATEGIES ON THE NATIONAL LEVEL

An issue debated during the preparation for Sweden's Work Environment Act of 1978 was whether research findings could provide a basis for any regulations applicable to organizational and psychosocial conditions of work. In the final instance, very little space was allocated to such matters, which prompted considerable disappointment in many quarters. The reaction of the labor unions was especially negative. It was at their prompting that greater attention came to be paid to the area of work organization, giving rise to a number of specific recommendations. One was to arrange further training for persons employed in the occupational health services (physicians, nurses, safety and hygiene engineers, physiotherapists, and occupational health psychologists). This was supported and implemented by experienced researchers at the Research Department of Sweden's National Board of Occupational Safety and Health in 1978. At that time, a decentralized occupational safety and health organization was built up in Sweden. Larger companies employed their own, smaller and middle-sized enterprises that jointly formed an occupational health services center. Financing came partly through the owner companies and partly from state subsidies.

The year 1993 became a turning point. It was decided that a significant amendment should be made to Sweden's Work Environment Act. Formal responsibility for the work environment was transferred from the Occupational Health and Safety Committees (where both employer and employees are represented) to the companies and their managers. This inevitably required greater attention to the views and problems of managerial staff. The interests of management had to be met with more respect, while the physical health and mental well-being of employees could not be neglected. The Occupational Health Services (OHS) no longer had the occupational health and safety committees as channels through which the company could be accessed.

A second (and dramatic) change was the removal of government financial support for OHS. The major and rapid expansion of OHS initiated and funded by the Social Democratic government during the 70s and 80s came to a halt. The benefits of the services were disputed, as was the need for a state-subsidized system. OHS was included in a general austerity program. The idea was that the need for OHS would be demonstrated by its capacity to survive in a supply and demand context.

A further dramatic change (especially for Sweden) was the rapidly increasing rate of unemployment. This had a multifaceted background, with many of the factors operating in the same direction:

- Attempts to reduce public-sector expenditure and, thereby, staffing levels;
- Efforts to increase productivity throughout the economy using ultramodern technology, which does not seem to increase the demand for labor;
- Substantial reductions in taxation, which may have given citizens greater individual freedom, but did not seem to have markedly affected employment opportunities; and
- Increased competition from other countries.

The OHS professionals became suddenly confronted with a previously inconceivable task: helping managers make personnel as dispensable as humanly possible. Managers, who are now the commercial clients of OHS, make no clear distinction between problems involving the development of a good and healthy work environment and those psychosomatic and psychosocial problems arising from personnel reduction.

INTERNAL CONTROL

As the role of the OHS became steadily weaker, strong efforts were put into developing and improving the regulation systems. A recent example of those efforts is a new ordinance on rehabilitation (July 1994). The employer now needs to be attentive to the needs for rehabilitation, take the necessary measures to satisfy the employees in this regard, and give financial support to these measures.

Another recent example is the ordinance on internal control of the working environment. This ordinance lays down special requirements concerning the manager's way of planning, directing, and following up on activities from a work environment viewpoint. The ordinance applies to all employers, whatever the size of the enterprise. Among other things, it mandates that employers have a work environment policy, action plans, a clear allocation of duties, and routines for mapping out risks and investigating accidents and ill-health (Statute Book of the Swedish National Board of Occupational Safety and Health Ordinance 1992). The Swedish Employers' Association has compiled complementary handbooks and guidelines for managing internal control procedures. To facilitate the implementation of internal control for small enterprises, the Swedish Institute of Production Engineering Research has published a manual. A ten-point system offers a step-wise procedure to be followed by a team consisting of the chief executive, production manager, personnel manager, and safety representative (see Table 27-1).

Useful tools for internal control are also being developed by researchers and research and development (R&D) experts, especially to support the management of companies where the demands on safety control are extremely high. Examples of this can be found in process industries, in the Swedish nuclear power companies, and where the near-accidents and risk patterns have not been mapped out particularly well, as in highly automated production at manufacturing companies (Backström and Döös 1996). The internal control procedure is usually participative in that manager, supervisor, operator, and production engineer are all involved in the assessments.

IMPACT OF THE SWEDISH FOUNDATION OF WORK LIFE

Many companies interested in ergonomic and organizational improvements apply for financial support from the Swedish Foundation of Work Life. This national foundation was established at the end of the 80s to support technological and managerial innovations and rehabilitation activities initiated by employers. Its financial basis came from a temporary six-month fee of 1.5% of the salary from all employees in Sweden, a fee originally imposed by the government to slow the rate of inflation. In total, $1.5 million were spent on projects in different regions of Sweden. The funds, intended only for practical improvements, were mainly distributed to workplaces seeking grants for implementation of programs related to changing work conditions. The financial support made up, on average, one third of the companies' actual costs. Almost 25,000 workplace programs (of which 75% were in the private sector) received grants and, at the end of the period, more than 12,000 minor contributions were

Table 27-1. The ten-points yearly program for managers of small enterprises*

	Corresponding to NBOSH 1992:6 paragraphs
1. Identify deficiencies and define concrete goals and measures for improvement.	5 and 6
2. Specify operational procedures for time budgeting, decision making, documentation, information activities, rehabilitation, and risk management.	7
3. Work out a systematic action plan containing targets, means, goals, and outcomes aimed at the current year; decide criteria for effect evaluation.	7
4. Who will be in charge of what? Find a suitable distribution of the tasks and duties to be carried out.	8
5. Register near-accidents and accidents and where they have occurred in the plant. Also list the occupational injuries (small as well as severe) and in what way they were related to work.	12
6. Trace if lack of knowledge and training contributed to apparent security and safety problems. Infer corresponding elements in the regular introduction program of the plant and/or in further training if it proves necessary.	9 and 10
7. Make an annual survey of absenteeism, turnover, and accidents in the plant to explore positive and/or negative trends. If necessary, try to find measures to turn the situation in a positive direction.	12 and 13
8. Compare the current action plan with those from previous years to find out if there are any measures that should be followed up or re-evaluated.	13
9. If some workstations should be given special attention due to risks or problems with work and/or machine design, take time to carry out a workplace analysis (on a single operations level concerning the kind of risks, protective devices, safety routines) to be able to suggest adequate measures for improvement.	11 and 12
10. Collect and study the national ordinances and recommendations relevant to the kind of production in your company.	

*Source: Swedish Institute of Production Engineering Research (1993). Publication no. 92831, 2nd Edition (free interpretation). (Based on National Board of Occupational Safety and Health [NBOSH] ordinance 1992:6—Internal Control of the Work Environment.)

used for disseminating knowledge and experience among companies. One fifth of them concerned traditional ergonomic efforts, another fifth rehabilitation, and slightly more than 50% aimed at changes to improve overall efficiency of the work organization. The smaller private enterprises (fewer than 50 employees) received about three times more money per employee than did the larger companies.

In 1995, before this program was terminated, a number of evaluations were carried out by Swedish researchers. The most extensive of these consisted of 1,200 projects on efforts to improve work organization (Gustavsen et al. 1996). This nationwide program has had a long-term effect on managers' awareness of keeping ergonomic standards as high as possible.

ERGONOMICS AS A PROFESSIONAL FIELD

OHS revealed the need for an improved level of personnel education and training, demonstrated by a substantial number of OHS employees becoming registered as European Ergonomists. Also, there is now a tendency for more academics to be involved. Sweden started a national registration of ergonomists in the early 1990s, but it has now joined the criteria agreed by the Center for Registration of European Ergonomists (CREE), which also includes a code of conduct. Recently, several one-year academic educational programs in ergonomics have been created, based on the CREE criteria.

The number of researchers in the field of ergonomics graduating from Swedish universities is continually increasing. More ergonomists are getting prominent jobs in industry or are becoming consultants. They deal with production, product design, and management of ergonomic change programs. In recent years, leading Swedish industries have strongly emphasized ergonomics in the design of their products and created or strengthened the broad competence of ergonomics groups. These groups focus on strategic issues of product development and service tasks, partly due to the customer focus of the companies. Thus, the ergonomics profession is going through a time of increased professionalism and recognition.

ERGONOMIC EQUIPMENT FOR INDUSTRIAL PRODUCTION

Swedish products have long been characterized by quality, ergonomics, functional design, and safety. Such products include Atlas Copco tools, Bacho hand tools, BT fork lift trucks, Husqvarna chain saws, Hörnell welding visors, and, of course, cars and trucks from Volvo, Saab, and Scania.

Earlier, company strategy was often to price standard products lower than products with improved ergonomic features. Ergonomic features were often added to the standard product at a later stage, which increased both costs and the number of variants in production.

A new trend can be seen where several companies have inverted the philosophy. New products are now designed with ergonomics integrated from the beginning. The new ergonomic product is the volume product and manufactured in large numbers, which often reduces costs and increases competitiveness in the marketplace. The old nonergonomic standard solutions can be offered to customers who prefer them. However, some companies have increased the price, even to a level above the cost for the new ergonomic volume product, to avoid a pricing policy that counteracts the new strategy.

Substantial R&D efforts have been directed toward improving the design of products and equipment for use in production. One example is hand tools. A conceptual model of influences in hand tool use has been developed by Sperling and collaborators (Sperling et al. 1993) (see Figure 27-1).

For evaluating hand tool work, a cube model considering force, precision, and time has been developed (Sperling et al. 1993) (see Figure 27-2). This model is readily applicable, and emphasizes several important influences on the ergonomic situation. In another version of this cube model, the variable "precision" has been changed to "work posture," which may be more relevant in certain applications. Finally, this model is one example of how the field of ergonomics is expanding in Sweden and how the number of factors for consideration in creating good working conditions is increasing.

The need to integrate ergonomics, not only in product development and workplace design, but also in manufacturing strategies, is being recognized more and more today. Ergonomic implementations that aim at far-reaching improvements in working conditions and overall productivity need to be integrated into an organizational context (Winkel and Westgaard 1996).

CASES FROM THE MANUFACTURING AND SERVICE INDUSTRIES

Today, virtually all large Swedish companies have introduced team organization on a broad scale. The motive for establishing work forms that enhance the personnel's all-around development,

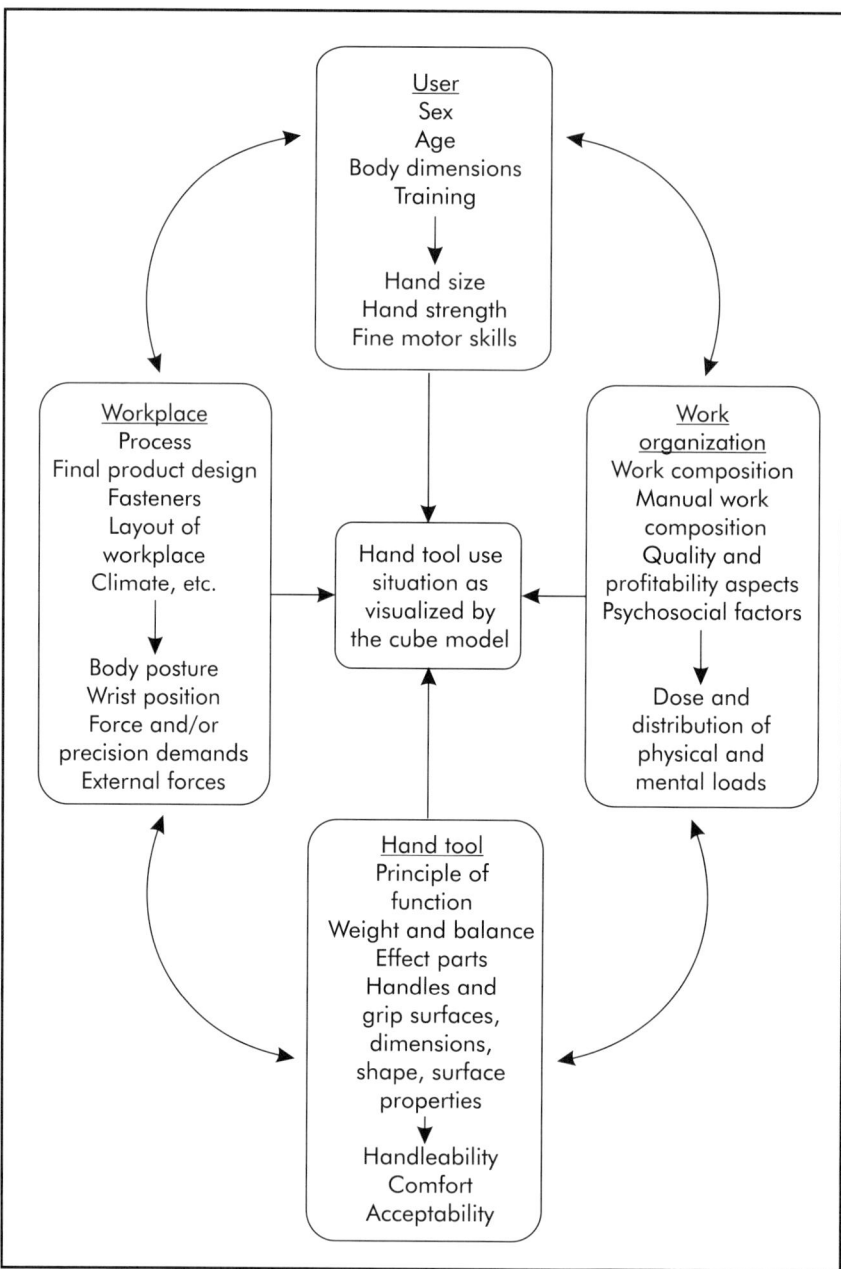

Figure 27-1 A model of influences affecting hand tool usage (Sperling et al. 1993). (Reprinted with permission from "Applied Ergonomics," 1993:24(3), Butterworth-Heinemann.)

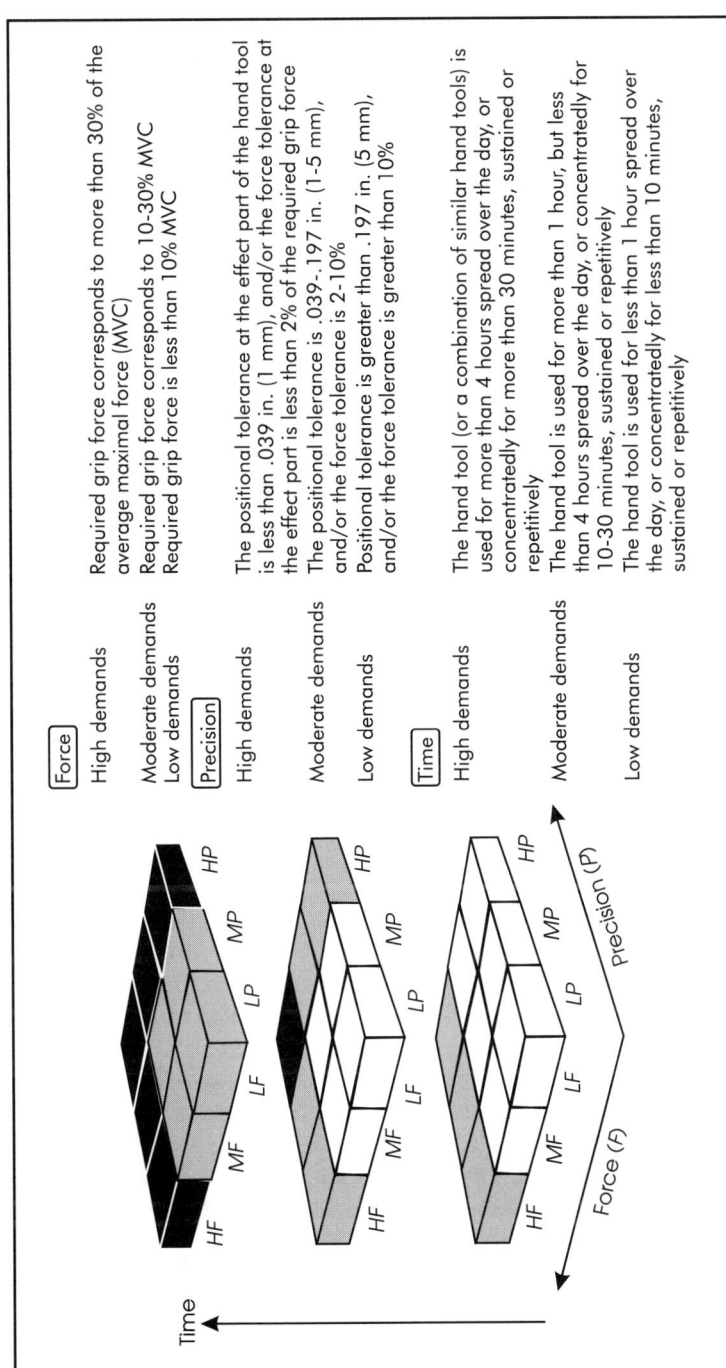

Figure 27-2 The cube model and proposed limits for acceptable and unacceptable work with hand tools. White subcubes = acceptable, grey subcubes = situations that must be investigated further, and black subcubes = unacceptable. L-low, M-medium, H-high (Sperling et al. 1993). (Reprinted with permission from "Applied Ergonomics," 1993:24(3), Butterworth-Heinemann.)

stem not only from work environment and ergonomic issues, but also from company strategic issues (Gabrielsson 1996). Following are some examples of companies that have particularly inspired and accelerated the development of new models for industrial work.

THE CASE OF ABB

In 1990, the management of ABB Sweden initiated a corporate-wide customer focus program named T50. This program is the most well known in Sweden, and has offered inspiration and ideas to many other Swedish firms. The objective of the program was to satisfy customers and motivate employees. The means to reach these goals was primarily by exacting time objectives, that is, all lead-times within ABB Sweden were to be halved by 1993 at the latest. This, management considered, could only be achieved through continuous decentralization of responsibilities and power, which, in turn, required a focus on personnel skills development so as to create increasingly independent and multi-skilled employees.

Management focused on lead time since it is closely related to quality, productivity, and profitability and is easily understood and communicated. Since the program was launched, the T50 concept has been refined. From a focus on time only, the program was developed to use a more complete "tool box." The vision of T50 was a program with a beginning, but no end.

Creating the T50 Program

At the outset in 1990, ABB Sweden arranged two awareness meetings attended by 400 company executives and union representatives. The purpose was to initiate a transformation process and mark the starting point of the T50 program. Highlighted were such problems as the increase in costs without any corresponding increase in productivity, the high turnover of personnel, absenteeism, and the difficulties experienced in recruiting young people to industrial work.

The awareness meetings were then repeated, thus reaching 2,000 managers and supervisors. Later, various other management seminars were held. Information activities were aimed at employees in Purchasing, Quality, Personnel, Marketing Communication, Sales, and R&D. Each group's specific role in the T50 program was explained.

The T50 program involves all ABB-companies, including manufacturing companies with short-order cycles and fairly standardized products, as well as project-oriented organizations with order cycles of several years, involving a large number of suppliers. Common features considered necessary for success were: commitment and involvement from top management; an engaging vision; clear objectives; a widespread sense of urgency (awareness of the need to improve); communication of the vision and objectives; training program and qualified support; feeling of ownership by those most affected by the change program; order of the process—"top down-bottom up"; and integration with key suppliers and supply management.

Each company was to assign one member of the senior management team as the person responsible for T50. This person also would be the liaison officer between the company and the central T50 team. Every third month, the T50 team arranged a one-day seminar for those in charge to facilitate an exchange of experience. Further, it was recommended that the strategic plans should reflect the potential of the T50 program, and the budget for the coming year should include costs and revenues related to the program. T50 was also placed permanently on the agenda of company board meetings.

The T50 program was primarily based on a massive information campaign on both the corporate and company level, appropriate training programs, and several company-run projects in a pilot or full-scale format. The different projects had to be centrally authorized to be recognized as T50 projects. The ABB companies could also apply to become T50 companies. For authorization, certain requirements had to be fulfilled, as shown in Table 27-2.

Target-oriented Teams

T50 projects often included the formation of target-oriented teams, and ABB Sweden had more than 1,000 of them by 1994. The use of target-oriented teams implied a need to develop team members' knowledge and skills, including methods for total quality management. Also needed was training in basic capabilities such as material planning, quality control, packing and transport routines, as well as human relations subjects (for example, how to relate in a group and how to handle conflicts). Implementation of target-oriented teams also demanded that corporate and company management learn about team dynamics, leadership in team-oriented organizations, and how to propel continuous improvement forward. New roles had to be designed for first-line managers whose role was changing to that of coach, whose main responsibility was to facilitate the work of others.

Table 27-2. Requirements for authorization of T50 projects and T50 companies

T50 Project
• Clear, far-reaching goals with respect to customer value and total cycle time • A major unit/division fully involved in a plan for expanding a pilot project • Skills development of employees included • Well-defined metrics • Focus on early results
T50 Company
• Time reduction > 50% since June 1990 • On-time deliveries continuously at 95-100% • Customer focus program showing results • Supply management showing results • Inventories reduced by 30% • Productivity improvement > 20% in 50% of the company's operations • Benchmarking of 50% of the company's operations • Company is based on target-oriented teams • Program for continuous improvement involves the majority of employees • Program/objectives for the development of leadership and employees' skills and attitudes

The Machines Division at ABB Industrial Systems AB—One Example of the T50 Project

To get a flow-oriented production organization based on target-oriented teams, division management decided to rearrange the whole workshop, an area of 200,000 ft^2 (18,580 m^2). In 1992, a project group undertook a preliminary study on existing production layout and processes. They identified and quantified different improvement potentials: shorter and faster material flows, adjustment of the production layout to future demands of target-oriented teams, reduction of utilized area, and an increase of the personnel's understanding of the production process. Also important, but not quantified, were ergonomic work improvement possibilities regarding lifting devices, noise and lighting conditions, and ventilation. Plans were made to make lunchrooms and group or meeting rooms more available for the target-oriented teams.

The project group decided the principal solutions for the final layout. They also determined the investment plan and scheduled activities. The principal solutions were then broken down and adjusted to the unit level where planning groups, consisting of a production leader, work group representative, safety representative, and production technician, carried out the final detailed solution. If necessary, the planning groups could get support from other function groups.

Members in the target teams had great influence on the final layout of the unit. Their participation created good solutions and also made the actual changes work very smoothly. The total layout project was paid back within 1.8 years.

The physical work environment was substantially improved, especially at one unit where workers' musculoskeletal disorders had been a great problem. The total transportation distance of product units was reduced from 8,900 to 2,700 ft (2,713 to 823 m) and the number of moves calling for the use of overhead cranes was reduced from 19 to two. Also, the new production layout became the necessary structure for utilizing the potential of the target-oriented teams.

AUTOMOTIVE CASES—VOLVO AND SAAB

Sweden's two car industries, Volvo and Saab, have become internationally well established, and export most of their production. Largely due to the sharp decline in sales around 1990, production in Sweden is now concentrated in their main plants in Göteborg and Trollhättan.

Car manufacturing has a high incidence of ergonomic problems due to the many physically and mentally monotonous tasks. The rate of musculoskeletal problems is high in all kinds of assembly work. Volvo, as well as Saab, has worked for decades to reform manual industrial work. They have experimented with shop floor reorganization and work groups to better meet the workers' needs for job involvement and skill development. Both companies are continuing these efforts, trying to obtain efficient production and good working conditions. They combine sociotechnology with aspects of Toyota's production philosophy. Groups working toward goals, multiskilling, a participative approach, continuous improvements, learning organizations, customer focus, and a very strong emphasis on change processes are commonly adopted (Sandberg 1995). Today's focus is on integrating company performance and human factors and ergonomics demands on work conditions.

Volvo and Saab make use of a similar concept in their production, namely KLE and QLE/H respectively, which means quality, timeliness, and economics. The sequence of these three terms is the order of priority, and for each factor there is a set of key variables to measure performance on the group level and to stepwise aggregate these variables to company level (see Table 27-3). Human resources are emphasized as a facilitating factor or the means of achieving the overall goals. The concepts are comprehensive, including, in addition to the previously mentioned factors, leadership development, and a flat organization structure with decentralized responsibility, authority, improved information, and communication.

Both Volvo and Saab have achieved substantial improvements in many of the key variables. Several reasons can be identified. Team organization with multiskilling increases flexibility and reduces vulnerability due to personnel changes. Work teams responsible for planning their own job rotation can more easily compensate when individual members are unable to perform certain tasks due to physical problems.

The use of human key variables has also increasingly been put into practice in other Swedish companies. Competence matrixes and other charts can be applied easily and aid decisions about human resource development.

Table 27-3. Commonly-used key variables

- Number of deficiencies
- Deficiency ratings
- Customer complaints

- Lead time
- Material turnover rate
- Delivery precision

- Working hours per car
- Value of products in progress
- Capital turnover rate

- Sickness, absenteeism
- Number of work injuries
- Number of improvement suggestions

- Competence matrixes—give work force and plant flexibility and vulnerability
- Investments in personnel competence development
- Motivation indexes
- Personnel turnover rate

SOCIOTECHNICAL JOB DESIGN IN SAWMILLS

Sawmills, a dominant industry in Sweden for centuries, are based on automated technology and can be described as process industries. Over a period of 25 years, the concept of a sociotechnical design of sawmills was developed by Ager (1993). This included idea development, development of supplier equipment, machine manufacturers, and consultants' know-how, with active collaboration from the sawmill industry. The concepts have gradually been introduced in several sawmills. To a large extent, five of the most recently-built Swedish sawmills have adopted this concept, which has gained a high degree of acceptance from union and company representatives. The main components of sociotechnical design in sawmills are described in Table 27-4.

A development model for the organizational change process from a traditional sawmill to a sociotechnical sawmill has been proposed by Ager (1993) (Figure 27-3). This model has been implemented in several sawmills.

SERVICE INDUSTRY—TOOLS FOR ERGONOMIC IMPROVEMENTS

Many Swedish companies realize the multifaceted background of ergonomic problems at the workplace and try to solve them by using so-called broad strategies. Depending on the problem and the scope of freedom for workers' participation in and their influence upon the planning of improvement measures, we find more or less top-down and bottom-up approaches. The effectiveness of some strategies have been demonstrated and documented.

Following is a typical example of how an ergonomic problem was solved, not exclusively by physical treatment and rehabilitation programs, but by changing a number of organizational conditions (Westlander 1995).

From 1989 to 1992, one service division of Swedish Telecom (now Telia AB), underwent a program of comprehensive organizational change; major efforts were made to improve the work environment of the approximately 300 telephone operators employed by the Manual Services Division.

One of the Telecom Group's head office specialists on occupational health issues, the project initiator, joined the managerial staff of Telecom Services to work out a program to find means and forms for solving the prevailing health-related work environment deficiencies. Over a period of three years, activities at Telecom Services were to be restructured with the aim of eliminating a series of work environment problems, including the operators' strong physical ties to the workstation (since the services could only be provided at a computer terminal), the intrinsic repetitiveness of the work, and inadequate premises. Taken as a whole, these problems were regarded as being the cause of extremely high absenteeism (relatively speaking), which was in itself a central problem from the perspective of company management.

The planned change was based on a package of several, dissimilar measures (a broad and diverse strategy) and favorable effects of different kinds were envisaged. The proposal involved:

- Changes in the work organization with a view to achieving more variation in telephone operators' work and greater scope for physical mobility;

Table 27-4. Characteristics of sociotechnology as applied in Swedish sawmills*

- Teamwork in autonomous production groups
  The group is formed around an organically-integrated part of the chain of operations, preferably along the material flow. The group has a rather high degree of autonomy regarding planning, production control, and internal organization of work. Certain administrative and maintenance tasks and a shared responsibility are delegated to the group. The group members are encouraged to develop multiskilling.

- Ager 1993, free translation aggregation or concentration of personnel
  By adapting material flow and buildings, as well as by decentralized maintenance, personnel are brought together as closely as possible. The purpose is to improve contacts, cooperation, and job rotation. Nonstraight material flows (U- or L-shaped) and a shared control room as a work center are examples of how to achieve this.

- Development of work roles with sufficient work content
  This means creation of functions, job descriptions, and a work organization, which both fulfills the demands regarding good job content and also suits the group of persons to be involved.

- Integration of processes and activities
  This principle means the creation of compact production departments including several operations, a separate flow and high degree of autonomy regarding planning, organization, and maintenance. The purpose is to support group organization and obtain a better overview, more holistic work, and a faster flow of material.

- Selected automation adapted to human characteristics
  Emphasis is put on the allocation of tasks to humans and machines, while considering the content of the operators' work. Also, the operators should be freed from unqualified and direct surveillance of material and machines. This will result in reduced manning, faster production, reduced costs, and elimination of dangerous or heavy tasks.

- Adaptable workplaces, adjustable to the individual
  To support variation of posture and job rotation, the workplaces should be designed to be adaptable, giving freedom of choice, mobility, and self-pacing. One example is to provide buffer areas for work to be performed instead of a machine-paced flow of work.

- Decentralized maintenance
  The production personnel perform some maintenance, and the integration and collaboration with the maintenance personnel are strengthened. Production disturbances are reduced and the possibility of increased numbers of personnel in critical tasks when needed is increased.

- Reward systems for quality and skill development
  The salary system and other rewards have to promote product quality and the utilization of raw material and production equipment. Skill development, cooperation, and shouldering responsibility should also be stimulated by the reward system.

*Source: Ager 1993, free translation

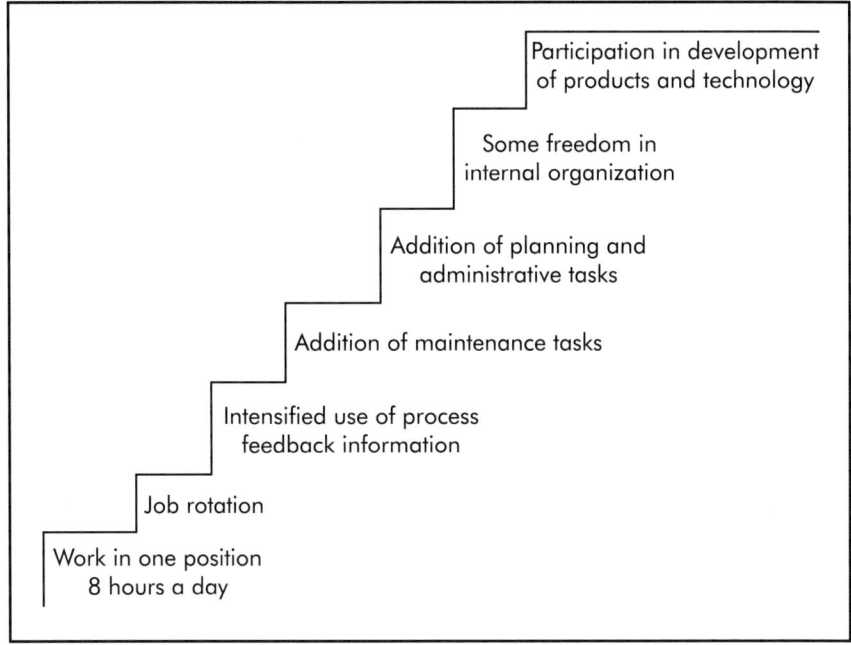

Figure 27-3 The development steps as a model for personnel and production organization development (Ager 1993).

- The combining of operator work with so-called service tasks so that there would be greater personal flexibility with regard to working hours;
- A changed organizational structure with the aim of promoting a sense of a group community, thus providing social contact between telephone operators at work;
- The rebuilding of the telephone operators' offices according to a new architectural design so that the physical environment was adapted to the new organizational form, thereby satisfying external physical conditions for the realization of an intragroup community; and
- More long-term development of computer facilities, which would allow a multiplicity of tasks to be undertaken at all workstations.

The implementation was continuously evaluated by an independent research team. The findings of the final evaluation indicated that, of the measures encompassed by the broad effort made to promote change and improvement of work-related health (release from musculoskeletal and visual discomfort from video display terminal [VDT] work), those concerning the work premises and the group form of organization were most successful. They were fully implemented. In particular, the new offices and their ergonomic design had the favorable effects that management desired.

CLOSING WORDS

The Scandinavian countries have, in many aspects, similar views on how occupational health should be taken into consideration in the continuous development and modernization of working conditions. This chapter concentrates on features of special interest to plant managers since they are in especially important positions for surveying and promoting worker well-being.

That is why we have given an overview of Swedish strategies in the form of legislation, institutionalization of health promotion, inspection, and intervention systems, as well as financial support systems. We have also found it important to describe how the professionalization of ergonomists is proceeding.

The selection of companies are meant to illustrate ambitious endeavors on the local level. The company cases described should be regarded as a few examples of many organizations going in the same direction.

It is a growing endeavor in Sweden and in the other Nordic countries to use the now extensive research findings from the various occupational health disciplines contributing to the field of ergonomics. There is also growing consciousness that strong efforts should be put on transforming this knowledge into practical methods easily accessible to and suitable for managers.

REFERENCES

Ager, B. *Creation and Implementation of Sociotechnical Design Concepts in the Swedish Sawmill Industry 1970-1993.* Doctoral thesis,TRITA-TRD R-93-1. Stockholm, Sweden: Department of Manufacturing Systems, Wood Technology and Processing, Royal Institute of Technology, 1993. In Swedish.

Backström, T., and Döös, M. "The Riv Method: A Participative Risk Analysis Method and its Application." *New Solutions,* 1996.

Gabrielsson, Å. *Industrial Work of the Future—Integration of Company Performance and Employee Development.* LiU-Tek-Lic-1996:13. Linköping, Sweden: Linköping Institute of Technology, 1996.

Gustavsen, B., Ekman-Philips, M., Wikman, A., and Hofmaier, B. *Concept-driven Development and the Organization of the Process of Change—An Evaluation of the Swedish Working Life Fund.* Amsterdam, The Netherlands: John Benjamin, 1996.

Sandberg, Å., ed. *Enriching Production—Perspectives on Volvo's Uddevalla Plant as an Alternative to Lean Production*. Avebury, England: Ashgate Publishing Limited, 1995.

Sperling, L., Dahlman, S., Wikström, L., Kilbom, Å., and Kadefors, R. "A Cube Model for the Classification of Work with Hand Tools and the Formulation of Functional Requirements." *Applied Ergonomics* 1993, 24(3): 212-220.

Statute Book of the Swedish National Board of Occupational Safety and Health Ordinance. *Internal Control of the Working Environment*. Solna, Sweden: AFS, 1992.

Westlander, G. "Means, Goals, and Outcomes of a Comprehensive Occupational Health Program for Telecommunication Workers." *International Journal of Health Services* 1995, 25(2): 313-332.

Winkel, J., and Westgaard, R., eds. "Achieving Ergonomics Impact through Management Intervention." Special Issue, *Applied Ergonomics* 1996: 27(2).

CHAPTER 28

ERGONOMICS AND TQM

Holger Luczak
Professor and Director

Kai Krings*
Research Assistant

Stefan Gryglewski
Research Assistant

Georg Stawowy
Research Assistant

Institute of Industrial Engineering and Ergonomics
Aachen, Germany

TQM PHILOSOPHY

Total quality management (TQM) is based on a broader notion of quality than that of older quality assurance concepts. In the TQM context, quality is not only focused on the product, but on the whole business, including all external relations and internal processes. One principle underlying the TQM philosophy is customer orientation (Figure 28-1). The idea of the customer as purchaser or user of a product or a service is broadened by the shareholder, the employee, and society as a whole.

According to the Japanese Institute of Standards, quality conforms to the requirements of the customer. *Quality*, defined this way, is based on the needs and expectations of customers and covers several dimensions.

The customer orientation applies to external, as well as internal customers. A purchaser customer, for example, has requirements with regard to product, styling, design, sales, and marketing, after sales service, and price. Internal customers pay attention to job satisfaction, leadership, education and training, supportive structure, communication, reward and recognition, and performance appraisal. Shareholders require profit. Society emphasizes environmental friendliness and savings for the national economy.

In terms of defining quality characteristics, the relationship between the enterprise and the customer is bidirectional. Anticipation and continuous adaptation of requirements necessitates intensive communication. Furthermore, the supplier influences the quality criteria by:

---

*Currently at: Schott Glass, Mainz, Germany.

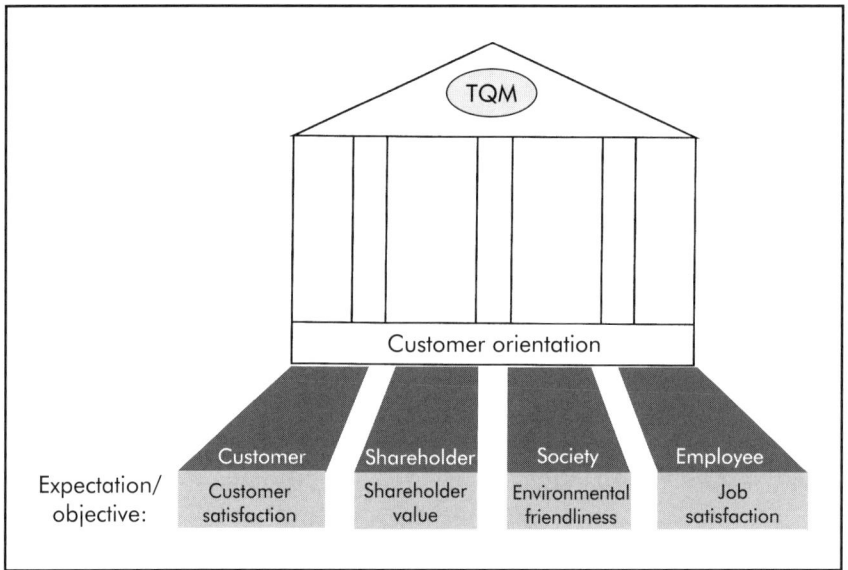

Figure 28-1 The objectives of TQM.

- Choosing markets, products, and services;
- Influencing the market and customer requirements (for example, marketing); and
- Weighing customer interests of purchaser, work force, shareholder, and society.

Defining and weighing quality standards also has implications for company ethics. Is the primary goal of the business exclusively profit-making or is there some consideration for the employees' interests?

To identify all internal activities of a company and its miscellaneous customer-supplier interrelations, it is necessary to structure them in terms of business processes, with individuals or groups as internal suppliers and customers.

The system of the business as a whole is a network of all employees interacting through supplier and customer relationships as the basic element of quality orientation—structured by organizational aspects (Figure 28-2).

This can only be achieved by involving all members of staff and management. It demands commitment across all hierarchies to continuously improve quality, as well as a business culture that fosters active participation (Grant et al. 1994). "TQM provides an

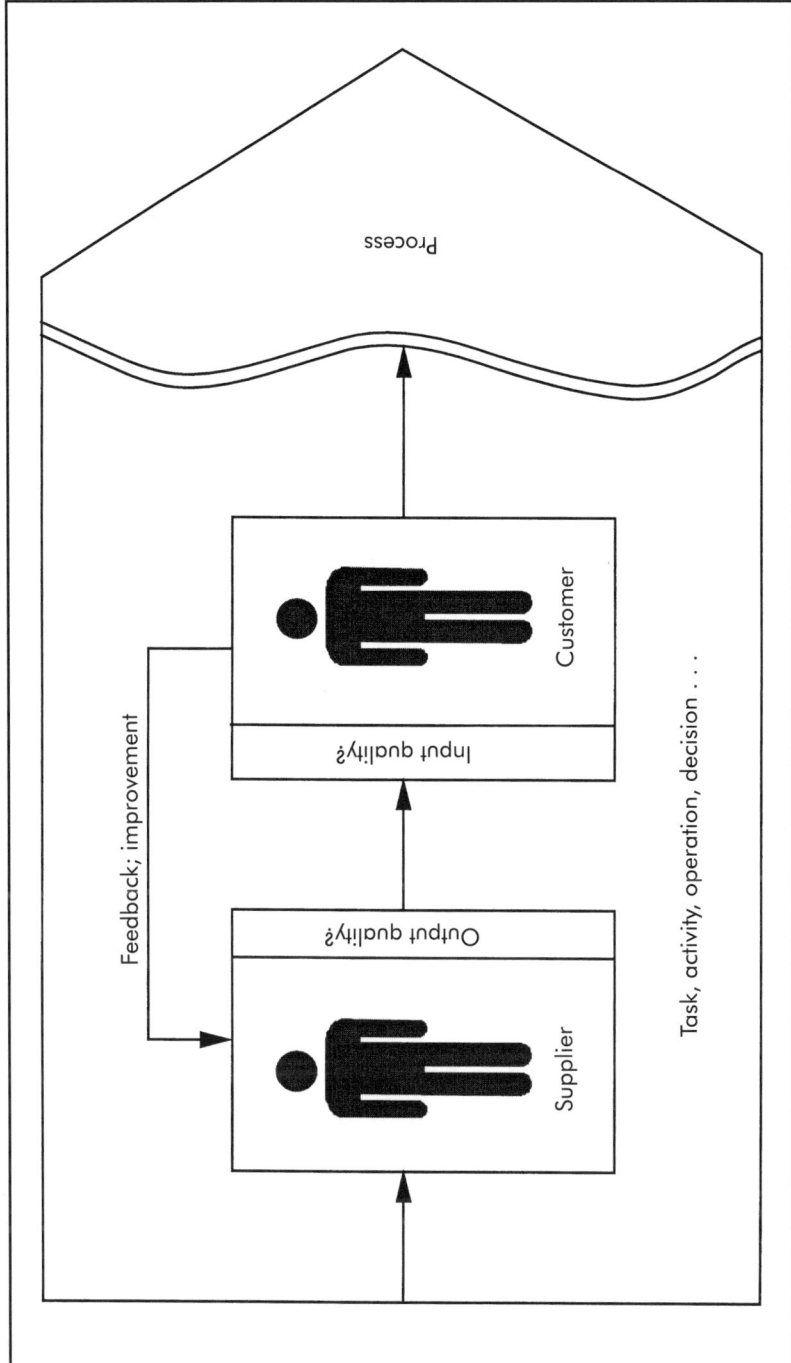

Figure 28-2 Process orientation and internal/external supplier and customer relationship.

environment where fear is eliminated, where all employees take pride in their work, where they feel part of the same team, and where the goals of the organization are their own" (Logothetis 1992).

In comparison to quality assurance and quality management concepts, total quality management regards input and output, and the entire business process from supplier to customer. Therefore, it must consider all elements of the system and more dimensions of quality (Figure 28-3).

To fulfill the TQM demand of a holistic approach, the philosophy makes use of different philosophies, strategies, and methods of improving quality (Kamiske 1994; Logothetis 1992) (Figure 28-4).

OBJECTIVES OF ERGONOMICS

Ergonomics covers several approaches to working life. Microergonomics regards isolated aspects of the individual and work environment. Macroergonomic techniques focus on a holistic understanding of individuals and the human relations among them. Macroergonomics takes into account the unity of motivation, needs, wants, qualification, social aspects, as well as the communication, participation, and interaction of the enterprise's employees. It can be viewed as a kind of ergonomics-TQM relationship.

Ergonomics, with its different focus points, demands a systematic approach to unify different scientific disciplines in terms of an efficient conversion and contribution to enterprise development. An ergonomical design process should be (Wojda and Hacker 1995):

- Holistic and system-oriented;
- Objective-oriented and prospective;
- Content- and process-oriented; and
- Situated.

Regarding these necessary process characteristics, the objectives of ergonomics are to:

- Design a working environment adapted to the individual's physiological needs;
- Cultivate and maintain performance; and
- Ensure employee acceptance toward working conditions.

ERGONOMICS AS A CHANGE AGENT

Within the course of time, the definition of quality has changed, thereby changing the methods and tools to achieve quality.

The objectives of recent quality management concepts are reflected by criteria systems to evaluate quality management (QM)

Ergonomics and TQM

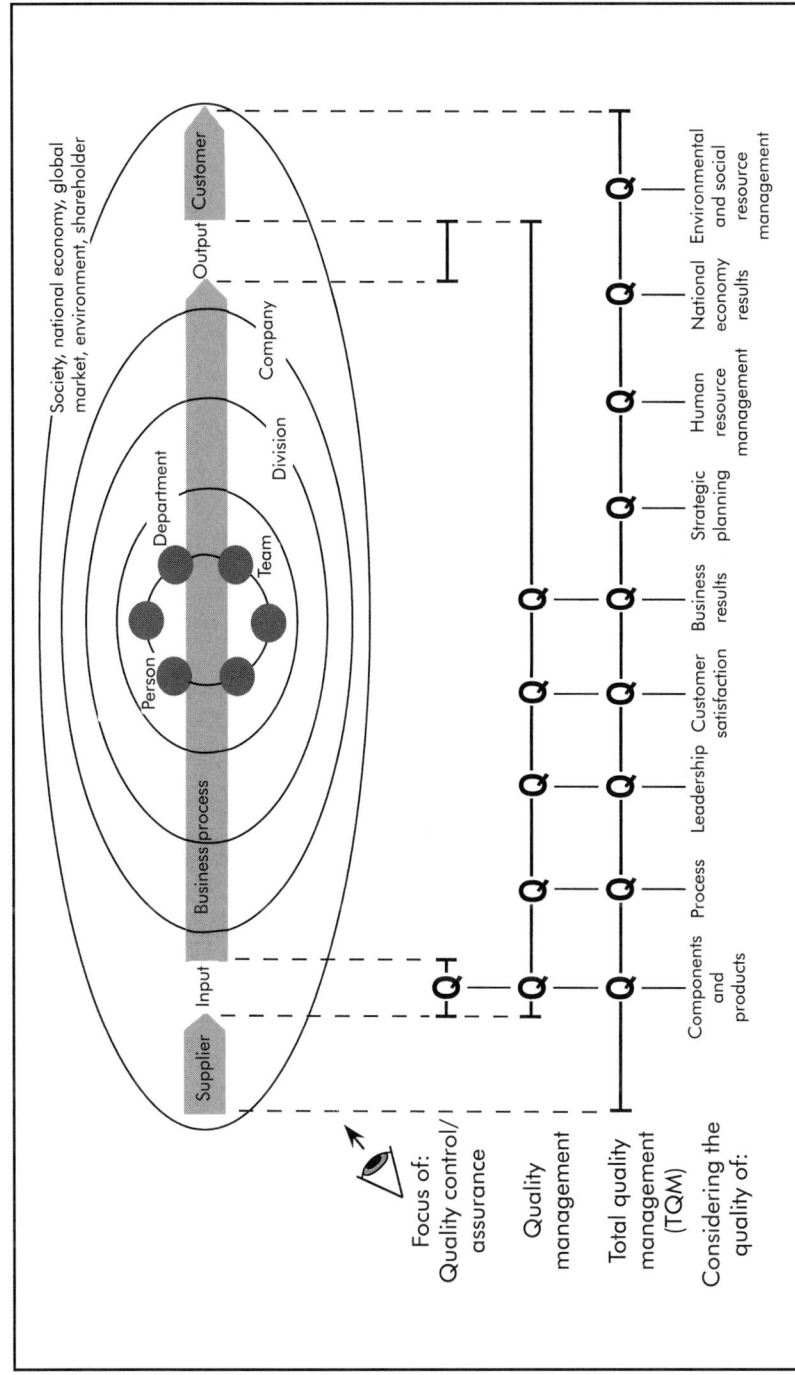

Figure 28-3 Dimension and range of quality concept.

Ergonomics in Manufacturing

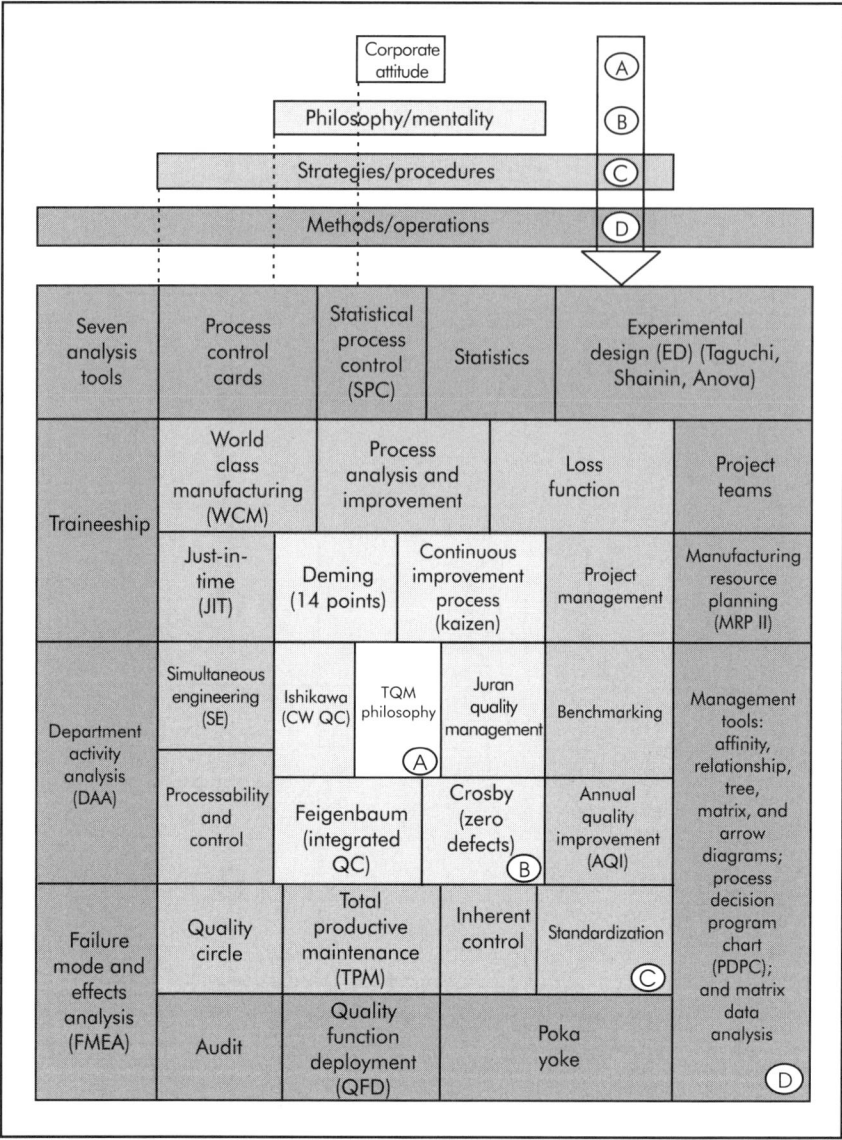

Figure 28-4 From TQM philosophy to methods for quality improvement.

efforts such as ISO 900X, the Malcolm Baldrige Quality Award (MBQA), and the European Quality Award (EQA). While criteria and their weighing differ among these evaluation systems, all three consider quality as product quality, process quality, and quality of working life. In comparison to the MBQA, the EQA puts more emphasis on the effects on society, therefore reflecting the tendency toward more holistic quality approaches.

The MBQA evaluates the achieved quality using seven criteria. Figure 28-5 illustrates the close relationship between the objectives of ergonomics and TQM. Black boxes symbolize a direct fit of goal, while white boxes indicate a correspondence to the preconditions or result. As shown in the matrix, the concept of ergonomics with its objectives covers all criteria, as defined in the MBQA. Therefore, quality management, in terms of a holistic and process-oriented quality notion, and ergonomics can be understood as different approaches with the same goals. TQM is contributing to process management, whereas ergonomics is providing appropriate content.

Having the improvement of quality in mind, the management and work force can choose an appropriate combination of quality philosophies, strategies and methods, and well-established ergonomical tools and methods like those mentioned in previous chapters of this book (Figure 28-6). But, the reasons are not evident for every quality deficiency. Some cannot be determined with well-modeled, analytical tools. For example, a system's complexity or

Correspondence between criteria of Malcolm Baldrige Award (columns) and goals of ergonomics (rows)	Leadership	Process management	Human resource evaluation	Strategic planning	Information and analysis	Customer focus and satisfaction	Business results
Working environment adapted to the individual's physiological needs		□	□	□	□		
Cultivating and maintaining performance	■	■	■	□	■	■	■
Ensuring employees' acceptance toward working conditions	■	■	■	□	□	□	□

■ Direct correspondence
□ Indirect correspondence

Figure 28-5 Correspondence between the goals of ergonomics and MBQA criteria.

the unexplored causal connections of physiological and psychological aspects can make it impossible to bring into play well-defined procedures, especially when conditions and strains are changing (Figure 28-7).

In these cases, the ability of staff to adapt to their working environment is important. This includes the ability to set long-term personal goals and change personal conduct, adapting it to altered situations and physical feedback. The continuous development of the capacity to act gives a direct link to a broad notion of health (Eckardstein et al. 1995; Ottawa Charter for Health Promotion 1986). In this sense, within a TQM context, there is a direct fit between ergonomics and occupational safety and health.

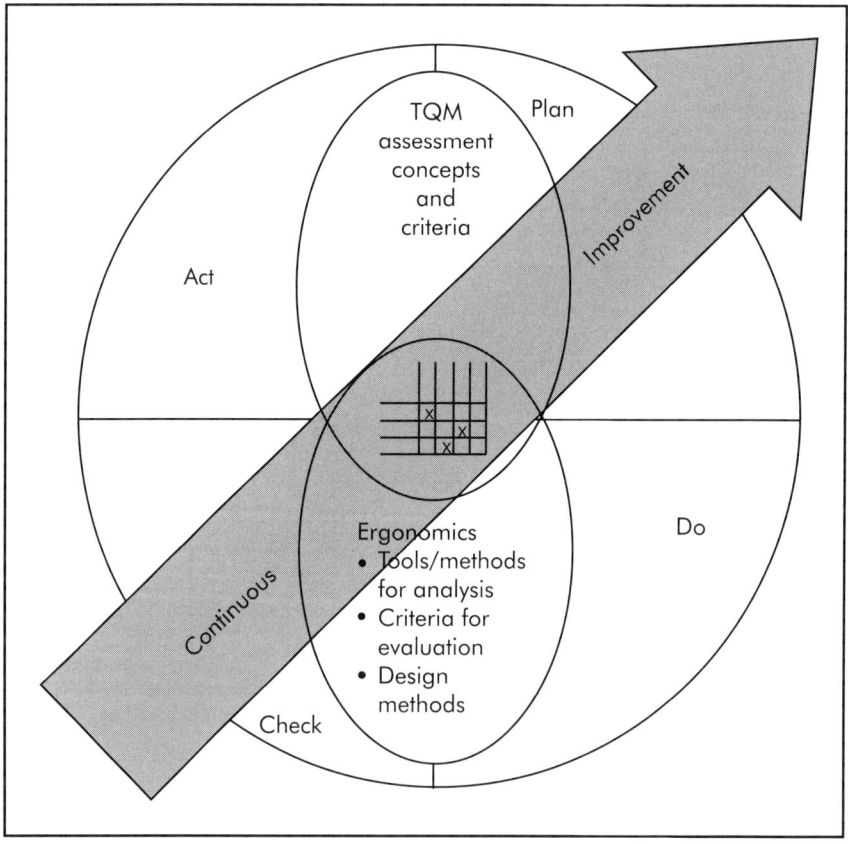

Figure 28-6 Analytical approach in well-modeled fields of ergonomics.

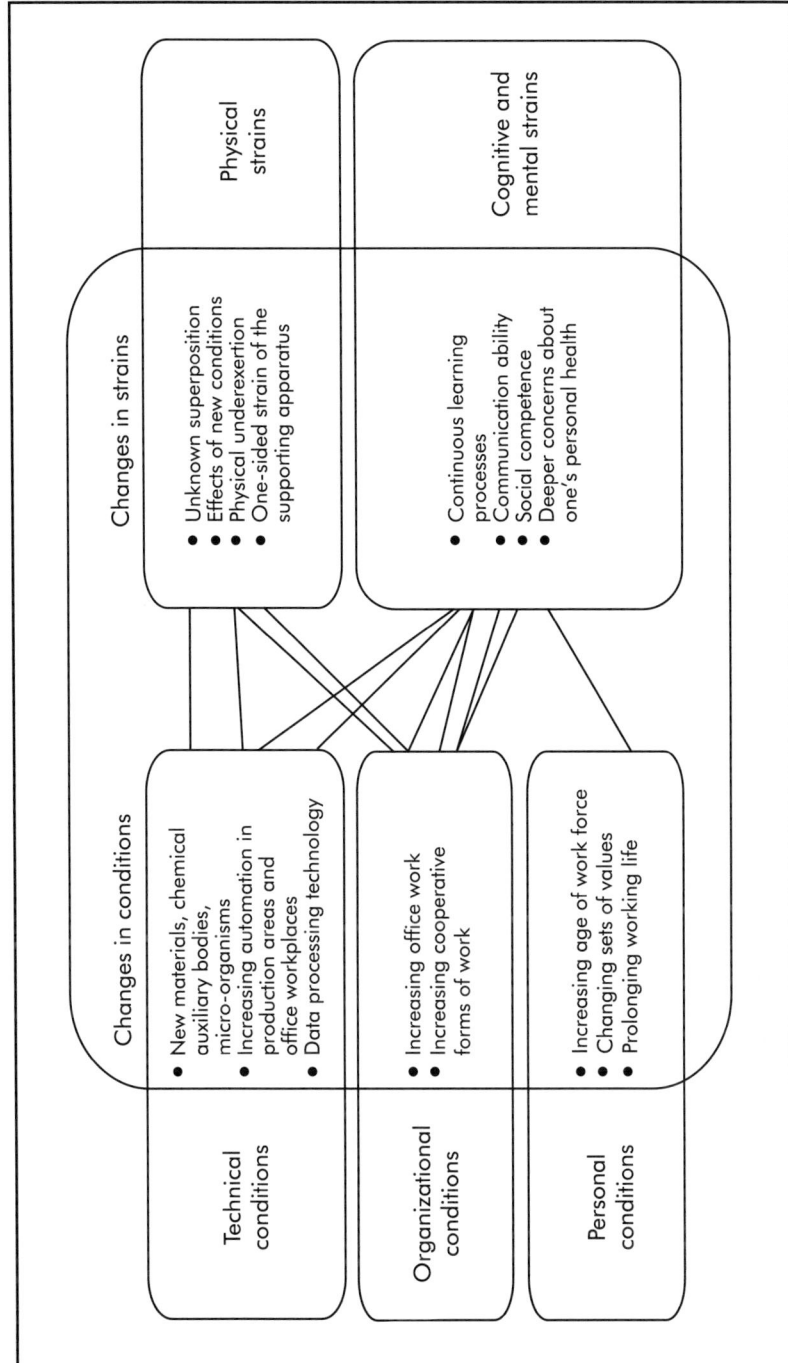

Figure 28-7 Changing conditions and strains.

For the individual staff member, as for the entire company, risk avoidance must be coupled with the use of personal development potential, which can only be achieved by appropriate processes (participation, organizational development, and continuous improvement). Such a holistic approach ensures the company's ability to adapt to changes in the conditional framework (Figure 28-8).

HEALTH PROMOTION IN TQM

Based on various dimensions of work-related human health and safety, the occupational safety and health approach has an impact on several quality dimensions:

1. The promotion of preventive occupational safety and health reduces the susceptibility of production processes to disruptions caused by accidents and sickness-related loss of working hours.

2. Through the long-term support of a healthy and capable work force, occupational safety and health ensures sustainable human resources management.

3. Health is a sine qua non for work force performance, and thus for the continuous improvement process, the basic principle behind the TQM philosophy.

OCCUPATIONAL SAFETY AND HEALTH AS A QUALITY TARGET

Even though the mutual dependency of safety and process security, as well as health and the entire quality of a company has often been pointed out in recent years, these aspects do not usually appear explicitly in the criteria systems for the evaluation of QM systems. Nevertheless, there is the necessity to integrate occupational safety and health promotion in TQM concepts (Krause et al. 1993; Rine 1994; Graham 1995).

ISO 900x generally contains few staff-oriented criteria. They relate to the necessity for professional and nonprofessional staff development. References to work safety and health promotion are missing.

Correspondingly, the Deming Application Prize only considers staff development with respect to education and dissemination criteria (Deming 1986). The Malcolm Baldrige Quality Award, however, also considers well-being and morale as a criterion for the quality category "human resource utilization" (National Institute of Standards and Technology 1995). These criteria correspond to the definition of health, which goes well beyond the absence of illness. The European Quality Award also focuses, to a considerable

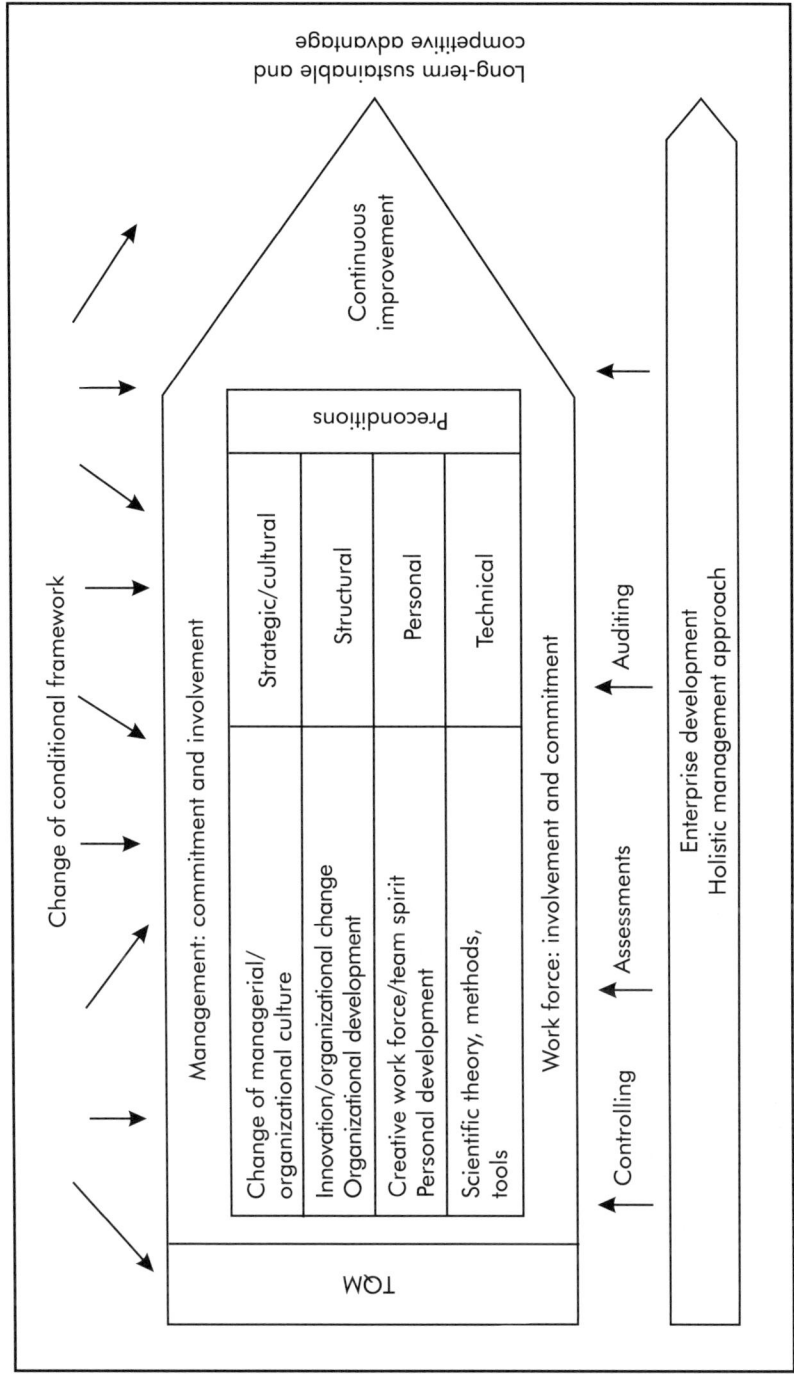

Figure 28-8 TQM in enterprise development.

extent, on human resource management, but generally remains more ambiguous in terms of the specific criteria (European Foundation for Quality Management 1994). In addition to the MBQA, and like the Deming Prize, the European Quality Award contains the criterion of social responsibility, another factor applicable to occupational safety and health.

Taking the aforementioned quality awards as an indicator of company focus, it becomes evident that the focus is on personnel development, whereas health promotion, in the sense of a continuous process, seems to be neglected. Although the relevance of personnel satisfaction, motivation and values, company climate, and other factors describing employee psychological well-being is generally recognized in recent QM concepts, related strains and direct impacts are rarely operationalized or systematically recorded.

The European Union passed a scope of directions that obligates companies to analyze their workplaces in terms of occupational safety and health to initiate appropriate improvements (European Commission 1993). The nations of the European Union must realize these directions by converting them into a national law. Thus, occupational safety and health is no longer just an ethical request or a company goal embedded in quality management undertakings, but a legal commitment.

The objective of personnel development is to foster technical, management, and social skills according to workplace demands. It is attempted through training to achieve continuous improvement of employee competencies and performance. But these measures must be supported and complemented by efforts that extend the work force's interests and abilities. The objectives of these additional measures, in contrast to personality development, have to be independent of job-related requirements. All staff members should be healthy, feel at ease at the workplace, and enjoy their work. They are presented under the general term, "personnel caring" (Kastner 1991), the essential complement to personnel development.

On the other hand, like other internal quality criteria, personnel caring cannot, from a business-economic point of view, be a genuine corporate aim. It can only be established as a recognized quality criterion if it contributes to the genuine aims of competitiveness and profitability. But its value should not be underestimated.

Personal Caring in TQM

Personal caring is the comprehension and sensible design of all activities related to the preservation of the work force, their intrinsic work motivation, social support, and well-being. The central point of interest of personal caring is healthy behavior and perception, and self-motivated continuation within the organization (Kastner 1991). A company cannot survive if a substantial part of its work force becomes sick or disabled. The value of the employees' knowledge and performance potential is often higher than the economic value of an infrastructure and this more than justifies investments for preservation and support. Good employees are scarce, valuable, and increasingly indispensable. Methods of short-term hiring and firing become more and more inappropriate, considering the increasing complexity of many tasks and the high cost of staff development. Caring for human resources is caring for the company resources. Without the first, performance improvements are hard to achieve.

An especially strong participation-oriented occupational safety and health approach may lead to the perception that occupational safety and health is primarily a task for the lower and middle levels of the hierarchy. Such a view does not do justice to the fostering role of occupational safety and health. It diminishes its significance compared to other corporate aims. This subsequently causes problems during the implementation phase when important decision makers are not involved in the processes and structural change cannot be implemented. The results of such an occupational safety and health promotion approach are mere attempts to influence behavior. Because appropriate occupational safety and health promotion measures involve management behavior, the organizational climate, as well as the underlying principles of the work organization, it must be part of the strategic quality dimensions. It is a well-studied fact that the commitment and role modeling of upper management constitutes the most important factor for the success of health programs (Curtis 1995).

Medical checkups, as part of company social services, may lead to an improvement in employee job satisfaction and work effectiveness. The perception of health as an achievable goal leads to the attitude that health is indispensable for a quality (working) life.

The emphasis here is not necessarily on physical fitness. Physical illness or disability are often smaller burdens to an enterprise than psychological and social disorders. Heavy physical impairments, such as heart attacks, cancer, or paralysis do not always impair productive performance, while moods, personal insecurity, depression, and psychosomatic conditions tend to be significant obstacles.

TQM for Occupational Safety and Health

When personnel caring is established as a criterion for business quality, we can turn around and ask what TQM principles and methods can teach us in order to achieve occupational safety and health. Like all internal services, occupational safety and health must be included in the process of continuous improvement. For this purpose, the multiple criteria and instruments of quality management can be applied. Presently, the TQM-like managing of occupational safety is being considered (Krause 1995; Senecal 1994; Curtis 1995). Figure 28-9 shows an exemplary translation of MBQA quality categories into occupational safety and health requirements.

The attitude concerning health-promoting behavior follows the hierarchical company structure. In organizations where executive staff members receive regular medical checkups and health counseling, and where they display healthy behavior, the call for company-wide health care from all levels follows. "Our boss wants to be in shape. So do we!"

Executive staff must prove its commitment to occupational safety and health, refusing to tolerate hazardous or unhealthy behavior, even when it means forfeiting a certain degree of productivity. Leadership style must contribute to a health-promoting operational climate. For this purpose, executive managers need a repertoire of appropriate actions to lend staff a competent, helpful, and performance-promoting hand (especially in problem situations), enabling self-motivation, instead of using orders and reprimands. Possible parameters for a health-promoting operational climate and good work motivation go far beyond the easily recordable sickness and fluctuation figures. They can be established through observation, conversations with staff, or interviews. Positive indications, for example, are friendliness, patience, humor, and the existence of a time culture.

Health is influenced on all levels of the work environment, from the chemical and physical work environment, work tasks, and even up to the level of company organization. However, early indications of health impairments, which may take the form of indisposition, missing work ethics, passivity, or fear of excessive demands, can, if recorded, serve as an early warning system and provide valuable

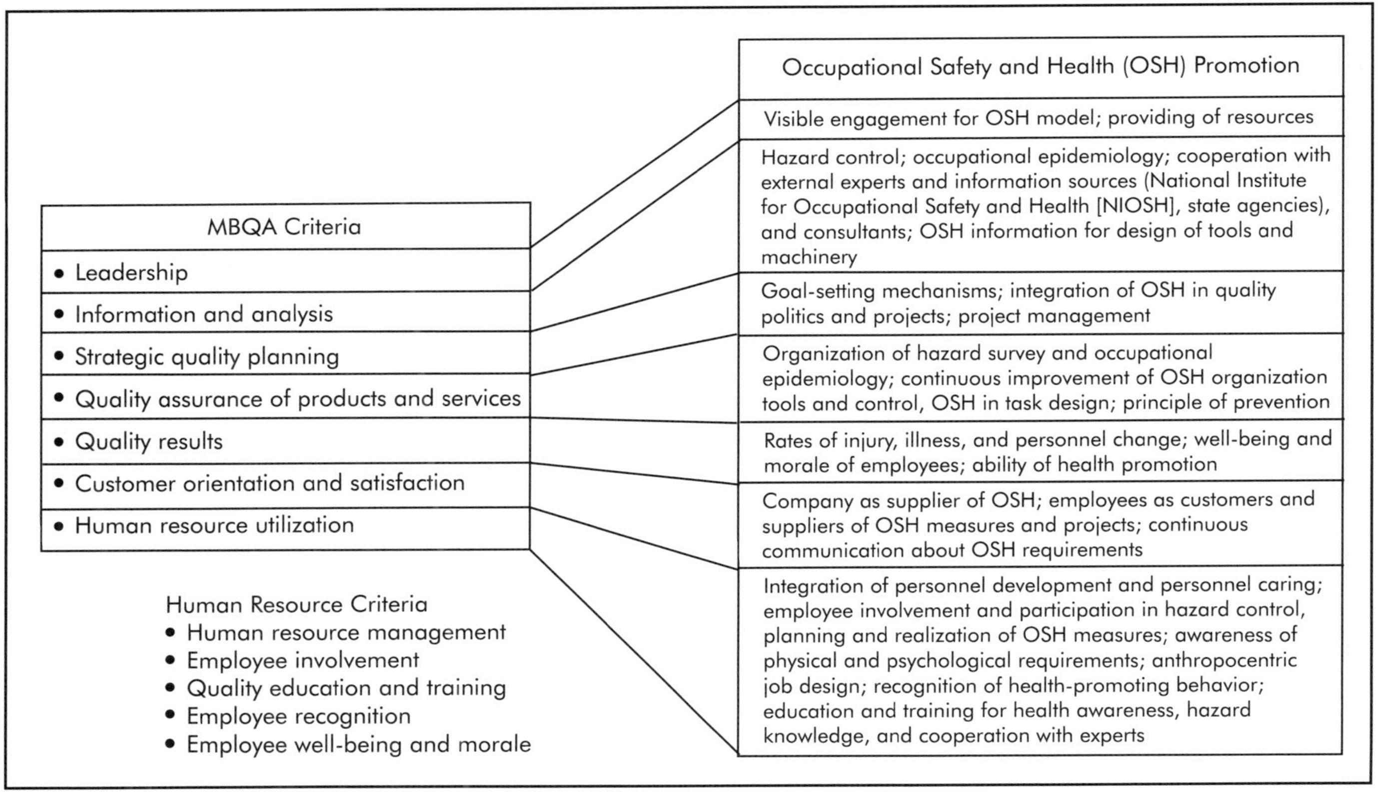

MBQA Criteria	Occupational Safety and Health (OSH) Promotion
• Leadership	Visible engagement for OSH model; providing of resources
• Information and analysis	Hazard control; occupational epidemiology; cooperation with external experts and information sources (National Institute for Occupational Safety and Health [NIOSH], state agencies), and consultants; OSH information for design of tools and machinery
• Strategic quality planning	Goal-setting mechanisms; integration of OSH in quality politics and projects; project management
• Quality assurance of products and services	Organization of hazard survey and occupational epidemiology; continuous improvement of OSH organization tools and control, OSH in task design; principle of prevention
• Quality results	Rates of injury, illness, and personnel change; well-being and morale of employees; ability of health promotion
• Customer orientation and satisfaction	Company as supplier of OSH; employees as customers and suppliers of OSH measures and projects; continuous communication about OSH requirements
• Human resource utilization	Integration of personnel development and personnel caring; employee involvement and participation in hazard control, planning and realization of OSH measures; awareness of physical and psychological requirements; anthropocentric job design; recognition of health-promoting behavior; education and training for health awareness, hazard knowledge, and cooperation with experts

Human Resource Criteria
- Human resource management
- Employee involvement
- Quality education and training
- Employee recognition
- Employee well-being and morale

Figure 28-9 Malcolm Baldrige Quality Award categories applied to occupational safety and health.

hints for coordination of man-machine systems, work-organization, or the operational climate.

Occupational safety and health promotion integrated into the context of TQM must comprise strategic planning, setting and operationalizing of aims, creating appropriate methods and instruments, organizing the acquisition of knowledge, continuous qualification of staff, efficient measures, measurements for the evaluation of such measures, control, and implementation.

1. Identify risks and potentials. Here, a number of very different methods of analysis, ranging from risk analysis to task, activity, and requirement analyses, can make a contribution. The complexity can generally be controlled only through participation-oriented procedures.

2. Evaluate risks and potentials. Risks for health and health promotion will always exist, even with good work designs. Thus, in terms of action-directed design concepts, priorities must be set guided by legal conditions (mandatory regulations), aims of the company, and preferences of the personnel. To succeed, this requires a target agreement process. The involvement of all persons in the target-finding process is an important precondition to adjust personal and organizational aims. Depending on this evaluation, different design options arise.

3. Minimize risks and utilization of potentials. The selected design measures must be able to predict the expected effects as precisely as possible. Here, costs and benefits need to be documented with sufficient precision.

4. Control measures. To get the expected result, measures must be implemented as planned. Project management must ensure the control of schedules and interim results. This is absolutely necessary because there can be interdependencies between measures for health promotion and other measures. Interim results show if the development is on the right track.

5. Improve measures. Flexible adjustment in those cases where the expected results do not show or where follow-up measures are to further improve results can lead to continuous improvement processes.

ECONOMICS AND HEALTH PROMOTION

By 1985, two-thirds of all American 500+ companies had implemented health-promotion programs. Over 80% were convinced that the benefits outweighed the costs. Estimated savings were in the region of $14 for every dollar invested in health promotion. But up to now, it has not been possible to forecast the long-term economic effects of health programs (Figure 28-10). In the meantime, control procedures have been developed for many internal business programs (Otte 1994). In general, these procedures recommend methodic demand planning, project management, and continuous evaluation.

Direct factors/ monetary quantifiable	• Continued payment of wages
Indirect factors/ monetary quantifiable	• Labor losses • Availability of machines and installations • Quality of products (especially scrap) • Lead times
Indirect factors/ difficult to quantify in monetary terms	• Corporate flexibility (for example, staffing flexibility) • Image of the enterprise on markets, including the labor market • Motivation, willingness for cooperation, and working atmosphere

Figure 28-10 Imposed effects of occupational safety and health promotion (Thiehoff 1992).

Ergonomic designs and measures of work safety and health promotion require investment. Decisions on investment measures are derived from investment planning procedures. If the hardly quantifiable evaluational criteria of safety- and health-promoting work design were taken into account at the stage of investment decisions, potential expensive mending measures for unsafe production facilities or ergonomic design deficiencies could be avoided. Besides strictly parameter-oriented procedures, there are so-called extended economic appraisal procedures that help evaluate, for example, nonmonetary criteria on the basis of use-value analyses (Gottschalk 1989; Zangenmeister 1994). Establishing operational target quantities allows later comparisons of the planned and realized conditions.

The disadvantage of such planning procedures is that they are very labor-intensive, involving the participation of various specialist disciplines in the planning process. However, this ensures that various target quantities are taken into account.

Analogous to financial accounting systems, some businesses increasingly turn to social accounting. Health reports, expenses for work safety, and health promotion are presented as company policy and made transparent internally and in external relations. One objective is to increase the company's prestige. In general, this reflects in the fluctuation rate and quality of applicants.

Cost

If personnel health is understood as a quality aim of TQM, the question regarding the cost benefit of health-promoting efforts has to be phrased differently. In general, whether quality makes economic sense and can be verified becomes a question of belief rather than of cost calculations (Yavas 1995). In those cases where quality costs are recorded and analyzed, strong doubts exist about the calculation procedures and the extent to which companies are comparable (Coopers & Lybrand 1994). In terms of health promotion, the problem of quantifying and balancing costs and benefits carries greater weight for four reasons:

1. Informational and emotional overload is often only measurable with unjustifiable costs. Even where data is available, there are no generally-accepted scales that determine, for example, acceptable or even personality-promoting strain intensities.

2. Usually, the relationship of weighing the quality dimension of health against other quality dimensions is the result of agreement processes.

3. The system boundaries for balancing are unclear. Each staff member is not only part of the company, but also belongs to other social groups such as the family and society, which also have costs and benefits. Which of them should be included in the balance? Conceivably, it would be an approach similar to Taguchi's quality definition, which includes all the costs for society. However, because of its complexity, this seems impractical for everyday business life.

4. The time scales of health-promoting effects, such as a longer working life, are much longer than the time scales of operational control (a business year) or the time span of responsibility cycles (managerial contracts), and cannot be effectively integrated in the given target system.

Operational work safety cannot be the principal task of only one function holder or department. It incorporates common commitments that demand continuous improvement of all processes, such as personnel training. It is not a requirement for single individuals or groups, but part of an overall value system.

This means that operational health promotion cannot be permanently implemented as regular programs and projects. This would be as expensive as quality assurance by inspection, performance evaluation, frequent quality audits, and capital investment in new machinery. Instead, the aim of operational health promotion must be to establish health conscious behavior as a company value and become effective in individually organized processes. In these terms, occupational safety and health consciousness can be implemented as a process, not a program.

ORGANIZATIONAL DEVELOPMENT THROUGH HEALTH PROMOTION

AREAS OF CONFLICT

Organizational development (OD) has the task of initiating long-term alteration processes in structures, rules, and ways of behavior in organizations. Many organizational development projects focus unilaterally on efficient organization patterns and normative precepts for the staff's behavior. Employees are to work as much and as effectively as possible and can improve their performance through training. In the context of lean management and business process re-engineering, companies often try to organize a small core of high-performing, healthy staff members in self-reliant teams, under acute pressure to perform. The associated health risks are ignored as long as possible.

The vision of staff members with unlimited dedication and steadily increasing potentials ignores reality. It cannot lead to the envisioned individually-organized process. On the personal side, the consequence is a great number of illnesses caused by stress. On the operational side, the consequence is unsuccessful re-engineering projects.

THE PROBLEM OF APPLYING TQM

The success of TQM is essentially dependent on the applied OD methods (Hawley 1995). In many cases, the business models of TQM and OD do not correspond. Thus, the direction model of the business determines the measures used to run it and research changes.

The process model, too, can be based on the concept of the company as a "trivial machine" (Foerster 1984), where a functional relationship between input and output exists and can be planned and controlled. In this model, the internal structure of the system "business company" is understood as the product of rational planning. The system's complexity plays an important role only in the sense that the complexity and dynamics of the planning processes exceed management's capacity. Therefore, the company's staff, with its expertise and external facilitators, must be involved in the planning process. For instance, quality circles, often aimed at improving some situation within a given concept, sometimes apply poor decision-making competencies (Vaziri 1987). This leads to participative Taylorism (Bradley and Hill 1987).

However, new concepts of organizational development conceive the company as a social system (Ackhoff 1994; Lee and Freedman 1984; Luhmann 1988). In this concept, the organization is not composed of working individuals; it is a network of transactions. The transactions are determined by the internal structures, rules, and media of communication.

Within these concepts, social systems have the quality of self-organization. This means they have the ability to change structures, rules, and communication media by themselves (Hayek 1967; Foerster 1984). Thereby, self-organization is not localized, but is dependent upon the outcome of the acts of all members of the system and their dynamic interaction. Intervention in such a system takes place in the form of target-oriented communication. However, given the inner complexity and dynamics, the effect cannot be precisely forecasted (Grossmann and Scala 1993).

Not only are entire companies social systems, but so are their departments and working groups. The customer-supplier relationship is, therefore, a relationship between social systems. What conditions have to be fulfilled by the internal and external interactions of all social systems within an organization so they can cooperate in preserving and improving the entire system?

First of all, because a social system cannot be driven by external forces in the right way, the systems must have the ability of organizing according to objectives—the ability of self-development (Probst 1987). On the level of the single member, this fits the given definition of health. People have to agree with the permanent changes imposed by TQM. Therefore, they should not be overcharged by the speed of change. Studies prove that unclarity about tasks and high-change dynamics are the main strain factors causing psychological disorders (Eckardstein et al. 1995).

Secondly, the entire goal must fit the internal goals of the single systems or single member. In terms of TQM, this is the problem of creating a common vision of quality objectives and an overall commitment to continuous improvement. Organizational development approaches that favor employee involvement, such as leading by contexts or the sociocratic approach, try to solve this problem by participation (Willke 1989; Zeleny 1989).

The underlying idea is that communication among members of an organization can affect the coordination of collective acting. In concrete terms, this means the promotion of staff competence and knowledge, creating room for autonomy, as well as fostering social support. Involving staff in reorganization processes and design decisions is a central factor of the success of such concepts. An exchange of expectations, requirements, information, and interests should result in the joint definition and realization of objectives. This can only be realized if the adjustment between different parts of the organization follows a common rationality. However, it has long been known that decision making in companies is characterized by boundaries (Simon 1957; Cyert and March 1963), causing conflicts that cannot simply be dissolved via communication. Therefore, organizational development has to manage compromises that allow integration of different objective systems. Partly formalized approaches have to be established, such as project organization, which gives some stability and time for learning and experiencing the negotiation and realization of compromises. This can result in a healthy culture without fear of unfairness and overcharge.

CONCLUSION

It might be concluded that occupational safety and health can be successful only if it focuses on influencing the contexts for self-organization of single employees and groups. To promote occupational safety and health, companies must:

- Create organizational preconditions, which involve organizational concepts, such as health circles, restructuring of work tasks, and extension of personal autonomy;
- Define and qualify requirements;
- Improve internal cooperation among various groups, like safety experts and industrial medicine doctors;

- Enhance cooperation with external organizations, such as health insurers and professional associations;
- Establish staff participation in all phases of the design processes; and
- Define evaluation measures that can be transformed into business objectives.

In the framework of organizational development and continuous improvement, work groups can be formed to specifically deal with questions of work safety and health promotion (Nieder and Susen 1994). Organizational development, understood as a participation-oriented process, ensures the collection of design knowledge from various knowledge-holders, including staff. Work groups are able to extract and evaluate avenues toward preventive work safety and health promotion.

Often, these kinds of project groups are called health circles (Nieder and Susen 1994), whose basic function is to enable staff to commit to personal health. Health circles are integrated into organizational development processes. The most important precondition for the success of health circle work is communication among equal-status members of the circle. An example for organizing occupational safety and health via health circles is shown in Figure 28-11.

The setting-up of such groups as a part of the TQM concept in the initial phase elucidates the significance of occupational safety and health as a goal of quality. In the long run, occupational safety and health has to be embedded in a participative organizational development process with congruent group and work tasks (Figure 28-12).

Only the integration of occupational safety and health, or generally of personnel caring, in work tasks can establish the concept as a long-term business value and reduce the costs incurred (Table 28-1).

REFERENCES

Ackhoff, R.L. *The Human Corporation: Integrating Work, Play, and Learning.* New York: Oxford University Press, 1994.

Bradley, K., and Hill, S. "Quality Circles and Managerial Interests." *Industrial Relations* 1987, 21: 68-82.

Coopers & Lybrand, European Foundation for Quality Management (ERGM). *Economic Aspects of Quality.* Panorama of EU Industry '94, European Commission, ed. Luxemburg: Amt für amtliche Veröffentlichunger der Europäischen Gemeinschaft, 1994.

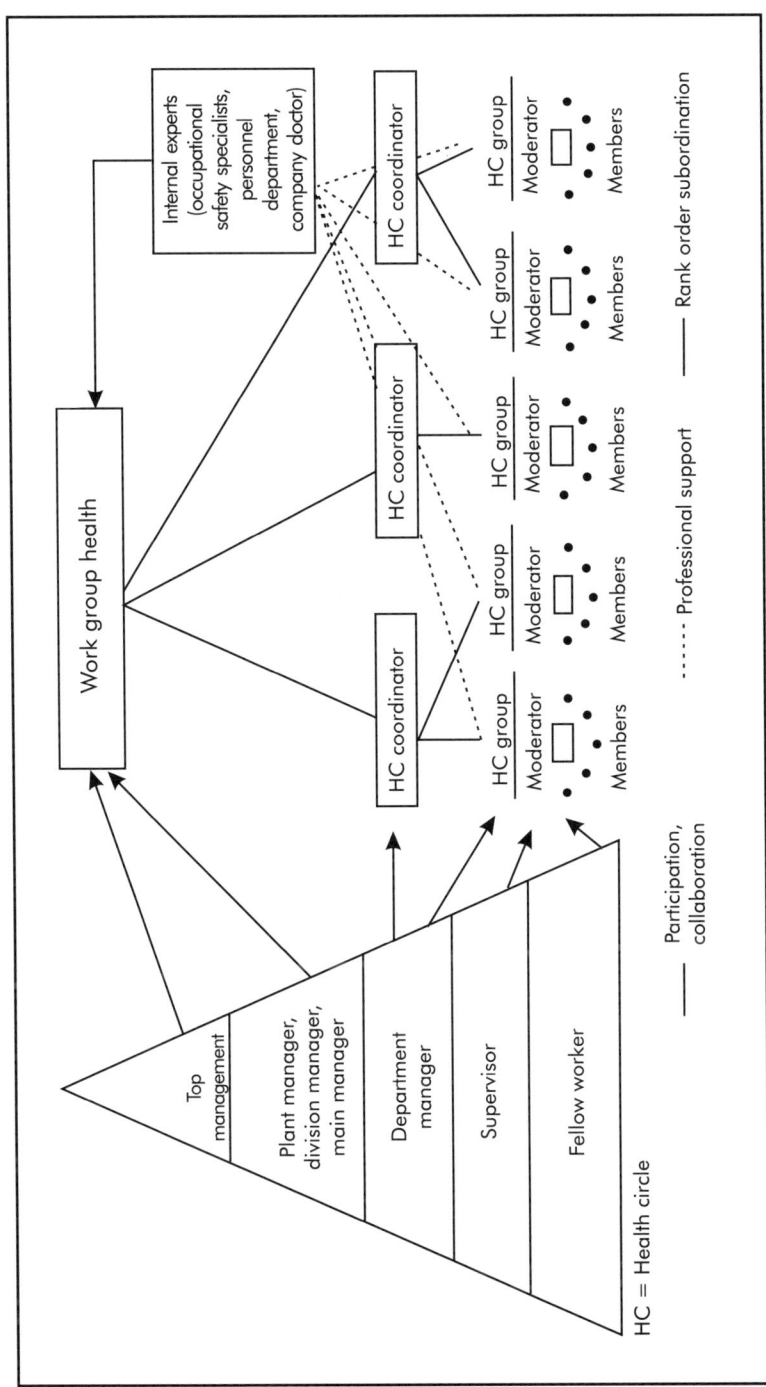

Figure 28-11 Organizational structure for promoting occupational safety and health with health circles (Gerpott and Schreiber 1996).

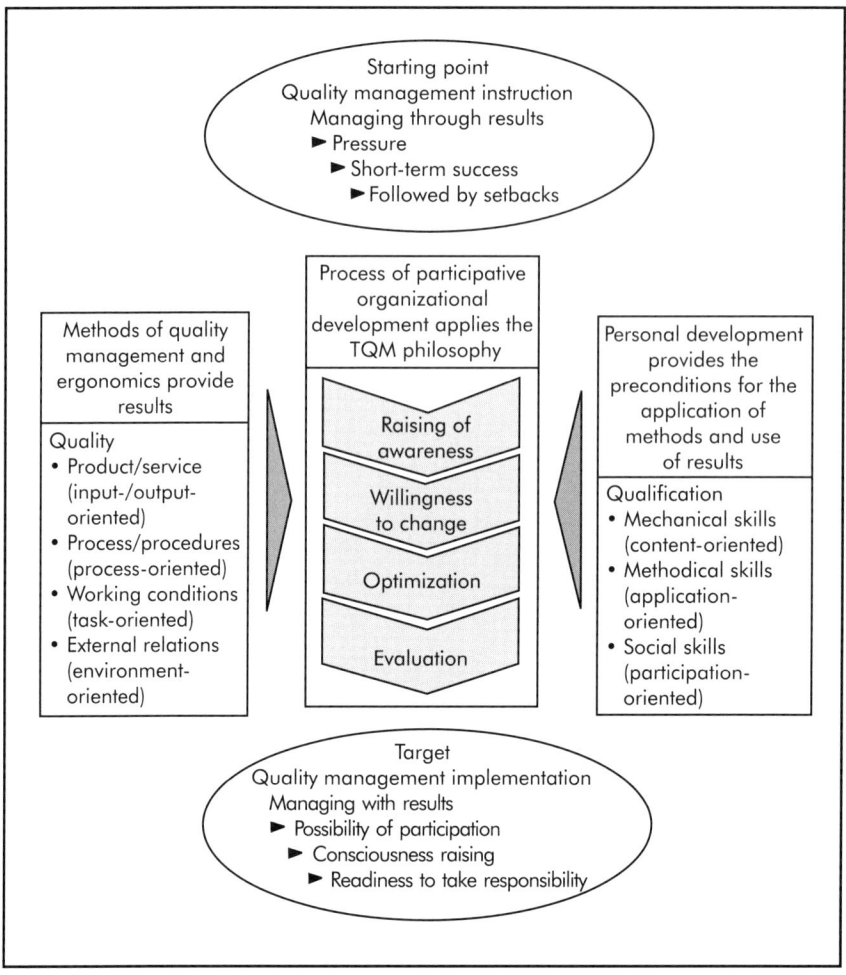

Figure 28-12 Participative organizational development (Luczak et al. 1995).

Curtis, S.L. "Safety and Total Quality Management." *Professional Safety* January 1995: 18-20.

Cyert, R.M., and March, J.G. *A Behavioral Theory of the Firm.* Englewood Cliffs, NJ: Prentice-Hall, 1963.

Deming, W.E. *Out of Crisis.* Cambridge, MA: MIT Center of Advanced Engineering Study, 1986.

Eckardstein, D.V., Lueger, G., Niedl, K., and Schuster, B. *Psychische Befindensbeeinträchtigungen im Betrieb.* München & Mering, Germany: Rainer Hampp Verlag, 1995.

European Commission, ed. *Soziales Europa: Europa für Sicherheit und Gesundheitsschutz am Arbeitsplatz.* Luxemburg: European Commission, 1993.

Table 28-1. Focal points of occupational safety and health on several organizational characteristics

Organizational Model Characteristics	OSH as Special Function	OSH Promotion through Programs and Projects	OSH as a Component of TQM in a Learning Organization
Vision	Process safety, avoidance of illness, freedom from impairment	OSH promotion, job satisfaction, avoidance of indirect hazards	OSH process imminent, training
Leadership	Standards, safety inspections, behavioral requirements	Standards, training	Standards, agreement of goals, team learning
Prosociality (trust, acceptance, tolerance)	Hierarchical expert-lay relation	Expert culture advises lay culture, human relations concepts	Cooperation of experts and lays in teams
Proactivity	Inherent technological safety, legal responsibility	Preventive TOP concepts for safety	Responsibility for the entire process in the work groups
Participation	Poor participation	Participative hazard control programs	Common vision finding and development of methods
Flexibility	Formalized approaches, narrow margins, rigid coupling to the values and aims of the company	Motivation of the employees for new, participative concepts, wider margins for decisions and maneuvering, positive attitude of the employees	New ideas and solutions are work tasks, wide margins for decisions, and maneuvering through complete tasks
Qualifying (personal competence)	Occupational qualification of experts	Broader competencies of employees, social and methodological training for participation	Integration of working and learning, higher redundancy of occupational qualifications, personal growth

European Foundation for Quality Management, ed. *The European Quality Award 1995-Bewerbungsbroschüre*. Brüssel, Belgium: European Foundation for Quality Management, 1994.

Foerster, H.V. H. Ulrich and G.J.B. Probst, eds. "Principles of Self-organization in a Socio Managerial Context." *Self-organization and Management of Social Systems*. Heidelberg, Germany: Springer, 1984.

Gerpott, T.J., and Schreiber, K. "Erfolgreiche Implementierung von Gesundheitszirkeln." *Personal* 1996, 2: 72-77.

Gottschalk, B. Wissenschaftliche Begleitung in der Umsetzung erweiterter Wirtschaftlichkeitsrechnungen. *Schriftenreihe Bundesanstalt für Arbeitsschutz*. FB 598. Dortmund, Germany: 1989.

Graham, S. "Safety and Quality: Forge the Ties that Bind." *Safety and Health* 1995, 151: 40-41.

Grant, R.M., Shani, R., and Krishnan, R. "TQM's Challenge to Management Theory and Practice." *Sloan Management Review* 1994, 35: 25-35.

Grossmann, R., and Scala, K. *Health Promotion and Organizational Development: Developing Settings for Health*. Regional Office for Europe of the WHO, document ICP/HSC 663, 1993.

Hawley, J.K. "Where's the Q in TQM?" *Quality Progress* October 1995: 63-64.

Hayek, F.A.V. *Studies in Philosophy, Politics, and Economics*. London, England: Rootledge & Keegan, 1967.

Kamiske, G.F. *Die hohe Schule des TQM*. Berlin, Germany: Springer, 1994.

Kastner, M. "Personalpflege." *Personalmanagement: Denken und Handeln im System*. M. Kastner and B. Gertenberg, eds. München, Germany: Quintessenz, 1991.

Krause, T., Durbin, T.J., and DiPiero, D.A. "Making Safety and Quality Work Together." *Occupational Hazards* 1993, 55: 55-58.

Krause, T.R. "Safety and Quality: Two Sides of the Same Coin." *Quality Progress* October 1995, 27: 51-55.

Lee, R.J., and Freedman, A.M. *Consultation Skills Reading*. Alexandria, VA: Pfeiffer and Company, 1984.

Logothetis, N. *Managing for Total Quality: From Deming to Taguchi and SPC*. New York: Prentice Hall, 1992.

Luczak, H., Otzipka, J., Flachsenberg, U., and Krings, K. "Qualitätsmanagement und Personalentwicklung." *Zeitschrift für Arbeistwissenschaft* 1995, 49: 149-156.

Luhmann, N. *Soziale Systeme: Grundriß einer allgemeinenr Theorie*. Frankfurt, Germany: Suhrkamp, 1988.

National Institute of Standards and Technology (NIST), ed. 1995. *Malcolm Baldrige National Quality Award 1995*. Gaithersburg, MD: NIST, 1995.

Nieder, P., and Susen, B. "Betriebliche Gesundheitsföderung und Organisationsentwicklung." *Personalführung* August 1994: 696-701.

Ottawa Charter for Health Promotion. *Health Promotion* 1986, 1(4):Iii-v.

Otte, R. *Gesundheit im Betrieb; Leistung durch Wohlbefinden*. Frankfurt, Germany: Frankfurter, Allgemeine Zeitung, Verlagsbereich Wirtschaftsbücher, 1994.

Probst, G.J.B. "Selbstorganisation und Entwicklung." *Die Unternehmung* 1987, 41(4): 242-255.

Rine, F. "Safety and Quality." *Safety and Health* 1994, 149: 30-31.

Senecal, P. "TQM: Putting People on the Path to Safety." *Occupational Hazards* November 1994, 56: 47-48.

Simon, H.A. *Models of Man: Social and Rational.* New York: John Wiley & Sons, 1957.

Thiehoff, R. *Erweiterte Wirtschaftlichkeitsrechnung-ein Beitrag zur ganzheitlichen Investitionsplanung.* Bundesarbeitsblatt Buch: Prävention im Betrieb. Arbeitsbedingungen gesundheitsgerecht gestalten. Bonn, Germany: BMA, 1992.

Vaziri, M.T. "Productivity Improvement through Quality Control Circles: A Comparative Approach." *Leadership and Organizational Development Journal* 1987, 8: 17-19.

Willke, H. R. Eschenbach, ed. "Controlling als Kontextsteuerung: Zum Problem dezentralen Enscheidens in vernetzten Organisationen." *Supercontrolling: Vernetzt denkenzielgerichtet entscheiden.* Wien, Germany: ServiceFachverlag, 1989.

Wojda, F., and Hacker, W. "Arbeitswissenschaft und Qualitätsmanagement." *Zeitschrift für Arbeitswissenschaft* 1995, 49: 126-130.

Yavas, B.F. "Quality Management Practices Worldwide: Convergence or Divergence." *Quality Progress* October 1995: 57-61.

Zangenmeister, C. "Erweiterte Wirtschaftlichkeitsanalyse." *Fortschrittliche Betriebsführung/Industrial Engineering* 1994, 43: 63-71.

Zeleny, M. "Socioscraty." *Human Systems Management* 1989, 8: 245-248.

INDEX

A

acceleration, 369
accommodation standard, 456
aircraft checklists, 404
American Industrial Hygiene
 Association, 441
Americans with Disabilities
 Act, 158
analysis, 19
ANSI-2365
 development, 317
 issues and concerns, 323
 structure and content, 320
arthritis, 71
anthropometric data, 44
anthropometry, 439
anthropomorphic models, 450
antivibration tools, 378
AT&T Global Information
 Solutions, 34
attributes of measurement, 400
audit assessment, 268
audit design, 401
auditing methods, 399
awkward posture
 health effects, 168
 posture analysis
  methods, 170
 posture checklists, 170
 posture classification, 177
 productivity effects, 169
 statistical sampling, 173
 work posture, 167

B

back belt, 452
back injury, preventing
 job design, 245
 personnel selection, 244
 personnel training, 244
 back support belts
  effect of, 250
  industrial, 247
  Japanese, 248
  OK-1 MODEL, 247
  orthotic devices, 246
  ProFlex™, 248
  weight-lifting, 246
behavioral failure, 462
benefits and costs, 414
biomechanical
 hazard, 352
 models, 451
Board of Certification of
 Professional Ergonomists,
 441
British Ergonomics Society, 441
Bundesanstalt für Arbeitsschutz
 und Arbeitsmedizin, 470

C

carpal tunnel syndrome, 71,
 299, 446
 causes, 301
 symptoms, 301
 treatment, 303
Center for Registration of
 European Ergonomists,
 490
checklists, 95, 359, 449
cognitive failure, 461
cognitive performance, 439
COMBIMAN, 54
computer-aided design for
 ergonomics, 93

computer-based job
    analysis, 361
computerized human modeling,
    54
crew chief, 57
cumulative trauma disorder
    incidence and costs, 44,
        305, 445
    in industry, 307
    prevention and control, 314
    risk factors, 309
    surveillance methods, 311
    what they are, 306
CYBERMAN, 54

D

Deere and Company, 36
DeQuervain's disease, 71
DeQuervain's tenosynovitis,
    291
    causes, 293
    how it develops, 292
    symptoms, 292
    treatment, 293
design, 455
discomfort surveys, 332
documentation, 366
down ramp series, 422
dynamic muscle work, 122

E

Eastman Kodak, 5
education program, 268
employee involvement, 354,
    385
epicondylitis, 71
ErgoCop™, 97
ergonomic
    circular grinding
        machine, 477
    design examples, 475

grinding workplace, 479
hand drill, 477
hazards, 352
job design, 153
ergonomics
    committee, 358, 365
    coordinator, 385
    evaluations, 398
    policy, 353
ergonomics working
    position—sitting
        checklist, 400
European Organization for
    Standardization, 88
European Quality Award, 514
evaluating the process
    employee surveys, 24
    ergonomics process and
        projects log
        management, 24
    facility plant organization
        evaluations, 24
    injury and illness record
        analysis, 23
exercise, 271
eye height, 50

F

facilitators, 461
fitness program, 271
fitting the task, 460
Fraunhofer-Gesellschaft, 471
functional capacity
    assessment, 273

G

Germany, 466, 468
Gesellschaft für
    Arbeitswissenschaft, 471
gradient, 421

H

hand-arm vibration, 371
hand tool design
   grip shape, 78
   grip size, 77
hand tools, 65
hazard prevention, 388
heavy dynamic muscle work, 127
Human Factors and Ergonomics Society, 441
human factors program, 392
Humanisierung des Arbeitslebens Action Program, 470

I

impact points of back injury prevention
   after the hire, 265
   after the injury, 267
   prior to hire, 265
Industrial Ergonomics Technical Group, 441
inspection costs, 427
Institute of Industrial Engineers, 441
ischemia, 71
isometric testing, 154

J

JACK, 57
job analysis, 281
job improvement survey, 354
job load and hazard analysis, 91
job placement assessment, 282

L

L.L. Bean, 39
lateral clearance, 49
lateral epicondylitis, 297
   causes, 298
   symptoms, 297
   treatment, 299
localized musculoskeletal discomfort, 136
location of the inspection station, 429

M

macroergonomics, 39, 508
maintaining working postures, 135
Malcolm Baldrige Quality Award, 514, 519
management commitment, 383
MANNEQUIN, 58
manual material handling, 128
manufacturing workstation, 43
mathematical models
   back injuries, 189
   carrying, 200
   job design, 189
   lifting, 190
   lowering, 192
   manual material handling, 189
   pulling, 198
   pushing, 196
   structure, 189
maximum holding time, 138
measurement criteria, 24
meatpackers' guidelines, 444, 456
medical management, 388
microergonomics, 508
monitoring progress, 23
muscular overload, 124
musculoskeletal
   disorders, 34, 153, 446
   injuries, 327, 353

N

narratives, 359
NIOSH lift equation, 450
number of ergonomists
    needed, 3
NYNEX, 38

O

Occupational Safety and
    Health Administration,
    352, 505
occupational safety and health
    promotion, 520
orthotic devices, 452
Ovako Working Posture
    Analysis System, 94
overuse injuries, 445

P

participative organizational
    development, 528
participatory ergonomics
    assessing readiness, 10
    benefits, 6
    explanation of, 5
    getting started, 13
    moving toward, 11
performance, 455
peritendinitis, 294
    causes, 295
    symptoms, 295
    treatment, 296
personal caring, 517
physical failure, 461
physical load, 121
Place of Work Rules, 467
Position Analysis
    Questionnaire, 400
preventing physiological
    overload, 131
products phase, 392
programs, 455

R

Raynaud's Disease, 71, 371
reach dimensions, 44
Red Wing Shoes, 35
Reichsausschuss für
    Arbeitsstudien, 466
repetitive strain injury, 445
repetitive work, 129
resonance, 370
revised NIOSH lifting equation
    action limit, 205
    angle of asymmetry, 212
    asymmetry components, 218
    composite lifting index, 228
    coupling
        classification, 214
        component, 222
    dish-washing machine
        unload, 235
    distance components, 217
    duration of lifting, 214
    equation limitations, 215
    frequency
        component, 219
        independent lifting index,
            228
        independent
            recommended weight
            limit, 225
        of lifting, 213
    horizontal
        component, 216
        location, 212
    job-related intervention
        strategy, 231
    lifting
        duration, 220
        index, 212
        task, 212
    load weight, 212
    loading supply tools, 232
    measurement requirements,
        211

National Institute for Occupational Safety and Health, 205
neutral body position, 213
procedure, 223
   single-task, 225
   multi-task, 225
recommended weight limit, 206
single-task lifting index, 228
single-task recommended weight limit, 228
special frequency adjustment procedure, 221
vertical
   component, 217
   location, 212
   travel distance, 212
work practices guide for manual lifting, 205
right-handed versus left-handed operators, 80
risk assessment, 91

S

Saab, 497
Safety at Work Act, 467
safety management information systems, 89
SAMMIE, 58
Scandinavian countries, 485
Single European Act, 467
snook, 457
standards, 145
static muscle work, 122
step series, 421
Stockholm Workshop Scale, 373
strength evaluation
   comparing methods, 156
   selecting a method, 157
   structure, 16
testing
   isoinertial, 155
   isokinetic, 155
   isometric, 154
surveillance, 358
Sweden's Work Environmental Act of 1978, 486
Swedish Foundation of Work Life, 488
Swedish model, 486

T

T50 Program, 494
team parameters, 17
teams, 459
ten steps toward successful ergonomics, 1
tendonitis, 71
tenosynovitis, 71
Tokyo Marine, 36
total quality management, 354
training, 362, 389, 448
   characteristics of, 107
   environment, 109
   five phases
      delivery, 111
      development, 111
      integration and improvement, 115
      measurement, 113
      planning, 110
   job specific
      employees, 119
      health and safety professionals, 118
      managers, 117
      manufacturing engineers, 118
      medical and human resource professionals, 120

trainer/facilitator, 108, 461
TRANSOM JACK™, 99
trigger finger and thumb, 288
  causes, 289
  how it develops, 288
  symptoms, 289
  treatment, 290

U

Union Pacific, 37
upper extremities, 66, 351
usability analysis, 440

V

Verband für Arbeitsstudien und Betriebsorganisation, 471
vibration, 369
  antivibration tools, 378
  hand-arm, 371
  whole-body, 370
Volvo, 497

W

where the ergonomics function is placed, 1
whole-body vibration, 370
Work Analysis Checklist, 400
work height, 49
work-rest model, 146
work site analysis, 387
Works Council Constitution Act, 467
worksheets, 449
workstation
  dimensions, 48
  redesign, 32
worth of a project, 417
wrist position, 80